Signaling and Communication in Plants

Series Editors

František Baluška
Department of Plant Cell Biology, IZMB, University of Bonn, Kirschallee 1,
D-53115 Bonn, Germany

Jorge Vivanco
Center for Rhizosphere Biology, Colorado State University, 217 Shepardson Building,
Fort Collins, CO 80523-1173, USA

For further volumes:
http://www.springer.com/series/8094

Günther Witzany • František Baluška
Editors

Biocommunication of Plants

Springer

Editors
Günther Witzany
Telos - Philosophische Praxis
Bürmoos
Austria

František Baluška
Universität Bonn
Inst. Zelluläre und Molekulare
Botanik (IZMB)
Bonn
Germany

ISSN 1867-9048 e-ISSN 1867-9056
ISBN 978-3-642-23523-8 e-ISBN 978-3-642-23524-5
DOI 10.1007/978-3-642-23524-5
Springer Heidelberg Dordrecht London New York

Library of Congress Control Number: 2011945292

Printed on acid-free paper

Springer is part of Springer Science+Business Media (www.springer.com)

Preface

Why Biocommunication in Plants?

If we speak of biocommunication in plants, we first must clarify the terms of communication and signalling which are based on systems we define as languages and codes. We should rely on the recent results of the pragmatic turn in the philosophy of science of the last century, which clarify the conditions for generating correct sentences in science.

Biocommunication is defined as meaningful interaction between at least two living agents, which share a repertoire of signs (representing a kind of natural language) that are combined (according to syntactic rules) in varying contexts (according to pragmatic rules) to transfer content (according to semantic rules).

Contrary to all former concepts, these three levels of semiotic rules are complementary parts of any natural language or code-based system. According to Charles Morris, we cannot speak of language or signal-mediated communication if one of these three levels is missing. So the most recent definition of biocommunication is this: sign-mediated and rule-governed meaningful interactions that depend on a communally shared repertoire of signs, codes and rules. Importantly, these features are lacking in any abiotic physical interaction.

Additionally, we know that mathematical and mechanistic theories of language are not helpful in investigations on natural languages and real-life communication processes because such theories cannot explain typical features of living agents that communicate. These aspects are not formalizable as no algorithm is available for de novo-generation (innovation) of coherent/correct sentences/sequences. This means that no natural language or code speaks or codes by itself but needs living and experiencing agents (biological systems) that are competent in using such languages or codes.

In the biology of the twentieth century, the physiology of cells, tissues, organs, and organisms of all kingdoms was the mainstream direction in biological research. In the 1970s, an increasing use of "communication" as a metaphor also occurred in biology. During the last decade of this period, interest in communication (no longer being used as a metaphor) within and between organisms overtook that of the pure physiological understanding of organisms. Cell-to-cell communication now dominates contemporary cell biology, resulting in enormous knowledge about a great variety of signalling systems serving for organization / coordination of production, release, uptake, and processing of "information" within and between cells.

In parallel, the use of "language" as a metaphor increased from the middle of the twentieth century with growing knowledge about the genetic code. Most of the processes that evolve, constitute, preserve, store, and rearrange the genetic storage medium DNA are terms that were originally used in linguistics, such as nucleic acid language, genetic code, "codes without commas" (Francis Crick), coding, copying, translation, transcription, sequence homology. Meanwhile, the linguistic approach lost its metaphorical character and the similarity between natural languages/codes, and the genetic storage medium DNA are not only accepted but are adapted in epigenetics, bioinformatics, biolinguistics, protein linguistics, and biosemiotics.

The advantage of methodical adaptation of communication and linguistic terminology is in having appropriate tools for differentiation at specific levels, which is otherwise difficult to describe non-reductively by pure physiology. This means that language-like systems and communication processes occur at the bottom of living nature. Language and communication are not inventions of humans, nor are they (as often claimed) anthropomorphous adaptations to describe the non-human living nature. It is becoming obvious that every coordination and organization within and between cells, tissues, organs, and organisms needs meaningful signs: chemical molecules that serve as signals, symbols and codes for conveying essential messages that serve as vital indicators of environmental (both abiotic and biotic) conditions. Because no code codes itself, as no language speaks itself, these signs need to be sensed and interpreted in a correct context by biological agents, i.e., there must be subjects/ representatives of sign production and sign interpretation. This means that if sensing and contextual interpretation fails, this will then result in non-appropriate (non-adaptive) behaviour and can have even fatal consequences for cells, tissues, organs, and organisms.

The method of analyzing any part of a machine in detail to get a picture of its whole functional blueprint, which can then be used to reproduce or manipulate it, or to produce an even more perfect one (taking genetic engineering as an example); is still useful if we are dealing with machines. However, growing evidence of biological processes makes it doubtful whether investigating organisms with this mechanistic attitude will be useful in the future. Communication between cellular parts, cells, tissues, organs, and organisms is far from being a procedure which can be reduced to mechanistic input/output or cause/reaction descriptions. It is evident that communication processes within and between living organisms include a variety of circumstances and competences that must be fulfilled in parallel if communicative acts are to have successful consequences, such as common coordination.

First of all, no single organism is able to communicate as an emerging property. It must be a community, a society, or a swarm of organisms that each share an identity (group) and a competence to sense others as being part of their biological identity (self/nonself competence, kin recognition), even if this competence is shared genetically solely. To biocommunicate, it is necessary that an organism has some skills that serve as signs (signals, symbols), such as chemical molecules either produced directly by itself or as secondary metabolites or even molecules in the surroundings that are not produced by the organism but can still be manipulated, according to the organismal needs.

Secondly, organisms must share a competence to use these signs in a coherent manner, which means using these signs in a strict temporal and spatial context. In most cases, it is not just one signalling molecule but complex networks of signalling molecules and channels that are dynamically combined in a certain manner to transport messages (information) effectively. This represents a common feature of sign-use in biocommunication processes, which is called their correct combination or syntax.

Thirdly, organisms are part of ecological habitat in which they live together with other organisms of the same or related species, as well as with an abundance of nonrelated organisms of other kingdoms. This context exactly represents the natural history of organismic swarms or communities in which they – and this is only a recently experienced feature – evolved and developed certain abilities to appropriate response behaviours according to their survival. These include sensing, learning, and memory, which are the preconditions for faster adaptations.

Finally, the signalling molecules, which serve as signs, transfer messages with meanings (semantics). The informational (semantic) content, which is transported, triggers certain response behaviours by the same or related, or even unrelated, organisms. Interestingly, the signal sequence or signal content does not necessarily depict a single meaning, i.e., function can vary according to different situational contexts. This means that even identical signs can transport a variety of different messages according to different contextual needs and scenarios. This is important in very dense ecological habitats, for example, in the rhizosphere biology. The different uses of identical signs (sequences) enable the generation of dialects within same species that can transport messages, which are microecosphere-specific. These include sensitive self/nonself recognition between slightly differently adapted populations of the same species in the same ecological habitat.

Although sign-mediated interactions (i.e., communication processes) are very reliable in most cases, they do not function mechanistically in a strict sense. Syntax (combination), pragmatics (context), and semantics (content) must function in parallel to ensure and optimize coordination and thus survival of group members.

These semiotic rules do not function mechanistically but may be varied, deleted, or, in certain circumstances, generated de novo. Additionally, biosemiotic rules do not function by themselves but need semiotic subjects, i.e., living organisms that use and understand such rules. If no living organism is present, semiotic rules, signs, and communication are absent. Although highly conserved semiotic rules are modifiable, environmental circumstances, such as stress, trigger adaptive responses. In such cases, signals may transport new messages, which previously did not exist, broadening the communicative competences of organisms and their evolutionary capabilities. This is different in the case of abiotic (purely physical) processes, where semiotic (syntactic, pragmatic, semantic) rules of sign-use are not relevant as natural laws are sufficient alone. No biosemiotic rules are used or are necessary for water molecules to freeze into ice.

To give an answer to the question "Why biocommunication in plants": biocommunication in plants integrates both biology of plants and communicative competences of plants. It allows more coherent explanation and description of full range of behavioural capabilities of plants that cannot be covered by mechanistic or even reductionistic approaches. Natural communication assembles full range of signal-mediated interactions that are necessary to organize coordinations within and between cells, tissues, organs and organisms.

Bürmoos, Austria
Bonn, Germany

Günther Witzany
František Baluška

Contents

Key Levels of Biocommunication in Plants

Witzany Guenther

Abstract As in all organisms, the evolution, development, and growth of plants depend on the success of complex communication processes. These communication processes are primarily signal-mediated *interactions* and not simply an exchange of information. Therefore, identification of meaning functions of signaling molecules depends on coherent investigation of *interactional patterns* in which signaling occurs. These interactions involve active coordination and active organization of a variety of timely ordered steps and substeps conveyed by signs. A wide range of chemical substances and physical influences serve as signs. Different abiotic (water, light, gravity) or biotic influences (symbiotic interactions, attack, defense, mating, etc.) require different behaviors. Depending on the behavior, the core set of signs common to species, families, and genera of plants is variously produced, combined, and transported. This allows entirely different communication processes to be carried out with the same types of chemical molecules (e.g., auxin, see below), which optimizes energy cost (see below).

1 With Communication, Plants Coordinate Complex Interactions

Plants have long been considered metabolic growth automatons with very simple stimuli-response reactions based on input-output mechanics. Research in last decades completely changed this picture. We now know plants as highly sensitive organisms which actively sense their environment on different levels within their plant body (intraorganismic) and interact with same, related, and nonrelated plants (interorganismic); with nonplant organisms such as fungi, bacteria, and animals (transorganismic); and—additionally—with abiotic influences from the

W. Guenther (✉)
Telos-Philosophische Praxis, Bürmoos, Austria
e-mail: witzany@sbg.at

G. Witzany and F. Baluška (eds.), *Biocommunication of Plants*,
Signaling and Communication in Plants 14,
DOI 10.1007/978-3-642-23524-5_1, © Springer-Verlag Berlin Heidelberg 2012

environment such as nutrient and water availability, light, gravity, wind, and temperature. All these sensory data have to be processed, memorized, and compared with memorized information to generate appropriate response behavior. Information processing occurs in parallel as well as the response behavior is of tremendous complexity and involves decision, organization of appropriate signaling molecules for a variety of different signaling patterns, and a highly sophisticated coordination of all steps and substeps especially in the root zone and in root-stem communication. Biocommunication means there will be no coordination and organization of plant organisms without signaling processes (Fig. 1).

Plants assess their surroundings, estimate how much energy they need for particular goals, and then realize the optimum variant. Plants constantly take measures to control certain environmental resources. They perceive themselves and can distinguish between self and nonself. This capability allows them to protect their territory and promote kinship. They process and evaluate information and then modify their behavior accordingly. Successful communication processes allow the plants to prosper; unsuccessful ones have negative, potentially lethal repercussions. Intraorganismic communication involves sign-mediated interactions in cells (intracellular) and between cells (intercellular). Intercellular communication processes are crucial in coordinating growth and development, shape, and dynamics. Such

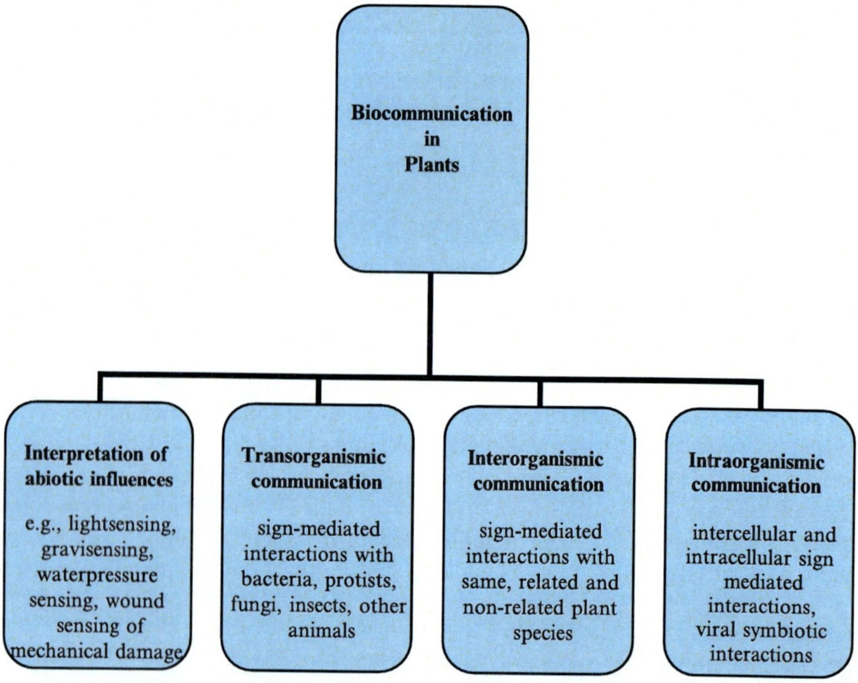

Fig. 1 Key levels of biocommunication in plants

communication must function on both the local level as well as between widely separated plant parts. This allows plants to react in a differentiated manner to its current developmental status and physiological influences (Witzany 2010).

2 Semiochemical Vocabulary: Transmitters, Hormones, RNAs, Reusable Elements

The chemical communication in and between plants is so complex that more than 20 different groups of molecules with communicatory function have currently been identified. Up to 100,000 different substances, known as secondary metabolites, are active in the root zone.

For synaptic neuronal-like cell-cell communication, plants use neurotransmitter-like auxin and presumably also neurotransmitters such as glutamate, glycine, histamine, acetylcholine, and dopamine. Alongside the classical phytohormones auxin, cytokinin, gibberellin, ethylene, and abscisic acid, the plant peptide hormone systemin has been noticed to be important; plants use this to systematically react to local injuries. In activating an effective defense response, a combination of systemin, jasmonate, and ethylene serves as signal molecules. The production (biosynthesis) of brassinolide hormones is important for cellular processes and development steps (Baluška et al. 2006; Baluška and Mancuso 2009).

Plant hormones control not only plant growth and development but also serve in communication within the same species, with related or unrelated plant species, and with insects. Beyond phytohormones, the chemical messenger substances include peptides such as phytosulphokine growth factors and RNAs.

MicroRNAs play an important role in intracellular communication during plant development, either in cleavage during translation/transcription or in preventing translation. MicroRNAs are apparently necessary for meristem function, organ polarity, vascular development, floral patterning, and hormone response. Many of them are developmentally or environmentally regulated. Small interfering RNA probably serves as a signal during early development. In later developmental phases, the RNAi-dependent epigenetic processes are reminded of this early development phase, for example the heterochromatin configuration. At any rate, these RNAs play important roles in chromatin regulation and therefore in epigenetic silencing (Dugas and Bartel 2004; Kidner and Martienssen 2005).

Small molecules and proteins that normally support important functions in plant immunity, such as nitric oxide and reactive oxygen species (ROS), have been identified as multiply reusable components of other biological processes. Nitric oxide (NO) is a substance that has a regulatory function in numerous signal processes such as germination, growth, reproduction, and disease resistance. The same is true for diverse species of ROS.

2.1 *Signaling Molecules Serve in More than One Communication Process*

Auxin is used in hormonal, morphogenic, and transmitter pathways. As an extracellular signal at the plant synapse, auxin serves to react to light and gravity. It also serves as an extracellular messenger substance to send electrical signals and functions as a synchronization signal for cell division. At the intercellular, whole plant level, it supports cell division in the cambium, and at the tissue level, it promotes the maturation of vascular tissue during embryonic development, organ growth as well as tropic responses and apical dominance. In intracellular signaling, auxin serves in organogenesis, cell development, and differentiation. Especially in the organogenesis of roots, for example, auxin enables cells to determine their position and their identity (Baluška and Mancuso 2009). These multiple functions of auxin demonstrate that identifying the momentary usage is extremely difficult because the context (investigation object of pragmatics) of use can be very complex and highly diverse.

2.2 *Interpretation and Response Behavior to Sensory Data of Inanimated Nature*

The entire configuration of a plant (morphogenesis) is partially determined by mechanical inputs, for example, wind and gravity. Responses to contact involve signal molecules and hormones along with intracellular calcium, reactive oxygen species, octadecanoids, and ethylene.

Another common feature is contact-related gene expression. Many of these genes code for calcium-binding proteins, cell wall changes, defense, transcription factors, and kinase proteins.

The detection of resources and their periodic, cyclic availability plays a key role in plant memory, planning, growth, and development.

Interpretation processes in the plant body are highly sensitive. In taller growing plants, for example, the water balance places enormous demands on cell wall development and cell wall structures, which must adapt to the (often extreme) pressures involved in storage and pressure distribution. A sophisticated and multileveled feedback and feedforward system guarantees a plant-compatible water balance even under extreme environmental conditions. Plants are especially sensitive to light and have various receptors for UV, blue, green, red, and far red light. The angle of the light, combined with sensation of the growth of adjoining plants, is decisive in enabling plants to coordinate their growth with respect to the optimal light angle and shade avoidance. The roots receive constant signals from the aboveground parts of the plant for specific growth orientations (Baluška et al. 2006; Baluška and Mancuso 2009).

3 Transorganismic Communication

Sign-mediated interactions with organisms belonging to other species, genera, families, and organismic kingdoms are vital for plants and are coordinated and organized in parallel. They are almost always symbiotic or parasitic and range from mutually beneficial via neutral, up to damaging behaviors. The different forms of symbiotic communication require very different behaviors from the participating partners. This involves large numbers of complementary direct and indirect defense behaviors.

A limited number of chemical messenger substances are available to maintain and simultaneously conduct the communication between (a) root cells (of three different types), (b) root cells and microorganisms, (c) root cells and fungi, and (d) root cells and insects. The communication process in the root zone is generally trans-, inter-, and intraorganismic and requires a high communicative competence in order to be successfully interactive on all three levels and to distinguish messenger molecules from (similar) molecules not being part of messages.

A special type of plant synapse resembles the immunological synapse of animal cells and allows plants to respond to pathogen and parasite attacks as well as to establish stable symbiotic interactions with rhizobia bacteria and fungal mycorrhiza. Electrical signals can reinforce chemical signals or overcome short-distance responses of fungal mycelia that can be present on root surfaces. Some rhizobia bacteria are taken up in plant cells via phagocytosis during symbiotic interactions with roots of leguminous plants. The symbiotic relationship between legumes and rhizobial bacteria leads to the formation of nitrogen-binding nodules in the root zone. Nod factor signaling and thigmotropic responses of root hairs overlap here as well (Walker et al. 2002).

Today, several hundred species of fungi colonize more than 100,000 different plant species. This type of cohabitation requires symbiotic signaling. Roots develop from rhizomes in order to provide better conditions for mycorrhizal fungi, which in turn supply plants with better nutrients. For the fungus, the relationship is either balanced or predatory. Endophytic fungi, however, live in plants without triggering disease symptoms. Similar to the symbiosis between plants and mycorrhizal fungi, the symbiosis between asexual endophytes and grasses also represents a type of complementary parasitism (Witzany 2011).

Plants, insects, and microbes share a particular repertoire of signals. Some are therefore also employed strategically. Thus, plants also use insect hormones (prostaglandins) for specific defense behavior. Signal theft is common. Because plants can detect their own signals, they can presumably also detect similar signals that are used in communication between insects.

4 Interorganismic Communication in Plants

Plants can distinguish between self and nonself. Thus, defense activities are initiated against foreign roots in order to protect the plant's own root zone against intruders. The individual sphere of a root, along with its symbiotic partners, requires certain fundamental conditions in order to survive and thrive. When these prerequisites are threatened by the roots of other plants, substances are produced and released in the root zone that hinder this advance. Such defense activities are also deployed as antimicrobial substances against the microflora in the root zone.

Plant roots produce a wide range of chemical substances: (a) some enable species-specific interactions; (b) many of these substances are released tens of centimeters into the surroundings; (c) these substances have strong but not necessarily negative effects on animals, bacteria, viruses, and fungi; (d) released substances have a defensive function against other plants; and (e) many substances have absorptive characteristics that reduce the negative effects of substances. Plants use biotic signals to inform each other about the presence, absence, and identity of neighboring plants, growth space, growth disturbances, and competition (Dunn and Handelsman 2002; Fleming 2005; Baluška et al. 2006).

5 Intraorganismic Communication

5.1 Intercellular Communication

Short-distance communication differs considerably from long-distance communication. As a rule, both complement each other. Intercellular communication in the root zone (in the soil) differs from that in the stem region aboveground. Both are necessarily coordinated with one another in order to enable life in these different habitats. Intercellular communication informs other plant parts about events in specific organs or regions of the plant (especially in large plants), for example, sugar production in leaves, the reproduction in flowers, and resource utilization by the roots (Baluška et al. 2006).

Plant cells are connected by plasmodesmata. These connecting channels enable the flow of small molecules as well as ions, metabolites, and hormones, and allow the selective exchange (size exclusion limit) of macromolecules such as proteins, RNAs, and even cell bodies. The plasmodesmata impart plants with a cytoplasmic continuum known as the symplasm. But plasmodesmata are more than mere transport channels; they also regulate and control the exchange of messenger substances in a very complex manner. In symplastic signaling, the intercellular communication of plants differs fundamentally from that in other organismic kingdoms. It integrates various communication types such as local and long-distance communication. Beyond symplastic communication (especially in the meristem, where new tissues are produced), plants also exhibit the receptor-ligand communication typical of animals. While receptor-ligand communication determines stomatal patterning in the epidermis of mature leaves, trichome patterning is mediated by symplastic signaling (Baluška et al. 2006; Baluška and Mancuso 2009).

For long-distance signaling movement, proteins play an important role. Movement proteins convey information bearing RNA from the stem and leaves to the remote roots and flowers. The movement protein allows the mRNA to enter the plasmodesmata tunnel, into the phloem flow. Once it has entered this transport system, it can relatively rapidly reach all parts of the plant. These RNAs can control the levels of other proteins. The level contains information for local tissues, for example, about the general physical condition of the plant, the season, or the presence of dangerous enemies.

Plasmodesmata are prerequisites for intercellular communication in higher plants. In embryogenesis, they are an important information channel between fetal and maternal tissue. The further the development of the embryo, the more reduced the cell-cell communication between embryo and maternal tissue.

Cell-cell communication via direct transmission of transcription factors plays a central role in root radial and epidermal cell patterning as well as in shoot organogenesis. The cellular organization of the roots is determined during the plant's embryonic development and is controlled by intercellular communication.

There are about 1,000 known protein kinases/phosphatases, numerous secondary messengers, and many thousands of other proteins. Through their life cycles and their growth zones, plants develop a life history of environmental experience that they can pass on to later generations and, should they themselves grow to be several hundred years old, utilize themselves. Even small plants store stress experiences in their memories and then use these memories to coordinate future activities (Baluška et al. 2006; Baluška and Mancuso 2009).

5.2 *Intracellular Communication*

Intracellular communication in plants takes place between the symbiogenetically assimilated unicellular ancestors of the eukaryotic cell, mainly between the cell body and cell periphery. It transforms and transmits external messages into internal messages that exert a direct (epigenetic) influence on the DNA storage medium and trigger genetic processes; this leads to the production of signal molecules that generate a response behavior.

Reports on the transfer of mitochondrial genes between unrelated plant species caused some surprise. While gene transfer is an extremely rare event in animals and fungi, it is common between plant mitochondria. Variations in repetitive DNA that manifest themselves as variation in the nuclear DNA complex have far-reaching ecological and life history consequences for plants.

The function of a eukaryotic cell depends on successful communication between its various parts. Plastids send signals to regulate nuclear gene expression and thus to reorganize macromolecules in response to environmental influences. It has been shown that microRNAs regulate certain developmental processes such as organ separation, polarity, and identity, and that they define their own biogenesis and function. Eukaryotic genomes are regionally divided into transcriptionally active euchromatin and transcriptionally inactive heterochromatin.

Epigenetic changes can take place without changes in genomes, for example, through various inactivations and activations of genetic datasets via chromatin remodeling, transposon/retrotransposon release, DNA methylation, novel transcription, histone modification, and transcription factor interactions. Various stress situations in plants are known to cause transposon movements, and bacterial infections or UV stress can cause chromosomal rearrangements, that is, changes in higher-order regulation levels that control the transcription processes of the protein-coding DNA. Repetitive DNA is present in two syntactic combinations: tandem repeats and dispersed repeats. Tandem repeats consist of sequences that can contain several thousand copies of elements that are dispersed throughout the genome. Pericentromeric sequences consist of a central repetitive nucleus flanked by moderately repetitive DNA. Telomeric and subtelomeric sequences consist of tandem repeats at the physical end of the chromosomes. Retroelements and transposable elements are involved in replication and reinsertion at various sites in complex processes: These include activation of excision, DNA-dependent RNA transcription, translation of RNA into functioning proteins, RNA-dependent DNA synthesis (reverse transcription), and reintegration of newly produced retroelement copies into the genome (Villarreal 2005; Witzany 2010).

5.3 Viral Symbiotic Interactions

Via endocytosis, however, bacteria, viruses, and viroids interfere with this intracellular communication and can support, disrupt, or even destroy it. Intracellular communication offers viruses the opportunity to integrate certain genetically coded abilities of the host into their own genome or to integrate their own genetic datasets into the host genome. The ability of viruses to integrate different genetic datasets probably plays a major role in symbiogenetic processes. The eukaryotic cell is composed of a multicompetent nucleus as a basic building block of life and a cell periphery consortium that was symbiogenetically the ancestor of other endosymbionts. Interestingly, both the nucleus and viruses have several similar features and capabilities: They both lack the protein synthesis pathways and the fatty acid producing pathways. Viruses were probably very important in the evolution of eukaryotic cells because they were able to conduct cell-cell union. There are strong reasons too, that the eukaryotic nucleus is of viral origin (Villarreal 2005; Roossinck 2010).

Many DNA viruses have encoded numerous nucleic acid metabolisms that are very similar to cell proteins. Examples include DNA polymerases, ribonucleotide reductase subunits, DNA-dependent RNA polymerase II subunits, DNA topoisomerase II, thymidylate synthase, helicases, and exorbinuclease.

One of the interaction processes between plant viruses and their host organisms creates a defense level against foreign genetic parasites. Plant viruses code for silencing suppressors in order to act against host RNA silencing, and some of these suppressors effect microRNA multiplication and hinder plant development. But also some viroids play a symbiotic role. Despite their small size and their noncoded genome, viroids can multiply, systematically spread from cell to cell, and trigger symptoms in the host (Roossinck 2010).

References

Baluška F, Mancuso S (2009) Signaling in plants. Springer, Berlin

Baluška F, Mancuso S, Volkmann D (2006) Communication in plants. Springer, Berlin/Heidelberg

Dugas DV, Bartel B (2004) MicroRNA regulation of gene expression in plants. Curr Opin Plant Biol 7:512–520

Dunn AK, Handelsman J (2002) Toward an understanding of microbial communities through analysis of communication networks. Antonie van Leeuwenhoeck 81:565–574

Fleming AJ (2005) Intercellular communication in plants, Ann Plant Rev. Blackwell Publishing, Oxford

Kidner CA, Martienssen RA (2005) The developmental role of micro RNA in plants. Curr Opin Plant Biol 8:38–44

Roossinck MJ (2010) Lifestyles of plant viruses. Philos Trans R Soc Lond B Biol Sci 365:1899–1905

Villarreal LP (2005) Viruses and the evolution of life. ASM Press, Washington, DC

Walker TS, Bais HP, Grotewold E, Vivanco JM (2002) Root exudation and rhizosphere biology. Plant Physiol 132:44–51

Witzany G (2010) Biocommunication and natural genome editing. Springer, Dortrecht

Witzany G (2011) Biocommunication in soil microorganisms. Springer, Berlin, New York

Information, Noise and Communication: Thresholds as Controlling Elements in Development

Anthony Trewavas

Abstract Organisms are dependent on the continual transmission of information both within cells and from outside them. Information is concerned with the conveyance of signals that require both a transmitter and a receiver able to decide what is sent. Accuracy in transmission is degraded by noise, and the evidence that shows noisiness in genetic circuitry is described. Reliable noise coupled with positive feedback constructs probabilistic thresholds amongst a population. In higher plants, stochastic distribution of thresholds enables quantitative variation amongst cells, tissues or plants to variable strengths of signals. It is the function of information to be communicated, but the gel structure of the cytoplasm together with the ordering by structured water might instead increase noise in transmission by interfering with the necessary movement of molecules in signal transduction. To reduce potential noise in signal transmission and transduction, it is suggested that abrupt phase transitions in microdomains of the cytoplasmic gel structure are induced by cytoplasmic calcium, amongst other signals. Plasmodesmata also contain actin gels, and communication between cells may simply be controlled by abrupt gel phase transitions. Two threshold phenomena are thus seen in plant cells important during development. The first involves noise and positive feedback; the second, gel phase transition.

1 What Is Information?

Biological information is conveyed by particular sequences of signals and messages that originate within the cell or outside it. Information theory, first propounded by Shannon and Weaver (1949), stated that the information content of any message

---

A. Trewavas (✉)
Institute of Molecular Plant Science, University of Edinburgh, Edinburgh, Scotland
e-mail: trewavas@ed.ac.uk

G. Witzany and F. Baluška (eds.), *Biocommunication of Plants*,
Signaling and Communication in Plants 14,
DOI 10.1007/978-3-642-23524-5_2,

was determined by the probability of occurrence of a particular message as against others. Shannon and Weaver (1949) were concerned with the accuracy of transmission of messages through phone lines to a receiver and with maintaining the secrecy of transmission.

The critical components of information conveyance are firstly a transmitter of information and secondly an interpreter of the information transmitted. Following Aristotle's implied meaning towards information as 'surprise', it was argued that the more surprising the contents of the message, the greater its information content and in turn the lower the probability of its occurrence.

If, for example, the message transmitted was 'the sky is blue', little information is being conveyed since the information is of little surprise. If, on the other hand, the conveyed message is 'the sky is green and black striped', that is most certainly surprising, containing new unexpected information, thus increasing its novelty and in turn its information content. Surprise does suggest rarity, and rare things by definition occur infrequently and thus with low probability. Low probability messages are associated with strong constraints on the information transferred. These constraints can, in many cases, be related to the degree of detail in the message; the greater the detail, the greater the likely information content. The message 'the sky is deep blue interlaced with aeroplane vapour trails, there is a light warm wind and the smell of honeysuckle in the air' increases the information content too and constrains or limits the described scene compared to the first simple message, 'the sky is blue'.

Shannon drew attention to the possible relation of information content to entropy (Vedral 2010). Highly ordered systems have low entropy; disordered ones, high entropy. In terms of messages, comprehensible messages are very ordered whereas disordered messages can be uninterpretable. There are about half a million words in the Oxford English Dictionary. Only certain discrete combinations of words and in a particular order out of a truly enormous number of possible word combinations provide sensible information. For a five-word combination, there are at least 10^{25} possibilities. Five-word messages that make sense to a human receiver are probably of the order of a few thousand. Genuine messages are therefore by definition rare and note also the specific elements of interpretation that have to be present in the English-speaking receiver. A light signal is interpreted differently by a seedling stem compared to their leaves.

However, there are intrinsic problems with trying to determine the information content of any biological signal. If, for example, a plant growing in laboratory conditions, experiences a change in light intensity, that is expected information because such variations are normal for any growing plant. Variable cloud cover and sun specks lead to unexpected changes in light intensity. If, on the other hand, the change in light intensity is accompanied by a change in temperature, water availability and humidity, then the information content will be higher and may indicate the progress towards evening.

2 Noise in Transmission of Information Degrades Accuracy in Response

One of the major concerns of Shannon and Weaver (1949) was to try and estimate the accuracy of transmission of messages down phone lines. The degradation of information transmission is called noise. The effect of noise is usually to jumble or omit perceived words, and meaningless messages are more probable than meaningful ones. Thus, the relationship that Shannon developed, equating information to entropy. What he indicated is that noise in message transmission is disordering. Thus, if noise occurs in cellular messages, this may have serious consequences for either survival or interpretation of external signals.

2.1 Noise Is Likely Inevitable in Living Systems

Living cells use many thousands of chemical reactions and other molecular interactions. There is inevitable noise in such processes since many reactions are probabilistic, requiring two or more molecules to come together in a crowded cytoplasmic environment. Later in this chapter, I will indicate how the structure of the cytoplasm may interfere in these necessary events and increase unwanted biological noise. Life survives because the tendency of randomising processes at the single molecule level is however countermanded by correcting statistical forces. That is, a larger number of molecules working together tend, on average, to counteract individual stochastic events. Many control circuits have been constructed in cells to offset or reduce noise. Negative feedback is the commonest, providing information to the earlier part of the circuit to try and modulate or stabilise throughput. But one hazard of negative feedback is the delay in response and that, in itself, often makes the process noisy. Feedback really requires instant effects if it is to reduce noise substantially.

The simplest circuitry perhaps involves gene activation, transcription, translation and that immediately introduces probabilistic events that can destabilise control. DNA during transcription can change its structure; proteins necessary for transcription can drop off or change conformation and become non-functional for periods of time. In other transduction circuits, signalling complexes have to be formed from large numbers of soluble proteins aggregating together; delays and failures in construction must inevitably be common. Channels for ion signals, detected using patch clamp, are observably noisy. Noise is endemic, and the problem that arises is how individual living cells can manage and survive within that framework of noise.

2.2 Evidence for Noise in Genetic Circuitry

The evidence for noise in transcription/translation is extensive, and a variety of single bacterial, yeast and cultured cell systems have been used to demonstrate its presence. The methods developed to demonstrate noise have all marked milestones in technical achievement. Suitable fluorescent probes with some superb microscopy have enabled comparison of copy numbers of specific proteins and mRNAs between individual cells from the same culture. Lango and Hasty (2006) list 25 papers that have used this technology. The ultimate has been the imaging of the synthesis of individual protein and mRNA molecules.

The most common detection of noise has been to compare copy numbers of both specific mRNA and specific proteins between single cells. Greater noise between individual cells is to be expected in proteins that are expressed in small rather than large abundances, and this has proved usually to be the case (Federoff and Fontana 2002). Elowitz et al. (2002) defined two kinds of noise in protein copy numbers/cell that they observed. Intrinsic noise was defined as the variation in expression between two identical genes in the same cell. Extrinsic noise was considered global within the cell reflecting, for example, variations in polymerase numbers or other regulatory proteins affecting many transcription events. Intrinsic noise disappeared more quickly than extrinsic noise when cells were followed through cell cycles.

Lack of correlation between a specific mRNA level and its protein product are considered to originate from the differential stability of both; mRNA in bacterial cells, for example, decays stochastically within a few minutes, proteins are far more stable (Taniguchi et al. 2010). Transcription rates, regulatory dynamics and genetic factors all contribute to the amplitude of noise (Elowitz et al. 2002). Rosenfeld et al. (2005) measured the quantitative relation between transcription factor concentration and the rate of protein production from the downstream gene (so-called gene regulation function) and observed how the ratio between these two fluctuated dynamically, thus limiting the accuracy of genetic circuitry. Textbook models that picture transcription factors binding to DNA and protein synthesis continuing in an orderly level thereafter are clearly very misleading.

The range in copy number of a single protein species between individual cells can be enormous. Careful measurements using a technique that could measure individual protein molecules indicated up to 15-fold variation (Taniguchi et al. 2010). If the genetic circuitry incorporated positive feedback at some stage in their control sequence, then noise itself was sufficient to enable the induction of two distinct phenotypes. To and Maheshri (2010) introduced a promoter with a single binding site for an effector molecule or seven binding sites for the same effector in a system with positive feedback in the control loop and showed that noise was able to induce bistable states without any change at all in effector concentration; some cells were spontaneously switched on, others not.

2.3 *Noise Can Spontaneously Induce Polarity and Ensure Each Cell Is Effectively a Unique Phenotype*

Similar and significant results of To and Maheshri (2010) were found in the establishment of yeast polarity in an unpolarised cell. Membrane-bound signalling molecules able to recruit from a cytoplasmic pool with positive feedback and in limited cytoplasmic copy number (and thus noisy), spontaneously established a site of polarity in yeast (Altschuler et al. 2008). Such results have obvious significance for the establishment of polarity in many stages of plant development. Deterministic models would not of course predict this unexpected outcome that must result from noise-induced variations in the conformation of either the promoter or the membrane-binding protein, in this case, CDC42. In *E. coli*, a single-chance event, the spontaneous dropping of a repressor from DNA in the *lac* system can introduce a bistable condition in which lactose floods into the cell and switches on the lactose metabolising system (Pearson 2008).

Noise in an upstream gene due to transcription factor variation can be transmitted to downstream genes (Pedraza and van Oudernardene 2005). Further observations of complexity were made when a number of different gene products were all imaged in single cells at the same time. Analysis of 11 genes altogether indicated that each cultured cell produced its own unique pattern of gene expression, thus generating individual phenotypes (Levsky et al. 2002).

2.4 *Transcription and Translation in Single Cells Takes Place in Brief Bursts*

One surprising feature that has emerged from observations of single mRNA or single protein molecule production is that synthesis takes place in bursts rather than continuously, thus again contradicting textbook models. By constructing a special technology for visualising individual mRNA molecules for a single gene, Golding et al. (2005) were able to image the production of single mRNA species and found that throughout the period of observation, the gene was active in bursts producing between 1 and 8 molecules each time, but synthesis only occupied 10% of the observation period. By imaging the appearance of single protein molecules in a single bacterial cell by fluorescence, Yu et al. (2006) observed patterns of stochastic bursts in synthesis with long periods of inactivity. There were usually only 1–2 bursts in synthesis/cell cycle, and the numbers of molecules/burst followed a simple power series. Synthesis of p53 in human cells oscillated with different frequencies between single cells after stimulation by radiation (Geva-Zatorsky et al. 2006). Bursting characteristics in synthesis obviously tends to randomise production in time.

With a delay between the synthesis and degradation of any molecule, Pedraza and Paulsson (2008) observed that a simple memory was created. Sigal et al. (2006), using human cells, observed that different proteins within one metabolic pathway showed less variation than between proteins in other pathways. They quantified the levels of some 20 different protein species and reported that the high or low noise variability could last at least between two cell cycles. Again, they indicate that this is a kind of molecular memory. The persistent memory for protein levels might induce cell individuality. Memory can only be present however if something has first been learnt. The learning mechanism involves the variable synthesis of specific proteins in this case, and such learning and memory capabilities are equally present in plant cells (Trewavas 1999).

Cells of the same type can again generate diverse physiological traits. A further study that labelled 2,500 proteins in yeast under different growth conditions found that there were dramatic specific-protein differences in noise that were correlated with function (Newman et al. 2006). However, these authors also reported that there was much greater noise in the proteins that respond to environmental signals, whilst those involved in protein synthesis were much quieter.

2.5 *Is Noise Useful or If Not, Can It Be Reduced?*

By engineering mutations into a control region of genes that confer antibiotic resistance in yeast, Blake et al. (2006) constructed two strains that differed in the noisiness of their expression. When incubated in a normally lethal concentration of antibiotic, the noisier strain survived much better. This is a kind of 'bet hedging' that noise can introduce to improve fitness. There will always be some variants that potentially can accommodate stressful circumstances better and thus ensure survival of the line. Noise must thus have value in variable environments. But on the other hand, noise will also cause cells to deviate from the optimum that they might have achieved in its absence. So noise may be useful only under certain less-than-usual circumstances. Clearly, there should be a trade-off between the control of noise and the need to optimise behaviour, and different organisms will alter the balance in this trade-off. Noise may also degrade biological signals and cause difficulties in perception and reduce appropriate sensitivity. But there may be ways around this by synthesising large numbers of critical proteins. Cells also get noisier as they get older, perhaps unsurprisingly.

Very low levels of electrical noise in neurons actually improved the response to weak signals (Collins et al. 1996). At that time, the phenomenon was called stochastic resonance. In these situations, a periodic signal inside cells that might normally be insufficient to be sensed is enhanced by the presence of noise. Elowitz et al. (2002) set up an oscillatory system using negative feedback on some of their gene circuitry and observed greater noise as a consequence. Proteins that respond to environmental signals are noisier than those that deal with protein synthesis that are relatively quiet (Newman et al. 2006).

So do cells have ways of reducing noise? The capacity of a cell to control its internal processes is obviously limited by information loss. Theoretical analyses indicate that there are considerable limits to the possibility of reducing noise. To decrease the standard deviation of protein distribution by half between separate cells would require an increase of 16-fold in numbers of signalling proteins (Lestas et al. 2010). Cells can use brute force when necessary to reduce noise resulting in regulatory genes being transcribed tens of thousands of times/cell cycle.

In a cascade (e.g. MAP kinase cascade), information is obviously progressively lost from upstream events. Information transfer in cascades will be limited by the component(s) made in the lowest copy or activity numbers. A five-step linear cascade in gene circuitry, for example, requires at least 25 more bursts of synthetic activity than a single step to maintain the same capacity to reduce noise. 'The mechanisms for preventing noise propagation such as time averaging or kinetic robustness to upstream changes cause a greater loss of information; mechanisms that minimise information losses such as all-or-none, non-linear effects actually increase noise'. 'Making a decent job is 16 times harder than a half decent job' (Lestas et al. 2010).

Parallel signal and control systems can instead improve noise suppression because each pathway contributes independent information about the upstream state. However, the loss of information is determined by the number and frequency of signalling events, not their nature. There are physical constraints on the sensitivity with which external signals can be sensed and low impact signals will only be perceived with greater noise than larger ones (Bialek et al. 2005).

3 Consequences of Signal and Genetic Circuitry Noise for Plant Growth and Development Control

3.1 *Relevance of Noise in Genetic and Transduction Circuitry for Plant Development*

There are a number of significant conclusions for plant growth and development that can be drawn from the above studies. The above information was of necessity gained on single-cell systems, and it clearly applies to single eukaryotic cells. There are several single-cell systems in plants and for which noise might contribute to understanding their behaviour. These are guard cells, the fertilised embryo, pollen tubes and root hairs. Lateral roots and maybe even leaves and buds may be in this category too because they potentially originate from single cells. These tissues surely use positive feedback as part of their behavioural response to inducing stimuli and to carry development and plasticity in responses forwards. If there are errors or extrinsic noise in the progenitor cell such as the fertilised embryo, it is certainly feasible that these noise variations will be continued in the final seed by epigenetic processes that it is now known, can last through generations (Molinier et al. 2006).

In an article entitled 'Reliable Noise', Levens and Gupta (2010) point out that statistical fluctuations (i.e. noise) involving a weak promoter of a transcription

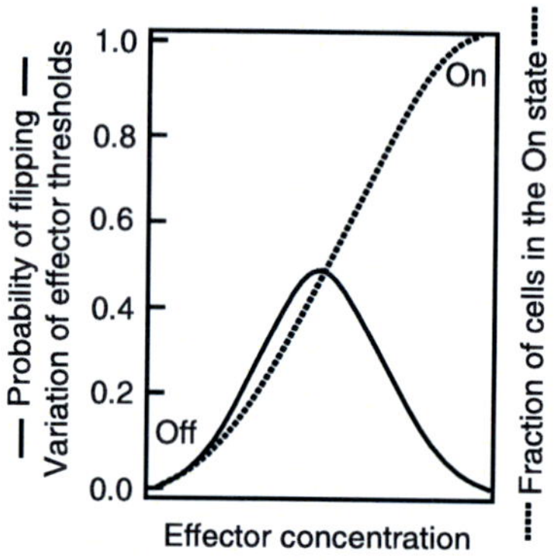

Fig. 1 The fraction of cells expressing a gene is a function of the concentration of an effector molecule. At very high or low effector concentrations, the population of cells has the gene either On or Off (as in the system used by To and Masheri 2010 and described in the text). At intermediate concentrations of the effector, some cells are 'On' and others are 'Off'. The right axis (*dotted line*) indicates the fraction of cells expressing the gene at different effector concentrations. The left axis (*continuous line*) represents the probability that a cell has flipped from the 'Off' state (no gene expression) to the full 'On' state of gene expression. The *continuous curve* also represents distribution of effector thresholds in the population of cells. Data redrawn from Levens and Gupta 2010

factor, can generate intrinsic noise. If the transcription factor is short lived, then the noise can be amplified by inducing extrinsic noise on each of the genes the transcription factor binds to; including, if so arranged, the original transcription factor itself. Dependent on the numbers of transcription sites as well as the potential variable strengths of promoters, different target genes may be tuned to switch to high output at different concentrations of the transcription factor. The consequence is a range of different phenotypes each with its own combination of gene products expressed to different degrees and responding differentially to a defined signal. Positive feedback of this kind can also fix the original gene into the 'on' position. Such stochastic switching will eventually generate a range of responses in unicellular organisms to a defined signal. Figure 1 summarises their thesis and is based on the observations of To and Maheshri (2010).

The mechanism described by Levens and Gupta (2010) in tissue responses is even more relevant if the initial gene(s) is concerned with controlling the synthesis of effectors. As indicated above, seeds could be an excellent example. Evidence that noise is an issue in plant cells and tissues and is observable between individual seeds was provided by Dahal et al. (1994) who reported variations of one enzyme up to a 1000-fold between individual seeds.

3.2 *Synchronising Effects of Signals in Plant Cells*

I have indicated previously that one of the most puzzling features of the effects of exogenously added plant growth regulators is that they appear to synchronise the responding tissues (Trewavas 1982, 1987, 1991). Synchronisation suggests an underlying probabilistic mechanism. The classic example is in cell division in which cells have to cross a threshold before commencing division with the thresholds varying stochastically (noisily) amongst individual cells (Smith and Martin 1973). The effect of increasing the size of a cell division stimulus is simply to enable those cells whose threshold has now been exceeded to enter the division cycle. The thresholds are not necessarily fixed however, but can be lowered by various environmental or hormonal triggers. Most crucially, a system using variable thresholds enables a dose response to be constructed to variations in the concentration of the inducing stimulus. Smith and Martin (1973) considered that the construction of the threshold involved positive feedback mechanisms and thus the introduction of noise.

Figure 2a, b are taken from Bradford and Trewavas (1994). The symbols of Fig. 2a represent data points of the germination against time of a null gibberellin mutant of tomato when placed in different concentrations of gibberellin. Crucially, the impact of increasing the exogenous gibberellin concentration is to induce more seeds to cross the threshold from dormancy to germination. But a further effect of increasing the gibberellin stimulus is to increase a faster rate of germination in those seeds whose threshold has been exceeded. The lines in Fig. 2a were calculated by Kent Bradford from a simple model that contains both a threshold and a time

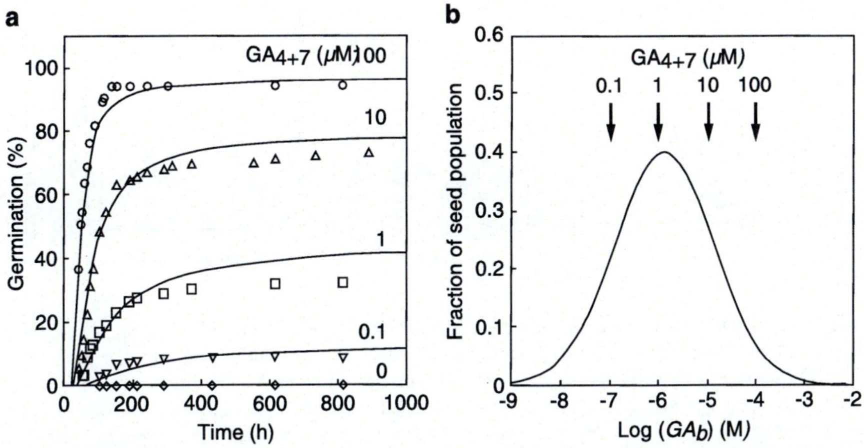

Fig. 2 Germination time courses and distribution of thresholds to gibberellin concentration in a population of tomato seeds. (**a**) Germination time courses (represented by symbols) of a GA-deficient mutant of tomato in different gibberellin concentrations from 0.1 μM to 100 μM. Increasing the gibberellin content of the medium increases the number of seeds germinating and shortens the time to germination too. The *solid lines* are the time courses predicted by a simple model incorporating both time and gibberellin concentration. (**b**) This graph shows the distribution of thresholds to gibberellin amongst the population of seeds. Only seeds with thresholds above the applied concentration will germinate. The extent to which gibberellin concentration exceeds the threshold increases the rate of germination. The distribution of thresholds is stochastic. Figures copied from Bradford and Trewavas (1994) with permission

component. Figure 2b indicates that there must be a Poisson (stochastic) distribution in thresholds amongst the individual seeds population. Note the similarity in character of response in Fig. 2b to Fig. 1 (probability in flipping and effector threshold variation axis).

The prediction here is that extrinsic noise in the fertilised cell is then stabilised by positive feedback and epigenetic processes, so that this initial noise variation is carried through to the mature seed. There is clearly a long-term memory in operation.

Figure 3a, b are modified from Gilroy and Trewavas (2001). Figure 3a reports the numbers of individual cereal aleurone cells that synthesise α amylase as gibberellin concentration in the medium is increased. There is clearly population variation in the thresholds of individual cells, as more cells cross their gibberellin threshold, more amylase is synthesised. The data shown as *triangles* and *filled circles* are plotted on the template of Fig. 1. The distribution of thresholds and thus the probability of cells to synthesise amylase is similar in character to Fig. 2b and to

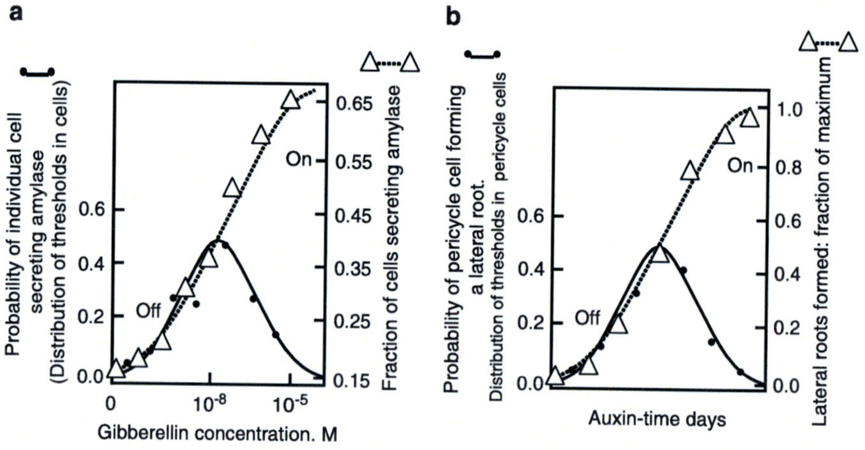

Fig. 3 Examples in plant cells showing that populations of tissue cells exhibit a stochastic variation in thresholds to inducing stimuli. (**a**) Reports the effect of variation in gibberellin concentration (from zero then 10^{-11} to 10^{-5} M in tenfold steps) on the numbers of individual aleurone cells synthesising amylase. The *triangles* represent the fraction of aleurone cells synthesising amylase, and the *closed circles* the probability of aleurone cells to synthesise amylase and thus the distribution of thresholds amongst the cell population. The symbols are plotted on the template of Fig. 1 and indicate the strong similarity in behaviour to Fig. 1. (**b**) Reports the effect of various lengths of time of treatment of root segments with auxin on numbers of lateral roots formed. Segments were incubated in auxin for variable periods of time and then removed and further incubated for a total period of 6 days in the absence of auxin. Lateral root numbers were then estimated. Lateral roots are formed from the pericycle. The triangles represent the fraction of lateral roots formed against the maximum number plotted against the total auxin treatment time in days. The *closed circles* represent the variation in thresholds to auxin amongst the population of pericycle cells and thus the probability that lateral roots will be formed. Again the data have been plotted on the template of Fig. 1 and indicates likely similarity in mechanism. Original data for (a, b) are to be found in Gilroy and Trewavas (2001)

Fig. 1. The distribution of thresholds is stochastic and likely resulting from the stochastic variation of noise coupled with a positive feedback mechanism during aleurone cell development.

Figure 3b shows numbers of lateral roots formed against auxin-time as the inducing stimulus. Root segments were incubated in auxin for variable periods of time before estimates of all treatments for lateral root number after 6 days. Again, the actual data are plotted as the symbols of *triangles* and *closed circles* on the template of Fig. 1. Since lateral roots are generally thought to develop from a single pericycle cell, the variation in thresholds again looks stochastic and presumably results from noise plus positive feedback during root and in particular pericycle cell development. As more cells cross their thresholds as the exposure to auxin increases, more lateral roots are formed.

By including this kind of mechanism involving noise and positive feedback in critical proteins, cells and tissues exhibit a dose-dependent response to the strengths of environmental or hormonal signals and to their duration. Other aspects of development where this mechanism may control is in seed dormancy breakage, leaf drop related to water deprivation in trees, bud break, root hairs, guard cells, etc. (Trewavas 1987, 2003). These examples indicate the importance of the threshold in understanding these phenomena.

4 Communication if Information Starts Within the Individual Plant Cell

4.1 Stochastic Responses Are Observed in Individual Plant Cells In Situ

One way to reduce noise is to use parallel changes that meet at some point and the result then averaged. It could potentially be seen as a basic reason why organisms became multicellular some two billion years ago, each cell receiving information and interpreting it with the necessary input of noise. With appropriate interaction, the noise level could be reduced. But this noise reduction will only work if the information from both cells is adequately and quickly transferred between cells and the subsequent response then being the average between the two cells. Does this actually happen in plant tissues?

That the stochastic, probabilistic response found in single cells above could be observed in single cells in situ in tissues was clearly shown by Nick et al. (1993). These authors used a microbeam of red light to switch on anthocyanin synthesis. They observed that there was considerable spottiness in response with patches of cells of varying sizes being switched on when using intermediate levels of illumination. They considered that the spottiness resulted from positive feedback in the transduction processes. Variation in individual cell thresholds is indicated. With saturating levels of red light, all cells respond. However, over the longer term, they observed a much slower inhibitory response that stopped anthocyanin synthesis.

Leaf patchiness in guard cell responses to closing signals is well established. However, the responses of individual cells to exogenously applied abscisic acid look distinctly stochastic and similar to the probabilistic response described above (Trewavas 2003). Variation in thresholds between individual guard cells is again implied. The speed with which patches of guard cell apertures change in response to closing signals however, suggests potential patch interaction issuing from another quicker source. A vapour phase–closing signal from the mesophyll is indicated (Sibbersen and Mott 2010). Excess short-term water loss from the leaf causes short-term stomatal closure by vapour phase signals. Prolonged water loss generates a slower ABA-dependent signal and now ABA-dependent closure and lasting for a much longer period.

4.2 Two Signals in Plant Development Change?

If this situation in guard cell closure can be generalised, and I consider it can be, the suggestion is that at least two signals are communicated in many aspects of plant behaviour. Growth regulators do not act as the initial inducers of behavioural change but as later signals that prolong and deepen the cellular change enabling its continuation for much longer periods of time and presumably reflecting the strength and depth of the signal. Certainly, recognition of this potential would mediate previous controversy based on the observable speed of cellular change as against the slower kinetics of changes in growth regulator concentration (Firn and Digby 1980). Perhaps a simple analogy from paper photography might suggest what is going on. Changes in development or behaviour are initially induced like the developer in photography; plant hormones act more like the fixative.

5 The Gel Nature of the Cytoplasm Provides for an Alternative Set of Threshold-Controlled Changes

5.1 Cytoplasm Is an Organised, Highly Structured Network

The cell is a highly structured entity. Although the basic outlines of the kinds of organelles, their structure, function and behaviour have been reasonably clear for many years, there is an area that is rarely referred to. The molecular structure of the cytoplasm is unclear apart from the generalised statement that some or all of it is gel-like in nature. Communication between cells and within cells is changed by the perception of the nature of cytoplasmic gels and their behaviour. In this context, then another controlling, threshold phenomenon appears separate from that indicated above.

The evidence for a defined structure of the cytoplasm at the molecular level comes from at least six sources.

1. The remarkable experiments of Zalokar (1960) and later Kaempner and Miller (1968). These authors respectively centrifuged whole cells of either a *Neurospora* hypha or the alga *Euglena gracilis*. Centrifugal segregation was accomplished in *Euglena*, for example, into the common fractions of starch grains, nucleus and large organelles, ER and a cytoplasmic soluble fraction. However, no macromolecules or enzyme activity were detected in the soluble fraction of the alga or fungal hyphae despite the retention of viability (Srere 2000). These observations confirmed earlier suppositions from the 1930s that cytoplasmic proteins are not free in the cytoplasm but attached to large subcellular structures that can be easily centrifuged.
2. The second indicator of structure comes from evidence for metabolons, integrated entities of enzymes that are responsible for metabolic pathways (Burbulis and Winkel-Shirley 1999; Winkel 2004). Metabolons encompass all the major metabolic pathways. The metabolon structure ensures that substrates in the pathway are not free but passed from one enzyme to another ensuring greater speed of metabolic output. Some metabolons may only transiently associate and may combine into different complexes. In signal transduction, large complexes of proteins are thought to form transiently around nucleation sites formed from PH or SH domains in membrane-bound proteins.
3. Polyribosomes have been shown to be localised to specific cytoplasmic regions and mislocalisation alters the phenotype (Luby-Phelps 2000). Even when cells were heavily permeabilised, enabling molecules of 400,000 molecular weight to penetrate, very few proteins were observed to leak out, indicating binding to the cellular contents.
4. Using two hybrid methods, large-scale networks of protein-protein interactions and co-expression networks in yeast and plant cells have been reported (Costanzo et al. 2010; Ficklin et al. 2010; Mutwil et al. 2010; Yu et al. 2008). These networks exhibit the typical small world, or scale free, network structure constructed of hubs and connectors.
5. Much of the cytoplasm is penetrated throughout by a network of microtubules and microfilaments and intermediate filaments to which other proteins can attach themselves. A complex of note is the peripheral cytoskeleton found underneath and attached to the plasma membrane that is about 100 times thicker than the plasma membrane (Alberts et al. 1983). It is known that it is this structure, and not the internal cortical matrix, that is responsible for governing specific aspects of morphological development in *Acetabularia* (Briere and Goodwin 1988; Goodwin and Pateromichelakis 1979 Goodwin et al. 1983; Mandoli 1998).
6. The experiments by Ling (1992) examined what happened to the potassium in the cells when they were cut in half. Although potassium is thought to be soluble in the cytoplasm, Ling (1992) observed that potassium only leaked out when proteins started to do so as the cell died. Some kind of structured binding of potassium to protein is indicated.

All these data suggest that the cytoplasm is a complex integrated network with perhaps microdomains specific for particular functions. The description of the cytoplasm as a gel capable of transition to a sol is of long standing and owes much to observations of organisms like *Amoeba* whose pseudopodial behaviour is constructed by swift changes between gel and sol. What then is known about gel structure?

5.2 *The Design of Specific Synthetic Gels Is Intensely Researched*

While the gel structure in organisms is the subject of intense research, good understanding may be gained by investigating the behaviour of synthetic gels. Artificial or synthetic gels are loosely described as two-component systems of a semisolid nature, but rich in liquid. There is intense industrial interest in the construction of 'intelligent polymer gel systems' for biotechnology, medicine and environmental issues (see references in Chen and Hoffman 1995). Gels with particular properties for drug delivery or for DNA transformation with the aim of delivery across the plasma membrane and directly into the nucleus have been constructed (Pack et al. 2005). A gel whose volume oscillates controlled by a non-linear reaction involving redox oscillations has been reported (Yoshida et al. 1999). These properties indicate the potential for biological gels constructed in different ways to have biologically interesting properties.

The cytoplasm contains anywhere from 20% to 40% protein, and it is some of these proteins, actin is a good example, that are likely responsible for cytoplasmic gel structure. However, with many proteins in the cytoplasmic gel and capable of gel formation, there is room for the construction of gel microdomains with different properties. Whatever structure is present in the cytoplasm, it cannot be fixed but must be capable of being changed in order to accommodate development and the response to signals. Flexibility in gel structure and behaviour becomes essential.

5.3 *Synthetic Gels Indicate the Presence of Structured Water*

The most familiar synthetic gel is the culinary jelly constructed from partly degraded collagen (gelatin). Such gels are formed at 5% collagen to water. Other gels using different polymers can form with a 1/1,000, polymer/water ratio. Such gels maintain their shape even though composed 95% or more of water. The water must clearly be in a form different from ordinary liquid water.

Each water molecule is an electric dipole with a $\delta+$ charge on the proton and a $\delta-$ charge on the oxygen (Fig. 4a). H-bonds can form between different water molecules and enable the formation of non-covalent water structures and most

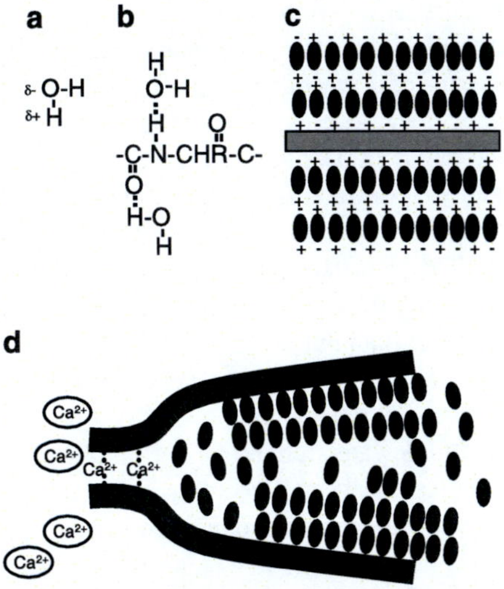

Fig. 4 Potential relation of structured water to gel formation in cells. (**a**) The water dipole, slight negative charge on the oxygen, slight positive charge on the hydrogen. (**b**) Potential hydrogen bonding of water molecules to the peptide bond. (**c**) On unfolded proteins, layers of water build up through initial hydrogen bonding to the peptide bond and then through hydrogen bonding to these vicinal water layers. The layers of structured water could be up to ten layers deep between adjacent polypeptide chains, thus linking them together in formation of a gel. (**d**) Ca^{2+} can cross-link adjacent polypeptide chains through negatively charged side chains and others and thus disrupt the structured water between them

certainly do so when ice forms. It was originally thought that unfolded, extended proteins would adopt a random coil configuration, but the three-dimensional structures of unfolded proteins, like partially degraded collagen, have turned out in contrast to be reasonably well defined. These configurations have been found to be stabilised by the interactions and structuring of water molecules around them. Although gelatin is an artificial gel, its structure has recently been clarified and may be representative of other unfolded globular proteins (Kozlov 1983; Carvagal and Lanier 2006).

Kozlov (1983) in early work indicated that water in gelatin existed in at least three distinguishable configurations. The first is now known to result from alignment of separate chains of collagen. In the proline-rich regions, the collagen molecules are cross-linked to adjacent chains through three or more water molecules. The first and third water molecules are hydrogen bonded through the carbonyl (−C=O) residues of the peptide bond of two adjacent chains of collagen. These two water molecules are then linked together by a third acting as a bridge (Carvagal and Lanier 2006). The second form of water is a tightly bound, usually single layer, of water molecules (vicinal water) responsible for hydration. There are several kinds known. Charged collagen side chains structure water around themselves. The water molecules structure initially through the dipoles and then to each other. Hydrophobic residues generate clathrate structures again around themselves.

However, if the protein is unfolded then in the open, neutral, polypeptide regions of collagen, water is attached through hydrogen bonds to the –imino (–NH) and (–C=O) carbonyl groups of many of the other peptide bonds (Fig. 4b). This second form of water does not freeze even at temperatures of −60°C.

The third more weakly bound water results from hydrogen bonding to the water molecules already attached to the open polypeptide regions and can form layers of structured-water attachment, four to even ten layers deep as intimated in Fig. 4c (Pollack 2001; Ling 2006). This structured water is in a form somewhere between the structure of ice and liquid, that is, in the structure expected of a gel. Not only will the viscosity be higher than pure water, but the diffusion rates of hydrated ions within structured water are proposed to be very much slower than in free liquid; rates of diffusion will be size-dependent.

Cytoplasmic, structured water (characterised as the restriction on freedom of motion of water molecules) has been detected with a variety of physical approaches such as NMR, frequency dielectric dispersion and quasi-elastic neutron scattering (Pollack 2001).

5.4 Structured Water in the Cytoplasm May Affect Ion Fluxes

Although charged molecules like K^+ or Cl^- could initially compete for the protein-charged groups as the gel is forming, the concentration of water is orders of magnitude higher. Thus, it is envisaged that initially it is water molecules that act to nucleate structured-water formation. Only later will hydrated K^+ or Cl^- penetrate structured water, bind to the charged protein side chains and remain held in the structured-water complex. If most cytoplasmic potassium is directly bound to the negatively charged, protein side chains inside the structured-water skin, then it may not be free in the conventional sense. Even when the plasma membrane is breached, potassium could remain bound until either the structured water is disorganised, and potassium becomes freely soluble, or the cell commences to lose both protein and potassium in agreement with observation (Ling 1992). Electrical integrity will thus be partly retained provided the structured water regions remain intact.

> The picture that emerges is that cell proteins exist in a semi-solid gel-like state and their water of hydration possesses unique solvent properties as a consequence of this organisation (Garlid 2000).

5.5 The Impact of Structured Water for Cytoplasmic Functioning

The presence of structured water does present problems for understanding cellular behaviour. Structured water will likely retard or inhibit direct interaction of cytoplasmic proteins, and yet, rapid transient protein-protein interactions are essential in our present understanding of signal transduction processes and indeed many other processes that will involve inevitably structured water. Ling (1992) calculated that if only 5% of cell proteins are in an unfolded state, then virtually all cellular water would be structured.

Unfolded proteins organise the water dipoles into a low entropy structure along the polypeptide surface. However, low entropy structures contain stored energy that could be used to drive certain molecular processes. Pollack (2001) considers that cells can use the energy implicit in structured water to drive various cellular processes such as secretion, vesicle transport and actin/myosin-controlled movements.

Culinary jelly will resist freezing at −15°C. The ordered or structured water that presumably pervades the whole of the gel prevents the formation of the normal ice structure, which is itself dependent on a strict arrangement of hydrogen bonding between water molecules. Such observations suggest that a particular cytoplasmic gel state may account for freezing resistance in plants. The accumulation of low molecular weight antifreeze molecules commonly thought to account for this property would only lower freezing temperatures by a few degrees. Antifreeze proteins in animals adopt the same protein configurations and presumably structure water around themselves as does gelatin (Carvagal and Lanier 2006).

The importance of the nature of water inside cells has been highlighted by various researchers, and I quote only a few. For example, Watterson (1987) pointed to the observations that indicated that unfolded proteins like filamentous actin must be surrounded by clusters of water molecules. These are tightly bound water molecules and cannot be removed osmotically (Ito et al. 1991). Actin gels can be formed at concentrations as low as 0.1% actin/water. When ATP is added, the gel exhibits large contractions in volume and expels water. Watterson (1987) hypothesised that other proteins (at least 60 are known) bind to actin by mimicking the topological structure of water around actin and removing the structured water as a complete entity.

Wiggins (2002) pointed to the evidence of microdomains in the cytoplasm and that the properties of water in these domains differ substantially from liquid water. She pointed to two different kinds of water inside cells: high density and low density. High-density water can participate directly in peptide or polynucleotide hydrolysis through increased free OH^- or with the locally high concentrations of protons and hydroxyl ions. Low-density water can energise the removal of water in hydrolysis reactions. If the cytoplasm is structured in this manner, then it is to be expected that discrete areas of cytoplasm will be demarcated to perform specific functions as a result of prior localised protein and enzyme activities.

Finally, Pollack (2001) has suggested that structured water may play an important role in enabling water to easily rise to the top of tall trees.

5.6 Gels and Phase Transition Cooperativity: conformational spread

If the cytoplasm has the characteristics of a complex gel-like state, then to understand how cellular properties can be altered requires understanding of the potential

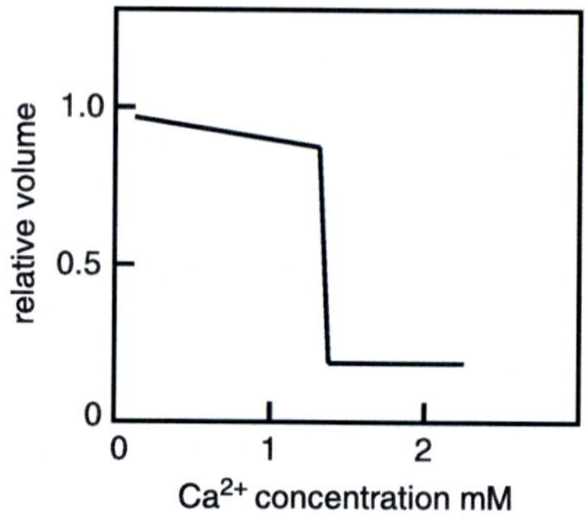

Fig. 5 Abrupt change in gel property with slight change in environment. Sodium polyacrylate gels were incubated in increasing concentrations of Ca^{2+} ions. At a critical concentration, a threshold is passed and the gel abruptly contracts. Data adapted from Tasaki and Byrne (1992)

changes in gel structure and behaviour. Again the properties of synthetic polymer gels are instructive (Pollack 2001).

Gelatin of course undergoes a phase change between two different states: liquid (sol) and solidified (gel), dependent on temperature and the conversion is reversible and usually abrupt. The transition in phase forms one of the basic properties of most non-covalently linked gels and the change can be induced by very subtle environmental alterations *once a threshold is exceeded* (e.g. Chen and Hoffman 1995; Pollack 2001). Figure 5 shows an example using Ca^{2+} on a synthetic gel; the change in volume is abrupt and reversible. Phase transition can increase the ion permeability of the gel 1,000-fold; it can shift solutes, increase the freedom of motion of water molecules and propel ions; some gels can oscillate in volume (Yoshida et al. 1999), and others act mechanically to propel a gel strip along another gel in response to an electric field (Pollack 2001; Pollack and Reitz 2001). Many of these properties are similar to the known capabilities of cells. Some conductive gels can oscillate their internal current when exposed to a constant current. Oscillations in plant cell electrical potential are not uncommonly reported (e.g. Shabala et al. 1997). An important corollary is that if gels retain their shape and the cytoplasm is largely a gel, what then is the real function of the plasma membrane? Clearly, it is not in traditional view as a bag to hold the contents in.

The threshold character and abruptness of phase change indicates the underlying mechanism relies on the cooperative behaviour amongst the constituent molecules. Two mechanisms of phase change can be envisaged. The first possibility is that described by Pollack (2001). Once a few non-covalent linkages in the gel polymer structure are unpicked, all the additional linkages rapidly collapse; the structure unzips as it were and collapses into a more stable state when a stimulus threshold is crossed. The originating factor here is surely noise in molecular structure and the low entropy structure that provides the energy for phase transition. This mechanism for gel phase transition argues that local structural change in a few linkages induces an electron cloud shift in a component polymer that then in turn induces and propagates an electron cloud shift along the whole molecule and then to other molecules. Alternatively, quantum coherence might explain the process. Figure 4d shows an example diagrammatically in which Ca^{2+} unpicks structured water by cross-linking polypeptide chains.

The second possibility is that described by Bray and Duke (2004) as Conformational Spread. They report the evidence that from a number of systems, for example, actin filaments and others, conformational changes can propagate through extended lattices of protein molecules. All these phenomena show high cooperativity (*narrow range of stimulus change between threshold and full response*). In the case of an actin gel, for example, the binding of gelsolin can solubilise actin filaments, changing gel characteristics. Binding of a single gelsolin molecule at one end propagates a conformational change along the whole actin molecule, and that may be sufficient to disrupt the structured water between adjacent molecules, thus breaking the actin gel structure. Conformational spread would continue disrupting the whole filamentous gel. Cofilin may work in similar fashion reducing filaments to monomers by conformational spread. Again, molecular noise would allow some gelsolin molecules to attach and initiate the process.

Perhaps, equally significant are the subtle environmental shifts that initiate transition cooperativity in synthetic gels. These are slight changes in pH, temperature, chemicals/biochemicals, salts, solvents and electrical and mechanical stimuli (Pollack 2001). This list is remarkably similar to summaries I constructed of environmental changes that induced bud and seed dormancy breakage, induced adventitious root formation, abscission or cell division (Trewavas 1992). Does cytoplasmic gel phase transition initiate these aspects of plant development? If conformational change enables critical proteins to now contact each other, might this not be sufficient to initiate new changes in development?

5.7 The Role of Ca^{2+} in Structured-Water Disorganisation and Signalling

Changes in cytoplasmic Ca^{2+} accompany many if not all signalling processes in plants. There are substantial amounts of Ca^{2+} in the cytoplasm in a bound form and these are probably several orders of magnitude higher than the 'free', resting Ca^{2+} detectable by fluorescence ratio imaging or aequorin (Gilroy and Trewavas 2001). Bound cytoplasmic Ca^{2+} may be involved in non-covalent, cross-linking of different protein molecules or of different regions of proteins. Such cross-linking, if present, will prevent the formation of structured water. The much larger, unbound but hydrated Ca^{2+} ion (compared to the hydrated K^+ ion) may also be mainly restricted to cellular regions free of structured water.

However, an increase in cytoplasmic Ca^{2+}, initiated by signalling, will act to initiate a phase transition in many areas of cytoplasmic gel. Actin gels illustrate the potential. Ca^{2+} is known to cause precipitation of actin, bundling of filamentous actin and initiate actin gel contraction in volume with concomitant expulsion of water (Bray 1992). The effect of Ca^{2+} is to disrupt the structured water around actin chains (Fig. 4d) and thus presumably to cross-link adjacent actin proteins through negatively charged side chains such as the carboxyl groups on aspartate and glutamate residues. However, other amino acid residues may be involved. Urry (1971) has indicated that Ca^{2+} binding in two proteins and no doubt many others takes place in areas rich in glycine residues increasing the potential binding sites available and quotes sulfhydryl groups as potential binding sites too. There are probably many proteins able to bind Ca^{2+}. The structure of the gel must be in some sort of dynamic state enabling some Ca^{2+} ions to penetrate the gel structure to initiate cross-linking. Once started, the whole structured-water complex is cooperatively destabilised using the energy available from the low entropy structure of structured water.

5.8 *Is There a Role for K^+ in Phase Transition?*

The common view of Ca^{2+} signalling is that signals open relevant channels in either the vacuole membrane or in the plasma membrane allowing the flow of Ca^{2+} down its electrochemical gradient into the cytoplasm. However, an alternative is to release Ca^{2+} from its bound form in the cytoplasm itself. A detailed compartmental analysis using washout procedures of $^{45}Ca^{2+}$ indicated the identified cytoplasmic compartment as having about 2 mM Ca^{2+} (Smart and Trewavas 1984). Many hundreds of measurements in plant cells place free cytoplasmic Ca^{2+} as at least four orders of magnitude lower. There are proteins that bind very large numbers of Ca^{2+} ions and these might be an explanation of these contradictory observations.

If signalling initiates a local disorganisation of structured water, then bound potassium will be effectively solubilised and could displace Ca^{2+} from these weakly bound cytoplasmic sites. Any signal that initially increases free cytoplasmic K^+ will, in turn, transiently increase cytoplasmic Ca^{2+}. But the intervention of the vacuole should ensure the removal of excess free cytoplasmic K^+ and see situation rapidly returned to what it was before. Any excess Ca^{2+} remaining will be mopped up by the activation of Ca^{2+}-dependent ATPases and sequestered into cellular stores before a slow return to the initial state. The effects of phase transition will be temporary overall, but the likelihood is that the new gel structure that is reconstructed will be different because of the metabolic and phosphorylation events that have occurred during the transition and the new environmental circumstances that have been sensed.

In both cases described above, where transient Ca^{2+} elevations are observed and structured water disorganised, the cytoplasmic volume should transiently increase; although, the additional water might be taken up by the vacuole or expelled to the wall. Interactions between different kinds of proteins that were previously hindered by structured water can now occur more freely. For example, Ca^{2+}-dependent protein kinases might more easily contact and phosphorylate protein substrates increasing their negative charge and thus increasing Ca^{2+} binding sites. As substrates increase their negative charge, they in turn could be cross-linked by Ca^{2+} providing larger cytoplasmic areas free of structured water and enabling further downstream signalling processes to continue.

5.9 *Could Phase Transitions Be Communicated Through The Plasmodesmata?*

Plasmodesmata are regarded as organelles that provide for cytoplasmic continuity between adjacent cells (Oparka 2005). Plasmodesmata are concerned with the potential transport of signals during host-pathogen interactions, predation signals and aspects of development that require communication between cells. The structure is complex, involving ER and protein bodies, and each pore is lined with plasma membrane. Early measurements indicated that plasmodesmatal pores would only allow passage of molecules less than 1 kDa (Erwee and Goodwin 1983). But viruses can pass through plasmodesmata using a movement protein and can open the size exclusion limit to molecules larger than 10 kDa. This increase in size exclusion limit can be also obtained by treatment with azide or anaerobic stress, that is, conditions that damage oxidative respiration (Oparka 2005). Molecule size movement can therefore be controlled, is dependent on ATP and can permit protein movement between adjacent cells.

The presence of actin and some other associated proteins in plasmodesmata has been known for some time (White et al. 1994; Faulkner et al. 2009). The realisation that actin and other proteins might form a gel in the plasmodesmata and that phase transitions in gel structure might explain changes in size exclusion limits seems not to be generally appreciated. Gels will of course allow the movement of small molecules by rapid diffusion but structured water in the gel will seriously retard the movement of proteins and larger molecules. The only way that larger molecules could pass would be to dismember the gel structure and thus release the inhibition on movement posed by structured water.

Ding et al. (1996) used fluorescent dextrans of varying sizes to detect permeability between cells and observed that cytochalasin D and profilin both now permitted molecules as large as 20 kDa to pass through the plasmodesmata. Actin filaments are in a dynamic state and cytochalasin D and profilin will dismember them. Concomitantly, structured water will be disrupted and break apart the gel structure. Movement of proteins is thus enabled. Cytochalasin D and profilin will initiate an actin gel phase transition. Azide and anaerobic stress will inhibit cellular ATP production, and thus, both these treatments can be expected to impair the dynamics of actin polymerisation into filaments and ensure structured water and gel disruption. Phalloidin, on the other hand, stabilises actin gel structure by cross-linking actin filaments. Ding et al. (1996) reported that phalloidin counteracted the opening of plasmodesmatal pore size by cytochalasin D and profilin. Potentially then, viruses increase the size exclusion limit by disrupting the actin gel structure and causing the breakdown of structured water that inhibits their movement between cells.

Increases in cytoplasmic Ca^{2+} have been shown to shut the plasmodesmatal valve (Erwee and Goodwin 1983; Tucker 1990). The effect of Ca^{2+} on actin gels is to cause the formation of a plug (Bray 1992). As actin gels contract, the volume diminishes, expelling some water. In the small plasmodesmatal pore, such phase transitions should either reduce its permeability or even completely plug it.

Could such changes in gel structure be communicated to adjacent cells? When phytochrome is activated by red light, transient increases in cytoplasmic Ca^{2+} have been observed (Shacklock et al. 1992). Nick et al. (1993) did indeed observe that red light effects were limited to individual cells or small clusters. Thus, the change in Ca^{2+} seems to be limited to the cell which senses the signal, by closing the size exclusion pore. The reason that Ca^{2+} shuts down the size exclusion limit is surely to ensure that further communication between cells must continue to operate through the wall as much of auxin movement is known to occur. Thus, the aim is temporary exclusion of movement of other soluble growth regulators. If under normal conditions the size exclusion limit is low and plasmodesmatal actin in the form of a gel, then there is the potential for gel phase transition induced by other signals to be communicated into adjacent cells dismembering local gel structure with consequences for transmission and influence beyond the responding cell.

6 Conclusions

Thresholds seem to be important elements in plant cell and tissue behaviour. Two ways have been suggested whereby threshold might be explained. The first of these is assumed to be positive feedback accompanied by noise in critical transcription factors. The second sees thresholds as developing from abrupt phase transitions in gels. These phase changes may be limited to micro-domains in the cytoplasm because one feature of Ca^{2+} signalling is its pronounced spatial character. The crucial issue here is that thresholds coupled with a probability of transition through the threshold provides for a simple way in which either a population of plants or tissues or cells from a plant exhibit a quantitative response to differing strengths of signals. More research on the threshold is now surely warranted.

References

Alberts B, Bray D, Lewis J, Raff M, Roberts K, Watson JD (1983) Molecular biology of the cell. Garland, New York

Altschuler SJ, Angenent SB, Wang Y, Wu LF (2008) On the spontaneous emergence of cell polarity. Nature 454:886–889

Bialek W, Setayeshgar S, Callan CG (2005) Physical limits to biochemical signalling. Proc Natl Acad Sci USA 102:10040–10045

Blake WJ, Balazsi G, Kohanski MA, Isaacs FJ, Murphy KF et al (2006) Phenotypic consequences of promoter transcriptional noise. Mol Cell 24:853–865

Bradford KJ, Trewavas AJ (1994) Sensitivity thresholds and variable time scales in plant hormone action. Plant Physiol 105:1029–1036

Bray D (1992) Cell movements. Garland, New York

Bray D, Duke T (2004) Conformational spread: the propagation of allosteric states in large multiprotein complexes. Annu Rev Biophys Biomol Struct 33:53–73

Briere C, Goodwin B (1988) Geometry and dynamics of tip morphogenesis in *Acetabularia*. J Theor Biol 131:461–475

Burbulis IE, Winkel-Shirley B (1999) Interactions among enzymes of the *Arabidopsis* flavonoid biosynthetic pathway. Proc Natl Acad Sci USA 96:12920–12934

Carvagal PA, Lanier TC (2006) The unfolded protein state revisited. In: Pollack GH, Cameron IL, Wheatley DN (eds) Water and the cell. Springer, Dordrecht, pp 235–252

Chen G, Hoffman AS (1995) Graft copolymers that exhibit temperature-induced phase transitions over a wide range of pH. Nature 373:49–52

Collins JJ, Imhoff TT, Grigg P (1996) Noise-enhanced tactile sensation. Nature 383:770

Costanzo M, Baryshnikova A, Bellay J, Kim Y et al (2010) The genetic landscape of a cell. Science 327:425–431

Dahal P, Still DW, Bradford KJ (1994) Mannanase activity in tomato endosperm caps does not correlate with germination timing. Plant Physiol 105:S-165

Ding B, Kwon M-O, Warnberg L (1996) Evidence that actin filaments are involved in controlling the permeability of plasmodesmata in tobacco mesophyll. Plant J 10:157–164

Elowitz MB, Levine AJ, Siggle ED, Swain PS (2002) Stochastic gene expression in a single cell. Science 297:1183–1186

Erwee MG, Goodwin PB (1983) Characterisation of the *Egeria densa* Planc. leaf symplast. Planta 158:320–328

Faulkner CL, Blackman LM, Collings DA, Cordwell SJ, Overall RL (2009) Anti-tropomyosin antibodies co-localise with actin filaments and label plasmodesmata. Eur J Cell Biol 88:357–369

Federoff N, Fontana W (2002) Small numbers of big molecules. Science 297:1129–1131

Ficklin SP, Luo F, Feltus FA (2010) The association of multiple interacting genes with specific phenotypes in rice using gene co-expression networks. Plant Physiol 154:13–24

Firn RD, Digby J (1980) The establishment of tropic curvatures in plants. Annu Rev Plant Physiol 31:131–148

Garlid KD (2000) The state of water in biological systems. Int Rev Cytol 192:281–302

Geva-Zatorsky N, Rosenfeld N, Itzkovitz S, Milo R, Sigal E, Dekel E, Yarnitsky T, Liron Y, Polak P, Lahav G, Alon U (2006) Oscillations and variability in the p53 system. Mol Syst Biol 2:0033

Gilroy S, Trewavas AJ (2001) Signal processing and transduction in plant cells: the end of the beginning. Nat Rev Mol Cell Biol 2:307–314

Golding I, Paulsson J, Zawilski SM, Cox EE (2005) Real time kinetics of gene activity in bacteria. Cell 123:1025–1036

Goodwin BC, Pateromichelakis S (1979) The role of electrical fields, ions and the cortex in the morphogenesis of *Acetabularia*. Planta 145:427–435

Goodwin BC, Skelton JL, Kirk-Bell SM (1983) Control of regeneration and morphogenesis by divalent cations in *Acetabularia meditteranea*. Planta 157:1–7

Ito T, Suzuki A, Stossel TP (1991) Regulation of water flow by actin-binding protein induced by actin gelation. Biophys J 61:1301–1305

Kaempner ES, Miller JH (1968) The molecular biology of *Euglena gracilis*: IV cellular stratification by centrifuging. Exp Cell Res 51:141–149

Kozlov PV (1983) The structure and properties of solid gelatin and the principles of their modification. Polymer 24:651–666

Lestas I, Vinnicombe G, Paulsson J (2010) Fundamental limits on the suppression of molecular fluctuations. Nature 467:174–178

Levens D, Gupta A (2010) Reliable noise. Science 327:1088–1091

Levsky JM, Shenoy SM, Pezo RC, Singer RH (2002) Single-cell gene expression profiling. Science 297:836–840

Ling GN (1992) A revolution in the physiology of the living cell. Kreiger Publishing Company, Florida

Ling GN (2006) A convergence of experimental and theoretical breakthroughs affirms the PM theory of dynamically structured water on the theory's 40th birthday. In: Pollack GH, Cameron IL, Wheatley DN (eds) Water and the cell. Springer, Dordrecht, pp 1–53
Longo D, Hasty J (2006) Dynamics of single cell gene expression. Mol Syst Biol 2:64–74
Luby-Phelps K (2000) Cytoarchitecture and physical properties of cytoplasm, volume, diffusion, intracellular surface area. Int Rev Cytol 192:189–220
Mandoli DF (1998) Elaboration of body plan and phase change during development of *Acetabularia*. Annu Rev Plant Physiol Plant Mol Biol 49:173–198
Molineir J, Ries G, Zipfel C, Hohn B (2006) Transgeneration memory of stress in plants. Nature 442:1046–1049
Mutwil M, Ysadel B, Schutte M, Loraine A, Ebenhoh O, Person S (2010) Assembly of an interactive correlation network for the *Arabidopsis* genome using a novel heuristic clustering algorithm. Plant Physiol 152:29–43
Newman JRS, Ghaemmaghari S, Ihmels J, Breslow DK, Noble M, DeRisi JL, Weismann JS (2006) Single cell proteomic analysis of *S. cerevisiae* reveals the architecture of biological noise. Nature 441:840–846
Nick P, Ehmann B, Furuya M, Schafer E (1993) Cell communication, stochastic cell responses and anthocyanin patterns in mustard cotyledons. Plant Cell 5:541–552
Oparka K (2005) Plasmodesmata, vol 18, Ann Plant Rev. Oxford, Wiley-Blackwell
Pack DW, Hoffman AS, Pun S, Stayton PS (2005) Design and development of polymers for gene delivery. Nat Rev Drug Discov 4:581–593
Pearson H (2008) The cellular hullabaloo. Nature 453:150–153
Pedraza JM, Paulsson J (2008) Effects of molecular memory and bursting on fluctuations in gene expression. Science 319:339–343
Pedraza JM, van Oudenaarden A (2005) Noise propagation in gene networks. Science 307:1965–1969
Pollack GH (2001) Cells, gels and the engines of life. Ebner and Sons, Seattle, WA
Pollack GH, Reitz FB (2001) Phase transitions and molecular motion in the cell. Cell Mol Biol 47:885–900
Rosenfeld N, Young JW, Alon U, Swain PS, Elowitz MB (2005) Gene regulation at the single cell level. Science 307:1962–1965
Shabala SN, Newman IA, Morris J (1997) Oscillations of H^+ and Ca^{2+} ion fluxes around the elongation region of corn roots and effects of external pH. Plant Physiol 113:111–118
Shacklock PS, Read ND, Trewavas AJ (1992) Cytosolic free calcium mediates red light-induced photomorphogenesis. Nature 358:753–755
Shannon CE, Weaver W (1949) The mathematical theory of communication. University of Illinois Press, Urbana
Sibbersen ED, Mott KA (2010) Stomatal responses to flooding of the intercellular air spaces suggest a vapor-phase signal between the mesophyll and the guard cells. Plant Physiol 153:1435–1442
Sigal A, Milo R, Cohen A, Geva-Zatorsky N, Klein Y, Liron Y, Rosenfeld N, Danon T, Perzov N, Alon U (2006) Variability and memory of protein levels in human cells. Nature 444:643–646
Smart CC, Trewavas AJ (1984) Abscisic acid induced turion formation in *Spirodela polyrrhiza* L. IV. Comparative ion flux characteristics of the turion and the vegetative frond and the effect of ABA during early turion development. Plant Cell Environ 7:521–539
Smith JA, Martin L (1973) Do cells cycle? Proc Natl Acad Sci USA 70:1263–1267
Srere PA (2000) Macromolecular interactions: tracing the roots. Trends Biochem Sci 25:150–153
Taniguchi Y, Choi PJ, Li G-W, Chen H, Babu M, Hearn J, Emili A, Xie XS (2010) Quantifying *E. coli* proteome and transcriptome with single molecule sensitivity in single cells. Science 329:533–538
Tasaki I, Byrne PM (1992) Discontinuous volume transitions in ionic gels and their possible involvement in the nerve excitation process. Biopolymers 32(8):1019–23

To T-L, Maheshri (2010 Noise can induce bimodality in positive transcriptional feedback loops without bistability. Science 327:1142–1145

Trewavas AJ (1982) Growth substance sensitivity: the limiting factor in plant development. Physiol Plant 55:60–72

Trewavas AJ (1987) Timing and memory processes in seed embryo dormancy—a conceptual paradigm for plant development questions. Bioessays 6:87–92

Trewavas AJ (1991) How do plant growth substances work? II. Plant Cell Environ 14:1–12

Trewavas AJ (1992) Growth substances in context: a decade of sensitivity. Biochem Soc Trans 20:102–108

Trewavas AJ (1999) How plants learn. Proc Natl Acad Sci USA 96:4216–4218

Trewavas AJ (2003) Aspects of plant intelligence. Ann Bot 92:1–20

Tucker EB (1990) Calcium-loaded 1,2-bis(2-aminophenoxy)ethane-N, N, N', N'-tetraacetic acid blocks cell-to-cell diffusion of carboxyfluorescein in staminal hairs of *Setcreasea purpurea*. Planta 182:34–38

Urry DW (1971) Neutral sites for calcium ion binding to elastin and collagen: a charge neutralisation theory for calcification and its relationship to atherosclerosis. Proc Natl Acad Sci USA 68:810–814

Vedral V (2010) Decoding reality: the universe as quantum information. Oxford University Press, Oxford

Watterson JG (1987) A role for water in cell structure. Biochem J 248:615–617

White RG, Badelt K, Overall RL, Vesk M (1994) Actin associated with plasmodesmata. Protoplasma 180:169–184

Wiggins PM (2002) Water in complex environments such as living systems. Phys A 314:465–491

Winkel BSJ (2004) Metabolic channelling in plants. Annu Rev Plant Biol 55:85–107

Yoshida R, Kokufuta E, Yamaguchi T (1999) Beating polymer gels coupled with a non-linear chemical reaction. Chaos 9:260–267

Yu J, Xiao J, Ren X, Lao K, Xie XS (2006) Probing gene expression in live cells one protein molecule at a time. Science 311:1600–1603

Yu H, Braun P, Yildirim MA, Lemmens I et al (2008) High quality binary protein interaction map of the yeast interactome network. Science 322:104–110

Zalokar M (1960) Cytochemistry of centrifuged hyphae of *Neurospora*. Exp Cell Res 19:114–132

Plant Hormones and Metabolites as Universal Vocabulary in Plant Defense Signaling

Dirk Balmer and Brigitte Mauch-Mani

Abstract Plants are sessile organisms exposed to a highly dynamic environment, and physiological flexibility including the rapid activation of suitable defense responses is crucial for their survival. Plants are confronted with an armada of pathogens and pests, and throughout the ongoing evolutionary arms race with these attackers, they have developed a sophisticated chemical signaling system, which allows them to activate highly specific and targeted defense responses. In this context, plant hormones and secondary metabolites play a pivotal role: they serve as signals in an intricate local and systemic communication network. This chapter presents recent insights into the vocabulary used by plants to fend off pathogens and pests.

1 Introduction

Despite a large variety of potential pathogens, only few are capable to successfully infect a particular plant species. The intricate self-protection system plants have developed during coevolution with their attackers makes disease the exception rather than the rule. Their defense barriers can only be overcome by specialized attackers. According to their lifestyle, plant pathogens are divided into biotroph and necrotrophs. Biotrophic pathogens obtain nutrients from living host cells; in contrast, necrotroph kill host cells to derive nourishment from dead tissue. Many pathogens, called hemibiotroph, exhibit both stages during their life cycle. The defense system of plants is multilayered and typically consists of preformed physical and chemical barriers as well as of inducible defenses. Phytoanticipin constitute the first layer of defense. They are products of secondary plant metabolism, synthesized during regular development, and stored in subcellular

D. Balmer • B. Mauch-Mani (✉)
Department of Biology, Laboratory of Molecular and Cell Biology, University of Neuchâtel, Neuchâtel, Switzerland
e-mail: Brigitte.Mauch@unine.ch

G. Witzany and F. Baluška (eds.), *Biocommunication of Plants*, Signaling and Communication in Plants 14,
DOI 10.1007/978-3-642-23524-5_3,

compartments (Morrissey and Osbourn 1999). Three main groups of such metabolites are known: phenolics, terpenes, and nitrogen-containing organic compounds (Walters 2010). A number of those compounds are toxic to pathogens. By preventing initial pathogen or pest entry, phytoanticipins provide additional time for the plant to set up inducible defenses. Another first layer of defense is induced upon recognition of conserved microbial features such as chitin, flagellin, and lipopolysaccharides (Göhre and Robatzek 2008). During this "innate immunity" response, plants perceive pathogen-associated molecular patterns (PAMPs) with the help of pattern recognition receptors (PRRs), leading to a PAMP-triggered immunity (PTI). Successful pathogens secrete effectors suppressing PTI, therefore promoting effector-triggered susceptibility (ETS). In turn, plants have resistance (R) proteins that recognize and attenuate pathogen-derived effectors, thus leading to an effector-triggered immunity (ETI; Jones and Dangl 2006). In induced plant defense, phytohormone and metabolites have a prominent role. Despite variations in quantity and blend between specific plants, tissues, and attackers, they participate in the fine-tuning and translation of induced defense signaling (Pieterse et al. 2009). Moreover, plants utilize hormones as a vocabulary facilitating local and systemic communication during disease management. The action of plant hormones during disease management follows the principle of Shannon and Weaver's (1949) classic model of communication. They defined communication as an interplay of four main parts: a *source* which is the origin of a given message, a *transmitter* that modulates a signal for the transport through a defined channel, a *receiver* which accepts the signal and transforms it to the message which is finally delivered to its *destination*. These four parts, namely source, transmitter, receiver, and destination, can consistently be applied to phytohormone-mediated signaling, such as defense reactions triggered by methyl salicylate (MeSA; Fig. 1). A locally infected plant

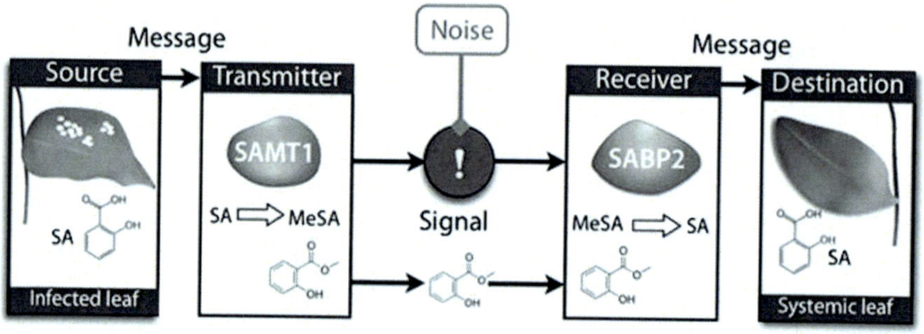

Fig. 1 Plant defense signaling follows the communication model of Shannon and Weaver (1949). Shannon and Weaver's model embodies an information *source*, *message*, *transmitter*, *signal*, *noise*, *receiver*, and *destination*. Methyl salicylate (MeSA)-triggered systemic defense is set up at a locally infected leaf that serves as source for the alarm message. Salicylic acid (SA) is induced and converted into MeSA by SA carboxyl methyltransferase 1 (SAMT1). SAMT1 acts as transmitter modifying the signal. MeSA then functions as mobile signal translocating to its destination, the noninfected systemic leaves. There, the message is perceived by salicylic acid-binding protein 2 (SABP2), which converts MeSA back into SA. SA then exerts its defense signaling function to immunize the systemic leaves. Some pathogens are able to manipulate the signaling cascade, thus acting as "noise" interfering with the message

part serves as source for a pathogen-specific alarm signal, which is often modified by cofactors and prepared for long-distance movement through the plant vascular system or in a volatile form through the air. The systemic tissue then perceives the alarm signal and decodes the message indicating the exact nature of the attack. This information allows the not-yet-infected tissue to turn on a defense reaction specifically adapted to the given stress. Recent advances in understanding the role of phytohormones have unveiled an extensive interplay between various hormones (Pieterse et al. 2009). Here, we present highlights and recent advances on the ability of chemicals to function as information carrier in an intricate semiochemical communication network modulating plant defense responses.

2 Plant Hormones Involved in Defense Signaling

Phytohormones are generally defined as "chemical regulators" produced by plants to regulate not only growth and development but also in response to biotic and abiotic stress. Six major plant hormone groups are distinguished: auxins (AUX), cytokinins (CK), gibberellins (GA), abscisic acid (ABA), ethylene (ET), and brassinosteroids (BR). Additional compounds such as jasmonic acid (JA), salicylic acid (SA), and systemin have been identified as hormone-like regulators of plant defense and development. The fact that various pathogens possess the ability to interfere with phytohormone signaling supports their pivotal role for defense. Some strains of the hemibiotrophic bacterial pathogen *Pseudomonas syringae* produce a phytotoxin called coronatine (COR). *P. syringae* uses COR to mimic JA signaling, thus downregulating SA-dependent defenses (Spoel and Dong 2008). In a Shannon and Weaver-type communication model (Fig. 1), COR functions as "noise," interfering with the signals and perturbing the messages sent by infected plant cells.

Hormonal signaling is based on key components such as receptors, protein interaction partners, and transcription factors, which are mostly conserved throughout higher plants (Bari and Jones 2009). Despite the variety of signal sources, channels, destinations, and signaling compounds, the hormones induced upon biotic stress share a common consequence of their action: they usually manipulate the expression of defense genes. For instance, out of 2,375 selected *Arabidopsis* genes, 705 messenger RNAs were found to be substantially changed upon SA, ET, methyl jasmonate (MeJA), and *Alternaria brassicicola* treatment (Schenk et al. 2000).

2.1 *Salicylic Acid*

SA belongs to the large group of phenolic plant compounds and plays a role not only in disease response but also in seed germination, cell growth, respiration, stomatal closure, senescence, thermo tolerance, and flowering (Vlot et al. 2009). In *Arabidopsis thaliana* and *Nicotiana benthamiana*, the majority of pathogen-induced SA is synthesized by isochorismate synthase (ICS; Vlot et al. 2009).

A SA-glucosyltransferase then converts most of the SA into *0-β*-glucoside (SAG; Dean and Delaney 2008). SAG is stocked in the vacuole, where it likely acts as storage form that can be converted back into SA when needed. SA is predominantly involved in defense against biotrophic pathogens. During defense communication, SA plays a role in both local and systemic resistance reactions. Locally, SA combats invading pathogens due to its natural antimicrobial properties (Murphy and Carr 2002). SA also functions as mediator of systemic acquired resistance (SAR). During SAR, a locally infected tissue emits phloem-mobile or airborne alarm signals to uninfected parts of the plants, thus rendering them more resistant against subsequent pathogen attack. Due to its presence in the phloem, SA was initially thought to be itself the signal mediating SAR. However, grafting experiments showed that SA is not required in the tissue *transmitting* the SAR signal, whereas it is indispensable in the systemic tissue *receiving* the SAR signal (Vernooij et al. 1994). In regard to the communication principle of Shannon and Weaver, SA seems therefore not to play a role as long-distance signal; it rather acts as a local communication mediator in infected cells and exerts a receiver-like function in noninfected tissue. The major role of SA during local disease management is the modification of cellular signaling pathways, mainly through the interaction with NPR1 (nonexpressor of PR1; Cao et al. 1997). NPR1 is present in the cytosol in a dimeric form. Accumulation of SA shifts the redox state inside the cell from oxidizing toward reducing conditions. Reduction of cysteine residues of NPR1 dimers leads to its monomerization. As a monomer, NPR1 translocates to the nucleus where it interacts with transcription factors such as TGAs and WRKYs to enhance defense gene expression (Vlot et al. 2009). Beside modification of NPR1 by shifting the redox state, SA also induces the expression of thioredoxins (TRX) that catalyze the monomerization of NPR1 (Tada et al. 2008). Therefore, NPR1 is the main "receiver" of the defense information delivered by SA, obtaining the signal via direct and indirect signal perception. Nevertheless, the true SA receptor is not yet known (Vlot et al. 2009).

2.2 Jasmonic Acid

Jasmonates are oxygenated fatty acids produced by the octadeconoid pathway (Staswick 2008). They are important for a variety of processes including pollen maturation, fruit development, photosynthesis, senescence, and root growth. Moreover, JA signaling is activated upon herbivore attack in a variety of different plant species and is crucial in regulating defense responses against necrotrophic pathogens and chewing insects (Pieterse et al. 2009). Furthermore, it also plays an important role during induced systemic resistance (ISR) mediated by nonpathogenic root-colonizing bacteria (Pieterse et al. 2009). Recently, the COP9 signalosome has been shown to regulate JA-dependent insect defense (Hind et al. 2011). Intriguingly, JA acts as a negative regulator of SA-dependent defenses (Bari and Jones 2009; Pieterse et al. 2009). Upon wounding of plant tissues, linoleic acid is released from membrane lipids of chloroplasts and incorporated into the octadeconoid pathway, where it is transformed into JA (Staswick 2008). JA can further be metabolized into various products including volatile MeJA, and it can conjugate with amino acids and sugars (Wasternack 2007).

Referring to the Shannon and Weaver model (Fig. 1), the source of JA as chemical regulator signal are membrane-derived lipids that are metabolized into jasmonates, which are then perceived by a COI1/JAZ co-receptor. Furthermore, JA has also been shown to be transmittable through the phloem into systemic tissues (Truman et al. 2007), therefore transporting a long-distance message to a destination tissue. Whether in its local or systemic destination, JA signaling drives the induction of defense-related genes. Further studies need to be undertaken to unveil how the products from JA-responsive genes contribute in detail in combatting disease.

2.3 Ethylene

The gaseous hormone ET is the major regulator of fruit ripening, seedling emergence, leaf and flower senescence, and organ abscission, but it contributes also to biotic stress signaling (van Loon et al. 2006). Both 1-aminocyclopropane-1-carboxylic acid (ACC) synthase and ACC oxidase, important enzymes in ET biosynthesis, are induced upon pathogen infection, wounding, and light stress (Wang et al. 2002). ET moves by diffusion from its original site of synthesis to systemic tissues. There, it is perceived by a family of membrane-localized receptors. The role of ET during pathogen defense signaling is rather ambiguous. ET contributes to basal resistance in *Arabidopsis* against *Pseudomonas syringae* pv. *syringae* and *Xanthomonas campestris* pv. *vesicatoria* (Ton et al. 2006). In contrast, the proliferation of the bacterial leaf pathogen *Pseudomonas syringae* pv. *glycinea* is impaired on mutants lacking the capacity to produce ET (Weingart et al. 2001). Often, disease symptoms are enhanced after ET treatment, probably due to the ET-triggered induction of senescence (van Loon et al. 2006). Recent findings suggest that ET plays a pivotal role during early defense reactions. Nitric oxide (NO) interacts with SA to regulate ET production mediating the hypersensitive response, a cell-death phenomenon associated with rapid localized resistance to pathogens (Mur et al. 2008). ET signaling is also involved in ISR (Pieterse et al. 2009). Due to the ambiguous mode of action, it can be assumed that ET does not play a role as message carrier itself during defense communication, rather acting as fine-tuning mediator in the cross talk of other major hormonal pathways (Sect. 4). In fact, ET is known to interact synergistically with both the JA and SA signaling network (Pieterse et al. 2009). Unlike other chemical regulators such as SA and JA, ET affects all developmental stages; the fluctuating effect of ET during defense communication therefore depends strongly on the age of the plant, the type of pathogen, and the environmental conditions.

2.4 Auxin

Auxins are the main chemical regulators of growth and cell differentiation in plants. They are principally occurring as indole-3-acetic acid (IAA). IAA is synthesized from two distinct pathways: one, using L-tryptophan as main precursor and another tryptophan-independent pathway (Buchanan et al. 2002). The majority of IAA *in planta* is synthesized in meristems, young leaves, and developing fruits and seeds. From its original site of biosynthesis, IAA is transported by nonpolar and polar transport mechanisms (Buchanan et al. 2002). Beside its crucial role in plant development, recent studies indicate that auxin also contributes to pathogen defense signaling in a rather ambivalent manner. The auxin-responsive gene GH3 has been shown to modulate SA and auxin signaling during *Pseudomonas syringae* infection in *Arabidopsis* (Zhang et al. 2007). *Arabidopsis* auxin-signaling mutants are more susceptible to the necrotrophic fungi *Plectosphaerella cucumerina* and *Botrytis cinerea* (Llorente et al. 2008). In contrast, treatment of *Arabidopsis* with an SA analogue resulted in the global repression of auxin-response genes, suggesting that the SA pathway inhibits auxin signaling to enhance pathogen resistance (Wang et al. 2007). Similarly, a plant microRNA (miR393) was discovered to contribute to antibacterial resistance in *Arabidopsis* by downregulation of TIR1, thus repressing auxin-responsive genes (Navarro et al. 2006). Hence, auxins seem to attenuate plant defense responses rather than to act as defense mediating signaling compound. It is known that pathogens are able to manipulate auxin signaling to promote disease (Padmanabhan et al. 2008). Taken together, auxins are believed to act as either negative or positive modulators of defense responses by affecting the catabolism of other hormonal pathways and the plant physiology in general.

2.5 Abscisic Acid

ABA is an isoprenoid phytohormone mainly involved in regulating seed germination, leaf senescence, and stomatal aperture and plays a crucial role in response to water and salt stress (Wasilewska et al. 2008). ABA is a phloem-mobile and long-distance signal synthesized primarily in vascular tissues (Nambara and Marion-Poll 2005). The role of ABA during pathogen defense is highly multifaceted and depends on the specific stage of defense and type of attacker (Ton et al. 2009). Generally, ABA is believed to act as a negative regulator of defense responses. ABA-deficient mutants or mutants impaired in ABA synthesis show increased resistance to different pathogens (Cao et al. 2011). Conversely, exogenous application of ABA can favor disease development (de Torres-Zabala et al. 2007). Different pathogens are known to produce ABA and thus interfere with host defense (Cao et al. 2011). However, ABA can also positively regulate defense responses (Mauch-Mani and Mauch 2005). The closure of stomata, which can serve as entry point for attacking bacteria, is triggered by ABA (Melotto et al. 2006). Moreover, ABA treatment mediates resistance against some necrotrophs. This ABA-induced resistance is based on ABA-dependent priming for deposition of callose-containing cell wall reinforcement against penetration by pathogens (Ton and Mauch-Mani 2004). Taken together, ABA acts as positive and negative chemical regulator of plant defense. During the initial phase of invasion, ABA positively regulates resistance through mediation of stomatal closure. In the subsequent early stage of invasion, ABA enhances resistance against fungi and oomycetes by triggering callose deposition but also diminishes resistance by inhibiting reactive oxygen species (ROS) generation and callose accumulation upon bacterial infection. Finally, during late defense reactions, ABA generally inhibits defense responses by suppressing JA, ET, and SA-dependent signaling (Ton et al. 2009).

2.6 *Brassinosteroids, Cytokinins, and Gibberellin*

Brassinosteroids, cytokinins, and gibberellin play rather minor roles in defense responses; only few studies are providing evidence that these classical phytohormones contribute to plant immune reactions. BR, known for their involvement in seed germination, cell division, flowering, and senescence, have been shown to enhance resistance of tobacco against TMV in an SA-independent manner (Nakashita et al. 2003). Similarly, exogenous application of BR on potato plants enhances their resistance against *Phytophthora infestans* (Krishna 2003). Components of BR signaling participate in early defense responses, as *Arabidopsis* mutants of the BR-receptor BRI1-associated receptor kinase 1 (BAK1) exhibit higher susceptibility to bacterial and fungal pathogens (Kemmerling et al. 2007), and BAK1 interacts with the flagellin-sensing transmembrane receptor kinase flagellin-sensitive 2 (FLS2) to initiate PAMP-triggered immunity during early pathogen perception (Chinchilla et al. 2007). Taken together, BR seems to play an indirect role during defense responses by influencing other hormonal pathways and by PAMP-triggered immunity (Bari and Jones 2009).

In turn, the roles of CK during defense responses are less understood. Mainly involved in stem cell control, vascular differentiation, chloroplast biogenesis, seed development, and shoot and root growth, CK was recently shown to contribute to pathogen responses. Disease symptoms of *Arabidopsis* roots against *Plasmodiophora brassicae* were found to be increased by CK (Siemens et al. 2006), and *Agrobacterium tumefaciens* enhances CK production in *Arabidopsis* plastids to induce tumor formation (Sakakibara et al. 2005). Therefore, CK seems to have rather disease-promoting effects, although its role in defense against different types of attackers is poorly understood.

In contrast, the growth-promoting hormone GA has been found to exert positive and negative effects on plant defense responses. GA stimulates plant growth by degradation of DELLA proteins, which negatively regulate plant growth. DELLA proteins regulate defense responses in *Arabidopsis* by altering SA- and JA-dependent immunity (Navarro et al. 2008). Hence, *Arabidopsis* DELLA mutants showed higher susceptibility to the necrotrophic pathogens *A. brassicicola* and *B. cinerea*, whereas the resistance against the biotrophs *Pst* DC3000 and *Hyaloperonospora arabidopsidis* was enhanced. Consequently, GA seems to be implicated in promoting resistance to biotrophs and susceptibility to necrotrophs. However, the mechanism of GA-regulated defense is still largely unexplored.

2.7 Systemin

Systemin is a plant peptide hormone playing an exclusive role following wounding in the *Solanaceae*. During herbivore attack, systemin is cleaved from its precursor prosystemin and stored in the cytoplasm (Ryan and Pearce 2001). Its local and systemic induction triggers the activation of proteinase inhibitor (PI). PIs prevent the uptake of essential amino acids in the insect midgut, causing developmental defects (Chen et al. 2005). Following perception, synthesis of JA and the expression of defense-related genes are activated (Kandoth et al. 2007). Grafting experiments have shown that neither systemin nor JA was required in the systemic tissue acquiring the signal, indicating that systemin acts at the local site of infection to facilitate the production of a long-distance and probably JA-derived signal (Li et al. 2002). Furthermore, overexpression of prosystemin resulted in an enhanced release of volatiles and synthesis of PIs upon herbivore attack in tomato, implicating that systemin and JA are regulating herbivore-induced systemic volatile emission (Degenhardt et al. 2010). So far, the exact role of peptide hormones in the regulation of plant defenses remains elusive.

3 Systemic Defense Signals

Following local events leading to the buildup of a defensive state, a signal has to be generated and transmitted to systemic plant parts. Induction of SAR follows PTI or ETI-mediated pathogen recognition and is associated with increased levels of SA and pathogen-related proteins (PR) in local and systemic tissues (Jones and Dangl 2006). At the root level, various microorganisms can trigger a systemic defense induction, as observed for ISR, or rhizobacteria-and-mycorrhiza-induced resistance (Pieterse et al. 2009). Moreover, during systemic wound response, herbivore-infected plants emit volatile signals to set up an indirect defense by attracting predatory insects (Heil and Silva Bueno 2007).

Systemic resistance represents an example of an intricate communication system, mediated by a series of mobile signals. Despite the major advances in recent years and the identification of multiple long-distance chemical signals, the exact nature of specific mobile signals remains elusive and controversial (Vlot et al. 2008). Recent studies proposed methyl salicylate (MeSA) as a critical SAR signal (Park et al. 2007). In TMV-infected tobacco leaves, SA carboxyl methyltransferase

1 (SAMT1) converts SA into MeSA, which is biologically inactive and volatile. MeSA can then act as a phloem-mobile or airborne signal immunizing noninfected systemic tissues. There, it is converted back to SA by salicylic acid-binding protein 2 (SABP2) (Park et al. 2007). However, MeSA is not essential for SAR expression in *Arabidopsis* (Attaran et al. 2009). Jasmonates are also accepted as mobile defense signals. Volatile methyl jasmonate (MeJA) functions as phloem- and xylem-mobile signal during systemic wound responses (Thorpe et al. 2007). SAR is compromised in jasmonate-deficient *Arabidopsis* mutants, suggesting a signaling role for JA during SAR (Truman et al. 2007). Nonetheless, the role of JA during SAR is highly debatable and likely conditional, depending on the experimental system and the applied pathogen dose (Shah 2009).

Furthermore, azelaic acid has been identified as a SAR-eliciting factor (Jung et al. 2009). Elevated levels of azelaic acid were found in petiole exudates of SAR-triggered plants, and locally applied radiolabelled azelaic acid was recovered in distant leaves, confirming its systemic nature. Its local application did not alter SA levels or SA-dependent gene expression (Jung et al. 2009). Recently, glycerol-3-phosphate (G3P) was discovered to act as critical mobile signal for SAR in *Arabidopsis* and soybean (Chanda et al. 2011). *Arabidopsis* G3P biosynthesis mutants are unable to induce SAR, and G3P derivates are translocated to distal tissues with the help of the lipid-transfer protein DIR1. Green leaf volatiles (GLVs) are also known to act as systemic defense signals, predominantly in response to wounding or herbivore attack (Heil and Silva Bueno 2007). They prime plants for enhanced induction of JA-dependent defenses during wounding and herbivore attack. Overall, recent studies suggest the presence of multiple mobile defense signals for systemic resistance. Beside MeSA, MeJA, azelaic acid, glycerol-3-phosphate, and GLVs, a variety of additional chemical regulators such as ET, ABA, sugars, and peptide hormones are likely to also contribute in systemic resistance. The nature of a specific signal strongly depends on the transport channel (vascular or airborne), on the plant species and its lifestyle, and on the type of attacker. Nevertheless, systemic defense highlights the plant's capability to apply a complex communication network with distinct *signal sources*, *channels*, and *signal receivers* according to Shannon and Weaver (1949).

4 Signal Cross Talk

In contrast to animals, plants do not possess cells that are exclusively specialized in immune reactions. In order to adapt their defense to a continuously changing environment, they fine-tune the cross talk of the different chemical regulators involved in defense signaling (Pieterse et al. 2009). Genome-profiling experiments with *Arabidopsis* hormone mutants revealed the presence of an extensive and pliable network between the three main chemical regulators SA, JA, and ET (Glazebrook et al. 2003). For instance, the interaction of SA and JA is normally antagonistic, due to trade-offs between SA-mediated resistance against biotrophs and JA-mediated resistance against necrotrophs. In *Arabidopsis*, JA-dependent defenses activated upon caterpillar feeding were suppressed by the SA-mediated defense reaction triggered by infection with the biotrophic pathogen *Hyaloperonospora arabidopsidis* (Koornneef et al. 2008). Similarly, exogenous application of SA diminishes the expression of JA-responsive genes such as *PDF1.2* and *VSP2*. However, the interaction between SA and JA is dose-dependent as simultaneous treatment with low doses of SA and JA was shown to trigger synergistic effects on SA- and JA-responsive genes (Schenk et al. 2000). The suppression of the JA pathway is mediated by NPR1, the master regulator of the SA pathway. The SA-driven suppression of JA-responsive genes does not require nuclear localization of NPR1, indicating that cytosolic NPR1 is mediating negative effects on JA-signaling by a yet unknown mechanism (Spoel et al. 2003). ET modulates the NPR1-dependent JA-SA antagonism by potentiating the SA-dependent expression of *PR1* and rendering the JA-suppressing effects independent of NPR1 (Leon-Reyes et al. 2009). Often, ET interacts with JA in a synergistic manner (Pieterse et al. 2009). The expression of the JA-responsive gene *PDF1.2* requires the concomitant activity of JA and ET signaling cascades (Penninckx et al. 1998). Both JA and ET treatment induces the expression of the ET-responsive transcription factors ERF1 and ORA59, indicating that JA and ET signaling share nodes of convergence (Pré et al. 2008). ET also interacts with SA-dependent defenses. In tobacco, ET is indispensable for the activation of SAR upon TMV infection (Verberne et al. 2003). The extensive cross talk between SA and ET has also been corroborated with the finding that the expression of SA-responsive genes was heavily affected in *Arabidopsis* mutants impaired in ET signaling (Glazebrook et al. 2003).

Beside the interaction of the major three defense hormones SA, JA, and ET, it is also known that other chemical regulators participate in the defense cross talk. ABA is known to generally attenuate SA- and JA/ET-dependent defense responses. In *Arabidopsis*, ABA inhibits the expression of JA and ET-responsive genes (Anderson et al. 2004). Moreover, ABA was demonstrated to interact antagonistically with SAR (Yasuda et al. 2008). Conversely, the activation of SAR inhibited the expression of ABA-responsive genes. Auxins are also known to affect the SA-JA-ET signaling network. The auxin responsive factors ARF6 and ARF8 have been demonstrated to promote jasmonic acid production (Nagpal et al. 2005), and auxin signaling enhances susceptibility of *Arabidopsis* to *P. syringae* (Navarro et al. 2006). Furthermore, both GA and brassinosteroids were shown to interact with the SA-JA-ET signaling network. DELLA proteins, the main regulators of GA signaling, were demonstrated to promote susceptibility to biotrophs and resistance to necrotrophs (Navarro et al. 2008). Similarly, brassinosteroids also interact with multiple hormones. They are known to affect ET biosynthesis, enhance auxin signaling, and interact antagonistically with ABA (Zhang et al. 2009). In spite of the advances acquired over the past years, the majority of the mechanism underlying hormone cross talk remains to be elucidated.

5 Concluding Remarks

During the past years, much has been learned regarding the role of phytohormones during plant defense responses. Chemical regulators of plant growth were shown to be also orchestrating pathogen and pest defense. Although general roles of phytohormones in immune responses are known, the dissection of mechanisms triggering signal generation, transport, and reception remains a challenge. Moreover, large-scale genomic analysis unveiled the presence of an intricate communication system driven by a multilayered cross talk of phytohormones and metabolites. Advances in the field of metabolomics and system biology will help to dissect this extensive network and lead to the discovery of novel blends of alarm signals. A better understanding of the hormone- and metabolite-triggered plant defense communication will also impact the development of disease and pest resistance in crops.

Acknowledgements The authors were supported by the National Centre of Competence in Research (NCCR) "Plant Survival" and by SNF Grant 31003A-120197, both research programs of the Swiss National Science Foundation. We also thank Ana Slaughter for critical reading of the manuscript. We apologize to our colleagues who could not be cited in this review due to space restrictions.

References

Anderson JP, Badruzsaufari E, Schenk PM, Manners JM, Desmond OJ, Ehlert C, Maclean DJ, Ebert PR, Kazan K (2004) Antagonistic interaction between abscisic acid and jasmonate-ethylene signaling pathways modulates defense gene expression and disease resistance in Arabidopsis. The Plant Cell 16:3460–3479

Attaran E, Zeier TE, Griebel T, Zeier J (2009) Methyl salicylate production and jasmonate signaling are not essential for systemic acquired resistance in Arabidopsis. The Plant Cell 21:954–971

Bari R, Jones JD (2009) Role of plant hormones in plant defense responses. Plant Molecular Biology 69:473–488

Buchanan RB, Gruissem W, Jones RL (2002) Biochemistry and molecular biology of plants. American Society of Plant Physiologists, Rockville

Cao H, Glazebrook J, Clark JD, Volko S, Dong X (1997) The *Arabidopsis* NPR1 gene that controls systemic acquired resistance encodes a novel protein containing ankyrin repeats. Cell 88:57–63

Cao FY, Yoshioka K, Desveaux D (2011) The roles of ABA in plant-pathogen interactions. J Plant Res, in press (doi: 10.1007/s10265-011-0409-y)

Chanda B, Xia Y, Mandal MK, Yu K, Sekine KT, Gao QM, Selote D, Hu Y, Stromberg A, Navarre D, Kachroo A, Kachroo P (2011) Glycerol-3-phosphate is a critical mobile inducer of systemic immunity in plants. Nat Genet, in press (doi:10.1038/ng.789)

Chen H, Wilkerson CG, Kuchar JA, Phinney BS, Howe GA (2005) Jasmonate-inducible plant enzymes degrade essential amino acids in the herbivore midgut. Proc Natl Acad Sci USA 102:19237–19242

Chinchilla D, Zipfel C, Robatzek S, Kemmerling B, Nürnberger T, Jones JD, Felix G, Boller T (2007) A flagellin-induced complex of the receptor FLS2 and BAK1 initiates plant defence. Nature 448:497–500 de Torres-Zabala M, Truman W, Bennett MH, Lafforgue G, Mansfield JW, Egea PR, Bögre L, Grant M (2007) *Pseudomonas syringae* pv. *tomato* hijacks the Arabidopsis abscisic acid pathway to cause disease. EMBO J 26:1434–1443

Dean JV, Delaney SP (2008) Metabolism of salicylic acid in wild-type, ugt74f1 and ugt74f2 glucosyl-transferase mutants of *Arabidopsis thaliana*. Physiologia Plantarum 132:417–425

Degenhardt DC, Refi-Hind S, Stratmann JW, Lincoln DE (2010) Systemin and jasmonic acid regulate constitutive and herbivore-induced systemic volatile emissions in tomato, Solanum lycopersicum. Phytochemistry 71:2024–2037

Glazebrook J, Chen W, Estes B, Chang HS, Nawrath C, Métraux JP, Zhu T, Katagiri F (2003) Topology of the network integrating salicylate and jasmonate signal transduction derived from global expression phenotyping. The Plant Journal 34:217–228

Göhre V, Robatzek S (2008) Breaking the barriers: microbial effector molecules subvert plant immunity. Annual Review of Phytopathology 46:189–215

Heil M, Silva Bueno JC (2007) Within-plant signaling by volatiles leads to induction and priming of an indirect plant defense in nature. Proc Natl Acad Sci USA 104:5467–5472

Hind SR, Pulliam SE, Veronese P, Shantharaj D, Nazir A, Jacobs NS, Stratmann JW (2011) The COP9 signalosome controls jasmonic acid synthesis and plant responses to herbivory and pathogens. The Plant Journal 65:480–491

Jones JD, Dangl JL (2006) The plant immune system. Nature 444:323–329

Jung HW, Tschaplinski TJ, Wang L, Glazebrook J, Greenberg JT (2009) Priming in systemic plant immunity. Science 324:89–91

Kandoth PK, Ranf S, Pancholi SS, Jayanty S, Walla MD, Miller W, Howe GA, Lincoln DE, Stratmann JW (2007) Tomato MAPKs LeMPK1, LeMPK2 and LeMPK3 function in the systemin-mediated defense response against herbivorous insects. Proc Natl Acad Sci USA 104:12205–12210

Kemmerling B, Schwedt A, Rodriguez P, Mazzotta S, Frank M, Qamar SA, Mengiste T, Betsuyaku S, Parker JE, Müssig C, Thomma BP, Albrecht C, de Vries SC, Hirt H, Nürnberger T (2007) The BRI1-associated kinase 1, BAK1, has a brassinolide-independent role in plant cell-death control. Current Biology 17:1116–1122

Koornneef A, Leon-Reyes A, Ritsema T, Verhage A, Den Otter FC, van Loon LC, Pieterse CM (2008) Kinetics of salicylate-mediated suppression of jasmonate signaling reveal a role for redox modulation. Plant Physiology 147:1358–1368

Krishna P (2003) Brassinosteroid-mediated stress responses. J Plant Growth Regul 22:289–297

Leon-Reyes A, Spoel SH, De Lange ES, Abe H, Kobayashi M, Tsuda S, Millenaar FF, Welschen RA, Ritsema T, Pieterse CM (2009) Ethylene modulates the role of NONEXPRESSOR OF PATHOGENSIS-RELATED GENES1 in cross talk between salicylate and jasmonate signaling. Plant Physiology 149:1797–1809

Li L, Li C, Lee GI, Howe GA (2002) Distinct roles for jasmonate synthesis and action in the systemic wound response of tomato. Proc Natl Acad Sci USA 99:6416–6421

Llorente F, Muskett P, Sánchez-Vallet A, López G, Ramos B, Sánchez-Rodríguez C, Jordá L, Parker J, Molina A (2008) Repression of the auxin response pathway increases *Arabidopsis* susceptibility to necrotrophic fungi. Molecular Plant 1:496–509

Mauch-Mani B, Mauch F (2005) The role of abscisic acid in plant-pathogen interactions. Current Opinion in Plant Biology 8:409–414

Melotto M, Underwood W, Koczan J, Nomura K, He SY (2006) Plant stomata function in innate immunity against bacterial invasion. Cell 126:969–980

Morrissey JP, Osbourn AE (1999) Fungal resistance to plant antibiotics as a mechanism of pathogenesis. Microbiology and Molecular Biology Reviews 63:708–724

Mur LA, Laarhoven LJ, Harren FJ, Hall MA, Smith AR (2008) Nitric oxide interacts with salicylate to regulate biphasic ethylene production during the hypersensitive response. Plant Physiology 148:1537–1446

Murphy AM, Carr JP (2002) SA has cell-specific effects on Tobacco mosaic virus replication and cell-to-cell movement. Plant Physiology 128:552–563

Nagpal P, Ellis CM, Weber H, Ploense SE, Barkawi LS, Guilfoyle TJ, Hagen G, Alonso JM, Cohen JD, Farmer EE, Ecker JR, Reed JW (2005) Auxin responsive factors ARF6 and ARF8 promote jasmonic acid production and flower maturation. Development 132:4107–4118

Nakashita H, Yasuda M, Nitta T, Asami T, Fujioka S, Arai Y, Sekimata K, Takatsuto S, Yamaguchi I, Yoshida S (2003) Brassinosteroid functions in a broad range of disease resistance in tobacco and rice. The Plant Journal 33:887–898

Nambara E, Marion-Poll A (2005) Abscisic acid biosynthesis and catabolism. Annual Review of Plant Biology 56:165–185

Navarro L, Dunoyer P, Jay F, Arnold B, Dharmasiri N, Estelle M, Voinnet O, Jones JD (2006) A plant miRNA contributes to antibacterial resistance by repressing auxin signaling. Science 312:436–439

Navarro L, Bari R, Achard P, Lisón P, Nemri A, Harberd NP, Jones JD (2008) DELLAs control plant immune responses by modulation the balance of jasmonic acid and salicylic acid signaling. Current Biology 18:650–655

Padmanabhan MS, Kramer SR, Wang X, Culver JN (2008) Tobacco mosaic virus replicase-auxin/indole acetic acid protein interactions: reprogramming the auxin response pathway to enhance virus infection. Journal of Virology 82:2477–2485

Park SW, Kaimoyo E, Kumar D, Mosher S, Klessig DF (2007) Methyl salicylate is a critical mobile signal for plant systemic acquired resistance. Science 318:113–116

Penninckx IA, Thomma BP, Buchala A, Métraux JP, Broekaert WF (1998) Concomitant activation of jasmonate and ethylene response pathways is required for induction of a plant defensin gene in *Arabidopsis*. The Plant Cell 10:2103–2113

Pieterse CM, Leon-Reyes A, Van der Ent S, Van Wees SC (2009) Networking by small-molecule hormones in plant immunity. Nature Chem Biol 5:308–316

Pré M, Atallah M, Champion A, De Vos M, Pieterse CM, Memelink J (2008) The AP2/ERF domain transcription factor ORA59 integrates jasmonic acid and ethylene signals in plant defense. Plant Physiology 147:1347–1357

Ryan CA, Pearce G (2001) Polypeptide hormones. Plant Physiology 125:65–68

Sakakibara H, Kasahara H, Ueda N, Kojima M, Takei K, Hishiyama S, Asami T, Okada K, Kamiya Y, Yamaya T, Yamaguchi S (2005) Agrobacterium tumefaciens increases cytokinin production in plastids by modifying the biosynthetic pathway in the host plant. Proc Natl Acad Sci USA 102:9972–9977

Schenk PM, Kazan K, Wilson I, Anderson JP, Richmond T, Somerville SC, Manners JM (2000) Coordinated plant defense responses in Arabidopsis revealed by microarray analysis. Proc Natl Acad Sci USA 97:11655–11660

Shah J (2009) Plants under attack: systemic signals in defence. Current Opinion in Plant Biology 12:459–464

Shannon CE, Weaver W (1949) The mathematical theory of communication. University of Illinois Press, Champaign

Siemens J, Keller I, Sarx J, Kunz S, Schuller A, Nagel W, Schmülling T, Parniske M, Ludwig-Müller J (2006) Transcriptome analysis of Arabidopsis clubroots indicate a key role for cytokinins in disease development. Molecular Plant–Microbe Interactions 19:480–494

Spoel SH, Dong X (2008) Making sense of hormone crosstalk during plant immune response. Cell Host & Microbe 3:348–351

Spoel SH, Koornneef A, Claessens SM, Korzelius JP, Van Pelt JA, Mueller MJ, Buchala AJ, Métraux JP, Brown R, Kazan K, van Loon LC, Dong X, Pieterse CM (2003) NPR1 modulates cross-talk between salicylate- and jasmonate-dependent defense pathways through a novel function in the cytosol. The Plant Cell 15:760–770

Staswick PE (2008) JAZing up jasmonate signaling. Trends in Plant Science 13:66–71

Tada Y, Spoel SH, Pajerowska-Mukthar K, Mou Z, Song J, Wang C, Zuo J, Dong X (2008) Plant immunity requires conformational changes of NPR1 via S-nitrosylation and thioredoxins. Science 321:952–956

Thorpe MR, Ferrieri AP, Herth MM, Ferrieri RA (2007) 11C-imaging: methyl jasmonate moves in both phloem and xylem, promotes transport of jasmonate, and of photoassimilate even after proton transport is decoupled. Planta 226:541–551

Ton J, Mauch-Mani B (2004) Beta-amino-butyric acid-induced resistance against necrotrophic pathogens is based on ABA-dependent priming for callose. The Plant Journal 38:119–130

Ton J, Pieterse CMJ, van Loon LC (2006) The relationship between basal and induced resistance in Arabidopsis. In: Tuzun S, Bent E (eds) Multigenic and induced systemic resistance in plants. Springer, New York, pp 197–224

Ton J, Flors V, Mauch-Mani B (2009) The multifaceted role of ABA in disease resistance. Trends Plant Sci 14:310–317

Truman W, Bennett MH, Kubigsteltig I, Turnbull C, Grant M (2007) Arabidopsis systemic immunity uses conserved defense signaling pathways and is mediated by jasmonates. Proc Natl Acad Sci USA 104:1075–1080

van Loon LC, Geraats BP, Linthorst HJ (2006) Ethylene as a modulator of disease resistance in plants. Trends in Plant Science 11:184–191

Verberne MC, Hoekstra J, Bol JF, Linthorst HJ (2003) Signaling of systemic acquired resistance in tobacco depends on ethylene perception. The Plant Journal 35:27–32

Vernooij B, Friedrich L, Morse A, Reist R, Kolditz-Jawhar R, Ward E, Uknes S, Kessmann H, Ryals J (1994) SA is not the translocated signal responsible for inducing SAR but is required in signal transduction. The Plant Cell 6:959–965

Vlot AC, Klessig DF, Park SW (2008) Systemic acquired resistance: the elusive signal(s). Current Opinion in Plant Biology 11:436–442

Vlot AC, Dempsey DA, Klessig DF (2009) Salicylic acid, a multifaceted hormone to combat disease. Annual Review of Phytopathology 47:177–206

Walters DR (2010) Plant defense: warding off attack by pathogens, herbivores, and parasitic plants. Blackwell Publishing Ltd, Oxford

Wang KL, Li H, Ecker JR (2002) Ethylene biosynthesis and signaling networks. The Plant Cell 14 (Suppl):S131–151

Wang D, Pajerowska-Mukhtar K, Culler AH, Dong X (2007) Salicylic acid inhibits pathogen growth in plants through the repression of the auxin signaling pathway. Current Biology 17:1784–1790

Wasilewska A, Vlad F, Sirichandra C, Redko Y, Jammes F, Valon C, Frei dit Frey N, Leung J (2008) An update on abscisic acid signaling in plants and more. Molecular Plant 1:198–217

Wasternack C (2007) Jasmonates: an update on biosynthesis, signal transduction and action in plant stress response, growth and development. Annals of Botany 100:681–697

Weingart H, Ullrich H, Geider K, Völksch B (2001) The role of ethylene production in virulence of *Pseudomonas syringae* pvs. *glycinea* and *phaseolicola*. Phytopathology 91:511–518

Yasuda M, Ishikawa A, Jikumaru Y, Seki M, Umezawa T, Asami T, Maruyama-Nakashita A, Kudo T, Shinozaki K, Yoshida S, Nakashita H (2008) Antagonistic interaction between systemic acquired resistance and the abscisic acid-mediated abiotic stress response in Arabidopsis. The Plant Cell 20:1678–1692

Zhang ZQ, Li Q, Li Z, Staswick PE, Wang M, Zhu Y, He Z (2007) Dual regulation role of GH3.5 in salicylic acid and auxin signaling during *Arabidopsis-Pseudomonas syringae* interaction. Plant Physiology 145:450–64

Zhang S, Wei Y, Lu Y, Wang X (2009) Mechanisms of brassinosteroids interacting with multiple hormones. Plant Signaling & Behavior 4:1117–1120

Gravity Sensing, Interpretation, and Response

Miyo Terao Moirta, Moritaka Nakamura, and Masao Tasaka

Abstract Because higher plants spend their sessile lives at the site of their germination, they rely on a number of strategies to ensure their survival in response to environmental stimuli. One of the stimuli to which plants can respond is gravity. Here, we describe recent findings with regard to the plant's response to gravity. We put specific emphasis on the molecular mechanism of gravitropism, which is a well-studied response to gravity. Since the direction and the magnitude of gravity are relatively constant on the surface of the Earth, gravitropism can be regarded as a posture adjustment, triggered by sensing the tilt of organs relative to the direction of gravity. Recent studies that combined molecular genetics and cell biological approaches in *Arabidopsis thaliana* have contributed to understand the mechanism of gravitropism.

1 Introduction

All living organisms evolve under the Earth's gravity in various ways. Both animals and plants sense the direction and magnitude of gravity and respond to these by regulating their growth and development. The responses to gravity appear to have critical effects upon terrestrial plants. To stand upright against the gravitational force, plants have developed a tough cell wall and wood tissue for structural support. Meanwhile, plants also utilize gravity as a directional cue to regulate the direction of their growth so as to be in a suitable position for absorption of water or nutrients, photosynthesis, reproduction, and morphogenesis. This chapter will provide a general introduction to plant responses to gravity, followed by a discussion of gravitropism, which is a response to gravity that has been well studied at the molecular level.

M.T. Moirta (✉) • M. Nakamura • M. Tasaka
Graduate School of Biological Sciences, Nara Institute of Science and Technology, Takayama/Ikoma/Nara, Japan
e-mail: mimorita@bs.naist.jp; mo-nakam@bs.naist.jp; m-tasaka@bs.naist.jp

G. Witzany and F. Baluška (eds.), *Biocommunication of Plants*, Signaling and Communication in Plants 14,
DOI 10.1007/978-3-642-23524-5_4,

2 Plant Responses Are Influenced by Gravity

It has been suggested that gravitational force influences various aspects of the cells and organs of plants, including their metabolism, intracellular architectures, cell growth, directional growth of organs, and development. Unlike other environmental signals, gravity is constitutive and is difficult to counteract in the laboratory setting. As such, technical difficulties have impeded the effort to investigate the response to gravity. Further development of space-based research, such as at the International Space Station, should enable a better understanding of the responses of all organisms to gravity (Correl and Kiss 2008). At the same time, efforts to alleviate the technical difficulties of research on the ground have greatly facilitated the study of gravitational responses. An advantage of ground-based experiments is the ability to assess their reproducibility more easily than space-based experiments. The study of responses to hypergravity using a centrifuge is one technique to investigate plant responses to gravity.

When dicot seedlings are grown under various hypergravity conditions, the rate of growth of shoots is affected by changes in the extensibility of the cell wall. This response is correlated to the cortical microtubule array, which is involved in the arrangement of cellulose microfibrils in the cell wall (Skagen and Iversen 1999; Matsumoto et al. 2010). Consistent with this, it has been reported that hypergravity affects cell wall components, resulting in the reinforcement of the cell wall (*Elodea* at 80 g; Chen et al. 1980; cress hypocotyl at 135 g; Hoson et al. 1996). In addition, the expression of genes involved in cell wall modification is also altered by hypergravity (*Arabidopsis* hypocotyl at 300 g; Zenko et al. 2003). This type of response to hypergravity, termed "gravity resistance," may reflect resistance of the plant cell against mechanical stress so as to support the plant body. An inhibitor of stretch-activated channel (Gd^{3+}) blocked this growth response to 300 g, implying that the stretch-activated channel is involved in sensing hypergravity stimulation (Soga et al. 2004, 2005). Gravity resistance involves sensing the magnitude of gravitational force or mechanical pressure, whereas gravitropism involves sensing the directional change of gravity. The shoot of an *Arabidopsis* mutant lacking gravity-sensing tissue for gravitropism showed a normal gravity resistance response, suggesting that the gravity perception mechanism of gravity resistance differs from that of gravitropism (Soga et al. 2004). The molecular mechanisms of sensing, signaling, and response to gravity resistance remain to be elucidated fully.

Peg formation in *Cucurbitaceae* plants is a unique form of gravimorphogenesis that represents another well-studied example of the plant response to gravity evaluated on the ground (Darwin 1880). *Cucurbitaceae* seedlings form a protruded tissue, called a peg, at one side of the boundary between the root and epicotyl just after germination. When flat cucumber seeds are placed horizontally, a peg forms at the lower side of the boundary (Witztum and Gersani 1975; Takahashi 1997). Meanwhile, vertically positioned seeds with the radicle pointing downward, clinoratated (random positioned) seeds, or seeds germinated under microgravity in space-flight experiments form one peg on each side of the boundary in a bilaterally symmetric manner (Takahashi et al. 2000). This suggests that gravity is not an essential signal for the development of the peg and that cucumber seeds have the potential to develop pegs on both sides. Thus, cucumber seedlings respond to gravity by suppressing peg formation at the upper side of the boundary of horizontally placed seeds. Although there is limited information on the molecular mechanism of peg formation because of the difficulty of molecular genetics in the cucumber, the response of cucumber seedlings is reminiscent of gravitropism. First, seeds perceive the directional information of gravity. Second, amyloplast sedimentation in the direction of gravity is observed in the tissue of the responding organ (Takahashi 1997). Third, auxin plays important roles. Expression of auxin-regulated gene was induced at the lower side and reduced at the upper side of the boundary region of horizontally placed seedlings. Exogenous application of auxin can induce the formation of a peg at the upper side, suggesting that each boundary region has the potential to form a peg (Kamada et al. 2000). These findings imply that the suppression of peg formation is caused by a decrease in auxin concentration or in the auxin response at the upper side, while peg formation is induced by an increase in the auxin concentration or response at the lower side. Sequence of the cucumber genome and progress in molecular genetic study of gravitropism using model plants may contribute to the identification of similarities and differences between these responses to gravity and provide new insights into plant responses to the directional cue of gravity.

2.1 Gravitropism

Gravitropism is a form of plant movement that is under continuous control with regard to the orientation and juxtaposition of the various parts of the plant body in response to gravity. In general, plant shoots grow upward (negative gravitropism), whereas roots grow downward (positive gravitropism). In higher plants, it has been thought that the relative directional change of gravity is suspected in specialized cells called statocytes, followed by signal conversion from the physical information into physiological information within the statocytes. The signal is subsequently transmitted to neighboring cells and other tissues, which leads to differential cell growth between the lower and upper flanks of the responsive organ (Morita 2010).

2.1.1 Starch Statolith Hypothesis

Since the direction and magnitude of gravity are almost constant across the surface of the Earth, gravitropism can be regarded as a posture adjustment triggered by sensing the tilt of organs relative to the direction of gravity (Boonsirichai et al. 2002; Tasaka et al. 1999). Since gravity acts upon mass, a number of organisms use relatively heavy cellular components, called statoliths or otoliths, to sense the direction of gravity. At the end of nineteenth century, it was observed that starch grains, which accumulate within particular plastids known as amyloplasts, sink in the direction of gravity within specific cells in the gravity-responding organs of higher plants. These observations led to the widely accepted starch statolith hypothesis, which holds that the sedimentation of the amyloplast toward the gravity vector within specific cells (statocytes) acts as the probable trigger for the directional sensing of gravity (Sack 1991, 1997). Genetic studies using a model plant, *Arabidopsis thaliana*, also support this hypothesis. The *phosphoglucomutase* (*pgm*) mutant is impaired in starch synthesis and exhibits a reduced gravitropic response in all graviresponsive organs (Caspar and Pickard 1989; Kiss et al. 1989, 1997; Weise and Kiss 1999).

2.1.2 Statocyte (Gravity Sensing Cell)

Studies using *Arabidopsis* have identified the cells responsible for sensing gravity. In *Arabidopsis* roots, the root cap comprises four tiers of columella cells and lateral root cap cells (Fig. 1). Columella cells contain sedimented amyloplasts. Genetic manipulation to remove the root cap abolishes root gravitropism (Tsugeki and Fedoroff 1999). Laser ablation of specific cells within the root cap has shown that the inner cells of the second tier of columella cells contribute greatly to root gravitropism (Blancaflor et al. 1998). These studies strongly suggest that columella

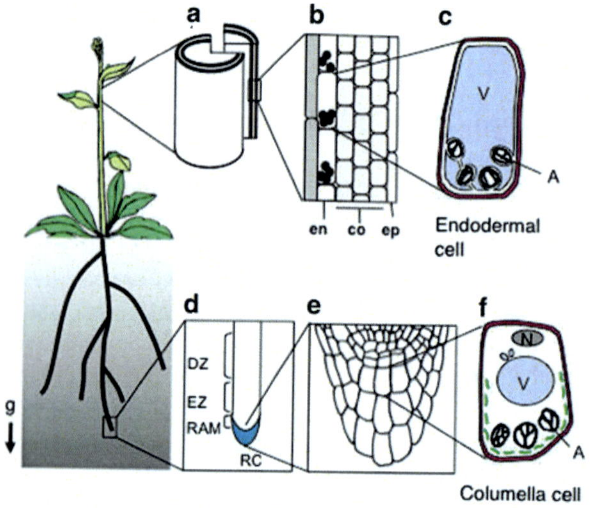

Fig. 1 Statocytes in *Arabidopsis thaliana*. *Upper* Schematic structure of shoot statocyte. (**a**) Stem tissue. One layer of endodermis is arranged cylindrically. (**b**) Arrangement of the epidermis (ep), the cortex (co), and the endodermis (en). (**c**) The endodermal cell. Starch-accumulating amyloplasts sedimented in the direction of gravity. *V* vacuole, *A* amyloplast. *Lower* Schematic structure of root statocyte. (**d**) Arrangement of the root cap (RC), the root apical meristem (RAM), elongation zone (EZ), and differentiation zone (DZ). (**e**) Root cap structure. (**f**) Schematic structure of the columella cell. *N* nucleus, *V* vacuole, *green lines* endoplasmic reticulum. Starch-accumulating amyloplasts sedimented in the direction of gravity. *g* direction of gravitational force

cells in the root cap are the gravity-sensing cells. The subcellular structure of root columella cell has the following characteristics: They contain relatively abundant cytoplasm and small vacuoles. The nucleus and ER are localized in a polarized configuration at the upper side and the periphery of the cell, respectively (Sack 1991; Konings 1995). Amyloplasts, which are derived from proplastids in columella initial cells, contain large starch granules but do not have an organized thylakoid membrane structure or photosynthetic pigments (Sack 1991).

In the gravity-responding region of *Arabidopsis* shoots, the epidermis, cortex, and endodermis surrounding vascular tissue and pith are arranged concentrically from the outside to the inside (Fig. 1). *Shoot gravitropism (sgr)1/scarecrow (scr)* and *sgr7/short-root (shr)* mutants exhibit no gravitropic response in their shoots (Fukaki et al. 1996). *SGR1/SCR* and *SGR7/SHR* are members of the *GRAS* gene family and encode transcription factors essential for the formation of the endodermis both in shoots and in roots, indicating that the endodermis is essential for shoot gravitropism (Fukaki et al. 1998; Wysocka-Diller et al. 2000; Helariutta et al. 2000). Since the endodermal cells of *Arabidopsis* shoots contain sedimentable amyloplasts, endodermal cells probably are statocyte (Tasaka et al. 1999; Morita and Tasaka 2004). Meanwhile, the roots of *sgr1/scr* and *sgr7/shr* mutants exhibit nearly normal gravitropism (Fukaki et al. 1996), indicating that the endodermal cells have little or no role in root gravitropism and that the statocytes of the roots and shoots have a distinct developmental origin. A recent study of shoot gravitropic mutants lacking orthologs of *SCR* (*weeping*) and *SHR* (*weeping2*) in Japanese morning glory (*Pharbitis nil*) indicated the importance of the endodermis for shoot gravitropism, as has been shown in *Arabidopsis* (Kitazawa et al. 2005, 2008).

Both the developmental origin and subcellular structure of endodermal cells are significantly different from that of root columella cells. The polarity of the nucleus and ER was unclear in the *Arabidopsis* endodermis. The most prominent feature of the shoot statocyte is a large central vacuole that occupies most of the cell volume (Fig. 2, Clifford and Barclay 1980; Sack 1987; Saito et al. 2005). Cytoplasm exists in transvacuolar strands and in a narrow space between the vacuolar and plasma membranes. Amyloplasts in endodermal cells are almost completely enclosed by a vacuolar membrane with only a thin layer of cytoplasm (Clifford et al. 1989; Saito et al. 2005). The endodermal amyloplasts are not likely orthotypical amyloplasts but are likely to be chloroplasts that specifically accumulate starch. Chlorophyll autofluorescence is observed in amyloplasts in endodermal cells as well as in chloroplasts in the neighboring cortical cells in *Arabidopsis* inflorescence stems (Fig. 2, Morita et al. 2006). Since the term "amyloplast" has been used for the starch-accumulating leucoplast, it is technically suitable to describe plastids in the columella cell but not those in the endodermal cell. A new term may be required to better describe the chloroplastic amyloplast in shoot statocytes.

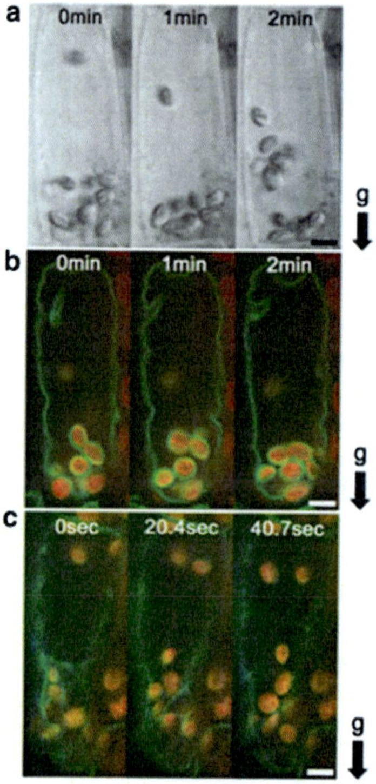

Fig. 2 Shoot endodermal cell contains dynamically moving amyloplasts and a vacuole. Live-cell imaging of an endodermal cell. The samples were hold vertically during observation by using a vertical stage microscope. (**a**) Sequential blight-field images showing amyloplast movement. (**b**) Sequential confocal images showing dynamic movements of amyloplasts (*red*) vacuolar membrane (*green*). Amyloplasts and vacuolar membrane are visualized by autofluorescence and vacuolar marker protein GFP-VAM3/SYP22 expressed under the endodermis specific *SCR* promoter. (**c**) Sequential confocal images showing dynamic amyloplasts (*red*) and F-actin (*green*) movements. Amyloplasts and F-actin are visualized by autofluorescence and GFP-mTalin expressed under the 35S promoter. *g* direction of gravitational force. Scale Bar=5μm.

2.1.3 Amyloplast Movement

Amyloplast movement toward the direction of gravity is likely to be important to sense the direction of gravity in all organs. The presentation time is defined as the minimum period of exposure to a gravity stimulus placed horizontally at 1 g that is required to elicit a gravitropic response. Amyloplast sedimentation occurs toward the new bottom of the statocyte upon gravitational stimulation within the presentation time (Sack et al. 1984, 1985). More correlative evidence has been provided by unique experiments with high-gradient magnetic fields (HGMFs) (Kondrachuk and Hasenstein 2001). HGMF was adopted to mimic a gravitational field to exploit differences between the diamagnetic susceptibilities of starch and the cytoplasm. HGMF was able to induce amyloplast relocalization, resulting in organ curvature similar to that of the gravitropic response (Kuznetsov and Hasenstein 1996). Since HGMF does not affect root gravitropism in starchless mutants, HGMF is unlikely to act on substances other than starch (Weise et al. 2000). In addition, several mutants lacking shoot gravitropism contain amyloplasts that fail to sediment but which disperse within the endodermal cell (*zigzag (zig)/sgr4* and *sgr2*, see below; Morita et al. 2002). The suppressor mutations that partially suppress the gravitropic phenotype of the *zig/sgr4* mutant partially restore amyloplast sedimentation in the endodermal cell (Niihama et al. 2005, 2009; Hashiguchi et al. 2010). These studies strongly support the idea that amyloplast movement toward the direction of gravity is important and is probably the key event triggering gravity sensing.

Although the word statolith refers to a "stationary stone," the behavior of amyloplasts differs considerably from that of the ideal statolith, particularly in shoot statocytes (Clifford and Barclay 1980; Sack and Leopold 1985; Saito et al. 2005). Live-cell imaging of endodermal cells in *Arabidopsis* stems revealed that amyloplasts exhibit continuous dynamic and complicated movements (Fig. 2, Saito et al. 2005; Nakamura et al. 2011). Recently, genetic studies demonstrated that the intracellular environment of the statocyte has considerable effects upon amyloplast movement.

Genes responsible for *Arabidopsis sgr2, zig/sgr4, sgr3, and sgr8/gravitropism defective (grv)2/katamari (kam)2* mutants that exhibit little or reduced shoot gravitropism encode proteins that have been implicated in vesicle transport to vacuoles (Kato et al. 2002; Morita et al. 2002; Yano et al. 2003; Silady et al. 2004). Endodermal amyloplasts exhibit little movement and do not sediment in these mutants. As mentioned above, endodermal amyloplasts pass through the narrow cytoplasmic space enclosed by the vacuolar membrane. Interestingly, these mutants show normal root gravitropism, probably due to relatively small vacuoles and abundant cytoplasm in root columella cells. This finding indicates that normal vacuolar function is required for amyloplast sedimentation, which is an important feature of shoot statocytes.

In addition to the vacuole, filamentous actin (F-actin) is involved in amyloplast movement (Fig. 2, Sack et al. 1986; Yamamoto and Kiss 2002; Hou et al. 2004; Saito et al. 2005). Most amyloplasts sediment with the direction of gravity, whereas a few amyloplasts exhibit saltatory movement in *Arabidopsis* endodermal cells (Saito et al. 2005). A recent study of *sgr9* mutant characterized by reduced gravitropism provides an interpretation for the complicated amyloplast dynamics found in the endodermal cell in the context of the interaction between amyloplasts and F-actin (Nakamura et al. 2011). Endodermal amyloplasts in this mutant exhibit dynamic movement but fail to sediment with the direction of gravity. Amyloplasts sometimes form a cluster that is abnormally entangled with F-actin in *sgr9* plants, whereas such clustered amyloplasts have never been found in wild-type plants. Inhibition of F-actin formation nullified both the effect of *sgr9* mutation on amyloplast sedimentation and the gravitropic response, suggesting excess interaction between amyloplasts and F-actin in the mutant. Thus, *Arabidopsis* endodermal amyloplasts appear to be in a dynamic equilibrium between sedimentation and saltatory movements, and this equilibrium is principally the result of interaction between the amyloplasts and F-actin in wild-type plants. According to this model, F-actin promotes the saltatory movements, whereas the SGR9, which is a RING-type E3 ligase localized to amyloplasts, may impede the interaction between amyloplasts and F-actin, allowing the amyloplasts to sediment in the direction of gravity (Nakamura et al. 2011).

2.1.4 Plastid-Based Gravity Sensing

Amyloplasts sedimentation is likely to be due to the dense accumulation of starch granules. The *pgm* mutant, however, exhibits a reduced but significant gravitropic response in both roots and shoots (Caspar and Pickard 1989; Kiss et al. 1989, 1997; Weise and Kiss 1999). Amyloplasts are unlikely to sediment to the bottom of cells in the mutant statocytes. The residual gravitropic response observed in the *pgm* mutant suggests that while starch is necessary for a full gravitropic response, its presence is not absolutely essential for sensing gravity. The extent of reduction in gravitropism is positively correlated with the reduction in starch content, suggesting that the mass of the amyloplast (starch) indeed affects the magnitude of the gravitropic response (Kiss et al. 1996; Sack 1991). As discussed in detail by Sack (1991, 1997), amyloplasts lacking starch, i.e., plastids, can act as susceptors and trigger a residual gravity response in starchless mutants. Thus, not starch but the plastid itself may act as a statolith, or a redundant gravity-sensing system may exist for the gravitropism observed in higher plants.

2.1.5 Intracellular Signaling

In columella cells, the ER is localized to the periphery of the cell. Since the ER represents an intracellular Ca^{2+} reservoir in general, it has been hypothesized that contact between amyloplasts and the peripheral ER could trigger release of Ca^{2+} stored in the ER as a possible gravity-sensing mechanism (Perbal and Driss-Ecole 2003). Recent research using electron micrography employing high-pressure freezing and freeze-substitution methods revealed close contact between amyloplasts and the cortical ER (Leitz et al. 2009). Unfortunately, a significant change in cytosolic Ca^{2+} concentration ($[Ca^{2+}]_c$) in columella cells in response to gravistimulation has not been observed (Legue et al. 1997). To date, there is insufficient evidence to support this attractive hypothesis. In contrast to $[Ca^{2+}]_c$, transient alkalization in the cytosol (pH_c) upon gravistimulation was detected in columella cells of *Arabidopsis* and in shoot statocytes of maize (Scott and Allen 1999; Fasano et al. 2001; Johannes et al. 2001). Consistent with this finding, artificial increase of the proton concentration in *Arabidopsis* columella cells using caged protons and UV irradiation partially inhibited root gravitropism (Fasano et al. 2001). In addition, a transient increase in pH_c was not detected in the *pgm* mutant, suggesting that starch-containing dense amyloplasts are required for subsequent cytosolic alkalization in wild-type columella cells (Fasano et al. 2001). However, the molecular mechanism of the graviresponsive transient increase in pH_c remains to be elucidated. The relationship between amyloplast displacement and the transient increase of pH_c is also undetermined.

The *ARG* (altered response to gravity)*1* gene, encoding a J-domain protein localized to endomembrane organelles, is also required for the transient increase in pH_c (Boonsirichai et al. 2003). *arg1* exhibits a reduced gravitropic response in the hypocotyls and in the roots (Sedbrook et al. 1999; Fukaki et al. 1997). *ARG1* and its paralog *ARL2* function in the root columella cells in root gravitropism (Boonsirichai et al. 2003; Harrison and Masson 2008). The presence of a J-domain implies that these proteins function as molecular chaperones, but their sites of action are unclear. Interestingly, PIN3 redistribution within the columella cells upon gravistimulation (see below) also fails to occur in *arg1* mutant. These findings suggest that ARG1 plays a role in the early processes of gravity signal transduction, which may modulate PIN3 redistribution. The relationship between the transient increase in pH_c and PIN3 redistribution is intriguing but yet to be defined.

2.1.6 Road to Organ Response

Auxin is an important and well-studied plant hormone that was identified as a substance that promotes cell elongation upon organ curvature during the phototropic response. According to the theory of Cholodony and Went, lateral transport of auxin within the responding organ is induced by directional stimulation by gravity (or light in phototropism), and the resulting asymmetric auxin distribution between the lower and upper sides induces organ curvature (Fig. 3). The auxin flow in the *Arabidopsis* root has been extensively studied (Petrásek and Friml 2009). Auxin derived from the shoot is usually transported rootward (toward the root tip) through vascular tissue and the central cylinder to root columella cells largely owing to the function of PIN1 and PIN4 (Gälweiler et al. 1998; Friml et al. 2002a). At the root cap, auxin is transported back from the columella toward the shoot through the lateral root cap and the epidermis, largely owing to the function of AUX1 and PIN2,

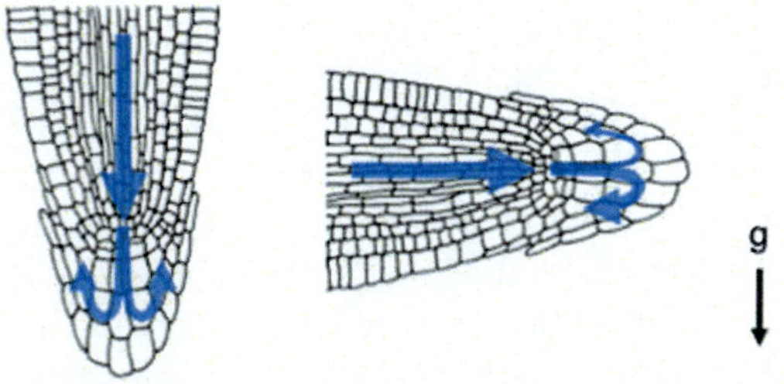

Fig. 3 Auxin flow in *Arabidopsis* root tip. Auxin flow in root tip before (*left*) and after (*right*) gravistimulation is schematically indicated. *Blue arrows* indicate auxin flow. Auxin is accumulated at the lower flank of the root after gravistimulation

and reaches the elongation zone (Chen et al. 1998; Luschnig et al. 1998; Müller et al. 1998; Utsuno et al. 1998; Swarup et al. 2005). It has been demonstrated that the coordinated function of auxin influx (AUX1, etc.) and efflux (PIN1, PGP1, etc.) transporters plays important roles in the establishment of the asymmetric distribution of auxin during root gravitropism (reviewed in Zazímalová et al. 2010).

Regulation of the flow and distribution of auxin are crucial in various developmental processes as well as tropism. Developmental cues and/or external signals determine the intracellular localization of PIN family proteins thereby directing intercellular auxin flow. Thus, the mechanism involved in regulation of intracellular localization of PIN family proteins has been extensively studied (reviewed in Grunewald and Friml 2010). Intracellular trafficking mechanisms, such as the GNOM-dependent pathway (Steinmann et al. 1999; Geldner et al. 2003), clathrin-dependent endocytosis (Dhonukshe et al. 2007, 2008), the retromer-dependent pathway (Jaillais et al. 2006, 2007; Kleine-Vehn et al. 2008), and targeting to the lumen of lytic vacuoles probably via the action of ESCRT (Spitzer et al. 2009), are all involved in control of PIN polarity and degradation. Regulation of PIN phosphorylation by PINOID Ser/Thr kinase and protein phosphatase 2A is known to be important for polar localization of PIN proteins (Christensen et al. 2000; Benjamins et al. 2001; Friml et al. 2004; Michniewicz et al. 2007). A recent study demonstrated that *MACCHI-BOU4/ENHANCER of PINOID/NAKED PINS IN YUC MUTANTS 1 (MAB4/ENP/NPY1)* and its paralogous genes encoding nonphototropic hypocotyl 3 (NPH3)-like proteins are required for regulation of PIN protein level and polar localization (Furutani et al. 2011). Interestingly, the proteins exhibit polar localization, which is almost identical to PIN polarity, although their molecular function is unclear. A number of mutants deficient in these factors exhibit gravitropic defects in the root, as well as developmental and morphological defects. The identity of the specific PIN responsible for the impaired gravitropic phenotype is unclear, as is the specific process in the gravitropic response that is affected a particular mutation in most cases.

During root growth, auxin is nearly equally distributed to the radially arranged epidermis through the lateral root cap, whereas the distribution at the columella cells becomes unequal in response to gravistimulation, leading to accumulation of auxin at the lower flank of the root. It is proposed that the directional signal sensed in the columella cells may be converted to a directional regulation of auxin flow. PIN3, which is expressed both in the root columella cells and in the shoot endodermis, is an ideal candidate as a regulator of lateral auxin flow upon gravistimulation (Friml et al. 2002b). Although the genetic contribution of PIN3 to gravitropism is not so obvious, it has been explained as arising from the genetic redundancy of PIN family genes (Kleine-Vehn et al. 2010). In the root, PIN3, which is distributed uniformly in the cell, moves to the lower side of the cell in response to gravistimulation (Friml et al. 2002b; Harrison and Masson 2008). A recent study suggests that redistribution of PIN3 in columella cells upon gravistimulation requires the activity of the GNOM-dependent trafficking pathway and that at least a fraction of the PIN3 might be redistributed via endosome-based translocation from one side of the cell to the other (transcytosis) (Kleine-Vehn et al. 2010). This might allow roots to rapidly change their growth (approx. 10–15 min; Mullen et al. 2000; Fasano et al. 2001), although the temporal relationship between PIN3 redistribution and root response remains to be elucidated. However, elucidation of the regulatory mechanism governing PIN distribution within statocytes may provide a clue to understanding the signal transduction of gravity directional sensing.

In the phototropic response, when the hypocotyl is irradiated by unidirectional light, the level of membrane-localized PIN3 protein in the endodermal cell on the lit side of the hypocotyl is reduced compared to that on the shaded side (Ding et al. 2011). This response requires phototropin and the GNOM-dependent trafficking pathway and is disturbed by excess PID activity. It is unclear whether PIN3 is redistributed in response to the directional light signal within a cell or if the regulation of the level of PIN3 protein in the lit side of cells is elicited by the signal. However, it is intriguing that different external directional signal may target the same PIN protein.

Although there is limited knowledge of long-distance signaling other than auxin transport, proton flux along the root tip during the gravitropic response has been reported (Zieschang and Sievers 1994; Monshausen and Sievers 2002). Asymmetric pH changes were observed at the surface of gravistimulated roots by proton-selective microelectrodes. The surface pH changes occurred at the root cap and progressed shootward (basipetally) to the elongation zone. A recent imaging technique demonstrates an ionic response during the gravitropic response with high spatiotemporal resolution (Monshausen et al. 2011). Roots show a highly dynamic pH pattern during vertical growth that is modified during gravistimulation. The root surface is acidified at the upper flank, whereas it is alkalized at the lower flank within 3 min after gravistimulation, which is much faster than the growth response, supporting the pH-dependent acid growth related to tropic curvature. In addition to proton flux, gravistimulation triggers asymmetric change in $[Ca^{2+}]_c$ in the root epidermis. The precise roles of the dynamics of surface pH and epidermal $[Ca^{2+}]_c$ induced by the AUX1-dependent auxin influx in the gravitropic response are not yet known. Elucidation of the temporal linkage between gravity-induced increases in pH_c and PIN3 redistribution occurred within the columella cells, and this long-distance ionic signaling may provide a clue to understanding the tropic organ response.

3 Conclusions

In gravitropism, sedimentation of a specific plastid, the amyloplast, is used as a statolith that provides directional information within the statocytes in each organ. Sedimentation of the amyloplast might also be utilized in gravimorphogenesis for directional information. In addition to the statolith, asymmetric auxin distribution in the responding organs may link the directional information provided by the statolith to the organ response in gravimorphogenesis. Thus, gravitropism and gravimorphogenesis possibly share similar signaling module(s), and this raises the intriguing possibility of an evolutional relationship between these directional organ responses.

In gravitropism, although it is clear that amyloplast displacement is important for triggering the directional cue, the identity of the gravity sensor remains unknown. The directional information of gravity, which is sensed locally in statocytes, is expanded to the response at the organ level. Auxin is likely to be a key carrier of information in this process. Understanding the signal conversion mechanism from the directional information to auxin flow is a critical issue, and this provides a clue to close in upon the gravity sensor.

References

Benjamins R, Quint A, Weijers D, Hooykaas P, Offringa R (2001) The PINOID protein kinase regulates organ development in *Arabidopsis* by enhancing polar auxin transport. Development 128:4057–4067

Blancaflor EB, Fasano JM, Gilroy S (1998) Mapping the functional roles of cap cells in the response of *Arabidopsis* primary roots to gravity. Plant Physiol 116:213–222

Boonsirichai K, Guan C, Chen R, Masson PH (2002) Root gravitropism: an experimental tool to investigate basic cellular and molecular processes underlying mechanosensing and signal transmission in plants. Annu. Rev. Plant. Biol. 53: 421–447

Boonsirichai K, Sedbrook JC, Chen R, Gilroy S, Masson PH (2003) Altered response to gravity is a peripheral membrane protein that modulates gravity-induced cytoplasmic alkalinization and lateral auxin transport in plant statocytes. Plant Cell 15:2612–2625

Caspar T, Pickard BG (1989) Gravitropism in a starchless mutant of *Arabidopsis*: implications for the starch-statolith theory of gravity sensing. Planta 177:185–197

Chen N, Siegel SM, Siegel BZ (1980) Gravity and land plant evolution: experimental induction of lignification by simulated hypergravity and water stress. Life Sci Space Res 18:193–198

Chen R, Hilson P, Sedbrook J, Rosen E, Caspar T, Masson PH (1998) The *Arabidopsis thaliana AGRAVITROPIC 1* gene encodes a component of the polar-auxin-transport efflux carrier. Proc Natl Acad Sci USA 95:15112–15117

Christensen S, Dagenais N, Chory J, Weigel D (2000) Regulation of auxin response by the protein kinase PINOID. Cell 100:469–478

Clifford PE, Barclay GF (1980) The sedimentation of amyloplasts in living statocytes of the dandelion flower stalk. Plant Cell Environ 3:381–386

Clifford PE, Douglas S, McCartney GW (1989) Amyloplast sedimentation in shoot statocytes having a large, central vacuole: Further interpretation from electron microscopy. J Exp Bot 40:1341–1346

Correl MJ, Kiss JZ (2008) Space-based research on plant tropism. In: Gilroy S, Masson PH (eds) Plant tropisms. Blackwell Publishing, Oxford, pp 161–182

Darwin C (1880) The power of movement in plants. John Murray, London

Dhonukshe P, Aniento F, Hwang I, Robinson DG, Mravec J, Stierhof YD, Friml J (2007) Clathrin-mediated constitutive endocytosis of PIN auxin efflux carriers in *Arabidopsis*. Curr Biol 17:520–527

Dhonukshe P, Tanaka H, Goh T, Ebine K, Mähönen A, Prasad K, Blilou I, Geldner N, Xu J, Uemura T, Chory J, Ueda T, Nakano A, Scheres B, Friml J (2008) Generation of cell polarity in plants links endocytosis, auxin distribution and cell fate decisions. Nature 456:962–966

Ding Z, Galván-Ampudia CS, Demarsy E, Łangowski Ł, Kleine-Vehn J, Fan Y, Morita MT, Tasaka M, Fankhauser C, Offringa R, Friml J (2011) Light-mediated polarization of the PIN3 auxin transporter for the phototropic response in *Arabidopsis*. Nat Cell Biol 13:447–452

Fasano JM, Swanson SJ, Blancaflor EB, Dowd PE, Kao TH, Gilroy S (2001) Changes in root cap pH are required for the gravity response of the *Arabidopsis* root. Plant Cell 13:907–921

Friml J, Benková E, Blilou I, Wisniewska J, Hamann T, Ljung K, Woody S, Sandberg G, Scheres B, Jürgens G, Palme K (2002a) AtPIN4 mediates sink-driven auxin gradients and root patterning in *Arabidopsis*. Cell 108:661–673

Friml J, Wiśniewska J, Benková E, Mendgen K, Palme K (2002b) Lateral relocation of auxin efflux regulator PIN3 mediates tropism in *Arabidopsis*. Nature 415:806–809

Friml J, Yang X, Michniewicz M, Weijers D, Quint A, Tietz O, Benjamins R, Ouwerkerk P, Ljung K, Sandberg G, Hooykaas PJ, Palme K, Offringa R (2004) A PINOID-dependent binary switch in apical–basal PIN polar targeting directs auxin efflux. Science 306:862–865

Fukaki H, Fujisawa H, Tasaka M (1996) *SGR1, SGR2, SGR3*: Novel genetic loci involved in shoot gravitropism in *Arabidopsis thaliana*. Plant Physiol 110:945–955

Fukaki H, Fujisawa H, Tasaka M (1997) The *RHG* gene is involved in root and hypocotyl gravitropism in *Arabidopsis thaliana*. Plant Cell Physiol 38:804–810

Fukaki H, Wysocka-Diller J, Kato T, Fujisawa H, Benfey PN, Tasaka M (1998) Genetic evidence that the endodermis is essential for shoot gravitropism in *Arabidopsis thaliana*. Plant J 14:425–430

Furutani M, Sakamoto N, Yoshida S, Kajiwara T, Robert HS, Friml J, Tasaka M (2011) Polar-localized NPH3-like proteins regulate polarity and endocytosis of pin-formed auxin efflux carriers. Development 138:2069–2078

Gälweiler L, Guan C, Müller A, Wisman E, Mendgen K, Yephremov A, Palme K (1998) Regulation of polar auxin transport by AtPIN1 in *Arabidopsis* vascular tissue. Science 282:2226–2230

Geldner N, Anders N, Wolters H, Keicher J, Kornberger W, Muller P, Delbarre A, Ueda T, Nakano A, Jurgens G (2003) The *Arabidopsis* GNOM ARF-GEF mediates endosomal recycling, auxin transport, and auxin dependent plant growth. Cell 112:219–230

Grunewald W, Friml J (2010) The march of the PINs: developmental plasticity by dynamic polar targeting in plant cells. EMBO J 29:2700–2014

Harrison BR, Masson PH (2008) ARL2, ARG1 and PIN3 define a gravity signal transduction pathway in root statocytes. Plant J 53:380–392

Hashiguchi Y, Niihama M, Takahashi T, Saito C, Nakano A, Tasaka M, Morita MT (2010) Loss-of-function mutations of retromer large subunits suppress the phenotype of *zig* mutant that lacks Qb-SNARE VTI11. Plant Cell 22:159–172

Helariutta Y, Fukaki H, Wysocka-Diller J, Nakajima K, Jung J, Sena G, Hauser MT, Benfey PN (2000) The *SHORT-ROOT* gene controls radial patterning of the *Arabidopsis* root through radial signaling. Cell 101:555–567

Hoson T, Nishitani K, Miyamoto K, Ueda J, Kamisaka S, Yamamoto R, Masuda Y (1996) Effects of hypergravity on growth and cell wall properties of cress hypocotyls. J Exp Bot 47:513–517

Hou G, Kramer VL, Wang YS, Chen R, Perbal G, Gilroy S, Blancaflor EB (2004) The promotion of gravitropism in *Arabidopsis* roots upon actin disruption is coupled with the extended alkalinization of the columella cytoplasm and a persistent lateral auxin gradient. Plant J 39:113–125

Jaillais Y, Fobis-Loisy I, Miége C, Rollin C, Gaude T (2006) AtSNX1 defines an endosome for auxin-carrier trafficking in *Arabidopsis*. Nature 443:106–109

Jaillais Y, Santambrogio M, Rozier F, Fobis-Loisy I, Miége C, Gaude T (2007) The retromer protein VPS29 links cell polarity and organ initiation in plants. Cell 130:1057–1070

Johannes E, Collings DA, Rink JC, Allen NS (2001) Cytoplasmic pH dynamics in maize pulvinal cells induced by gravity vector changes. Plant Physiol 127:119–130

Kamada M, Fujii N, Aizawa S, Kamigaichi S, Mukai C, Shimazu T, Takahashi H (2000) Control of gravimorphogenesis by auxin: accumulation pattern of *CS-IAA1* mRNA in cucumber seedlings grown in space and on the ground. Planta 211:493–501

Kato T, Morita MT, Fukaki H, Yamauchi Y, Uehara M, Niihama M, Tasaka M (2002) SGR2, a phospholipase-like protein, and ZIG/SGR4, a SNARE, are involved in the shoot gravitropism of *Arabidopsis*. Plant Cell 14:33–46

Kiss JZ, Hertel R, Sack FD (1989) Amyloplasts are necessary for full gravitropic sensitivity in roots of *Arabidopsis thaliana*. Planta 177:198–206

Kiss JZ, Wright JB, Caspar T (1996) Gravitropism in roots of intermediate-starch mutants of *Arabidopsis*. Physiol Plant 97:237–244

Kiss JZ, Guisinger MM, Miller AJ, Stackhouse KS (1997) Reduced gravitropism in hypocotyls of starch-deficient mutants of *Arabidopsis*. Plant Cell Physiol 38:518–525

Kitazawa D, Hatakeda Y, Kamada M, Fujii N, Miyazawa Y, Hoshino A, Iida S, Fukaki H, Morita MT, Tasaka M, Suge H, Takahashi H (2005) Shoot circumnutation and winding movements require gravisensing cells. Proc Natl Acad Sci USA 102:18742–18747

Kitazawa D, Miyazawa Y, Fujii N, Nitasaka E, Takahashi H (2008) Characterization of a novel gravitropic mutant of morning glory, *weeping2*. Adv Space Res 42:1050–1059

Kleine-Vehn J, Leitner J, Zwiewka M, Sauer M, Abas L, Luschnig C, Friml J (2008) Differential degradation of PIN2 auxin efflux carrier by retromer-dependent vacuolar targeting. Proc Natl Acad Sci USA 105:17812–17817

Kleine-Vehn J, Ding Z, Jones AR, Tasaka M, Morita MT, Friml J (2010) Gravity-induced PIN transcytosis for polarization of auxin fluxes in gravity-sensing root cells. Proc Natl Acad Sci USA 107:22344–22349

Kondrachuk AV, Hasenstein KH (2001) The effects of HGMFs on the plant gravisensing system. Adv Space Res 27:1001–1005

Konings A (1995) Gravitropism of roots: an evaluation of progress during the last three decades. Botanica Acta 44:195–223

Kuznetsov OA, Hasenstein KH (1996) Intracellular magnetophoresis of amyloplasts and induction of root curvature. Planta 198:87–94

Legue V, Blancaflor E, Wymer C, Perbal G, Fantin D, Gilroy S (1997) Cytoplasmic free Ca^{2+} in *Arabidopsis* roots changes in response to touch but not gravity. Plant Physiol 114:789–800

Leitz G, Kang BH, Schoenwaelder ME, Staehelin LA (2009) Statolith sedimentation kinetics and force transduction to the cortical endoplasmic reticulum in gravity-sensing *Arabidopsis* columella cells. Plant Cell 21:843–860

Luschnig C, Gaxiola RA, Grisafi P, Fink GR (1998) EIR1, a root-specific protein involved in auxin transport, is required for gravitropism in *Arabidopsis thaliana*. Genes Develop 12:2175–2187

Matsumoto S, Kumasaki S, Soga K, Wakabayashi K, Hashimoto T, Hoson T (2010) Gravity-induced modifications to development in hypocotyls of *Arabidopsis* tubulin mutants. Plant Physiol 152:918–926

Michniewicz M, Zago M, Abas L, Weijers D, Schweighofer A, Meskiene I, Heisler M, Ohno C, Zhang J, Huang F, Schwab R, Weigel D, Meyerowitz EM, Luschnig C, Offringa R, Friml J (2007) Antagonistic regulation of PIN phosphorylation by PP2A and PINOID directs auxin flux. Cell 130:1044–1056

Monshausen GB, Sievers A (2002) Basipetal propagation of gravity-induced surface pH changes along primary roots of *Lepidium sativum* L. Planta 215:980–988

Monshausen GB, Miller ND, Murphy AS, Gilroy S (2011) Dynamics of auxin-dependent Ca^{2+} and pH signaling in root growth revealed by integrating high-resolution imaging with automated computer vision-based analysis. Plant J 65:309–318

Morita MT (2010) Directional gravity sensing in gravitropism. Annu Rev Plant Biol 61:706–720

Morita MT, Tasaka M (2004) Gravity sensing and signaling. Curr Opin Plant Biol 7:712–718

Morita MT, Kato T, Nagafusa K, Saito C, Ueda T, Nakano A, Tasaka M (2002) Involvement of the vacuoles of the endodermis in the early process of shoot gravitropism in *Arabidopsis*. Plant Cell 14:47–56

Morita MT, Sakaguchi K, Kiyose S, Taira K, Kato T, Nakamura M, Tasaka M (2006) A C2H2-type zinc finger protein, SGR5, is involved in early events of gravitropism in *Arabidopsis* inflorescence stems. Plant J 47:619–628

Mullen JL, Wolverton C, Ishikawa H, Evans ML (2000) Kinetics of constant gravitropic stimulus responses in *Arabidopsis* roots using a feedback system. Plant Physiol 123:665–670

Müller A, Guan C, Gälweiler L, Tänzler P, Huijser P, Marchant A, Parry G, Bennett M, Wisman E, Palme K (1998) *AtPIN2* defines a locus of *Arabidopsis* for root gravitropism control. EMBO J 17:6903–6911

Nakamura M, Toyota M, Tasaka M, Morita MT (2011) An *Arabidopsis* E3 ligase SHOOT GRAVITROPISM 9 modulates the interaction between statoliths and F-Actin in gravity sensing. Plant Cell 23:1830–1848

Niihama M, Uemura T, Saito C, Nakano A, Sato MH, Tasaka M, Morita MT (2005) Conversion of functional specificity in Qb-SNARE VTI1 homologues of *Arabidopsis*. Curr Biol 15:555–560

Niihama M, Takemoto N, Hashiguchi Y, Tasaka M, Morita MT (2009) *ZIP* genes encode proteins involved in membrane trafficking of the TGN-PVC/vacuoles. Plant Cell Physiol 50:2057–2068

Perbal G, Driss-Ecole D (2003) Mechanotransduction in gravisensing cells. Trends Plant Sci 8:498–504

Petrášek J, Friml J (2009) Auxin transport routes in plant development. Development 136:2675–2688

Sack FD (1987) The structure of the stem endodermis in etiolated pea seedlings. Can J Bot 65:1514–1519

Sack FD (1991) Plant gravity sensing. Int Rev Cytol 127:193–252

Sack FD (1997) Plastids and gravitropic sensing. Planta 203:S63–S68

Sack FD, Leopold AC (1985) Cytoplasmic streaming affects gravity-induced amyloplast sedimentation in maize coleoptiles. Planta 164:56–62

Sack FD, Suyemoto MM, Leopold AC (1984) Kinetics of amyloplast sedimentation in gravistimulated maize coleoptiles. Planta 161:459–464

Sack FD, Suyemoto MM, Leopold AC (1985) Amyloplast sedimentation kinetics in gravistimulated maize roots. Planta 165:295–300

Sack FD, Suyemoto MM, Leopold AC (1986) Amyloplast sedimentation and organelle saltation in living corn columella cells. Am J Bot 73:1692–1698

Saito C, Morita MT, Kato T, Tasaka M (2005) Amyloplasts and vacuolar membrane dynamics in the living graviperceptive cell of the *Arabidopsis* inflorescence stem. Plant Cell 17:548–558

Scott AC, Allen NS (1999) Changes in cytosolic pH within *Arabidopsis* root columella cells play a key role in the early signaling pathway for root gravitropism. Plant Physiol 121:1291–1298

Sedbrook JC, Chen R, Masson PH (1999) *ARG1 (altered response to gravity)* encodes a DnaJ-like protein that potentially interacts with the cytoskeleton. Proc Natl Acad Sci USA 96:1140–1145

Silady R, Kato T, Lukowitz W, Sieber P, Tasaka M, Somerville C (2004) The *gravitropism defective 2 (grv2)* mutants of *Arabidopsis* are deficient in a protein implicated in endocytosis in *Caenorhabditis elegans*. Plant Physiol 136:3095–3103

Skagen EB, Iversen TH (1999) Simulated weightlessness and hyper-g results in opposite effects on the regeneration of the cortical microtubule array in protoplasts from *Brassica napus* hypocotyls. Physiol Plant 106:318–325

Soga K, Wakabayashi K, Kamisaka S, Hoson T (2004) Graviperception in growth inhibition of plant shoots under hypergravity conditions produced by centrifugation is independent of that in gravitropism and may involve mechanoreceptors. Planta 218:1054–1061

Soga K, Wakabayashi K, Kamisaka S, Hoson T (2005) Mechanoreceptors rather than sedimentable amyloplasts perceive the gravity signal in hypergravity-induced inhibition of root growth in azuki bean. Funct Plant Biol 32:175–179

Spitzer C, Reyes FC, Buono R, Sliwinski MK, Haas TJ, Otegui MS (2009) The ESCRT-related CHMP1A and B proteins mediate multivesicular body sorting of auxin carriers in *Arabidopsis* and are required for plant development. Plant Cell 21:749–766

Steinmann T, Geldner N, Grebe M, Mangold S, Jackson C, Paris S, Gälweiler L, Palme K, Jurgens G (1999) Coordinated polar localization of auxin efflux carrier PIN1 by GNOM ARF GEF. Science 286:316–318

Swarup R, Kramer EM, Perry P, Knox K, Leyser HM, Haseloff J, Beemster GT, Bhalerao R, Bennett MJ (2005) Root gravitropism requires lateral root cap and epidermal cells for transport and response to a mobile auxin signal. Nat Cell Biol 7:1057–1065

Takahashi H (1997) Gravimorphogenesis: gravity-induced formation of the peg in cucumber seedlings. Planta 203:S164–S169

Takahashi H, Fujii N, Kamada M, Higashitani A, Yamazaki Y, Kobayashi A, Takano M, Yamasaki S, Sakata T, Mizuno H, Kaneko Y, Murata T, Kamigaichi S, Aizawa S, Yoshizaki I, Shimazu T, Fukui K (2000) Gravimorphogenesis of *Cucurbitaceae* plants: development of peg cells and graviperception mechanism in cucumber seedlings. Biol Sci Space 14:64–74

Tasaka M, Kato T, Fukaki H (1999) The endodermis and shoot gravitropism. Trends Plant Sci 4:103–107

Tsugeki R, Fedoroff NV (1999) Genetic ablation of root cap cells in *Arabidopsis*. Proc Natl Acad Sci USA 96:12941–12946

Utsuno K, Shikanai T, Yamada Y, Hashimoto T (1998) *agr*, an agravitropic locus of *Arabidopsis thaliana*, encodes a novel membrane-protein family member. Plant Cell Physiol 39:1111–1118

Weise SE, Kiss JZ (1999) Gravitropism of inflorescence stems in starch-deficient mutants of *Arabidopsis*. Int J Plant Sci 160:521–527

Weise SE, Kuznetsov OA, Hasenstein KH, Kiss JZ (2000) Curvature in *Arabidopsis* inflorescence stems is limited to the region of amyloplast displacement. Plant Cell Physiol 41:702–709

Witztum A, Gersani M (1975) The role of polar movement of IAA in the development of the peg in *Cucumis sativus* L. Botanical Gazette 136:5–16

Wysocka-Diller JW, Helariutta Y, Fukaki H, Malamy JE, Benfey PN (2000) Molecular analysis of *SCARECROW* function reveals a radial patterning mechanism common to root and shoot. Development 127:595–603

Yamamoto K, Kiss JZ (2002) Disruption of the actin cytoskeleton results in the promotion of gravitropism in inflorescence stems and hypocotyls of *Arabidopsis*. Plant Physiol 128:669–681

Yano D, Sato M, Saito C, Sato MH, Morita MT, Tasaka M (2003) A SNARE complex containing SGR3/AtVAM3 and ZIG/VTI11 in gravity-sensing cells is important for *Arabidopsis* shoot gravitropism. Proc Natl Acad Sci USA 100:8589–8594

Zazímalová E, Murphy AS, Yang H, Hoyerová K, Hosek P (2010) Auxin transporters—why so many? Cold Spring Harbor Perspect Biol 2:a001552

Zenko C, Komatu K, Yokoyama R, Nishitani K, Kamisaka S (2003) Effect of hypergravity stimulus on *XTH* gene expression in *Arabidopsis thaliana*. Biol Sci Space 17:259–260

Zieschang HE, Sievers A (1994) Differential flank growth. Adv Space Res 14:135–144

Jasmonates in Plant Defense Responses

E. Wassim Chehab and Janet Braam

Abstract Plants constantly interact with a wide range of life-threatening organisms including herbivorous arthropods and pathogenic microbes. The plant fatty acid–derived jasmonates produced in response to biotic stresses are essential to survival. These oxylipins constitute part of the plant's sophisticated strategy to defend itself. Upon biotic attack, the increased accumulation of these metabolites diverts energy away from growth needs and channels it toward defense. The complex interplay between jasmonates and invader-specific elicitors provides the plant with gene expression regulatory potential to launch effective responses against the invaders. Such responses can be either direct, by producing molecules that are toxic to the invading organisms, or indirect, by attracting the natural enemies of such invaders. Jasmonates are also critical components in mediating the plant stress-induced systemic signal(s) to activate defense-related genes. The availability of jasmonate mutants has been crucial in identifying the roles these metabolites play in plant stress responses. In this chapter, we present an overview of jasmonate function in insect and pathogen defense, the cross talk between jasmonates and other phytohormones in fine-tuning such defenses, and the possible role these oxylipins play in mediating mechanoresponses.

1 Introduction

As sessile organisms, plants have evolved the ability to produce a chemical arsenal to effectively respond to diverse environmental challenges. Over the past three decades, oxylipin metabolites derived from membrane fatty acid metabolism have been recognized as playing central signaling roles in such responses (Conconi et al. 1996; Browse 2005; Glazebrook 2005; Howe and Jander 2008). These oxylipin

E.W. Chehab (✉) • J. Braam
Biochemistry and Cell Biology, Rice University, Houston, TX, USA
e-mail: ewchehab@rice.edu; braam@rice.edu

G. Witzany and F. Baluška (eds.), *Biocommunication of Plants*,
Signaling and Communication in Plants 14,
DOI 10.1007/978-3-642-23524-5_5,

compounds are known as jasmonates. Methyl jasmonate (MeJA) was the first oxylipin identified, originally extracted from *Jasminum grandiflorum* oil (Demole et al. 1962). Following the discovery of MeJA, various biochemical and genetic approaches on model plants identified many other jasmonates, their biosynthetic pathways and regulation, and their signaling mechanisms. In this chapter, we will discuss the powerful signaling capabilities of these molecules with focus on studies performed mainly in *Arabidopsis*.

2 Biosynthesis of Jasmonates

Vick and Zimmerman (1983) were the first to propose fatty acids as jasmonate precursors. Following their pioneering discovery, the details of the biosynthesis pathway were defined (Fig. 1) (Herms and Mattson 1992; Liechti and Farmer 2002; Wasternack 2007). In brief, external stimuli, through an undefined mechanism, activate type A phospholipases, such as DEFECTIVE IN ANTHER DEHISCENCE1 (DAD1) and DONGLE (DGL), releasing α-linolenic acid (α-LeA; 18:3) from chloroplast membrane glycerolipids (Ishiguro et al. 2001; Hyun et al. 2008). 13-Lipoxygenase (LOX) subsequently catalyzes the oxidation of the free α-LeA to 13-hydroperoxy-9,11,15-octadecatrienoic acid (13-HPOT). The latter is further metabolized by the CYP450 enzyme allene oxide synthase (AOS) to form the chemically unstable allene oxide 12,13-epoxyoctadecatrienoic acid (12,13-EOT), which is converted to a specific stereo-configured *(9S, 13S)*-12-oxo-phytodienoic acid (OPDA) by the enzyme allene oxide cyclase (AOC) (Ziegler et al. 2000). 13-HPOT is a common substrate for other CYP450 enzymes also involved in producing defense-related metabolites. Therefore, 13-HPOT catalysis by AOS is considered to be the first committed step in jasmonate formation. OPDA gets shuttled from the chloroplast to the peroxisomes through a transport mechanism partially dependent on the ATP-binding cassette (ABC) transporter COMATOSE (CTS), also known as PXA1 and PED3 (Zolman et al. 2001; Hayashi et al. 2002; Footitt et al. 2007). In the peroxisomes, OPDA is converted to 3-oxo-2-(2′[Z]-pentenyl)cyclopentane-1-octanoic acid (OPC-8:0) by OPDA reductase 3 (OPR3) (Sanders et al. 2000; Schaller et al. 2000; Stintzi and Browse 2000; Strassner et al. 2002). The carboxylic acid moiety of OPC-8:0 is then activated as CoA ester (OPC-8:CoA) by carboxyl-CoA-ligases, one of which has been identified as OPCL1 (Koo et al. 2006). OPC-8:CoA is then channeled for three rounds of β-oxidation to eventually yield jasmonic acid (JA) (Vick and Zimmerman 1983).

The β-oxidation steps are catalyzed by three core enzymes, acyl-CoA oxidase (ACX), the multifunctional protein (MFP; containing 2-*trans*-enoyl-CoA hydratase and L-2-hydroxy-acyl-CoA dehydrogenase activities), and 3-ketoacyl-CoA thiolase (KAT) (Thines et al. 2007). In *Arabidopsis*, there are five different ACXs. ACX1 is responsible for the production of 80% of wound-induced jasmonates (Cruz Castillo et al. 2004; Schilmiller et al. 2007); ACX1 and ACX5 act redundantly as evidenced by *axc1/5* double mutant, but not single mutants, showing severe JA deficiency

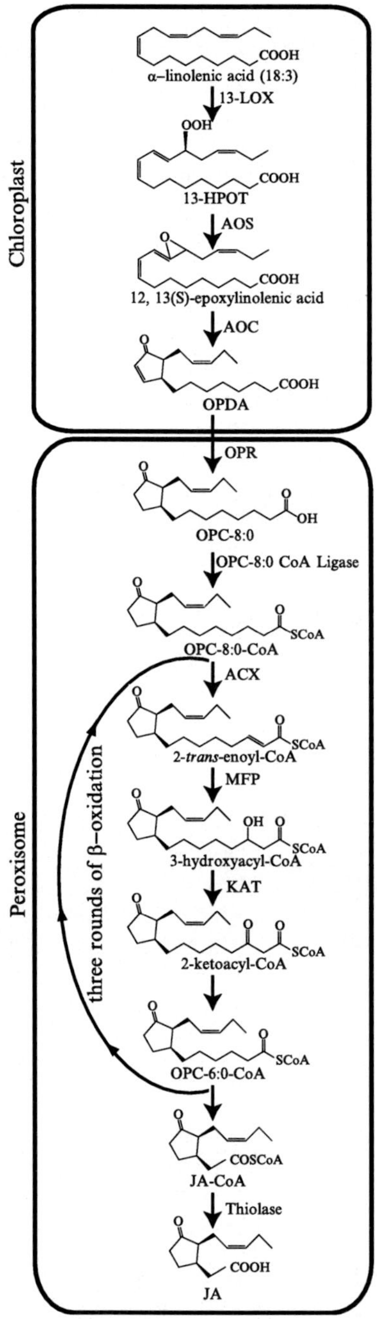

Fig. 1 *The biosynthetic pathway of jasmonates*. The first and second boxes include the reactions that take place in the chloroplast and peroxisomes, respectively

symptoms (Schilmiller et al. 2007). *Arabidopsis* also has two genes that encode MFPs. Mutants defective in one of the two *MFP* genes (*aim1*) show impairment in wound-induced JA production (Delker et al. 2007). The enzyme redundancy in the β-oxidation steps in *Arabidopsis* is also apparent by the presence of five *KAT* genes. KAT2 plays a role in wound-induced JA biosynthesis (Cruz Castillo et al. 2004). Of all the MFP and thiolase isozymes reported to date, no single MFP or thiolase gene mutation results in male sterility, a consequence of acute JA deficiency in *Arabidopsis* (Hayashi et al. 1998; Richmond and Bleecker 1999; Eastmond et al. 2000; Afitlhile et al. 2005; Schilmiller et al. 2007). Therefore, the specific contribution of all these family members to JA production remains to be determined.

The JA produced following β-oxidation can be derivatized to other metabolites with different biological activities such as tuberonic acid, cucurbic acid, *cis*-jasmone, methyl jasmonate (MeJA), and JA-Ile. In *Arabidopsis*, conjugation of the JA carboxylic acid group to isoleucine (Ile) is catalyzed by JAR1 (Staswick et al. 2002; Suza and Staswick 2008). Of all JA-derived metabolites identified to date, only JA-Ile has been shown to have direct signaling role.

3 Jasmonate Mutants

Table 1 summarizes a list of available mutants that have been essential for gaining evidence that jasmonates are important signaling molecules involved not only in plant defense (Albrecht et al. 1993; Howe et al. 1996; McConn and Browse 1996;

Table 1 *Arabidopsis* mutants essential in revealing the roles of jasmonates in plant biology

Mutants	Disrupted gene(s)	Phenotype	Affected process
fad3fad7fad8	*FAD3 FAD7* and *FAD8*	Male sterile	Desaturase activity
dad1	*Phospholipase A1*	Male sterile	Production of free α-LeA
dde1 (opr3)	*OPR3*	Male sterile	Conversion of OPDA to OPC8:0
dde2-2 (aos)	*AOS*	Male sterile	Conversion of 13HPOT to 12,13EOT
coi1	*COI1*	Male sterile	Protein degradation via SCF complex
jar1	*JAR1*	JA insensitive	JA conjugation mainly to Ile
acx1	*ACX1*	JA production deficiency	β-oxidation
acx5	*ACX5*	JA production deficiency	β-oxidation
cts (pxa) (ped3)	*COMATOSE*	JA production deficiency	OPDA transport to peroxisome
opcl1	*OPCL1*	JA production deficiency	CoA-ligase activity
cev1	*CeS3*	Constitutive JA response	Cell wall synthesis

Creelman and Mullet 1997; Staswick et al. 1998; Vijayan et al. 1998) but also in responses to abiotic stress (Parthier 1990), mechanotransduction (Falkenstein et al. 1991), and reproduction (Creelman and Mullet 1995; McConn and Browse 1996; Hause et al. 2000; Ishiguro et al. 2001). Severe *Arabidopsis* jasmonate mutants defective in synthesizing JA or incapable of perceiving this phytohormone are male sterile (Feys et al. 1994; Stintzi and Browse 2000; Park et al. 2002). Such an overt characteristic phenotype has helped researchers identify plants with defects in jasmonate biosynthesis or response.

Although the many enzymes involved in the JA biosynthetic pathway have been identified, only a few *Arabidopsis* male-sterile mutants have been isolated that are defective in JA biosynthesis. Failure to isolate additional mutants defective in JA synthesis is likely due to gene function redundancy. The *dad1* mutant has defective anther dehiscence and pollen grain maturity. *DAD1* encodes a chloroplast-specific phospholipase A1, which catalyzes the formation of free α-LeA (Ishiguro et al. 2001). However, the *DAD1* phospholipase is not essential for wound- and pathogen-induced jasmonate biosynthesis (Ellinger et al. 2010). Therefore, other lipases, such as DGL, also likely contribute to jasmonate formation (Hyun et al. 2008). *aos*, also known as *dde2-2*, has a disrupted *AOS* and therefore makes no JA (Park et al. 2002; von Malek et al. 2002). *opr3*, also known as *dde1*, has a T-DNA insertion in the second *OPR3* intron (Sanders et al. 2000; Stintzi and Browse 2000). Similar to *dad1*, both *aos* and *opr3* exhibit delayed pollen dehiscence and defects in anther filament elongation (Sanders et al. 2000; Stintzi and Browse 2000; Park et al. 2002; von Malek et al. 2002). *aos* and *opr3* also show defective resistance to herbivorous pests such as the common cabbage looper (*Trichoplusia ni*) and the fungal gnat (*Bradysia impatiens*) (Stintzi and Browse 2000; Chehab et al. 2008; Zhang and Turner 2008; Chehab et al. 2011). Surprisingly, there is a clear distinction in defense phenotypes between *aos* and *opr3*; while *aos* is highly susceptible to the necrotrophic fungi *Botrytis cinerea* and *Alternaria brassicicola*, *opr3* shows significant resistance to these pathogens (Stintzi and Browse 2000; Chehab et al. 2008; Rowe et al. 2010; Chehab et al. 2011). Although the ability of *opr3* to mount a defense response to necrotrophic fungi was originally interpreted to mean that OPDA is potentially sufficient for fungal defense, we recently showed that, upon fungal infection, *opr3* accumulates JA. Furthermore, our data are consistent with the idea that JA, and not OPDA, is the signal critical for fungal defense. Full-length, properly spliced *OPR3* transcripts accumulate in fungal-infected *opr3*, indicating that despite the large 17-kilobase intron insertion, *OPR3* transcripts can be properly spliced (Chehab et al. 2011). Thus, previous reports describing work with the *opr3* mutants and ascribing signaling function to OPDA must be reassessed because of the mutant's unexpected ability to accumulate JA. Furthermore, this finding offers a cautionary note to researchers working with intron-insertion mutations; splicing may be robust enough to generate low levels of intact mRNAs resulting in leaky mutations. Unraveling possible role(s) for OPDA in plant defense will require a true *opr3* null allele.

The ability of jasmonates or analogues, such as coronatine, to inhibit root growth has also been used to screen for jasmonate-insensitive mutants. Roots of *coronatine insensitive 1* (*coi1*) mutants are insensitive to coronatine and JA, and *coi1* flowers are male sterile due to an inability to produce viable pollen (Feys et al. 1994). Although *coi1* plants have their JA biosynthetic machinery intact, they are unable to perceive JA. Interestingly, *coi1* plants are highly susceptible to insect infestation and necrotrophic pathogen infection (Stintzi et al. 2001; Li et al. 2004; Reymond et al. 2004; Chen et al. 2005; Mewis et al. 2005; Paschold et al. 2007; Zarate et al. 2007). These findings led to the proposal, later confirmed by Sheard et al. (2010), that COI1 may be a JA receptor. *COI1* encodes an F-box protein that plays a critical role in JA perception through triggering protein degradation required for initiating JA responses (Devoto et al. 2005).

Similar to *coi1*, *jasmonate resistant 1* (*jar1*) was also isolated in a screen for JA-insensitive mutants. *jar1* was identified by failure of MeJA to inhibit root growth (Staswick et al. 1992). Mutants of *JAR1* are defective in conjugation of JA to Ile to generate JA-Ile (Staswick and Tiryaki 2004). Unlike auxin conjugation, whereby indole acetic acid conjugation to amino acids generates inactive hormone storage forms or intermediates to degradation (Bartel 1997), conjugation of JA is necessary to generate the active hormone form, JA-Ile. Although *jar1* has no detectable defects in reproduction, it exhibits reduced expression of JA-regulated genes and is susceptible to pathogens (Staswick et al. 1998).

4 Mechanism of JA Action

Characterization of the mutants described above provided profound insight into the physiological functions of jasmonate signaling. Biochemical approaches demonstrated that JA-Ile binds to COI1 (Sheard et al. 2010). COI1 functions similarly to the auxin receptor (Dharmasiri et al. 2005). COI1 associates with Cullin1, ASK1/ASK2, and Rbx1 to form an SCF complex (Xie et al. 1998; Xu et al. 2002; Liu et al. 2004; Ren et al. 2005). JA-Ile binding to SCF^{COI1} targets JA-ZIM (JAZ) domain proteins for ubiquitination and degradation by the 26S proteasome. The 12 *Arabidopsis* JAZ proteins are thought to be repressors that bind to and inhibit transcription factors, such as the bHLH transcription factor MYC2 (Chini et al. 2007; Thines et al. 2007; Yan et al. 2007; Fonseca et al. 2009; Seo et al. 2011; Song et al. 2011). Removal of the JAZ proteins enables JA-induced gene expression, generating the diverse plant responses (Fig. 2) (Chini et al. 2007; Thines et al. 2007; Yan et al. 2007; Fonseca et al. 2009; Sheard et al. 2010).

5 Jasmonates Orchestrate Plant Immunity to Biotic Stresses

Plant wounding and tissue consumption due to invasions from herbivore arthropods and pathogenic microbes constitute serious environmental challenges that threaten a plant's survival. The initial discovery of wound-induced genes (Green and Ryan 1972; Sanchez-Serrano et al. 1986) and the subsequent findings that the expression

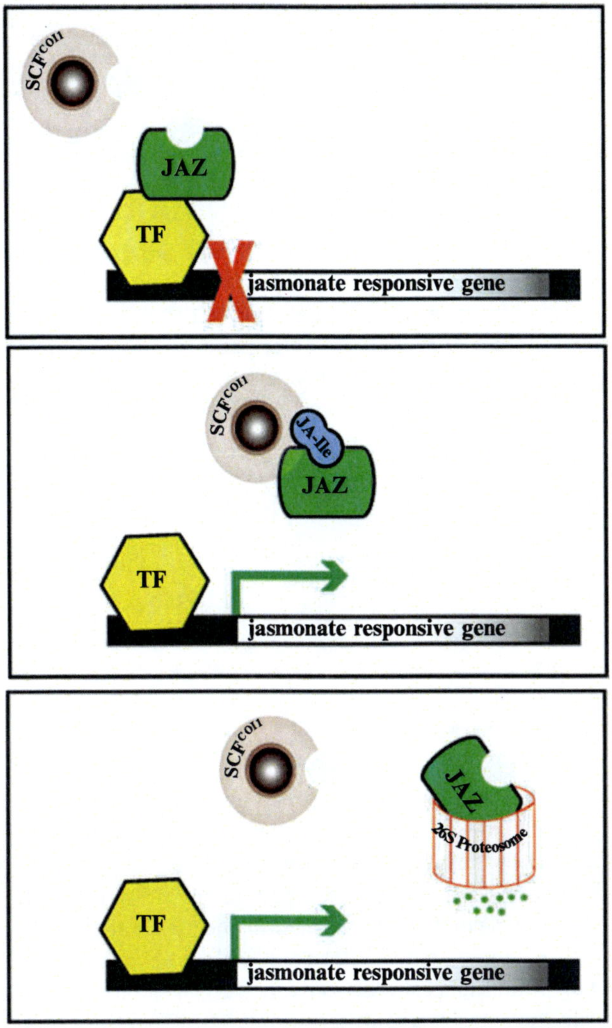

Fig. 2 *A model of SCF^{COI1} action in jasmonate signaling*. In the absence of jasmonates, JAZ proteins repressively bind transcription factors (TFs), thus inhibiting jasmonate-responsive gene expression. In the presence of JA-Ile, SCF^{COI1} binds to JAZ repressors and targets them for degradation via the 26 S proteasome, thus derepressing the TF and activating jasmonate-responsive gene expression

of many of these genes is induced by JA application were the initial studies implicating the jasmonate phytohormones in plant defense (Farmer et al. 1992; Farmer and Ryan 1992). Defenses regulated by jasmonates include the production of toxic molecules and/or the release of volatile compounds as airborne signals to attract predators that feed on the herbivores (Keinanen et al. 2001; Kessler and Baldwin 2002; Grubb and Abel 2006; Chehab et al. 2008; Rowe et al. 2010). Here, we discuss these defenses against herbivorous insects and pathogenic microbes.

5.1 *Immunity Against Insect Herbivores*

Evidence for the critical role of jasmonates in insect herbivore defense was obtained through bioassays on mutants compromised in the synthesis or perception of jasmonates. For example, the *Arabidopsis* triple mutant *fad3-2 fad7-2 fad8* with compromised desaturase activity exhibits a deficiency in α-LeA and thus is severely deficient in JA production (McConn and Browse 1996). Similar to *coi1*, the triple mutant has high susceptibility to soil gnats (*B. impatiens*) as compared to wild type (Stintzi et al. 2001). *Arabidopsis aos* mutants are susceptible to cabbage looper (*T. ni*) and aphid (*Myzus persicae*) infestation (Chehab et al. 2008; Chehab et al. 2011). Through the use of mutants in other plant model systems, such as tobacco (*Nicotiana attenuata*), it has been shown that jasmonates are also critical in plant defense against an array of other insect herbivores including thrips (*Frankliniella occidentalis*), beetles (*Diabrotica undecimpunctata*), and leafhoppers (*Empoasca sp.*) (Kessler et al. 2004).

5.1.1 Defensive Secondary Metabolites

One way jasmonates promote plant defense against insects is by activating secondary metabolic pathways. Some JA-induced secondary metabolites act by being toxic upon consumption. For example, glucosinolates produced in *Arabidopsis* act directly to deter herbivorous insect feeding (Grubb and Abel 2006; Rowe et al. 2010). Anthocyanins (Fang et al. 1998), oligolignol (Pauwels et al. 2008), and reactive oxygen species (Zhang and Xing 2008) can have similar effects. Alternatively, JA can trigger production of secondary metabolites that act indirectly in that they attract to the plant other insects that can help fend off the attacking herbivore. This tritrophic defense response involves JA-induced production of airborne distress signals that attract the natural enemies of the invading herbivorous arthropods. For example, aphid-infected *Arabidopsis* plants produce increased jasmonate levels which subsequently activate the production and release of volatile semiochemicals that attract aphid parasitoids (Birkett et al. 2000; Bruce et al. 2008). Thaler et al. (2002) also showed using tomato mutants the involvement of jasmonates in the production and release of the sesquiterpene β-caryophyllene and the monoterpenes α-pinene, β-pinene, 2-carene, and β-phellandrene, thought to attract caterpillar-predacious mites.

5.1.2 Defense Proteins

Jasmonates also induce the production of proteins that exert direct toxicity on herbivorous invaders. One such class of proteins is proteinase inhibitors, which disrupt the digestive process in the insect gut, thus thwarting the attack (Ryan 1990; Halitschke et al. 2003; Chen et al. 2005). These protease inhibitors inhibit gut proteases resulting in amino acid deficiencies that negatively affect the growth of the herbivore (Zavala et al. 2004; Lison et al. 2006). Another class of plant defense proteins includes enzymes that deplete consumed nutrients in the herbivore's midgut. For example, *ARGINASE* is highly upregulated in expression by jasmonates following herbivorous attack. As the insect ingests the leaf tissue, it also consumes the enzyme arginase which in turn digests arginine in the insect midgut and deprives the insect from this essential amino acid (Chen et al. 2004). Over the past few years, other jasmonate-induced defense proteins have been identified, such as polyphenol oxidase (PPO), leucine amino peptidase, and the acid phosphatase vacuolar storage protein 2 (VSP2). Interestingly, these defense proteins are relatively resistant to the insect proteases and other harsh conditions of the insect midgut (Felton et al. 1994; Constabel et al. 1995; Chen et al. 2005; Lison et al. 2006; Chen et al. 2007).

5.1.3 Transcriptional Regulation

Transcriptional profiling experiments show that the jasmonate-mediated production of secondary metabolites as well as the synthesis of the defense proteins are primarily mediated through transcriptional regulation (Reymond et al. 2000; Halitschke et al. 2001; Reymond et al. 2004; De Vos et al. 2005; Devoto et al. 2005; Major and Constabel 2006; Ralph et al. 2006). Depending on the type of the invading insect, distinct, yet overlapping, gene expression patterns can be observed (De Moraes et al. 2001; Heidel and Baldwin 2004). Therefore, it is likely that the combination of JA signaling coupled with attacker-derived signals tailor highly effective defense responses. Recent findings suggest that these signals depend on the invaders' feeding behaviors, known as guilds (Heidel and Baldwin 2004). For example, microarray analysis of *Arabidopsis* plants subjected to insects with different feeding guilds, the chewing cabbage worm (*Pieris rapae*) and the piercing thrips (*F. occidentalis*), shows that the majority of genes expressed are jasmonate-regulated; however, 61% of these genes had an expression pattern specific to one of the two attackers (De Vos et al. 2005). On the other hand, insects from the same feeding guild tend to evoke similar responses. For example, transcriptional profiling experiments of *Arabidopsis* plants infested with the chewing insects *P. rapae* and the Egyptian common leaf worm (*Spodoptera littoralis*) tend to induce nearly identical gene expression patterns (Reymond et al. 2004).

How plants sense the presence of feeding insects and initiate increased jasmonate production is still not fully known. Recent studies suggest that insect oral secretions, such as fatty acid-amino acid conjugates and peptides, might be perceived (Kessler and Baldwin 2002; Schmelz et al. 2006). Invader-induced host-derived elicitors, such as the cell wall–derived oligogalacturonic acid, are also known to increase JA production (Hu et al. 2003).

Jasmonate-mediated transcriptional regulation may also help defense-related plant energy allocation, especially since the energy cost associated with protecting the plant against invading pests is fairly large. Therefore, when a plant is subjected to herbivore or pathogen damage, it is in a survival dilemma: defend or grow. In such circumstances, jasmonates not only activate defense-related genes but also downregulate those genes involved in cell division, thus diverting more resources toward defense (Swiatek et al. 2002; Yan et al. 2007; Balbi and Devoto 2008; Pauwels et al. 2008; Zhang and Turner 2008).

All these findings emphasize that the jasmonate signaling pathway is not simply an on/off pathway, but instead an integrated and complex signaling network in which the defense response may be customized to the specific pest and modulated with growth regulation to attain optimal balance for plant survival.

5.2 *Pathogenic Microbial Defense*

5.2.1 Necrotrophs vs. Biotrophs

Pathogens can be generally divided into those that infect and feed off living tissue (biotrophs) and those that kill cells prior to feeding on them (necrotrophs) (Parbery 1996). In 1998, Vijayan and coworkers were the first to provide compelling evidence of the essential role jasmonates play in mediating plant defenses against pathogens. They showed that the necrotrophic fungus *Pythium mastophorum* infected and killed the *Arabidopsis* triple mutant *fad3-2 fad7-2 fad8* and *coi1* but not wild-type plants. Exogenous application of JA rescued the fungal resistance of *fad3-2 fad7-2 fad8* mutant but not that of *coi1*. This confirmed that the JA-mediated protection of the exogenously applied JA against the fungus was due to jasmonate-mediated defense signaling and not due to toxicity of JA on the fungus. Similar observations were also reported for the necrotrophic pathogens *Botrytis cinerea*, *Alternaria brassicicola*, and *Fusarium oxysporum* (Thomma et al. 1998; Berrocal-Lobo and Molina 2004; Chehab et al. 2008; Chehab et al. 2011). However, jasmonate does not mediate defense against biotrophic *Pseudomonas syringae*. Instead, salicylic acid (SA) is the phytohormone required for defense at least during the early stages of pathogenesis (Feys et al. 1994; Petersen et al. 2000; Kloek et al. 2001). Interestingly, JA and SA work antagonistically and reduce each other's responses (Niki et al. 1998; Kunkel and Brooks 2002; Traw et al. 2003; Cipollini et al. 2004; Bostock 2005; Koornneef and Pieterse 2008). Indeed, *P. syringae* produces coronatine, a JA-Ile analogue, thereby augmenting the JA signaling pathway and suppressing SA defense against parasitic growth.

Although plants respond to necrotrophic and biotrophic pathogens by activating different defense signaling mechanisms, JA and SA signaling share some common downstream responses; for example, production of camalexin, a primary *Arabidopsis* phytoalexin important for pathogen growth inhibition (Tsuji et al. 1992; Glazebrook and Ausubel 1994; Glazebrook et al. 1997), accumulates both in response to JA and

SA. Thus, common responses to different pathogens may be controlled by distinct regulatory networks; the mechanisms of this regulation remain to be elucidated.

5.2.2 Cross Talk of JA/SA

The molecular mechanisms responsible for the negative cross talk between SA and JA are not well understood. Repression of JA-induced gene expression by SA requires the function of (nonexpressor of PR genes1) NPR1 (Dong 2001; Pieterse and Van Loon 2004). Oxidized NPR1 forms oligomers and is localized in the cytosol (Mou et al. 2003). However, the redox state changes associated with SA production reduce NPR1. The resultant monomeric form is subsequently nuclear localized where it interacts with a class of basic domain/leucine zipper transcription factors to mediate the induction of SA-dependent genes (Despres et al. 2003; Mou et al. 2003; Spoel et al. 2003; Dong 2004). The transcriptional regulatory region of *NPR1* contains W-box binding sites for WRKY transcription factors. Interestingly, several WRKY transcription factors are also implicated in regulating SA-dependent defense responses as well as the SA/JA cross talk (Eulgem et al. 2000). WRKY70 is one of the few WRKYs demonstrated to play a role in the cross talk by positively regulating SA-mediated defenses and repressing JA responses (Journot-Catalino et al. 2006; Mao et al. 2007). Antisense suppression of *WRKY70* results in the activation of COI1-dependent genes, whereas overexpression of *WRKY70* results in the constitutive SA signaling and the suppression of jasmonate-response genes (Li et al. 2002; Li et al. 2004).

Mitogen-activated protein kinases (MAPKs) are also key players in JA/SA cross talk. MAPKs regulate plant responses to biotic challenges (Jonak et al. 2002). *Arabidopsis mpk4* mutants are JA insensitive, produce high levels of SA, and are resistant to *P. syringae* (Petersen et al. 2000).

5.2.3 Cross Talk of JA/ET

JA has been found to be conjugated to 1-aminocyclopropane-1-carboxylate (ACC), the precursor of ethylene. Although the function of this conjugated product is yet to be identified, its accumulation in plants may be relevant to the reported cross talk between JA and ethylene (ET) in regulating the expression of defense-related genes (Xu et al. 1994; O'Donnell et al. 1998; Penninckx et al. 1998; Rojo and Solano 2003). Ethylene and jasmonates can act in a synergistic or antagonistic manner depending on the stress encountered by the plant.

Pharmacological and mutant studies show that JA and ET act in synergy in plant defense against fungal pathogens (Pieterse et al. 1998; van Wees et al. 1999; Ellis and Turner 2001; Thomma et al. 2001; Berrocal-Lobo and Molina 2004). *PLANT DEFENSIN 1.2* (*PDF1.2*) and *ETHYLENE RESPONSE FACTOR 1 (ERF1)*, which encode an antimicrobial protein and a transcription factor, respectively, are highly induced upon infection with fungi, such as *A. brassicicola* (Penninckx et al. 1996; Penninckx et al. 1998). To achieve full expression of the two genes, the activation of both JA and ET signaling pathways is required (Penninckx et al. 1998). Lorenzo et al. (2003) demonstrated that ERF1 regulates the expression of *PDF1.2*. Therefore, upon pathogen infection, the JA and ET signaling pathways may converge to activate the expression of *ERF1*, which in turn regulates *PDF1.2* expression. Consistent with this notion, overexpressing *ERF1* results in the expression of defense-related genes that are responsive to both JA and ET (Lorenzo et al. 2003). Furthermore, overexpressing *ERF1* in *coi1* rescues expression of genes involved in fungal defense responses (Lorenzo et al. 2003). Therefore, the concerted action of JA and ET acts concomitantly to activate the defense responses against fungal pathogens.

On the other hand, antagonism between jasmonates and ET is also evident in wounding and insect herbivory responses (Rojo et al. 1999; Shoji et al. 2000; Lorenzo et al. 2004). As previously discussed, MYC2 is required for induction of expression of many JA-regulated genes. Expression of these genes responds to wounding and arthropod herbivory (Boter et al. 2004; Lorenzo et al. 2004; Dombrecht et al. 2007). Interestingly, wound-induced genes through the action of MYC2 are repressed by ERF1 (Lorenzo et al. 2004). Therefore, it appears that genes activated by JA but repressed by ET are part of the transcriptional response to insect herbivory attacks, whereas genes that require both phytohormones for full expression are more likely involved in protecting the plant against microbial pathogens.

6 Systemic Resistance

Biotic stress may not only launch defense responses at the wounding site but also systemic expression of defense-related genes and protection of healthy tissue from future attacks (Conrath et al. 2006; Frost et al. 2007; Ton et al. 2007; Chassot et al. 2008; Erb et al. 2008; Heil and Ton 2008; Vlot et al. 2008). Through the use of plant mutants defective in jasmonate synthesis or perception, it has been shown that these oxylipins regulate systemic resistance (Zhang and Baldwin 1997; Li et al. 2002; Thorpe et al. 2007). For example, tomato grafting experiments between wild-type and COI1-deficient plants show that response to jasmonates is necessary for recognizing the systemic wound signal in distal undamaged leaves but not required for production of the signal in damaged leaves (Li et al. 2002). Intact JA biosynthetic machinery only in the rootstock is required for the wound-induced systemic expression of JA-dependent genes in the unwounded distal leaves of the scion (Li et al. 2002; Lee and Howe 2003; Li et al. 2005). These findings as well as the ability of jasmonates to translocate through the vascular system indicate that JA and/or its related metabolites that are recognized by COI1 constitute part if not all of the systemic transmitted wound signal.

7 Thigmomorphogenesis and Jasmonates

Plants respond to repetitive touch or mechanostimulation by undergoing changes in growth that generally include a decrease in elongation growth and an increase in radial expansion (Braam 2005; Chehab et al. 2009). Although molecular responses to touch have been identified (e.g., Braam and Davis 1990; Braam 1992; Xu et al. 1995; Purugganan et al. 1997; Lee et al. 2005) and implications for touch-induced genes in mechanoresponses are reported (Sistrunk et al. 1994; McCormack and Braam 2003; Delk et al. 2005; Wang et al. 2011), there have been few insights into how thigmomorphogenesis is regulated. Over the past decade or so, some studies have implicated jasmonates in plant mechanoresponses. For example, Stelmach et al. (1998) showed that the application of coronatine on the common bean (*Phaseolus vulgaris*) causes physiological responses reminiscent of thigmomorphogenesis. Mechanically impeding root growth causes an increase in JA production and a temporary inhibition of root elongation. *Arabidopsis cev1* mutants with constitutively high levels of JA show thigmomorphogenetic-like phenotypes (Ellis et al. 2002). The physical impedance of *Bryonia dioica* causes elevation of intracellular MeJA levels in the tendrils. The application of MeJA, or its precursor 12-OPDA, on *B. dioica* elicits a coiling tendril response (Weiler et al. 1993). All these findings suggest jasmonates might be playing a role in linking the touch stimulus with the transduction pathway leading to the observed thigmomorphogenetic responses. However, further investigations are necessary to determine whether jasmonates are required for the mechanoresponsive pathway. Such a task can be achieved by utilizing jasmonate mutants defective either in their ability to synthesize jasmonates, such as *aos* (Park et al. 2002; Chehab et al. 2008), or in their ability to perceive jasmonates, such as *coi1* (Feys et al. 1994).

8 Concluding Remarks

Jasmonate-mediated defense responses to biotic attacks are crucial to the survival of plants. Since constitutive defense activation is energetically costly and in conflict with biotrophic pathogen defense, perhaps plants have evolved the JA regulatory pathway for switching on these responses only under appropriate conditions for optimal survival and growth. JA responses can be either direct or indirect but can be specific depending on the invading pest. Although major accomplishments have been achieved in understanding the mechanisms and regulation of jasmonate signaling, many unanswered questions remain to be resolved. The use of existing jasmonate mutants as well as the identification of new ones is crucial for further unraveling of the remaining mysteries of these powerful signaling molecules.

Acknowledgment Research in our lab related to this topic is based upon work supported by the National Science Foundation under Grant No. MCB 0817976 to JB

References

Afitlhile, M.M., H. Fukushige, M. Nishimura, and D.F. Hildebrand. 2005. A defect in glyoxysomal fatty acid beta-oxidation reduces jasmonic acid accumulation in Arabidopsis. *Plant Physiol Biochem* 43: 603–609.

Albrecht, T., A. Kehlen, K. Stahl, H.D. Knofel, G. Sembdner, and E.W. Weiler. 1993. Quantification of rapid, transient increases in jasmonic acid in wounded plants using a monoclonal antibody. *Planta* 191: 86–94.

Balbi, V., and A. Devoto. 2008. Jasmonate signalling network in *Arabidopsis thaliana*: crucial regulatory nodes and new physiological scenarios. *New Phytol* 177: 301–318.

Bartel, B. 1997. Auxin biosynthesis. *Annu Rev Plant Physiol Plant Mol Biol* 48: 51–66.

Berrocal-Lobo, M., and A. Molina. 2004. Ethylene response factor 1 mediates Arabidopsis resistance to the soilborne fungus *fusarium oxysporum*. *Mol Plant Microbe Interact* 17: 763–770.

Birkett, M.A., C.A. Campbell, K. Chamberlain, E. Guerrieri, A.J. Hick, J.L. Martin, M. Matthes, J.A. Napier, J. Pettersson, J.A. Pickett, G.M. Poppy, E.M. Pow, B.J. Pye, L.E. Smart, G.H. Wadhams, L.J. Wadhams, and C.M. Woodcock. 2000. New roles for cis-jasmone as an insect semiochemical and in plant defense. *Proc Natl Acad Sci USA* 97: 9329–9334.

Bostock, R.M. 2005. Signal crosstalk and induced resistance: straddling the line between cost and benefit. *Annu Rev Phytopathol* 43: 545–580.

Boter, M., O. Ruiz-Rivero, A. Abdeen, and S. Prat. 2004. Conserved MYC transcription factors play a key role in jasmonate signaling both in tomato and Arabidopsis. *Genes Dev* 18: 1577–1591.

Braam, J. 1992. Regulation of expression of calmodulin and calmodulin-related genes by environmental stimuli in plants. *Cell Calcium* 13: 457–463.

Braam, J. 2005. In touch: plant responses to mechanical stimuli. *New Phytol* 165: 373–389.

Braam, J., and R.W. Davis. 1990. Rain-, wind-, and touch-induced expression of calmodulin and calmodulin-related genes in Arabidopsis. *Cell* 60: 357–364.

Browse, J. 2005. Jasmonate: an oxylipin signal with many roles in plants. *Vitam Horm* 72: 431–456.

Bruce, T.J., M.C. Matthes, K. Chamberlain, C.M. Woodcock, A. Mohib, B. Webster, L.E. Smart, M.A. Birkett, J.A. Pickett, and J.A. Napier. 2008. cis-jasmone induces Arabidopsis genes that affect the chemical ecology of multitrophic interactions with aphids and their parasitoids. *Proc Natl Acad Sci USA* 105: 4553–4558.

Chassot, C., A. Buchala, H.J. Schoonbeek, J.P. Metraux, and O. Lamotte. 2008. Wounding of Arabidopsis leaves causes a powerful but transient protection against botrytis infection. *Plant J* 55: 555–567.

Chehab, E.W., R. Kaspi, T. Savchenko, H. Rowe, F. Negre-Zakharov, D. Kliebenstein, and K. Dehesh. 2008. Distinct roles of jasmonates and aldehydes in plant-defense responses. *PLoS One* 3: e1904.

Chehab, E.W., E. Eich, and J. Braam. 2009. Thigmomorphogenesis: a complex plant response to mechano-stimulation. *J Exp Bot* 60: 43–56.

Chehab, E.W., S. Kim, T. Savchenko, D. Kliebenstein, K. Dehesh, and J. Braam. 2011. Intronic T-DNA insertion renders Arabidopsis *opr3* a conditional JA producing mutant. *Plant Physiol* 156: 770–778.

Chen, H., B.C. McCaig, M. Melotto, S.Y. He, and G.A. Howe. 2004. Regulation of plant arginase by wounding, jasmonate, and the phytotoxin coronatine. *J Biol Chem* 279: 45998–46007.

Chen, H., C.G. Wilkerson, J.A. Kuchar, B.S. Phinney, and G.A. Howe. 2005. Jasmonate-inducible plant enzymes degrade essential amino acids in the herbivore midgut. *Proc Natl Acad Sci USA* 102: 19237–19242.

Chen, H., E. Gonzales-Vigil, C.G. Wilkerson, and G.A. Howe. 2007. Stability of plant defense proteins in the gut of insect herbivores. *Plant Physiol* 143: 1954–1967.

Chini, A., S. Fonseca, G. Fernandez, B. Adie, J.M. Chico, O. Lorenzo, G. Garcia-Casado, I. Lopez-Vidriero, F.M. Lozano, M.R. Ponce, J.L. Micol, and R. Solano. 2007. The JAZ family of repressors is the missing link in jasmonate signalling. *Nature* 448: 666–671.

Cipollini, D., S. Enright, M.B. Traw, and J. Bergelson. 2004. Salicylic acid inhibits jasmonic acid-induced resistance of *Arabidopsis thaliana* to spodoptera exigua. *Mol Ecol* 13: 1643–1653.

Conconi, A., M.J. Smerdon, G.A. Howe, and C.A. Ryan. 1996. The octadecanoid signalling pathway in plants mediates a response to ultraviolet radiation. *Nature* 383: 826–829.

Conrath, U., G.J. Beckers, V. Flors, P. Garcia-Agustin, G. Jakab, F. Mauch, M.A. Newman, C.M. Pieterse, B. Poinssot, M.J. Pozo, A. Pugin, U. Schaffrath, J. Ton, D. Wendehenne, L. Zimmerli, and B. Mauch-Mani. 2006. Priming: getting ready for battle. *Mol Plant Microbe Interact* 19: 1062–1071.

Constabel, C.P., D.R. Bergey, and C.A. Ryan. 1995. Systemin activates synthesis of wound-inducible tomato leaf polyphenol oxidase via the octadecanoid defense signaling pathway. *Proc Natl Acad Sci USA* 92: 407–411.

Creelman, R.A., and J.E. Mullet. 1995. Jasmonic acid distribution and action in plants: regulation during development and response to biotic and abiotic stress. *Proc Natl Acad Sci USA* 92: 4114–4119.

Creelman, R.A., and J.E. Mullet. 1997. Oligosaccharins, brassinolides, and jasmonates: nontraditional regulators of plant growth, development, and gene expression. *Plant Cell* 9: 1211–1223.

Cruz Castillo, M., C. Martinez, A. Buchala, J.P. Metraux, and J. Leon. 2004. Gene-specific involvement of beta-oxidation in wound-activated responses in Arabidopsis. *Plant Physiol* 135: 85–94.

De Moraes, C.M., M.C. Mescher, and J.H. Tumlinson. 2001. Caterpillar-induced nocturnal plant volatiles repel conspecific females. *Nature* 410: 577–580.

De Vos, M., V.R. Van Oosten, R.M. Van Poecke, J.A. Van Pelt, M.J. Pozo, M.J. Mueller, A.J. Buchala, J.P. Metraux, L.C. Van Loon, M. Dicke, and C.M. Pieterse. 2005. Signal signature and transcriptome changes of Arabidopsis during pathogen and insect attack. *Mol Plant Microbe Interact* 18: 923–937.

Delk, N.A., K.A. Johnson, N.I. Chowdhury, and J. Braam. 2005. CML24, regulated in expression by diverse stimuli, encodes a potential Ca^{2+} sensor that functions in responses to abscisic acid, daylength, and ion stress. *Plant Physiol* 139: 240–253.

Delker, C., B.K. Zolman, O. Miersch, and C. Wasternack. 2007. Jasmonate biosynthesis in *Arabidopsis thaliana* requires peroxisomal beta-oxidation enzymes—additional proof by properties of pex6 and aim1. *Phytochemistry* 68: 1642–1650.

Demole, E., E. Lederer, and D. Mercier. 1962. Isolement et détermination de la structure du jasmonate de méthyle, constituant odorant charactéristique de l'essence de jasmin. *Helv Chim Acta* 45: 645–685.

Despres, C., C. Chubak, A. Rochon, R. Clark, T. Bethune, D. Desveaux, and P.R. Fobert. 2003. The Arabidopsis NPR1 disease resistance protein is a novel cofactor that confers redox regulation of DNA binding activity to the basic domain/leucine zipper transcription factor TGA1. *Plant Cell* 15: 2181–2191.

Devoto, A., C. Ellis, A. Magusin, H.S. Chang, C. Chilcott, T. Zhu, and J.G. Turner. 2005. Expression profiling reveals COI1 to be a key regulator of genes involved in wound- and methyl jasmonate-induced secondary metabolism, defence, and hormone interactions. *Plant Mol Biol* 58: 497–513.

Dharmasiri, N., S. Dharmasiri, and M. Estelle. 2005. The F-box protein TIR1 is an auxin receptor. *Nature* 435: 441–445.

Dombrecht, B., G.P. Xue, S.J. Sprague, J.A. Kirkegaard, J.J. Ross, J.B. Reid, G.P. Fitt, N. Sewelam, P.M. Schenk, J.M. Manners, and K. Kazan. 2007. MYC2 differentially modulates diverse jasmonate-dependent functions in Arabidopsis. *Plant Cell* 19: 2225–2245.

Dong, X. 2001. Genetic dissection of systemic acquired resistance. *Curr Opin Plant Biol* 4: 309–314.

Dong, X. 2004. NPR1, all things considered. *Curr Opin Plant Biol* 7: 547–552.

Eastmond, P.J., M.A. Hooks, D. Williams, P. Lange, N. Bechtold, C. Sarrobert, L. Nussaume, and I.A. Graham. 2000. Promoter trapping of a novel medium-chain acyl-CoA oxidase, which is induced transcriptionally during Arabidopsis seed germination. *J Biol Chem* 275: 34375–34381.

Ellinger, D., N. Stingl, I.I. Kubigsteltig, T. Bals, M. Juenger, S. Pollmann, S. Berger, D. Schuenemann, and M.J. Mueller. 2010. DONGLE and DEFECTIVE IN ANTHER DEHISCENCE1 lipases are not essential for wound- and pathogen-induced jasmonate biosynthesis: redundant lipases contribute to jasmonate formation. *Plant Physiol* 153: 114–127.

Ellis, C., and J.G. Turner. 2001. The Arabidopsis mutant *cev1* has constitutively active jasmonate and ethylene signal pathways and enhanced resistance to pathogens. *Plant Cell* 13: 1025–1033.

Ellis, C., I. Karafyllidis, C. Wasternack, and J.G. Turner. 2002. The Arabidopsis mutant cev1 links cell wall signaling to jasmonate and ethylene responses. *Plant Cell* 14: 1557–1566.

Erb, M., J. Ton, J. Degenhardt, and T.C. Turlings. 2008. Interactions between arthropod-induced aboveground and belowground defenses in plants. *Plant Physiol* 146: 867–874.

Eulgem, T., P.J. Rushton, S. Robatzek, and I.E. Somssich. 2000. The WRKY superfamily of plant transcription factors. *Trends Plant Sci* 5: 199–206.

Falkenstein, E., B. Groth, A. Mithofer, and E.W. Weiler. 1991. Methyl jasmonate and linolenic acid are potent inducers of tendril coiling. *Planta* 185: 316–322.

Fang, Y., M.A.L. Smith, and M.F. Pdpin. 1998. Benzyladenine restores anthocyanin pigmentation in suspension cultures of wild *vaccinium pahalae*. *Plant Cell Tissue Org Cult* 54: 113–122.

Farmer, E.E., and C.A. Ryan. 1992. Octadecanoid-derived signals in plants. *Trends Cell Biol* 2: 236–241.

Farmer, E.E., R.R. Johnson, and C.A. Ryan. 1992. Regulation of expression of proteinase inhibitor genes by methyl jasmonate and jasmonic acid. *Plant Physiol* 98: 995–1002.

Felton, G.W., J.L. Bi, C.B. Summers, A.J. Mueller, and S.S. Duffey. 1994. Potential role of lipoxygenases in defense against insect herbivory. *J Chem Ecol* 20: 651–666.

Feys, B., C.E. Benedetti, C.N. Penfold, and J.G. Turner. 1994. Arabidopsis mutants selected for resistance to the phytotoxin coronatine are male sterile, insensitive to methyl jasmonate, and resistant to a bacterial pathogen. *Plant Cell* 6: 751–759.

Fonseca, S., J.M. Chico, and R. Solano. 2009. The jasmonate pathway: the ligand, the receptor and the core signalling module. *Curr Opin Plant Biol* 12: 539–547.

Footitt, S., D. Dietrich, A. Fait, A.R. Fernie, M.J. Holdsworth, A. Baker, and F.L. Theodoulou. 2007. The COMATOSE ATP-binding cassette transporter is required for full fertility in Arabidopsis. *Plant Physiol* 144: 1467–1480.

Frost, C.J., H.M. Appel, J.E. Carlson, C.M. De Moraes, M.C. Mescher, and J.C. Schultz. 2007. Within-plant signalling via volatiles overcomes vascular constraints on systemic signalling and primes responses against herbivores. *Ecol Lett* 10: 490–498.

Glazebrook, J. 2005. Contrasting mechanisms of defense against biotrophic and necrotrophic pathogens. *Annu Rev Phytopathol* 43: 205–227.

Glazebrook, J., and F.M. Ausubel. 1994. Isolation of phytoalexin-deficient mutants of Arabidopsis thaliana and characterization of their interactions with bacterial pathogens. *Proc Natl Acad Sci USA* 91: 8955–8959.

Glazebrook, J., M. Zook, F. Mert, I. Kagan, E.E. Rogers, I.R. Crute, E.B. Holub, R. Hammerschmidt, and F.M. Ausubel. 1997. Phytoalexin-deficient mutants of Arabidopsis reveal that PAD4 encodes a regulatory factor and that four PAD genes contribute to downy mildew resistance. *Genetics* 146: 381–392.

Green, T.R., and C.A. Ryan. 1972. Wound-induced proteinase inhibitor in plant leaves: a possible defense mechanism against insects. *Science* 175: 776–777.

Grubb, C.D., and S. Abel. 2006. Glucosinolate metabolism and its control. *Trends Plant Sci* 11: 89–100.

Halitschke, R., U. Schittko, G. Pohnert, W. Boland, and I.T. Baldwin. 2001. Molecular interactions between the specialist herbivore *manduca sexta* (Lepidoptera, sphingidae) and its natural host *nicotiana attenuata*. III. Fatty acid-amino acid conjugates in herbivore oral secretions are necessary and sufficient for herbivore-specific plant responses. *Plant Physiol* 125: 711–717.

Halitschke, R., K. Gase, D. Hui, D.D. Schmidt, and I.T. Baldwin. 2003. Molecular interactions between the specialist herbivore *manduca sexta* (Lepidoptera, sphingidae) and its natural host *nicotiana attenuata*. VI. Microarray analysis reveals that most herbivore-specific transcriptional changes are mediated by fatty acid-amino acid conjugates. *Plant Physiol* 131: 1894–1902.

Hause, B., I. Stenzel, O. Miersch, H. Maucher, R. Kramell, J. Ziegler, and C. Wasternack. 2000. Tissue-specific oxylipin signature of tomato flowers: allene oxide cyclase is highly expressed in distinct flower organs and vascular bundles. *Plant J* 24: 113–126.

Hayashi, M., K. Toriyama, M. Kondo, and M. Nishimura. 1998. 2,4-Dichlorophenoxybutyric acid-resistant mutants of Arabidopsis have defects in glyoxysomal fatty acid beta-oxidation. *Plant Cell* 10: 183–195.

Hayashi, M., K. Nito, R. Takei-Hoshi, M. Yagi, M. Kondo, A. Suenaga, T. Yamaya, and M. Nishimura. 2002. Ped3p Is a peroxisomal ATP-binding cassette transporter that might supply substrates for fatty acid beta-oxidation. *Plant Cell Physiol* 43: 1–11.

Heidel, A., and I.T. Baldwin. 2004. Microarray analysis of salicylic acid- and jasmonic acid-signalling in responses of *nicotiana attenuata* to attack by insects from multiple feeding guilds. *Plant Cell Environ* 2: 1362–1373.

Heil, M., and J. Ton. 2008. Long-distance signalling in plant defence. *Trends Plant Sci* 13: 264–272.

Herms, D.A., and W.J. Mattson. 1992. The dilemma of plants: to grow or defend. *Q Rev Biol* 6: 283–335.

Howe, G.A., and G. Jander. 2008. Plant immunity to insect herbivores. *Annu Rev Plant Biol* 59: 41–66.

Howe, G.A., J. Lightner, J. Browse, and C.A. Ryan. 1996. An octadecanoid pathway mutant (JL5) of tomato is compromised in signaling for defense against insect attack. *Plant Cell* 8: 2067–2077.

Hu, X., S. Neill, W. Cai, and Z. Tang. 2003. Hydrogen peroxide and jasmonic acid mediate oligogalacturonic acid-induced saponin accumulation in suspension-cultured cells of *panax ginseng*. *Physiol Plant* 118: 414–421.

Hyun, Y., S. Choi, H.J. Hwang, J. Yu, S.J. Nam, J. Ko, J.Y. Park, Y.S. Seo, E.Y. Kim, S.B. Ryu, W.T. Kim, Y.H. Lee, H. Kang, and I. Lee. 2008. Cooperation and functional diversification of two closely related galactolipase genes for jasmonate biosynthesis. *Dev Cell* 14: 183–192.

Ishiguro, S., A. Kawai-Oda, J. Ueda, I. Nishida, and K. Okada. 2001. The DEFECTIVE IN ANTHER DEHISCIENCE gene encodes a novel phospholipase A1 catalyzing the initial step of jasmonic acid biosynthesis, which synchronizes pollen maturation, anther dehiscence, and flower opening in Arabidopsis. *Plant Cell* 13: 2191–2209.

Jonak, C., L. Okresz, L. Bogre, and H. Hirt. 2002. Complexity, cross talk and integration of plant MAP kinase signalling. *Curr Opin Plant Biol* 5: 415–424.

Journot-Catalino, N., I.E. Somssich, D. Roby, and T. Kroj. 2006. The transcription factors WRKY11 and WRKY17 act as negative regulators of basal resistance in *Arabidopsis thaliana*. *Plant Cell* 18: 3289–3302.

Keinanen, M., N.J. Oldham, and I.T. Baldwin. 2001. Rapid HPLC screening of jasmonate-induced increases in tobacco alkaloids, phenolics, and diterpene glycosides in nicotiana attenuata. *J Agric Food Chem* 49: 3553–3558.

Kessler, A., and I.T. Baldwin. 2002. Plant responses to insect herbivory: the emerging molecular analysis. *Annu Rev Plant Biol* 53: 299–328.

Kessler, A., R. Halitschke, and I.T. Baldwin. 2004. Silencing the jasmonate cascade: induced plant defenses and insect populations. *Science* 305: 665–668.

Kloek, A.P., M.L. Verbsky, S.B. Sharma, J.E. Schoelz, J. Vogel, D.F. Klessig, and B.N. Kunkel. 2001. Resistance to pseudomonas syringae conferred by an *Arabidopsis thaliana* coronatine-insensitive (coi1) mutation occurs through two distinct mechanisms. *Plant J* 26: 509–522.

Koo, A.J., H.S. Chung, Y. Kobayashi, and G.A. Howe. 2006. Identification of a peroxisomal acyl-activating enzyme involved in the biosynthesis of jasmonic acid in Arabidopsis. *J Biol Chem* 281: 33511–33520.

Koornneef, A., and C.M. Pieterse. 2008. Cross talk in defense signaling. *Plant Physiol* 146: 839–844.

Kunkel, B.N., and D.M. Brooks. 2002. Cross talk between signaling pathways in pathogen defense. *Curr Opin Plant Biol* 5: 325–331.

Lee, G.I., and G.A. Howe. 2003. The tomato mutant spr1 is defective in systemin perception and the production of a systemic wound signal for defense gene expression. *Plant J* 33: 567–576.

Lee, D., D.H. Polisensky, and J. Braam. 2005. Genome-wide identification of touch- and darkness-regulated Arabidopsis genes: a focus on calmodulin-like and XTH genes. *New Phytol* 165: 429–444.

Li, L., C. Li, G.I. Lee, and G.A. Howe. 2002. Distinct roles for jasmonate synthesis and action in the systemic wound response of tomato. *Proc Natl Acad Sci USA* 99: 6416–6421.

Li, L., Y. Zhao, B.C. McCaig, B.A. Wingerd, J. Wang, M.E. Whalon, E. Pichersky, and G.A. Howe. 2004. The tomato homolog of CORONATINE-INSENSITIVE1 is required for the maternal control of seed maturation, jasmonate-signaled defense responses, and glandular trichome development. *Plant Cell* 16: 126–143.

Li, C., A.L. Schilmiller, G. Liu, G.I. Lee, S. Jayanty, C. Sageman, J. Vrebalov, J.J. Giovannoni, K. Yagi, Y. Kobayashi, and G.A. Howe. 2005. Role of beta-oxidation in jasmonate biosynthesis and systemic wound signaling in tomato. *Plant Cell* 17: 971–986.

Liechti, R., and E.E. Farmer. 2002. The jasmonate pathway. *Science* 296: 1649–1650.

Lison, P., I. Rodrigo, and V. Conejero. 2006. A novel function for the cathepsin D inhibitor in tomato. *Plant Physiol* 142: 1329–1339.

Liu, F., W. Ni, M.E. Griffith, Z. Huang, C. Chang, W. Peng, H. Ma, and D. Xie. 2004. The ASK1 and ASK2 genes are essential for Arabidopsis early development. *Plant Cell* 16: 5–20.

Lorenzo, O., R. Piqueras, J.J. Sanchez-Serrano, and R. Solano. 2003. ETHYLENE RESPONSE FACTOR1 integrates signals from ethylene and jasmonate pathways in plant defense. *Plant Cell* 15: 165–178.

Lorenzo, O., J.M. Chico, J.J. Sanchez-Serrano, and R. Solano. 2004. JASMONATE-INSENSITIVE1 encodes a MYC transcription factor essential to discriminate between different jasmonate-regulated defense responses in Arabidopsis. *Plant Cell* 16: 1938–1950.

Major, I.T., and C.P. Constabel. 2006. Molecular analysis of poplar defense against herbivory: comparison of wound- and insect elicitor-induced gene expression. *New Phytol* 172: 617–635.

Mao, P., M. Duan, C. Wei, and Y. Li. 2007. WRKY62 Transcription factor acts downstream of cytosolic NPR1 and negatively regulates jasmonate-responsive gene expression. *Plant Cell Physiol* 48: 833–842.

McConn, M., and J. Browse. 1996. The critical requirement for linolenic acid in pollen development, not photosynthesis, in an Arabidopsis mutant. *Plant Cell* 8: 403–416.

McCormack, E., and J. Braam. 2003. Calmodulins and related potential calcium sensors of Arabidopsis. *New Phytol* 159: 585–598.

Mewis, I., H.M. Appel, A. Hom, R. Raina, and J.C. Schultz. 2005. Major signaling pathways modulate Arabidopsis glucosinolate accumulation and response to both phloem-feeding and chewing insects. *Plant Physiol* 138: 1149–1162.

Mou, Z., W. Fan, and X. Dong. 2003. Inducers of plant systemic acquired resistance regulate NPR1 function through redox changes. *Cell* 113: 935–944.

Niki, T., I. Mitsuhara, S. Seo, N. Ohtsubo, and Y. Ohashi. 1998. Antagonistic effect of salicylic acid and jasmonic acid on the expression of pathogenesis-related (PR) protein genes in wounded mature tobacco leaves. *Plant Cell Physiol* 39: 500–507.

O'Donnell, P.J., M.R. Truesdale, C.M. Calvert, A. Dorans, M.R. Roberts, and D.J. Bowles. 1998. A novel tomato gene that rapidly responds to wound- and pathogen-related signals. *Plant J* 14: 137–142.

Parbery, D.G. 1996. Trophism and the ecology of fungi associated with plants. *Biol Rev* 71: 473–527.

Park, J.H., R. Halitschke, H.B. Kim, I.T. Baldwin, K.A. Feldmann, and R. Feyereisen. 2002. A knock-out mutation in allene oxide synthase results in male sterility and defective wound signal transduction in Arabidopsis due to a block in jasmonic acid biosynthesis. *Plant J* 31: 1–12.

Parthier, B. 1990. Jasmonates: hormonal regulators or stress factors in leaf senescence? *J Plant Growth Regul* 9: 57–63.

Paschold, A., R. Halitschke, and I.T. Baldwin. 2007. Co(i)-ordinating defenses: NaCOI1 Mediates herbivore-induced resistance in *nicotiana attenuata* and reveals the role of herbivore movement in avoiding defenses. *Plant J* 51: 79–91.

Pauwels, L., K. Morreel, E. De Witte, F. Lammertyn, M. Van Montagu, W. Boerjan, D. Inze, and A. Goossens. 2008. Mapping methyl jasmonate-mediated transcriptional reprogramming of metabolism and cell cycle progression in cultured Arabidopsis cells. *Proc Natl Acad Sci USA* 105: 1380–1385.

Penninckx, I.A., K. Eggermont, F.R. Terras, B.P. Thomma, G.W. De Samblanx, A. Buchala, J.P. Metraux, J.M. Manners, and W.F. Broekaert. 1996. Pathogen-induced systemic activation of a plant defensin gene in Arabidopsis follows a salicylic acid-independent pathway. *Plant Cell* 8: 2309–2323.

Penninckx, I.A., B.P. Thomma, A. Buchala, J.P. Metraux, and W.F. Broekaert. 1998. Concomitant activation of jasmonate and ethylene response pathways is required for induction of a plant defensin gene in Arabidopsis. *Plant Cell* 10: 2103–2113.

Petersen, M., P. Brodersen, H. Naested, E. Andreasson, U. Lindhart, B. Johansen, H.B. Nielsen, M. Lacy, M.J. Austin, J.E. Parker, S.B. Sharma, D.F. Klessig, R. Martienssen, O. Mattsson, A.B. Jensen, and J. Mundy. 2000. Arabidopsis map kinase 4 negatively regulates systemic acquired resistance. *Cell* 103: 1111–1120.

Pieterse, C.M., and L.C. Van Loon. 2004. NPR1: the spider in the web of induced resistance signaling pathways. *Curr Opin Plant Biol* 7: 456–464.

Pieterse, C.M., S.C. van Wees, J.A. van Pelt, M. Knoester, R. Laan, H. Gerrits, P.J. Weisbeek, and L.C. van Loon. 1998. A novel signaling pathway controlling induced systemic resistance in Arabidopsis. *Plant Cell* 10: 1571–1580.

Purugganan, M.M., J. Braam, and S.C. Fry. 1997. The Arabidopsis TCH4 xyloglucan endotransglycosylase. Substrate specificity, pH optimum, and cold tolerance. *Plant Physiol* 115: 181–190.

Ralph, S.G., H. Yueh, M. Friedmann, D. Aeschliman, J.A. Zeznik, C.C. Nelson, Y.S. Butterfield, R. Kirkpatrick, J. Liu, S.J. Jones, M.A. Marra, C.J. Douglas, K. Ritland, and J. Bohlmann. 2006. Conifer defence against insects: microarray gene expression profiling of Sitka spruce (*picea sitchensis*) induced by mechanical wounding or feeding by spruce budworms (*choristoneura occidentalis*) or white pine weevils (*pissodes strobi*) reveals large-scale changes of the host transcriptome. *Plant Cell Environ* 29: 1545–1570.

Ren, C., J. Pan, W. Peng, P. Genschik, L. Hobbie, H. Hellmann, M. Estelle, B. Gao, J. Peng, C. Sun, and D. Xie. 2005. Point mutations in Arabidopsis Cullin1 reveal its essential role in jasmonate response. *Plant J* 42: 514–524.

Reymond, P., H. Weber, M. Damond, and E.E. Farmer. 2000. Differential gene expression in response to mechanical wounding and insect feeding in Arabidopsis. *Plant Cell* 12: 707–720.

Reymond, P., N. Bodenhausen, R.M. Van Poecke, V. Krishnamurthy, M. Dicke, and E.E. Farmer. 2004. A conserved transcript pattern in response to a specialist and a generalist herbivore. *Plant Cell* 16: 3132–3147.

Richmond, T.A., and A.B. Bleecker. 1999. A defect in beta-oxidation causes abnormal inflorescence development in Arabidopsis. *Plant Cell* 11: 1911–1924.

Rojo, E., J. Leon, and J.J. Sanchez-Serrano. 1999. Cross-talk between wound signalling pathways determines local versus systemic gene expression in *Arabidopsis thaliana*. *Plant J* 20: 135–142.

Rojo, E., R. Solano, and J.J. Sanchez-Serrano. 2003. Interactions between signaling compounds involved in plant defense. *J Plant Growth Regul* 22: 82–98.

Rowe, H.C., J.W. Walley, J. Corwin, E.K. Chan, K. Dehesh, and D.J. Kliebenstein. 2010. Deficiencies in jasmonate-mediated plant defense reveal quantitative variation in *botrytis cinerea* pathogenesis. *PLoS Pathog* 6: e1000861.

Ryan, C.A. 1990. Protease inhibitors in plants: genes for improving defenses against insects and pathogens. *Annu Rev Phytopathol* 28: 425–449.

Sanchez-Serrano, J., R. Schmidt, J. Schell, and L. Willmitzer. 1986. Nucleotide sequence of proteinase inhibitor II encoding cDNA of potato (*solanum fuberosum*) and its mode of expression. *Mol Gen Genet* 203: 15–20.

Sanders, P.M., P.Y. Lee, C. Biesgen, J.D. Boone, T.P. Beals, E.W. Weiler, and R.B. Goldberg. 2000. The Arabidopsis DELAYED DEHISCENCE1 gene encodes an enzyme in the jasmonic acid synthesis pathway. *Plant Cell* 12: 1041–1061.

Schaller, F., C. Biesgen, C. Mussig, T. Altmann, and E.W. Weiler. 2000. 12-Oxophytodienoate reductase 3 (OPR3) is the isoenzyme involved in jasmonate biosynthesis. *Planta* 210: 979–984.

Schilmiller, A.L., A.J. Koo, and G.A. Howe. 2007. Functional diversification of acyl-coenzyme a oxidases in jasmonic acid biosynthesis and action. *Plant Physiol* 143: 812–824.

Schmelz, E.A., M.J. Carroll, S. LeClere, S.M. Phipps, J. Meredith, P.S. Chourey, H.T. Alborn, and P.E. Teal. 2006. Fragments of ATP synthase mediate plant perception of insect attack. *Proc Natl Acad Sci USA* 103: 8894–8899.

Seo, J.S., J. Joo, M.J. Kim, Y.K. Kim, B.H. Nahm, S.I. Song, J.J. Cheong, J.S. Lee, J.K. Kim, and Y.D. Choi. 2011. OsbHLH148, a basic helix-loop-helix protein, interacts with OsJAZ proteins in a jasmonate signaling pathway leading to drought tolerance in rice. *Plant J* 65: 907–921.

Sheard, L.B., X. Tan, H. Mao, J. Withers, G. Ben-Nissan, T.R. Hinds, Y. Kobayashi, F.F. Hsu, M. Sharon, J. Browse, S.Y. He, J. Rizo, G.A. Howe, and N. Zheng. 2010. Jasmonate perception by inositol-phosphate-potentiated COI1-JAZ co-receptor. *Nature* 468: 400–405.

Shoji, T., Y. Yamada, and T. Hashimoto. 2000. Jasmonate induction of putrescine N-methyltransferase genes in the root of *nicotiana sylvestris*. *Plant Cell Physiol* 41: 831–839.

Sistrunk, M.L., D.M. Antosiewicz, M.M. Purugganan, and J. Braam. 1994. Arabidopsis TCH3 encodes a novel Ca^{2+} binding protein and shows environmentally induced and tissue-specific regulation. *Plant Cell* 6: 1553–1565.

Song, S., T. Qi, H. Huang, Q. Ren, D. Wu, C. Chang, W. Peng, Y. Liu, J. Peng, and D. Xie. 2011. The jasmonate-ZIM domain proteins interact with the R2R3-MYB transcription factors MYB21 and MYB24 to affect jasmonate-regulated stamen development in Arabidopsis. *Plant Cell* 23: 1000–1013.

Spoel, S.H., A. Koornneef, S.M. Claessens, J.P. Korzelius, J.A. Van Pelt, M.J. Mueller, A.J. Buchala, J.P. Metraux, R. Brown, K. Kazan, L.C. Van Loon, X. Dong, and C.M. Pieterse. 2003. NPR1 Modulates cross-talk between salicylate- and jasmonate-dependent defense pathways through a novel function in the cytosol. *Plant Cell* 15: 760–770.

Staswick, P.E., and I. Tiryaki. 2004. The oxylipin signal jasmonic acid is activated by an enzyme that conjugates it to isoleucine in Arabidopsis. *Plant Cell* 16: 2117–2127.

Staswick, P.E., W. Su, and S.H. Howell. 1992. Methyl jasmonate inhibition of root growth and induction of a leaf protein are decreased in an *Arabidopsis thaliana* mutant. *Proc Natl Acad Sci USA* 89: 6837–6840.

Staswick, P.E., G.Y. Yuen, and C.C. Lehman. 1998. Jasmonate signaling mutants of Arabidopsis are susceptible to the soil fungus *pythium irregulare*. *Plant J* 15: 747–754.

Staswick, P.E., I. Tiryaki, and M.L. Rowe. 2002. Jasmonate response locus JAR1 and several related Arabidopsis genes encode enzymes of the firefly luciferase superfamily that show activity on jasmonic, salicylic, and indole-3-acetic acids in an assay for adenylation. *Plant Cell* 14: 1405–1415.

Stelmach, B.A., A. Muller, P. Hennig, D. Laudert, L. Andert, and E.W. Weiler. 1998. Quantitation of the octadecanoid 12-oxo-phytodienoic acid, a signalling compound in plant mechanotransduction. *Phytochemistry* 47: 539–546.

Stintzi, A., and J. Browse. 2000. The Arabidopsis male-sterile mutant, opr3, lacks the 12-oxophytodienoic acid reductase required for jasmonate synthesis. *Proc Natl Acad Sci USA* 97: 10625–10630.

Stintzi, A., H. Weber, P. Reymond, J. Browse, and E.E. Farmer. 2001. Plant defense in the absence of jasmonic acid: the role of cyclopentenones. *Proc Natl Acad Sci USA* 98: 12837–12842.

Strassner, J., F. Schaller, U.B. Frick, G.A. Howe, E.W. Weiler, N. Amrhein, P. Macheroux, and A. Schaller. 2002. Characterization and cDNA-microarray expression analysis of 12-oxophytodienoate reductases reveals differential roles for octadecanoid biosynthesis in the local versus the systemic wound response. *Plant J* 32: 585–601.

Suza, W.P., and P.E. Staswick. 2008. The role of JAR1 in jasmonoyl-L: isoleucine production during Arabidopsis wound response. *Planta* 227: 1221–1232.

Swiatek, A., M. Lenjou, D. Van Bockstaele, D. Inze, and H. Van Onckelen. 2002. Differential effect of jasmonic acid and abscisic acid on cell cycle progression in tobacco BY-2 cells. *Plant Physiol* 128: 201–211.

Thaler, J.S., A.L. Fidantsef, and R.M. Bostock. 2002. Antagonism between jasmonate- and salicylate-mediated induced plant resistance: effects of concentration and timing of elicitation on defense-related proteins, herbivore, and pathogen performance in tomato. *J Chem Ecol* 28: 1131–1159.

Thines, B., L. Katsir, M. Melotto, Y. Niu, A. Mandaokar, G. Liu, K. Nomura, S.Y. He, G.A. Howe, and J. Browse. 2007. JAZ repressor proteins are targets of the SCF(COI1) complex during jasmonate signalling. *Nature* 448: 661–665.

Thomma, B.P., K. Eggermont, I.A. Penninckx, B. Mauch-Mani, R. Vogelsang, B.P. Cammue, and W.F. Broekaert. 1998. Separate jasmonate-dependent and salicylate-dependent defense-response pathways in Arabidopsis are essential for resistance to distinct microbial pathogens. *Proc Natl Acad Sci USA* 95: 15107–15111.

Thomma, B.P., I.A. Penninckx, W.F. Broekaert, and B.P. Cammue. 2001. The complexity of disease signaling in Arabidopsis. *Curr Opin Immunol* 13: 63–68.

Thorpe, M.R., A.P. Ferrieri, M.M. Herth, and R.A. Ferrieri. 2007. 11C-imaging: methyl jasmonate moves in both phloem and xylem, promotes transport of jasmonate, and of photoassimilate even after proton transport is decoupled. *Planta* 226: 541–551.

Ton, J., M. D'Alessandro, V. Jourdie, G. Jakab, D. Karlen, M. Held, B. Mauch-Mani, and T.C. Turlings. 2007. Priming by airborne signals boosts direct and indirect resistance in maize. *Plant J* 49: 16–26.

Traw, M.B., J. Kim, S. Enright, D.F. Cipollini, and J. Bergelson. 2003. Negative cross-talk between salicylate- and jasmonate-mediated pathways in the wassilewskija ecotype of *Arabidopsis thaliana*. *Mol Ecol* 12: 1125–1135.

Tsuji, J., E.P. Jackson, D.A. Gage, R. Hammerschmidt, and S.C. Somerville. 1992. Phytoalexin accumulation in *Arabidopsis thaliana* during the hypersensitive reaction to *pseudomonas syringae* pv syringae. *Plant Physiol* 98: 1304–1309.

van Wees, S.C., M. Luijendijk, I. Smoorenburg, L.C. van Loon, and C.M. Pieterse. 1999. Rhizobacteria-mediated induced systemic resistance (ISR) in Arabidopsis is not associated with a direct effect on expression of known defense-related genes but stimulates the expression of the jasmonate-inducible gene atvsp upon challenge. *Plant Mol Biol* 41: 537–549.

Vick, B.A., and D.C. Zimmerman. 1983. The biosynthesis of jasmonic acid: a physiological role for plant lipoxygenase. *Biochem Biophys Res Commun* 111: 470–477.

Vijayan, P., J. Shockey, C.A. Levesque, R.J. Cook, and J. Browse. 1998. A role for jasmonate in pathogen defense of Arabidopsis. *Proc Natl Acad Sci USA* 95: 7209–7214.

Vlot, A.C., D.F. Klessig, and S.W. Park. 2008. Systemic acquired resistance: the elusive signal(s). *Curr Opin Plant Biol* 11: 436–442.

von Malek, B., E. van der Graaff, K. Schneitz, and B. Keller. 2002. The Arabidopsis male-sterile mutant dde2-2 is defective in the ALLENE OXIDE SYNTHASE gene encoding one of the key enzymes of the jasmonic acid biosynthesis pathway. *Planta* 216: 187–192.

Wang Y, Wang B, Gilroy S, Chehab EW, Braam J (2011) CML24 is involved in root mechanoresponses and cortical microtubule orientation in Arabidopsis. J Plant Growth Regul. doi: 10.1007/s00344-011-9209-9.

Wasternack, C. 2007. Jasmonates: an update on biosynthesis, signal transduction and action in plant stress response, growth and development. *Ann Bot* 100: 681–697.

Weiler, E.W., T. Albrecht, B. Groth, Z.Q. Xia, M. Luxem, H. Lib, L. Andert, and P. Spengler. 1993. Evidence for the involvement of jamonates and their octadecanoid precursors in the tendril coiling response of *bryonia dioica. Phytochemistry* 32: 591–600.

Xie, D.X., B.F. Feys, S. James, M. Nieto-Rostro, and J.G. Turner. 1998. COI1: an Arabidopsis gene required for jasmonate-regulated defense and fertility. *Science* 280: 1091–1094.

Xu, Y., P. Chang, D. Liu, M.L. Narasimhan, K.G. Raghothama, P.M. Hasegawa, and R.A. Bressan. 1994. Plant defense genes are synergistically induced by ethylene and methyl jasmonate. *Plant Cell* 6: 1077–1085.

Xu, W., M.M. Purugganan, D.H. Polisensky, D.M. Antosiewicz, S.C. Fry, and J. Braam. 1995. Arabidopsis TCH4, regulated by hormones and the environment, encodes a xyloglucan endotransglycosylase. *Plant Cell* 7: 1555–1567.

Xu, L., F. Liu, E. Lechner, P. Genschik, W.L. Crosby, H. Ma, W. Peng, D. Huang, and D. Xie. 2002. The SCF(COI1) ubiquitin-ligase complexes are required for jasmonate response in Arabidopsis. *Plant Cell* 14: 1919–1935.

Yan, Y., S. Stolz, A. Chetelat, P. Reymond, M. Pagni, L. Dubugnon, and E.E. Farmer. 2007. A downstream mediator in the growth repression limb of the jasmonate pathway. *Plant Cell* 19: 2470–2483.

Zarate, S.I., L.A. Kempema, and L.L. Walling. 2007. Silverleaf whitefly induces salicylic acid defenses and suppresses effectual jasmonic acid defenses. *Plant Physiol* 143: 866–875.

Zavala, J.A., A.G. Patankar, K. Gase, D. Hui, and I.T. Baldwin. 2004. Manipulation of endogenous trypsin proteinase inhibitor production in *nicotiana attenuata* demonstrates their function as antiherbivore defenses. *Plant Physiol* 134: 1181–1190.

Zhang, Z., and I.T. Baldwin. 1997. Transporter of [2-C-14]jasmonic acid from leaves to roots mimics wound-induced changes in endogenous jasmonic acid pools in *nicotiana sylvestris. Planta* 203: 436–441.

Zhang, Y., and J.G. Turner. 2008. Wound-induced endogenous jasmonates stunt plant growth by inhibiting mitosis. *PLoS One* 3: e3699.

Zhang, L., and D. Xing. 2008. Methyl jasmonate induces production of reactive oxygen species and alterations in mitochondrial dynamics that precede photosynthetic dysfunction and subsequent cell death. *Plant Cell Physiol* 49: 1092–1111.

Ziegler, J., I. Stenzel, B. Hause, H. Maucher, M. Hamberg, R. Grimm, M. Ganal, and C. Wasternack. 2000. Molecular cloning of allene oxide cyclase. The enzyme establishing the stereochemistry of octadecanoids and jasmonates. *J Biol Chem* 275: 19132–19138.

Zolman, B.K., I.D. Silva, and B. Bartel. 2001. The Arabidopsis pxa1 mutant is defective in an ATP-binding cassette transporter-like protein required for peroxisomal fatty acid beta-oxidation. *Plant Physiol* 127: 1266–1278.

Multitude of Long-Distance Signal Molecules Acting Via Phloem

Sylvie Dinant and Paula Suárez-López

Abstract As sessile organisms, plants use long-range signalling between organs in order to adapt to their environment. The phloem is an important pathway for such long-distance communication. It transports signals that trigger systemic defence responses to wounding, herbivory and infection by plant pathogens. It also plays a pivotal role for developmental transitions, such as floral induction and tuberization, in response to stimuli perceived by the leaves, and physiological adaptation to nutrient deprivation. The signals involved in these processes include hormones, metabolites, proteins and RNAs, transported by mass flow with the phloem translocation stream. Faster signals, such as electropotential waves, can be propagated by the phloem plasma membrane. Most recent studies showed that these signalling pathways can recruit combinations of signal molecules, and that additional steps, such as molecular 'hopping' and amplification, may occur within the phloem tissue. This provides a basis to explain how plants cope with multiple environmental stimuli to confer long-lasting effects against stresses and maintain plant growth and development.

S. Dinant (✉)
Institut Jean Pierre Bourgin, UMR1318, Institut National de la Recherche Agronomique, Versailles, France
e-mail: sylvie.dinant@versailles.inra.fr

P. Suárez-López
Centre for Research in Agricultural Genomics, CSIC-IRTA-UAB-UB, Barcelona, Spain
e-mail: paula.suarez@cragenomica.es

G. Witzany and F. Baluška (eds.), *Biocommunication of Plants*,
Signaling and Communication in Plants 14,
DOI 10.1007/978-3-642-23524-5_6,

1 Overview of Phloem Organization and Functions

1.1 The Phloem: A Pathway for Nutrient Allocation and Inter-Organ Communication

Higher plants are organized in specialized organs, which fulfil distinct functions in the uptake of nutrients and energy, storage of metabolites and adaptation to the environment. The exchanges of nutrients and information between organs occur in the vascular tissues, i.e. xylem and phloem, by long-distance transport of water, nutrients, metabolites and signal molecules. The phloem plays a key role in long-distance signalling for many developmental and environmental responses. For example, long-range induction of flowering is a classical case of signalling from the leaves to the shoot apical meristem (Zeevaart 2008). Systemic acquired protection against plant pathogens is another well-known example (Sticher et al. 1997). The propagation of gene silencing has also been shown to follow a similar pathway (Palauqui et al. 1997; Voinnet and Baulcombe 1997). Ultimately, the phloem was identified as a main route for the translocation of such systemic signals. A breakthrough in our comprehension of long-distance communication was the discovery that proteins and RNAs transported in the phloem can act as mobile signals (Lough and Lucas 2006).

Thus, the phloem is essential in a number of adaptation and developmental events that require a coordinated and integrated response of the whole plant. In this chapter, we will successively consider key cases of long-distance signalling *via* the phloem.

1.2 Anatomy and Biochemistry of the Phloem

1.2.1 Phloem Anatomy and Overall Functions

The emergence of the vascular tissue has been an early landmark of the evolution of land plants, with water uptake and transport from the roots carried out by the xylem and allocation of sugars resulting from carbon fixation by aerial organs carried out by the phloem (van Bel 2003a). Thus, the primary phloem function is the partitioning of carbohydrates produced as photosynthates from autotrophic to heterotrophic organs. Both the sieve elements (SEs), i.e. cells conducing phloem sap, and the companion cells (CCs) present a unique cellular organization (Sjölund 1997). The phloem is organized in functional zones specialized in loading, transport and unloading, and named 'collection phloem', 'transport phloem' and 'release phloem' (van Bel 2003a), with the transport phloem making up the major part of the phloem (van Bel 2003b). The driving force for long-distance transport in the sieve tubes makes use of a turgor gradient due to variations in photosynthate accumulation along the pathway that create a hydraulic pressure gradient (Thompson 2006; Knoblauch and Peters 2010). The phloem communications between organs follow an independent succession of vascular connections between source and sink organs, known as orthostichies, which depend on the plant phyllotaxy (Callos and Medford 1994; Orians 2005). This implies that not all sinks are equally supplied by source leaves. A consequence is that systemic signal molecules, such as salicylic acid (SA), move in large part with assimilate movement along an orthostichy and do not trigger a response in all sink leaves (Kiefer and Slusarenko 2003). Another well-established example of this vascular organization is the pattern of systemic colonization during viral infection, which also follows orthostichous phloem connections (Roberts et al. 2007).

1.2.2 Phloem Sap Composition

Phloem sap contains sugars, amino acids, organic acids, secondary metabolites, ions, peptides, hormones as well as a large range of macromolecules, including proteins, small RNAs and mRNAs (Turgeon and Wolf 2009; Dinant et al. 2010). The composition of the phloem sap and the supply in structural components of the SEs are controlled at the interface between the SE-CC complex (Sjölund 1997; Oparka and Turgeon 1999), with an integrated control of loading, lateral exchanges along the transport pathway and unloading (van Bel 2003a). The delivery of molecules from the CCs or adjacent parenchyma cells to the SEs takes place either through the apoplasm, based on a series of carriers and pumps, present on the plasma membrane of SEs and CCs (Lalonde et al. 2003; Dinant and Lemoine 2010), or through fields of specialized plasmodesmata at the CC-SE interface, constituting the plasmodesmata pore units (PPUs) (van Bel 2003a). Most macromolecules present in the SEs are synthesized in the CCs (Turgeon and Wolf 2009). The entry of macromolecules into the SE takes place *via* the plasmodesmata, whereas the loading of metabolites and hormones can follow either symplasmic or apoplasmic steps.

1.2.3 Methods to Analyze Phloem

One major difficulty in studying phloem activity is to sample phloem sap and to sample phloem cells (Sjölund 1997). Several methods are available, such as bleeding, stylectomy or EDTA-facilitated exudation, depending on plant species (Turgeon and Wolf 2009). These methods can be useful for the identification of phloem sap components, although they potentially cause artefacts (Dinant et al. 2010). Carbon isotope labelling has been used to follow the transport in the vasculature of various compounds, such as sugars, SA, methyl jasmonate (MeJA) or other substances (Minchin and Thorpe 1987; Kiefer and Slusarenko 2003; Rocher et al. 2006; Thorpe et al. 2007). Grafting has been widely used as experimental approach to test for the biological activity of a compound translocated in the phloem (Turnbull et al. 2002). Cold-girdling or split-root experiments are also interesting tools to confirm long-distance signalling. Magnetic resonance imaging has been developed for *in vivo* imaging of vascular tissues and can be used to measure phloem sap velocity (Windt et al. 2006; Mullendore et al. 2010). As for the isolation of phloem cells, laser microdissection has been successfully used (Nelson et al. 2006).

2 Long-Distance Signalling in Response to Biotic Stress

Given that plants are sessile, they cannot run away from threats or move in search of nutrients or favourable environments. In order to maximize fitness and survival, plants have evolved numerous strategies to perceive environmental signals and adapt their development to different habitats. This encompasses the perception of specific stresses by the different organs and the transmission of the information to the other parts of the plant. Several key cases of long-distance signalling *via* the phloem in response to biotic or abiotic stresses will be described in this section. One classical case is the systemic defence response triggered in the whole plant after an initial injury of the leaves caused by plant pathogens, pests or wounding. The second classical case of phloem long-distance signalling is initiated by nutrient deprivation in the soil. This induces a root-to-shoot signal involving the xylem, then a shoot-to-root phloem signal that allows the plant to maintain the nutrient homeostasis within the whole plant and to adapt rapidly its growth and its development to its environment.

2.1 Long-Distance Signalling to Wounding and Herbivory

2.1.1 Systemin and Jasmonates in Response to Wounding in Tomato

When a leaf is injured, resulting from herbivory or contact with a cutting surface, a systemic signal is transported to non-injured newly forming leaves (Wu and Baldwin 2010). This systemic response is associated with the production of protease inhibitors and the release of volatiles, as a defence mechanism against subsequent insect infestations. The long-distance signalling has been studied in details in tomato. It is initiated by the production of systemin, a small peptide of 18 amino acids, which is produced after cleavage of a propeptide, the prosystemin. Systemin was initially thought to be the systemic signal (Stratmann 2003). However, it is now well established that systemin-induced jasmonic acid (JA), or JA derivative, which moves systemically, represent the major signal molecules in wound response (Lee and Howe 2003; Li et al. 2003; Schilmiller and Howe 2005; Wasternack et al. 2006). This signalling pathway is propagated and amplified within the vascular tissues: the prosystemin is produced in the phloem parenchyma cells (Narváez-Vásquez and Ryan 2004), and the biosynthetic enzymes for the synthesis of jasmonates are present in the CC-SE complexes (Hause et al. 2000, 2003; Stenzel et al. 2003), which further confer to the phloem the ability to amplify the synthesis of jasmonates (van Bel and Gaupels 2004). A systemin-binding SR160/BRI1 receptor at the surface of cells was identified in *Solanum peruvianum* (Montoya et al. 2002; Scheer and Ryan 2002) and proposed to trigger the transduction pathway for the synthesis of jasmonates (Schilmiller and Howe 2005). However, its role has been controversial and recent studies suggested instead that the systemin receptor is a distinct although related BRI-like protein localized in the vascular tissues (Malinowski et al. 2009; Hind et al. 2010), yet to characterize. Hydrogen peroxide has also been proposed to constitute a secondary messenger in sink organs (Orozco-Cardenas et al. 2001).

2.1.2 Jasmonates and the JAZ Proteins

The action of systemin and jasmonates in long-distance signalling is unique to tomato. However, the role of JA and JA derivatives such as the JA-amino-acid conjugate jasmonyl-L-isoleucine (JA-Ile) in response to wounding or to herbivory has been generalized to other species. In *Arabidopsis*, it has been discovered that jasmonates, most likely as JA-Ile, interact with the CORONATIN-INSENSITIVE 1 (COI1) unit of the E3 ubiquitin ligase complex SCF-COI1 (Skip/Cullin/F-box–COI1). A third component of the jasmonate co-receptor complex is inositol pentakisphosphate (Sheard et al. 2010). This complex targets, for subsequent degradation by the 26S proteasome, the JAZ proteins (Thines et al. 2007), which are repressors of the JA-inducible genes (Kazan and Manners 2008; Staswick 2008). Whether this transduction pathway only acts in the leaves or whether it is also in action in the transport phloem to relay and amplify the signal(s) has not been investigated. Other JA derivatives, the JA metabolite *cis*-jasmone (CJ) and MeJA, have been proposed to be active in defence signalling (Birkett et al. 2000; Bruce et al. 2008; Wu et al. 2008), some of them, such as MeJA, being transported *via* the phloem (Thorpe et al. 2007).

2.1.3 Propagation of Electric Potential Waves in the Phloem

Other systemic signals have been proposed to act in wound responses, including oligosaccharides, reactive oxygen species (ROS), hydraulic signals, electrical signals or other plant hormones (Rhodes et al. 1996; Mancuso 1999; Wasternack et al. 2006; Fromm and Lautner 2007; Maffei et al. 2007; Heil and Ton 2008; Shah 2009; Zimmermann et al. 2009). The role of electric potential waves (EPWs) in long-distance signalling in response to wounding was shown in tomato (Rhodes et al. 1996) and further examined in *Vicia faba* and barley (Furch et al. 2007; Zimmermann et al. 2009). EPWs, which are relayed by Ca^{2+} influx, can propagate very rapidly in the phloem, in response to wounding or other stimuli, such as burning or cooling, which in turn triggers various responses in the SE (Furch et al. 2007, 2009, 2010; van Bel et al. 2011a). The propagation rates of EPWs vary from 5 to 200 cm min^{-1}, depending on EPW classes, which is much faster than those driven by components transported by phloem sap (Fromm and Lautner 2007; Zimmermann et al. 2009). The observation that the accumulation of JA-Ile occurs rapidly in distal leaves, as soon as 5 min after wounding (Koo et al. 2009), is consistent with such EPW propagation rates. The preferential transmission of electrical signals in the phloem has been proposed to result from the low electrical conductance of plasmodesmata in lateral direction and on the high degree of electrical coupling *via* the sieve pores in longitudinal direction (Kempers and van Bel 1997; Fromm and Lautner 2007).

2.1.4 The Emerging Action of Reactive Oxygen Species and RbohD

Another main component of rapid propagation of signalling in response to wounding is the accumulation of ROS produced by a *RESPIRATORY BURST OXIDASE HOMOLOG D* (*RBOHD*) gene in *Arabidopsis* (Miller et al. 2009). This pathway is independent of ethylene, JA or SA. It is triggered by wounding, heat, cold, high-intensity light or salinity stresses, at a propagation rate of 8.4 cm min^{-1}. This established that ROS accumulation along a systemic signal front is essential for long-distance signalling in plants (Miller et al. 2009). This signal propagates in the apoplasm of the vascular tissues. The presence of a complete antioxidant system in the phloem sap also suggests a tight control of oxidative stress in this compartment (Walz et al. 2002).

Airborne signals also participate in long-distance signalling to wounding or herbivory (Heil and Silva Bueno 2007), overcoming the restrictions resulting from the plant's orthostichy (Frost et al. 2007; Heil and Ton 2008). Altogether, these observations support the idea that multiple long-distance signalling systems operate, JA-acting either in a cell-autonomous or in a cell-non-autonomous signalling pathway (Heil and Ton 2008; Koo and Howe 2009).

2.2 *Systemic Response of Plants to Pathogen Attack*

2.2.1 SA, MeSA and SABP: An Integrated Pathway in Tobacco

The role of the phloem in the mounting of systemic defences in plant immune responses has been also investigated in details (Durrant and Dong 2004; Grant and Lamb 2006). During interactions with an avirulent plant pathogen, the recognition of pathogen-associated molecular patterns (PAMPs) by host cells first triggers a local response, known as hypersensitive response (HR) (Jones and Dangl 2006), then a general immune response, resulting from the generation by infected leaves of a long-distance signal transported *via* the phloem (Durrant and Dong 2004). This long-lasting response known as the systemic acquired resistance (SAR) is characterized by an enhanced resistance to plant pathogens in newly formed organs associated with the production of pathogenesis-related (PR) proteins and an oxidative burst (Sticher et al. 1997; Durrant and Dong 2004; Grant and Lamb 2006; Zhang and Zhou 2010). The triggering of this signalling pathway is associated with the production of SA. In tobacco, it has been demonstrated that the activity of a SAMT (SA methyl transferase 1) in inoculated leaves enables the production of MeSA from SA (Park et al. 2007). MeSA is then transported systemically and constitutes a critical signal for the establishment of the systemic response (Seskar et al. 1998). In the systemic tissues, MeSA is hydrolyzed into SA by the MeSA esterase activity of SA-binding protein 2 (SABP2) and this newly generated SA triggers SAR (Forouhar et al. 2005). This mechanism is also active in *Arabidopsis* and potato (Kumar and Klessig 2003; Park et al. 2007; Vlot et al. 2008b; Manosalva et al. 2010).

2.2.2 Lipid-Derived Molecules: Modulators or Relays?

The identity of the systemic mobile signal(s) for SAR is so far still unclear, and other signals have been identified (Vlot et al. 2008a). Plant pathogen interactions trigger locally the synthesis of a lipid transfer protein (LTP) DIR1 (Maldonado et al. 2002) and a plastid glycerolipid factor, dependent from the biosynthetic genes *FAD7*, *SFD1* and *SFD2* (Kachroo et al. 2001; 2004; Chaturvedi et al. 2008), which probably form a complex. Both MeSA and this DIR1-lipid complex, acting as mobile signals, are required for the systemic activation of SAR (Liu et al. 2011a). *DIR1* transcripts were found in the phloem companion cells (Ivashikina et al. 2003), providing support in favour of a role as a phloem-specific carrier of signal (van Bel and Gaupels 2004). Other compounds, such as terpenoids or peptides, which are released by the action of extracellular proteases, have also been implicated in systemic signalling (Durner and Klessig 1999; Suzuki et al. 2004; Xia et al. 2004; Rustérucci et al. 2007; Shah 2009). A recent study also showed the role of azelaic acid, a nine-carbon dicarboxylic acid, in priming systemic defences in *Arabidopsis* (Jung et al. 2009). Azelaic acid induces the expression of *AZI1*, a gene encoding a predicted secreted protease inhibitor/LTP, which modulates production and/or translocation of the mobile signal during SAR. These signal molecules would act together with MeSA. Alternatively, they may act as relays for the amplification of the initial signal(s).

2.2.3 Hormone Crosstalks and the Multifactorial Plant Immune System

Jasmonates have also been described as signals essential for establishing systemic immunity in response to *Pseudomonas syringae* (Truman et al. 2007). However, this model is still quite controversial (Shah 2009), since conflicting evidences indicated that neither MeSA nor jasmonate were essential as systemic signals for

SAR (Cui et al. 2005; Mishina and Zeier 2007; Attaran et al. 2009). There is also a large body of evidence of antagonist interplays with other hormones such as auxin or abscisic acid (Chen et al. 2007; Wang et al. 2007; Ding et al. 2008; De Torres Zabala et al. 2009; Fan et al. 2009a; Truman et al. 2010). It has been proposed that depending on the combination of pathogen attackers, complex hormone crosstalks are activated to fine-tune induced defences (Leon-Reyes et al. 2009, 2010; Makandar et al. 2010). Interestingly, most hormones have been identified in the phloem sap (Hoad 1995), including auxin, cytokinins, gibberellins, abscisic acid, 1-aminocyclopropane-1-carboxylic acid (the precursor of ethylene), MeJA and SA, with the exception of brassinosteroids and strigolactones. Overall, the idea is emerging that multiscale and multifactorial defence systems can operate proper temporal and spatial integration to confer lasting disease resistance and prevent unfavourable signal interactions to concomitantly defend against multiple pathogens (Bruce and Pickett 2007; Spoel et al. 2007; Parker 2009; Shah 2009).

2.2.4 Nitric Oxide and the Concept of Molecular 'Hopping'

Nitric oxide (NO) is also involved in signalling (Crawford and Guo 2005; Durner and Klessig 1999; Leitner et al. 2009). NO and S-nitrosothiols (SNO) are produced in the phloem CCs, in response to biotic and abiotic stresses, and have been proposed to be important signals, acting in the phloem cells downstream of SA (Rustérucci et al. 2007; Gaupels et al. 2008). One mode of NO action in the phloem would be through binding to some enzymes, thereby modifying their activity, which in turn would induce signal synthesis or activation (Gaupels et al. 2008). This led to the interesting model, proposed by van Bel and Gaupels (2004), that the role of the phloem, including production, release and distribution of signal molecules, may also encompass modulation and amplification of signals along the pathway, as observed in tomato for wound response (Wasternack et al. 2006). This concept, recently termed molecular 'hopping' by van Bel and co-workers (van Bel et al. 2011b), is based on long-standing observations that release/retrieval processes occur along the transport phloem pathway (Minchin and Thorpe 1987; Ayre et al. 2003; Hafke et al. 2005). It assumes a key role played by the CCs and phloem parenchyma cells, which are connected to the SEs by the PPUs, in relaying and/or amplifying signal(s).

2.2.5 Alternative Long-Distance Signalling Pathways

Several observations provide support in favour of additional long-distance pathways. First, the pattern of signalling does not always strictly follow phloem orthostichies (discussed in van Bel and Gaupels 2004). Root-to-shoot signalling has been shown to trigger systemic defences, such as induced systemic resistance (ISR) (van Loon et al. 1998). This is also supported by the observation that the pattern of sucrose distribution over the leaves, revealing phloem mass flow, overlapped only partially that of the SAR induction (Kiefer and Slusarenko 2003). One main additional pathway is phloem-to-xylem transfer of signals, since it was shown that SA transported *via* the phloem is redistributed upward in small amounts *via* the xylem (Rocher et al. 2006), and MeJA moves both in the phloem and in the xylem (Thorpe et al. 2007). Xylem can also transport from root-to-shoot a large range of nutrients, metabolites and hormones acting potentially as signals. In addition, airborne signals, including MeSA, MeJA and green leaf volatiles, directly contribute to these defence mechanisms (Farmer 2001; Frost et al. 2007; Shah 2009).

2.3 Phloem Conductivity in Response to Injury and Aphid Feeding

2.3.1 Sieve Element Occlusion in Response to Injury

Strikingly, the properties of transport in the phloem can be altered in response to biotic or abiotic stresses. A key case of such changes is the dispersion of protein bodies, named forisomes, observed in the SEs of *Vicia faba* in response to wounding or heating, which was associated with a transitory arrest of mass flow in sieve tubes (Furch et al. 2007; Thorpe et al. 2010). A similar process has been observed in *Cucurbita maxima* after burning of the leaf tip and was proposed to result from the aggregation of proteins in the vicinity of sieve plates (Furch et al. 2010). These rapid, reversible processes depend on the generation of an EPW and on Ca^{2+} influx (van Bel et al. 2011a). Their downstream effects on defence signalling are not known.

2.3.2 Manipulation of Phloem by Aphid Feeding

Aphid infestation induces defence mechanisms whose effects are defeated by aphids. Indeed phloem-feeding insects express 'decoy' defences and suppress the JA-regulated defences that affect insect performance (Thompson and Goggin 2006; Walling 2008; Giordanengo et al. 2010). During feeding, aphids inject in sieve tubes a saliva that contains compounds preventing occlusion of sieve elements (Will and van Bel 2006; Will et al. 2007, 2009). Aphid feeding can also induce in the phloem a systemic response potentially modifying transport properties. In response to aphid infestation of celery by *Myzus persicae*, it was shown that the transport phloem responded by a systemic transcriptional reprogramming, leading to multiple adjustments, potentially impacting metabolic pathways as well as phloem transport (Divol et al. 2005). The expression of several genes acting on cell wall modifications and water uptake was affected, which could modify the conductivity of the phloem tissue. These changes were specifically regulated by aphid infestation, since viral or bacterial infections led to a different response (Divol et al. 2005).

3 Long-Distance Signalling in Response to Nutrient Deficiency

In response to fluctuations in nutrient concentration, plants generate local and systemic signals in order to communicate the nutrient status to the whole plant and trigger adaptive responses (Forde 2002a; Schachtman and Shin 2007; Giehl et al. 2009; Chiou and Lin 2011). Recent advances in phosphate and nitrate homeostasis are illustrative examples of signalling in response to nutrient availability.

3.1 Response to Phosphate Starvation

3.1.1 Coordinated Root-to-Shoot and Shoot-to-Root Signalling

Plant cells maintain inorganic phosphate (Pi) concentrations, despite large variations of Pi availability in the soil. The complex regulation of Pi homeostasis involves local signalling, long-distance transport through the xylem and phloem, transcriptional and post-transcriptional gene control and several types of non-coding regulatory RNAs (Chiou and Lin 2011). Pi is acquired in roots through phosphate transporters encoded by PHT1 genes (Forde 2002a; Mudge et al. 2002; Misson et al. 2004; Shin et al. 2004). Once Pi status is sensed, local and systemic signals are triggered. It has been proposed that systemic signals are transported in the xylem from roots to shoots, which in turn generate secondary long-range signals that move to roots *via* the phloem (Chiou and Lin 2011). Under Pi deprivation, primary root growth is arrested and the number and length of lateral roots increase, a response that depends on local signalling (Linkohr et al. 2002; Svistoonoff et al. 2007). In addition, Pi starvation induces changes in gene expression to facilitate Pi uptake, remobilization and recycling (Chiou and Lin 2011). Pi uptake is noticeably regulated by long-distance signals (Liu et al. 1998; Burleigh and Harrison 1999; Thibaud et al. 2010).

3.1.2 The Role of Hormones and Sucrose

Several plant hormones are involved in Pi starvation responses, but they seem to affect mainly local responses, rather than systemic signalling (Chiou and Lin 2011). However, the recently identified hormones strigolactones might play a role in long-distance communication. Up-regulation of strigolactones by Pi deficiency contributes to changes in shoot architecture (Yoneyama et al. 2007; López-Ráez et al. 2008; Umehara et al. 2010; Kohlen et al. 2011). Moreover, strigolactones have been detected in *Arabidopsis* xylem sap, suggesting that these hormones act as root-to-shoot signals involved in Pi starvation responses (Kohlen et al. 2011). Split-root experiments have indicated the existence of systemic suppressors of phosphate starvation-induced genes when Pi is available to one portion of the roots (Liu et al. 1998; Burleigh and Harrison 1999; Franco-Zorrilla et al. 2005). Pi itself has been proposed to act as such mobile signal. However, down-regulation of one of these genes occurs before internal Pi levels increase, and, in addition, a reduction in Pi flow does not affect this down-regulation, suggesting that the systemic signal is not Pi (Burleigh and Harrison 1999; Thibaud et al. 2010). Another putative long-distance signal is sucrose, as Pi starvation leads to increased levels of sugars in leaves, and these sugars are transported in the phloem to roots (Chiou and Lin 2011). Conclusive evidence on the role of sucrose as a systemic signal has been hindered, however, by the difficulty in separating its signalling from its metabolic role.

3.1.3 A New Actor in the Landscape: miR399, a Major Signal in Pi Homeostasis

Recent studies on the role of a microRNA (miRNA), miR399, in phosphate homeostasis have shed light on the identity of the phloem-transmissible signal(s). MiR399, which is induced by Pi deficiency, down-regulates the levels of its target transcript *PHO2*, encoding a ubiquitin-conjugating E2 enzyme required to prevent over-accumulation of Pi in shoots (Delhaize and Randall 1995; Fujii et al. 2005; Aung et al. 2006; Bari et al. 2006; Chiou et al. 2006). *PHO2* and miR399 are expressed in the vasculature and play a role in the systemic regulation of Pi uptake and translocation (Aung et al. 2006; Bari et al. 2006; Chiou et al. 2006). In addition, this miRNA has been detected in the phloem sap of two plant species (Pant et al. 2008). All this suggested that miR399 might act as a systemic Pi homeostasis signal. Indeed, shoot-to-root movement of mature miR399 has been demonstrated independently by two research groups, using grafting experiments in *Arabidopsis* and tobacco (Lin et al. 2008; Pant et al. 2008). MiR399-overexpressing (miR399-OX) scions caused a down-regulation of *PHO2* in rootstocks, and both miR399-OX/wild-type- and wild-type/miR399-OX-grafted plants showed increased Pi levels in scions, indicating biological activity of transported miR399 molecules (Lin et al. 2008; Pant et al. 2008). Although these results have been obtained using miR399-OX plants and, therefore, confirmation that the same mechanism operates in wild-type plants is still needed, they strongly argue for a role of miR399 as a phloem-mobile signal in Pi homeostasis. The existence of additional, miR399-independent systemic signals triggered by vacuolar Ca^{2+}/H^{+} transporters has been recently pointed out (Liu et al. 2011b). Identification of these long-distance molecules awaits further research.

3.2 Nitrate Homeostasis

3.2.1 Nitrate Uptake and Root Architecture and the N Status of the Plant

In addition to being a nutrient source, nitrate also functions as a signal molecule, regulating gene expression (Stitt 1999; Liu et al. 2009; Castaings et al. 2011; Krouk et al. 2010a). NO_3^- homeostasis at the plant level is controlled by sensing of exogenous NO_3-, but also by systemic N signalling. One example is the N regulation of root architecture. This response involves (1) a local response, implicating the nitrate transceptor NRT1.1 (Remans et al. 2006) and the transporter NRT2.1 (Filleur et al. 2001), both acting on NO_3^- uptake and signalling (Little et al. 2005; Remans et al. 2006); (2) a root-to-shoot signalling event, involving cytokinins (Takei et al. 2001, 2002; Rahayu et al. 2005); and (3) a shoot-to-root signalling of the N status regulating nitrate uptake and root branching. Experiments with split-root systems have clearly demonstrated the existence of systemic controls on the specific repression of root NO_3- uptake systems and root branching by high N status of the plant and provided strong evidence that the regulatory signals arise in the shoot (Forde 2002a).

3.2.2 NO_3- and Amino Acids as Signals

Many investigations attempted to identify the nature of the systemic shoot-to root signal molecule. NO_3^- per se might be a signal since it is transported long-distance through the phloem by the nitrate transporters NRT1.7 and NRT1.9 (Fan et al. 2009b; Wang and Tsay 2011), although there is little evidence of a long-range role as signal molecule rather than metabolite. Because nitrate is assimilated into amino acids, it was proposed that the increase in the pools of amino acids, such as Gln, Glu and Asn, may provide a systemic signal of the N status of the plant to regulate root response and repression of NO_3- uptake (Cooper and Clarkson 1989; Forde 2002b; Miller et al. 2008; Forde and Walch-Liu 2009). However, conflicting data have been reported and did not always support this hypothesis. In the *hni* mutants, a class of mutants affected in systemic shoot-to-root response, there was an inverse correlation between amino acids levels and repression of the nitrate transporter *NRT2.1*, suggesting that amino acids are not involved as systemic signals (Girin et al. 2010). Because amino acids are also an N source and can be metabolized, it is unclear whether their effect on NO_3- uptake results from a role as signalling molecules or from a role in overall N supply. Uptake of N is also tightly coordinated with C assimilation in shoots, supported by the observation that NRT2.1 and NRT1.1, as well as other inorganic nutrient transporters, are regulated by sugars (Lejay et al. 1999, 2003; Liu et al. 2009). In addition, an uncharacterized oxidative pentose phosphate pathway–dependent sugar-signalling pathway has been recently identified (Lejay et al. 2008). Hormone control was also proposed to participate in the shoot-to-root long-range signalling; auxin acts directly on root architecture, in coordination with nitrate signalling, and the recent demonstration that NRT1.1 transports not only nitrate but also auxin establishes a connection between nutrient and hormone signalling (Guo et al. 2002; Krouk et al. 2010b). However, the nature of the signal(s) acting in shoot-to-root signalling of N status is still unknown.

3.2.3 Transduction of N Signal and Roles for miR167, miR169 and miR393

Several signalling components triggered by N status have been identified, and include sensors, such as the transceptor NRT1.1, kinases (CIPK8), ubiquitin ligases (NLA) and transcriptional factors or regulators, such as NLP7, LBD37/38/39 and the master clock control gene CCA1, acting on the control of nitrogen assimilation genes (Peng et al. 2007; Gutiérrez et al. 2008; Castaings et al. 2009; Hu et al. 2009; Rubin et al. 2009). Furthermore, in roots, this transduction pathway interplays with auxin signalling that also affects nitrate nutrition (Krouk et al. 2011). Several miRNAs have been identified in this feedback control, including miR393, miR167 or miR169 (Gifford et al. 2008; Vidal et al. 2010; Zhao et al. 2011). Interestingly, two of these miRNAs, miR167 and miR169, were detected in the phloem sap of pumpkin or rapeseed, suggesting a role in phloem long-distance signalling (Yoo et al. 2004; Buhtz et al. 2008, 2010). These findings and the recent demonstration of the systemic role of miR399 in phosphate starvation (Pant et al. 2008; Lin et al. 2008) may indicate a general role of miRNAs in long-range signalling in response to nutrient starvation (Yoo et al. 2004; Kehr 2009).

4 Long-Distance Signalling in Developmental Programs

Plant tissues and organs develop from meristems, which are usually sheltered to prevent their damage. This protection entails a trade-off: Meristems cannot directly detect many environmental signals. However, external cues are perceived by different parts of the plant body, like leaves or roots. Therefore, communication among different tissues and organs is essential to achieve coordinated development. Examples of cell-to-cell communication, long-distance signalling through the phloem and xylem and secretion of regulatory molecules to modulate development have been described (Giakountis and Coupland 2008; Lehesranta et al. 2010; Sieburth and Lee 2010; Urbanus et al. 2010; Domagalska and Leyser 2011; Proust et al. 2011; Van Norman et al. 2011). This section focuses on developmental processes regulated by long-range signals *via* the phloem and the mobile molecules that have been identified so far, as well as others that might be involved.

4.1 Plant Reproduction: The Identification of a Florigen Component

4.1.1 The Mysterious Florigen

The existence of long-distance signals regulating flowering was demonstrated in the 1930s, on the basis of grafting experiments between plants induced and non-induced to flower. These findings led to the concept of floral stimulus or 'florigen', a transmissible substance that induces flowering in all higher plants (reviewed by Lang 1952). Experimental evidence suggested that leaf-generated inhibitors of flowering also exist, and later on it was proposed that the floral stimulus must have a complex composition, including several different molecules (Bernier 1988). The mobile signal, simple or complex, is produced in leaves and is transported in the phloem to the shoot apical meristem, where flowers develop (Bernier 1988). Many different molecules have been postulated as components of the florigen, including sucrose, gibberellins (GAs), cytokinins, other plant hormones, certain amino acids, proteins, mRNAs, small RNAs and SA (Bernier 1988; Corbesier and Coupland 2005; Suárez-López 2005). Diverse biochemical and physiological approaches, however, failed to demonstrate, during decades, that these molecules are systemic flowering signals, except perhaps for GAs in a grass species (Bernier 1988; King and Evans 2003; Corbesier and Coupland 2005; Suárez-López 2005).

4.1.2 Evidence on the Major Role of FLOWERING LOCUS T

Molecular genetics experiments, however, pinpointed a possible florigen component. The description of the expression pattern of several flowering-time genes, the use of tissue-specific promoters to express these genes in the phloem or in the shoot apical meristem and the exploitation of classical grafting techniques have been crucial for this advance. Two major players in the photoperiodic regulation of flowering, the transcriptional regulator CONSTANS (CO) and the small globular protein FLOWERING LOCUS T (FT), were shown to be expressed in leaf vascular tissues of *Arabidopsis* plants, suggesting their possible involvement in long-distance signalling (Takada and Goto 2003; An et al. 2004). In addition, expression of *CO* specifically in phloem companion cells or in the minor veins of mature leaves, but not in the shoot apical meristem, was sufficient to complement the late-flowering phenotype of *co* mutants and to induce *FT* expression in the phloem (An et al. 2004; Ayre and Turgeon 2004). In wild-type *Arabidopsis*, *FT* mRNA is expressed mainly in the leaves and absent, or present at extremely low levels, in the shoot apex (Kobayashi et al. 1999; Takada and Goto 2003; Abe et al. 2005; Wigge et al. 2005; Corbesier et al. 2007). However, FT acts in the shoot apex through its interaction with the bZIP transcription factor FD, which is preferentially expressed in the shoot apex of *Arabidopsis* and maize (Abe et al. 2005; Wigge et al. 2005; Muszynski et al. 2006). Furthermore, two FT-like proteins, one of them highly homologous to FT, were detected in the phloem sap of *Brassica napus* (Giavalisco et al. 2006). All these results pointed to FT as a good candidate for a florigen component. The fact that FT is a small protein also fitted with the hypothesis of FT being mobile.

Finally, 70 years after Chailakhyan coined the term 'florigen' (Chailakhyan 1936), evidence that a molecule acts as a long-distance flowering signal has been obtained. Monocot and dicot plant species, as well as long-day, short-day and day-neutral plants, have been shown to use FT as a florigenic molecule (Corbesier et al. 2007; Jaeger and Wigge 2007; Lin et al. 2007; Mathieu et al. 2007; Tamaki et al. 2007). The first indication that a product of the *FT* gene might be part of the floral stimulus was obtained in tomato by showing that plants overexpressing *SINGLE FLOWER TRUSS* (*SFT*, the tomato *FT* orthologue) grafted onto *sft* mutant stocks rescue the late-flowering phenotype of these mutants (Lifschitz et al. 2006). Movement of the *SFT* mRNA could not be detected, indicating that either the SFT protein or a downstream target moves to the shoot apical meristem to induce flowering. In addition, the results of Lifschitz et al. (2006) suggested that *SFT*-stimulated signals are conserved in different plants.

Then, several landmark papers, using diverse approaches including comparison of the localization of endogenous *FT* mRNA and engineered FT proteins fused either to reporter proteins or to small tags, tissue-specific expression and tissue-specific silencing of FT, expression of non-mobile versions of this protein and grafting experiments to test the transmission of the effects of FT on flowering time provided strong evidence that *Arabidopsis* FT and rice Hd3a—an orthologue of FT—proteins move in the phloem to the shoot apical meristem (Corbesier et al. 2007; Jaeger and Wigge 2007; Mathieu et al. 2007; Tamaki et al. 2007). Experimental support for translocation of FT in the phloem and transmission of its effect was also obtained in cucurbits using heterografts between two cucurbit species (Lin et al. 2007). In this work, two FT-like proteins were detected in the phloem sap of *Cucurbita maxima* (Lin et al. 2007). Although movement of endogenous FT proteins from the leaves to the shoot apex has not been demonstrated yet, all these findings strongly support that FT is a component of florigen.

4.1.3 A Model for the Mode of Action of FT in *Arabidopsis*

Based on the results described above, a model for the regulation of *Arabidopsis* flowering by long-distance signals has been proposed. Inductive photoperiodic conditions perceived in the leaf lead to stabilization of CO, which induces *FT* transcription in the leaf phloem (An et al. 2004; Valverde et al. 2004). Once translated in the phloem CCs, the FT protein enters the phloem stream and moves to the shoot apical meristem, where it interacts with FD to activate the expression of at least one floral meristem identity gene, *APETALA1* (*AP1*) (Abe et al. 2005; Wigge et al. 2005; Corbesier et al. 2007; Jaeger and Wigge 2007; Lin et al. 2007; Mathieu et al. 2007; Tamaki et al. 2007). In the shoot apical meristem, FT also up-regulates the expression of *SOC1*, which is another gene involved in flowering-time control (Turck et al. 2008).

4.1.4 Role of FT Homologues in Other Species

At least part of this mechanism of regulation is conserved, with some variations, in rice, tomato and cucurbits (Yano et al. 2000; Izawa et al. 2002; Lifschitz et al. 2006; Lin et al. 2007; Tamaki et al. 2007). In several other species, CO and/or FT homologues are also involved in the regulation of flowering time, and the interaction between FT and FD has also been shown or suggested (Pnueli et al. 2001; Li and Dubcovsky 2008; Turck et al. 2008). However, in rice, an FD homologue has not been identified yet (Tsuji et al. 2011). Interestingly, recent results suggest that rice has at least two florigen components: Hd3a, which promotes flowering under short days, and RFT1—another FT-like protein highly similar to Hd3a—that promotes flowering, much later, under long days (Tamaki et al. 2007; Komiya et al. 2009). Three members of the FT protein family, but, intriguingly, not Hd3a and RFT1, have been detected in the phloem sap of rice, suggesting that other proteins of this family might also be mobile (Aki et al. 2008). In pea, evidence suggesting that two FT genes are also involved in long-distance promotion of flowering has recently been obtained (Hecht et al. 2011).

4.2 *Other Components Acting on Flower and Tuber Induction*

4.2.1 Other Transcription Factors Involved in Flowering

In addition to FT, several *Arabidopsis* FT homologues are also involved in flowering. TWIN SISTER OF FT (TSF) affects flowering partially redundantly with FT, might also act as a long-distance signal and, as mentioned above, is present in the phloem of *B. napus* (Michaels et al. 2005; Yamaguchi et al. 2005; Giavalisco et al. 2006; Mathieu et al. 2007). TERMINAL FLOWER 1 (TFL1) and BROTHER OF FT AND TFL1 (BFT) repress flowering and play redundant roles in inflorescence meristem development (Shannon and Meeks-Wagner 1991; Yoo et al. 2010). The presence of the TFL1 protein in places of the shoot apical meristem where its mRNA is not present suggests that TFL1 moves cell-to-cell (Conti and Bradley 2007). These results emphasize the ability of the small FT-like proteins to move.

Some plants require exposure to low temperatures to flower (vernalization). The MADS-box transcription factor FLC plays an important role in the response to vernalization (Amasino 2010). FLC levels are high in late-flowering *Arabidopsis* plants that respond to vernalization (Michaels and Amasino 1999; Sheldon et al. 1999). FLC represses the production of systemic signals (FT) in the leaves and prevents the shoot apical meristem from being competent to respond to these signals, by repressing FD and SOC1 expression, until plants are vernalized (Searle et al. 2006). When plants are vernalized, *FLC* mRNA and protein levels are reduced and flowering can occur (Michaels and Amasino 1999; Sheldon et al. 1999, 2000; Searle et al. 2006).

4.2.2 Role of Hormones in Systemic Induction of Flowering?

Several plant hormones affect the induction of flowering. Among them, GAs seem the most likely to act as mobile flowering signals. They affect flowering in many plants and can be transported in the phloem and xylem sap (Bernier 1988; Davis 2009; Mutasa-Göttgens and Hedden 2009). In the grass *Lolium temulentum*, when flowering is induced, the bioactive GAs, GA_5 and GA_6 increase at the shoot apex shortly after an increase of their GA_{20} precursor in leaves. Moreover, when labelled GA_5 is exogenously applied, it is transported to the shoot apex (King et al. 2001; King and Evans 2003). This suggests that GA_5, and perhaps GA_6, might act as florigenic molecules. It has been proposed that selective degradation of certain GAs just below the shoot apex restricts their access to the shoot apical meristem, but GA_5 is protected from this degradation, allowing this GA to reach the shoot apex and induce flowering (King et al. 2008). In *Arabidopsis*, levels of GA_4 and sucrose increase in the shoot apex before floral initiation under short days. These increases probably result from transport of GA_4 and sucrose produced outside the shoot apex (Eriksson et al. 2006). These results suggest that the florigenic GAs might be different in different species.

Cytokinins are also considered putative florigen components, and the results supporting this view have recently been reviewed (Bernier 2011). However, a recent report shows that cytokinins promote flowering and induce transcription of TSF in *Arabidopsis* leaves, suggesting that cytokinins might act upstream of long-distance signals (D'Aloia et al. 2011). Nevertheless, previous results indicated a direct effect of cytokinins at the shoot apical meristem, and therefore further research is required to show whether cytokinins act as mobile signals or not.

4.2.3 A Role for Sucrose

In addition to increasing at the shoot apex just before floral initiation, sucrose also increases rapidly in leaf phloem exudates (Corbesier and Coupland 2005; Eriksson et al. 2006). Mutants affected in starch synthesis or mobilization exhibit altered flowering times (Corbesier and Coupland 2005). Under certain conditions, sucrose can complement the late-flowering phenotypes of several mutants, including *co*, but not that of *ft* (Roldán et al. 1999; Ohto et al. 2001). Altogether, these observations suggest a long-range signalling role for sucrose, which would act downstream of *CO* and upstream or in parallel with *FT*. In addition, complex interactions between sucrose, cytokinin and GA signalling have been proposed (Périlleux and Bernier 2002; Suárez-López 2005). Given that GAs, sucrose and cytokinins affect many aspects of plant growth and development, it is difficult to demonstrate whether their effects on the systemic regulation of flowering are direct or indirect.

4.2.4 The Roles of miR172 and BEL5 in Tuber Formation

Tuber formation is a mode of vegetative reproduction regulated, like flowering, by long-distance signals generated in the leaves (Abelenda et al. 2011). The photoreceptor phytochrome B (PHYB) represses tuberization in potato, whereas the homeobox transcription factor StBEL5 and the microRNA 172 (miR172) promote tuber formation (Jackson et al. 1996; Chen et al. 2003; Banerjee et al. 2006; Martin et al. 2009). Movement of *StBEL5* mRNA through grafts correlates with tuber induction (Banerjee et al. 2006). Plants with reduced levels of PHYB tuberize earlier than wild-type plants and show reduced abundance of *StBEL5* transcript in leaves and increased abundance in stolons at early stages of tuber development, suggesting that PHYB might regulate *StBEL5* mRNA movement (Jackson et al. 1996; Martin et al. 2009).

Interestingly, PHYB affects miR172 levels in a similar way as it affects *StBEL5* mRNA. This, together with the presence of miR172 in vascular bundles and the transmission of its effect on tuberization through grafts, has led to the hypothesis that miR172 might be a long-distance signalling molecule or might regulate mobile signals (Martin et al. 2009). The role of miR172 in flowering-time control in several species and its detection in the phloem sap of *Brassica napus* are consistent with this hypothesis (Buhtz et al. 2008; Zhu and Helliwell 2011). Alternatively, it has been speculated that miR172 might function as a cell-to-cell signal mediating the effect of PHYB from the mesophyll on the expression of FT in the phloem in *Arabidopsis* (Abelenda et al. 2011).

A sucrose transporter, StSUT4, affects tuber induction, suggesting that sucrose plays a role in the systemic regulation of this process (Chincinska et al. 2008), but since sucrose is required to form starch, a major component of tubers, it is difficult to distinguish a metabolic from a signalling role of sucrose. Recently, it has been proposed that a potato FT homologue might be a mobile signal for tuberization, although results supporting this hypothesis have not been reported yet (Abelenda et al. 2011).

4.2.5 Still Unanswered Questions

Despite the impressive knowledge recently acquired on the systemic regulation of flowering, many questions are still unanswered. FT does not seem to be 'the' florigen, but a major florigen component, as other FT-like proteins also act as mobile flowering signals. It remains to be shown whether other types of molecules might play a similar role together with, or alternatively to, FT and its homologues. For example, GAs seem to play a systemic role in *L. temulentum*, but recent evidence suggests that FT might also affect flowering in this plant (King et al. 2006; Skøt et al. 2011).

In beet, two FT homologues play opposite roles in floral induction, but no evidence that any of the two is mobile has been reported so far (Pin et al. 2010). The CO/FT module also regulates seasonal growth cessation in trees (Böhlenius et al. 2006), but whether transport of FT is required for this process has not been tested yet. In order to understand fully the long-distance signalling process, it will be necessary to understand how FT is loaded into the phloem, transported and unloaded in target tissues, as well as the mechanisms that control the response of these tissues to the mobile signal. Part of this response is mediated by FD in the shoot apex, but FT promotes flowering both through FD-dependent and independent pathways, suggesting that additional genes are involved (Wigge et al. 2005).

4.3 *Vegetative Development and Morphogenesis*

4.3.1 Role of Long-Distance Transport of RNAs in Morphogenesis

Although flowering is a paradigm of systemic signalling in the field of plant development, other developmental events also involve long-distance signals. Leaf development was shown to be affected by a graft-transmissible RNA in tomato (Kim et al. 2001). mRNAs encoding other developmental regulators, as well as small RNAs that down-regulate the expression of developmental genes, have been detected in phloem sap, and some of them are transmissible through graft junctions, suggesting that RNAs can also act as long-distance signals for the control of plant development (reviewed in Lough and Lucas 2006). However, further demonstrations that RNAs act as mobile signals for developmental regulation have to be obtained (Kehr 2009; Turgeon and Wolf 2009).

4.3.2 Role of FT as a General Regulator of Plant Development

In addition to their role in flowering, FT proteins are involved in other developmental events. In tomato, SFT affects leaf development and maturation, stem growth and the formation of abscission zones through long-distance signalling (Shalit et al. 2009). Ectopic expression of rice *Hd3a* in vascular tissues, as well as overexpression of *Arabidopsis FT* or *FD*, affects vegetative traits, such as internode elongation or leaf development (Teper-Bamnolker and Samach 2005; Wigge et al. 2005; Tamaki et al. 2007). Taken together, these observations indicate roles of FT proteins beyond flowering and further point out FT as a general systemic regulator of plant development.

4.3.3 Role of PHYB and SPA1 in Response to Light Perception

At least two genes involved in light perception and signalling, *PHYB* and *SUPPRESSOR OF PHYA-105* (*SPA1*), regulate the production of FT through their effect on the stability of the CO protein (Valverde et al. 2004; Laubinger et al. 2006). PHYB influences other developmental events involving intercellular and inter-organ communication, as well as long-distance signalling for other processes, including tuberization and plant disease resistance (Jackson et al. 1998; Bou-Torrent et al. 2008; Griebel and Zeier 2008). SPA1 is required in the phloem to control not only flowering time but also seedling photomorphogenesis and leaf expansion, but SPA1 itself is not mobile, indicating that SPA1 affects non-cell-autonomous regulators of these processes (Ranjan et al. 2011). Identification of the mobile molecules acting downstream of PHYB and SPA1 to control vegetative development awaits further investigations.

4.3.4 Other Potential Long-Distance Signal Molecules Acting on Development

Long-range signalling is also involved in vascular development. In *Arabidopsis*, xylem expansion associated with hypocotyl and root secondary growth is promoted after floral induction and requires graft-transmissible signals (Sibout et al. 2008). Interestingly, low levels of FLC, a flowering-time regulator, correlate with xylem expansion (Sibout et al. 2008). Recent results suggest that GAs might be the mobile signal (Ragni et al. 2011), but the identity of this signal has not been proven yet. Given that FLC represses FT, it would be interesting to test whether FT plays a role in this process. Other hormones are also candidates for mobile signals regulating vegetative development. Auxin is known to affect many developmental events, and although there is evidence of transport of auxin in the phloem, the best-studied mechanisms of auxin transport do not involve this vascular conduit (Lehesranta et al. 2010; Peer et al. 2011).

5 Concluding Remarks

Long-distance signalling *via* the phloem has been shown during the past decade to recruit a variety of signal molecules, including hormones, peptides, macromolecules, nutrients and metabolites. These signals are involved in many developmental and adaptive processes. At least some phloem-mobile signals may be common to several processes, as is the case for FT. However, in most cases, the nature of these signals is still elusive, and the conclusive demonstration of a signalling role for candidate signal(s) is often controversial. In contrast, it is now well established that chemical signals act together with electrical signals acting faster in a long-distance range. Further, the concept of molecular 'hopping' for relay and amplification of signal molecules in the transport phloem opens up new avenues to address the mechanism of long-distance signalling in higher plants and needs to be further tested in the future.

A main issue in a near future will be to determine the molecular mechanisms coordinating the action of multiple signalling pathways acting in the phloem tissue. Are they based on crosstalks during signal transduction or are they regulated at a gene-network level, as proposed recently in the context of integration of hormone signalling (Jaillais and Chory 2011)? This points out that we need to improve our knowledge on gene expression networks acting in the phloem, which are still poorly characterized (Vilaine et al. 2003; Le Hir et al. 2008), and on the subsets of macromolecules, proteins, mRNAs and miRNAs, loaded into the sieve elements, translocated long-distance and acting non-cell autonomously. Another major exciting issue is the identification of the factors required for transport of macromolecules in the translocation stream. The recent discovery of the formation of large ribonucleoprotein complexes in the phloem sap suggests indeed that this process is highly regulated and might be involved in the specific transport of selected molecules (Ham et al. 2009; Ma et al. 2010; Li et al. 2011). Understanding how the mobile signals leave the phloem to reach their target tissues also requires further research.

References

Abe M, Kobayashi Y, Yamamoto S, Daimon Y, Yamaguchi A, Ikeda Y, Ichinoki H, Notaguchi M, Goto K, Araki T (2005) FD, a bZIP protein mediating signals from the floral pathway integrator FT at the shoot apex. Science 309:1052–1056

Abelenda JA, Navarro C, Prat S (2011) From the model to the crop: genes controlling tuber formation in potato. Curr Opin Biotechnol 22:287–292

Aki T, Shigyo M, Nakano R, Yoneyama T, Yanagisawa S (2008) Nano scale proteomics revealed the presence of regulatory proteins including three FT-like proteins in phloem and xylem saps from rice. Plant Cell Physiol 49:767–790

Amasino R (2010) Seasonal and developmental timing of flowering. Plant J 61:1001–1013

An H, Roussot C, Suárez-López P, Corbesier L, Vincent C, Piñeiro M, Hepworth S, Mouradov A, Justin S, Turnbull C, Coupland G (2004) CONSTANS acts in the phloem to regulate a systemic signal that induces photoperiodic flowering of *Arabidopsis*. Development 131:3615–3626

Attaran E, Zeier TE, Griebel T, Zeier J (2009) Methyl salicylate production and jasmonate signaling are not essential for systemic acquired resistance in *Arabidopsis*. Plant Cell 21:954–971

Aung K, Lin S-I, Wu C-C, Huang Y-T, C-l Su, Chiou T-J (2006) *pho2*, a phosphate overaccumulator, is caused by a nonsense mutation in a microRNA399 target gene. Plant Physiol 141:1000–1011

Ayre BG, Turgeon R (2004) Graft transmission of a floral stimulant derived from *CONSTANS*. Plant Physiol 135:2271–2278

Ayre BG, Keller F, Turgeon R (2003) Symplastic continuity between companion cells and the translocation stream: long-distance transport is controlled by retention and retrieval mechanisms in the phloem. Plant Physiol 131:1518–1528

Banerjee AK, Chatterjee M, Yu Y, Suh S-G, Miller WA, Hannapel DJ (2006) Dynamics of a mobile RNA of potato involved in a long-distance signaling pathway. Plant Cell 18:3443–3457

Bari R, Datt Pant B, Stitt M, Scheible WR (2006) PHO2, microRNA399, and PHR1 define a phosphate-signaling pathway in plants. Plant Physiol 141:988–999
Bernier G (1988) The control of floral evocation and morphogenesis. Ann Rev Plant Physiol Plant Mol Biol 39:175–219
Bernier G. My favourite flowering image: the role of cytokinin as a flowering signal. J Exp Bot, in press
Birkett MA, Campbell CAM, Chamberlain K, Guerrieri E, Hick AJ, Martin JL, Matthes M, Napier JA, Pettersson J, Pickett JA, Poppy GM, Pow EM, Pye BJ, Smart LE, Wadhams GH, Wadhams LJ, Woodcock CM (2000) New roles for cis-jasmone as an insect semiochemical and in plant defense. Proc Natl Acad Sci USA 97:9329–9334
Böhlenius H, Huang T, Charbonnel-Campaa L, Brunner AM, Jansson S, Strauss SH, Nilsson O (2006) CO/FT regulatory module controls timing of flowering and seasonal growth cessation in trees. Science 312:1040–1043
Bou-Torrent J, Roig-Villanova I, Martínez-García JF (2008) Light signaling: back to space. Trends Plant Sci 13:108–114
Bruce TJA, Pickett JA (2007) Plant defence signalling induced by biotic attacks. Curr Opin Plant Biol 10:387–392
Bruce TJA, Matthes MC, Chamberlain K, Woodcock CM, Mohib A, Webster B, Smart LE, Birkett MA, Pickett JA, Napier JA (2008) cis-jasmone induces *Arabidopsis* genes that affect the chemical ecology of multitrophic interactions with aphids and their parasitoids. Proc Natl Acad Sci USA 105:4553–4558
Buhtz A, Springer F, Chappell L, Baulcombe DC, Kehr J (2008) Identification and characterization of small RNAs from the phloem of *Brassica napus*. Plant J 53:739–749
Buhtz A, Pieritz J, Springer F, Kehr J (2010) Phloem small RNAs, nutrient stress responses, and systemic mobility. BMC Plant Biol 10:64
Burleigh SH, Harrison MJ (1999) The down-regulation of *Mt4*-like genes by phosphate fertilization occurs systemically and involves phosphate translocation to the shoots. Plant Physiol 119:241–248
Callos JD, Medford JI (1994) Organ positions and pattern formation in the shoot apex. Plant J 6:1–7
Castaings L, Camargo A, Pocholle D, Gaudon V, Texier Y, Boutet-Mercey S, Taconnat L, Renou J-P, Daniel-Vedele F, Fernandez E, Meyer C, Krapp A (2009) The nodule inception-like protein 7 modulates nitrate sensing and metabolism in Arabidopsis. Plant J 57:426–435
Castaings L, Marchive C, Meyer C, Krapp A (2011) Nitrogen signalling in *Arabidopsis*: how to obtain insights into a complex signalling network. J Exp Bot 62:1391–1397
Chailakhyan MK (1936) New facts in support of the hormonal theory of plant development. Dokl Biol Sci 13:79–83
Chaturvedi R, Krothapalli K, Makandar R, Nandi A, Sparks AA, Roth MR, Welti R, Shah J (2008) Plastid ω3-fatty acid desaturase-dependent accumulation of a systemic acquired resistance inducing activity in petiole exudates of *Arabidopsis thaliana* is independent of jasmonic acid. Plant J 54:106–117
Chen H, Rosin FM, Prat S, Hannapel DJ (2003) Interacting transcription factors from the three-amino acid loop extension superclass regulate tuber formation. Plant Physiol 132:1391–1404
Chen Z, Agnew JL, Cohen JD, He P, Shan L, Sheen J, Kunkel BN (2007) *Pseudomonas syringae* type III effector AvrRpt2 alters *Arabidopsis thaliana* auxin physiology. Proc Natl Acad Sci USA 104:20131–20136
Chincinska IA, Liesche J, Krügel U, Michalska J, Geigenberger P, Grimm B, Kühn C (2008) Sucrose transporter StSUT4 from potato affects flowering, tuberization, and shade avoidance response. Plant Physiol 146:515–528
Chiou T-J, Lin S-I (2011) Signaling network in sensing phosphate availability in plants. Annu Rev Plant Biol 62:185–206
Chiou TJ, Aung K, Lin SI, Wu CC, Chiang SF, Su CL (2006) Regulation of phosphate homeostasis by microRNA in Arabidopsis. Plant Cell 18:412–421

Conti L, Bradley D (2007) TERMINAL FLOWER1 is a mobile signal controlling *Arabidopsis* architecture. Plant Cell 19:767–778

Cooper HD, Clarkson DT (1989) Cycling of amino-nitrogen and other nutrients between shoots and roots in cereals—a possible mechanism integrating shoot and root in the regulation of nutrient uptake. J Exp Bot 40:753–762

Corbesier L, Coupland G (2005) Photoperiodic flowering of *Arabidopsis*: integrating genetic and physiological approaches to characterization of the floral stimulus. Plant Cell Environ 28:54–66

Corbesier L, Vincent C, Jang S, Fornara F, Fan Q, Searle I, Giakountis A, Farrona S, Gissot L, Turnbull C, Coupland G (2007) FT protein movement contributes to long-distance signaling in floral induction of *Arabidopsis*. Science 316:1030–1033

Crawford NM, Guo F-Q (2005) New insights into nitric oxide metabolism and regulatory functions. Trends Plant Sci 10:195–200

Cui J, Bahrami AK, Pringle EG, Hernandez-Guzman G, Bender CL, Pierce NE, Ausubel FM (2005) *Pseudomonas syringae* manipulates systemic plant defenses against pathogens and herbivores. Proc Natl Acad Sci USA 102:1791–1796

D'Aloia M, Bonhomme D, Bouché F, Tamseddak K, Ormenese S, Torti S, Coupland G, Périlleux C (2011) Cytokinin promotes flowering of Arabidopsis *via* transcriptional activation of the *FT* paralogue *TSF*. Plant J 65:972–979

Davis SJ (2009) Integrating hormones into the floral-transition pathway of *Arabidopsis thaliana*. Plant Cell Environ 32:1201–1210

De Torres ZM, Bennett MH, Truman WH, Grant MR (2009) Antagonism between salicylic and abscisic acid reflects early host–pathogen conflict and moulds plant defence responses. Plant J 59:375–386

Delhaize E, Randall PJ (1995) Characterization of a phosphate-accumulator mutant of *Arabidopsis thaliana*. Plant Physiol 107:207–213

Dinant S, Lemoine R (2010) The phloem pathway: new issues and old debates. C R Biol 333:307–319

Dinant S, Bonnemain J, Girousse C, Kehr J (2010) Phloem sap intricacy and interplay with aphid feeding. C R Biol 333:504–515

Ding X, Cao Y, Huang L, Zhao J, Xu C, Li X, Wang S (2008) Activation of the indole-3-acetic acid–amido synthetase GH3-8 suppresses expansin expression and promotes salicylate- and jasmonate-independent basal immunity in rice. Plant Cell 20:228–240

Divol F, Vilaine F, Thibivilliers S, Amselem J, Palauqui JC, Kusiak C, Dinant S (2005) Systemic response to aphid infestation by *myzus persicae* in the phloem of *Apium graveolens*. Plant Mol Biol 57:517–540

Domagalska MA, Leyser O (2011) Signal integration in the control of shoot branching. Nat Rev Mol Cell Biol 12:211–221

Durner J, Klessig DF (1999) Nitric oxide as a signal in plants. Curr Opin Plant Biol 2:369–374

Durrant WE, Dong X (2004) Systemic acquired resistance. Annu Rev Phytopathol 42:185–209

Eriksson S, Böhlenius H, Moritz T, Nilsson O (2006) GA_4 is the active gibberellin in the regulation of *LEAFY* transcription and *Arabidopsis* floral initiation. Plant Cell 18:2172–2181

Fan J, Hill L, Crooks C, Doerner P, Lamb C (2009a) Abscisic acid has a key role in modulating diverse plant–pathogen interactions. Plant Physiol 150:1750–1761

Fan S-C, Lin C-S, Hsu P-K, Lin S-H, Tsay Y-F (2009b) The *Arabidopsis* nitrate transporter NRT1.7, expressed in phloem, is responsible for source-to-sink remobilization of nitrate. Plant Cell 21:2750–2761

Farmer EE (2001) Surface-to-air signals. Nature 411:854–856

Filleur S, Dorbe M-F, Cerezo M, Orsel M, Granier F, Gojon A, Daniel-Vedele F (2001) An *Arabidopsis* T-DNA mutant affected in *Nrt2* genes is impaired in nitrate uptake. FEBS Lett 489:220–224

Forde BG (2002a) Local and long-range signaling pathways regulating plant responses to nitrate. Annu Rev Plant Biol 53:203–224

Forde BG (2002b) The role of long-distance signalling in plant responses to nitrate and other nutrients. J Exp Bot 53:39–43

Forde BG, Walch-Liu PIA (2009) Nitrate and glutamate as environmental cues for behavioural responses in plant roots. Plant Cell Environ 32:682–693

Forouhar F, Yang Y, Kumar D, Chen Y, Fridman E, Park SW, Chiang Y, Acton TB, Montelione GT, Pichersky E, Klessig DF, Tong L (2005) Structural and biochemical studies identify tobacco SABP2 as a methyl salicylate esterase and implicate it in plant innate immunity. Proc Natl Acad Sci USA 102:1773–1778

Franco-Zorrilla JM, Martin AC, Leyva A, Paz-Ares J (2005) Interaction between phosphate-starvation, sugar, and cytokinin signaling in Arabidopsis and the roles of cytokinin receptors CRE1/AHK4 and AHK3. Plant Physiol 138:847–857

Fromm J, Lautner S (2007) Electrical signals and their physiological significance in plants. Plant Cell Environ 30:249–257

Frost CJ, Appel HM, Carlson JE, De Moraes CM, Mescher MC, Schultz JC (2007) Within-plant signalling *via* volatiles overcomes vascular constraints on systemic signalling and primes responses against herbivores. Ecol Lett 10:490–498

Fujii H, Chiou TJ, Lin SI, Aung K, Zhu JK (2005) A miRNA involved in phosphate-starvation response in *Arabidopsis*. Curr Biol 15:2038–2043

Furch AC, Hafke JB, Schulz A, van Bel AJ (2007) Ca^{2+}-Mediated remote control of reversible sieve tube occlusion in *Vicia faba*. J Exp Bot 58:2827–2838

Furch AC, van Bel AJ, Fricker MD, Felle HH, Fuchs M, Hafke JB (2009) Sieve element Ca^{2+} channels as relay stations between remote stimuli and sieve tube occlusion in *Vicia faba*. Plant Cell 21:2118–2132

Furch ACU, Zimmermann MR, Will T, Hafke JB, van Bel AJE (2010) Remote-controlled stop of phloem mass flow by biphasic occlusion in *Cucurbita maxima*. J Exp Bot 61:3697–3708

Gaupels F, Furch AC, Will T, Mur LA, Kogel KH, van Bel AJ (2008) Nitric oxide generation in *Vicia faba* phloem cells reveals them to be sensitive detectors as well as possible systemic transducers of stress signals. New Phytol 178:634–646

Giakountis A, Coupland G (2008) Phloem transport of flowering signals. Curr Opin Plant Biol 11:687–694

Giavalisco P, Kapitza K, Kolasa A, Buhtz A, Kehr J (2006) Towards the proteome of *Brassica napus* phloem sap. Proteomics 6:896–909

Giehl RFH, Meda AR, von Wirén N (2009) Moving up, down, and everywhere: signaling of micronutrients in plants. Curr Opin Plant Biol 12:320–327

Gifford ML, Dean A, Gutierrez RA, Coruzzi GM, Birnbaum KD (2008) Cell-specific nitrogen responses mediate developmental plasticity. Proc Natl Acad Sci USA 105:803–808

Giordanengo P, Brunissen L, Rusterucci C, Vincent C, van Bel A, Dinant S, Girousse C, Faucher M, Bonnemain JL (2010) Compatible plant-aphid interactions: how aphids manipulate plant responses. C R Biol 333:516–523

Girin T, El-Kafafi E-S, Widiez T, Erban A, Hubberten H-M, Kopka J, Hoefgen R, Gojon A, Lepetit M (2010) Identification of Arabidopsis mutants impaired in the systemic regulation of root nitrate uptake by the nitrogen status of the plant. Plant Physiol 153:1250–1260

Grant M, Lamb C (2006) Systemic immunity. Curr Opin Plant Biol 9:414–420

Griebel T, Zeier J (2008) Light regulation and daytime dependency of inducible plant defenses in Arabidopsis: phytochrome signaling controls systemic acquired resistance rather than local defense. Plant Physiol 147:790–801

Guo FQ, Wang R, Crawford NM (2002) The *Arabidopsis* dual-affinity nitrate transporter gene *AtNRT1.1* (*CHL1*) is regulated by auxin in both shoots and roots. J Exp Bot 53:835–844

Gutiérrez RA, Stokes TL, Thum K, Xu X, Obertello M, Katari MS, Tanurdzic M, Dean A, Nero DC, McClung CR, Coruzzi GM (2008) Systems approach identifies an organic nitrogen-responsive gene network that is regulated by the master clock control gene *CCA1*. Proc Natl Acad Sci USA 105:4939–4944

Hafke JB, van Amerongen JK, Kelling F, Furch AC, Gaupels F, van Bel AJ (2005) Thermodynamic battle for photosynthate acquisition between sieve tubes and adjoining parenchyma in transport phloem. Plant Physiol 138:1527–1537

Ham BK, Brandom JL, Xoconostle-Cazares B, Ringgold V, Lough TJ, Lucas WJ (2009) A polypyrimidine tract binding protein, pumpkin RBP50, forms the basis of a phloem-mobile ribonucleoprotein complex. Plant Cell 21:197–215

Hause B, Stenzel I, Miersch O, Maucher H, Kramell R, Ziegler J, Wasternack C (2000) Tissue-specific oxylipin signature of tomato flowers: allene oxide cyclase is highly expressed in distinct flower organs and vascular bundles. Plant J 24:113–126

Hause B, Hause G, Kutter C, Miersch O, Wasternack C (2003) Enzymes of jasmonate biosynthesis occur in tomato sieve elements. Plant Cell Physiol 44:643–648

Hecht V, Laurie RE, Vander Schoor JK, Ridge S, Knowles CL, Liew LC, Sussmilch FC, Murfet IC, Macknight RC, Weller JL (2011) The pea *GIGAS* gene is a *FLOWERING LOCUS T* homolog necessary for graft-transmissible specification of flowering but not for responsiveness to photoperiod. Plant Cell 23:147–161

Heil M, Silva Bueno JC (2007) Within-plant signaling by volatiles leads to induction and priming of an indirect plant defense in nature. Proc Natl Acad Sci USA 104:5467–5472

Heil M, Ton J (2008) Long-distance signalling in plant defence. Trends Plant Sci 13:264–272

Hind SR, Malinowski R, Yalamanchili R, Stratmann JW (2010) Tissue-type specific systemin perception and the elusive systemin receptor. Plant Signal Behav 5:42–44

Hoad GV (1995) Transport of hormones in the phloem of higher plants. Plant Growth Regul 16:173–182

Hu H-C, Wang Y-Y, Tsay Y-F (2009) AtCIPK8, a CBL-interacting protein kinase, regulates the low-affinity phase of the primary nitrate response. Plant J 57:264–278

Ivashikina N, Deeken R, Ache P, Kranz E, Pommerrenig B, Sauer N, Hedrich R (2003) Isolation of *AtSUC2* promoter-*GFP*-marked companion cells for patch-clamp studies and expression profiling. Plant J 36:931–945

Izawa T, Oikawa T, Sugiyama N, Tanisaka T, Yano M, Shimamoto K (2002) Phytochrome mediates the external light signal to repress *FT* orthologs in photoperiodic flowering of rice. Genes Dev 16:2006–2020

Jackson SD, Heyer A, Dietze J, Prat S (1996) Phytochrome B mediates the photoperiodic control of tuber formation in potato. Plant J 9:159–166

Jackson SD, James P, Prat S, Thomas B (1998) Phytochrome B affects the levels of a graft-transmissible signal involved in tuberization. Plant Physiol 117:29–32

Jaeger KE, Wigge PA (2007) FT protein acts as a long-range signal in *Arabidopsis*. Curr Biol 17:1050–1054

Jaillais Y, Chory J (2011) Unraveling the paradoxes of plant hormone signaling integration. Nat Struct Mol Biol 17:642–645

Jones JDG, Dangl JL (2006) The plant immune system. Nature 444:323–329

Jung HW, Tschaplinski TJ, Wang L, Glazebrook J, Greenberg JT (2009) Priming in systemic plant immunity. Science 324:89–91

Kachroo P, Shanklin J, Shah J, Whittle EJ, Klessig DF (2001) A fatty acid desaturase modulates the activation of defense signaling pathways in plants. Proc Natl Acad Sci USA 98:9448–9453

Kachroo A, Venugopal SC, Lapchyk L, Falcone D, Hildebrand D, Kachroo P (2004) Oleic acid levels regulated by glycerolipid metabolism modulate defense gene expression in *Arabidopsis*. Proc Natl Acad Sci USA 101:5152–5157

Kazan K, Manners JM (2008) Jasmonate signaling: toward an integrated view. Plant Physiol 146:1459–1468

Kehr J (2009) Long-distance transport of macromolecules through the phloem. F1000. Biol Rep 1:131

Kempers R, van Bel AJE (1997) Symplasmic connections between sieve element and companion cell in the stem phloem of *Vicia faba* L. have a molecular exclusion limit of at least 10 kDa. Planta 201:195–201

Kiefer IW, Slusarenko AJ (2003) The pattern of systemic acquired resistance induction within the Arabidopsis rosette in relation to the pattern of translocation. Plant Physiol 132:840–847

Kim M, Canio W, Kessler S, Sinha N (2001) Developmental changes due to long-distance movement of a homeobox fusion transcript in tomato. Science 293:287–289

King RW, Evans LT (2003) Gibberellins and flowering of grasses and cereals: prizing open the lid of the "florigen" black box. Annu Rev Plant Biol 54:307–328

King RW, Moritz T, Evans LT, Junttila O, Herlt AJ (2001) Long-day induction of flowering in *Lolium temulentum* involves sequential increases in specific gibberellins at the shoot apex. Plant Physiol 127:624–632

King RW, Moritz T, Evans LT, Martin J, Andersen CH, Blundell C, Kardailsky I, Chandler PM (2006) Regulation of flowering in the long-day grass *Lolium temulentum* by gibberellins and the *FLOWERING LOCUS T* gene. Plant Physiol 141:498–507

King RW, Mander LN, Asp T, MacMillan CP, Blundell CA, Evans LT (2008) Selective deactivation of gibberellins below the shoot apex is critical to flowering but not to stem elongation of Lolium. Mol Plant 1:295–307

Knoblauch M, Peters WS (2010) Münch, morphology, microfluidics—our structural problem with the phloem. Plant Cell Environ 33:1439–1452

Kobayashi Y, Kaya H, Goto K, Iwabuchi M, Araki T (1999) A pair of related genes with antagonistic roles in mediating flowering signals. Science 286:1960–1962

Kohlen W, Charnikhova T, Liu Q, Bours R, Domagalska MA, Beguerie S, Verstappen F, Leyser O, Bouwmeester H, Ruyter-Spira C (2011) Strigolactones are transported through the xylem and play a key role in shoot architectural response to phosphate deficiency in nonarbuscular mycorrhizal host Arabidopsis. Plant Physiol 155:974–987

Komiya R, Yokoi S, Shimamoto K (2009) A gene network for long-day flowering activates *RFT1* encoding a mobile flowering signal in rice. Development 136:3443–3450

Koo AJK, Howe GA (2009) The wound hormone jasmonate. Phytochemistry 70:1571–1580

Koo AJK, Gao X, Daniel Jones A, Howe GA (2009) A rapid wound signal activates the systemic synthesis of bioactive jasmonates in Arabidopsis. Plant J 59:974–986

Krouk G, Crawford NM, Coruzzi GM, Tsay Y-F (2010a) Nitrate signaling: adaptation to fluctuating environments. Curr Opin Plant Biol 13:265–272

Krouk G, Lacombe B, Bielach A, Perrine-Walker F, Malinska K, Mounier E, Hoyerova K, Tillard P, Leon S, Ljung K, Zazimalova E, Benkova E, Nacry P, Gojon A (2010b) Nitrate-regulated auxin transport by *NRT1.1* defines a mechanism for nutrient sensing in plants. Dev Cell 18:927–937

Krouk G, Ruffel S, Gutiérrez RA, Gojon A, Crawford NM, Coruzzi GM, Lacombe B (2011) A framework integrating plant growth with hormones and nutrients. Trends Plant Sci 16:178–182

Kumar D, Klessig DF (2003) High-affinity salicylic acid-binding protein 2 is required for plant innate immunity and has salicylic acid-stimulated lipase activity. Proc Natl Acad Sci USA 100:16101–16106

Lalonde S, Tegeder M, Throne-Holst M, Frommer WB, Patrick JW (2003) Phloem loading and unloading of sugars and amino acids. Plant Cell Environ 26:37–56

Lang A (1952) Physiology of flowering. Ann Rev Plant Physiol 3:265–306

Laubinger S, Marchal V, Le Gourrierec J, Wenkel S, Adrian J, Jang S, Kulajta C, Braun H, Coupland G, Hoecker U (2006) *Arabidopsis* SPA proteins regulate photoperiodic flowering and interact with the floral inducer CONSTANS to regulate its stability. Development 133:3213–3222

Le Hir R, Beneteau J, Bellini C, Vilaine F, Dinant S (2008) Gene expression profiling: keys for investigating phloem functions. Trends Plant Sci 13:273–280

Lee GI, Howe GA (2003) The tomato mutant *spr1* is defective in systemin perception and the production of a systemic wound signal for defense gene expression. Plant J 33:567–576

Lehesranta SJ, Lichtenberger R, Helariutta Y (2010) Cell-to-cell communication in vascular morphogenesis. Curr Opin Plant Biol 13:59–65

Leitner M, Vandelle E, Gaupels F, Bellin D, Delledonne M (2009) NO signals in the haze: nitric oxide signalling in plant defence. Curr Opin Plant Biol 12:451–458

Lejay L, Tillard P, Lepetit M, Olive Francesc D, Filleur S, Daniel-Vedele F, Gojon A (1999) Molecular and functional regulation of two NO3– uptake systems by N- and C-status of *Arabidopsis* plants. Plant J 18:509–519

Lejay L, Gansel X, Cerezo M, Tillard P, Müller C, Krapp A, von Wirén N, Daniel-Vedele F, Gojon A (2003) Regulation of root ion transporters by photosynthesis: functional importance and relation with hexokinase. Plant Cell 15:2218–2232

Lejay L, Wirth J, Pervent M, Cross JM-F, Tillard P, Gojon A (2008) Oxidative pentose phosphate pathway-dependent sugar sensing as a mechanism for regulation of root ion transporters by photosynthesis. Plant Physiol 146:2036–2053

Leon-Reyes A, Spoel SH, De Lange ES, Abe H, Kobayashi M, Tsuda S, Millenaar FF, Welschen RAM, Ritsema T, Pieterse CMJ (2009) Ethylene modulates the role of *nonexpressor of pathogenesis-related genes1* in cross talk between salicylate and jasmonate signaling. Plant Physiol 149:1797–1809

Leon-Reyes A, Du Y, Koornneef A, Proietti S, Körbes AP, Memelink J, Pieterse CMJ, Ritsema T (2010) Ethylene signaling renders the jasmonate response of *Arabidopsis* insensitive to future suppression by salicylic acid. Mol Plant Microbe Interact 23:187–197

Li C, Dubcovsky J (2008) Wheat FT protein regulates *VRN1* transcription through interactions with FDL2. Plant J 55:543–554

Li C, Liu G, Xu C, Lee GI, Bauer P, Ling HQ, Ganal MW, Howe GA (2003) The tomato suppressor of *prosystemin-mediated responses2* gene encodes a fatty acid desaturase required for the biosynthesis of jasmonic acid and the production of a systemic wound signal for defense gene expression. Plant Cell 15:1646–1661

Li P, Ham B-K, Lucas WJ (2011) CmRBP50 Protein phosphorylation is essential for assembly of a stable phloem-mobile high-affinity ribonucleoprotein complex. J Biol Chem 286:23142–23149

Lifschitz E, Eviatar T, Rozman A, Shalit A, Goldshmidt A, Amsellem Z, Alvarez JP, Eshed Y (2006) The tomato *FT* ortholog triggers systemic signals that regulate growth and flowering and substitute for diverse environmental stimuli. Proc Natl Acad Sci USA 103:6398–6403

Lin M-K, Belanger H, Lee YJ, Varkonyi-Gasic E, Taoka K-I, Miura E, Xoconostle-Cázares B, Gendler K, Jorgensen RA, Phinney B, Lough TJ, Lucas WJ (2007) FLOWERING LOCUS T protein may act as the long-distance florigenic signal in the cucurbits. Plant Cell 19:1488–1506

Lin S-I, Chiang S-F, Lin W-Y, Chen J-W, Tseng C-Y, Wu P-C, Chiou T-J (2008) Regulatory network of microRNA399 and *PHO2* by systemic signaling. Plant Physiol 147:732–746

Linkohr BI, Williamson LC, Fitter AH, Leyser HM (2002) Nitrate and phosphate availability and distribution have different effects on root system architecture of *Arabidopsis*. Plant J 29:751–760

Little DY, Rao H, Oliva S, Fo D-V, Krapp A, Malamy JE (2005) The putative high-affinity nitrate transporter NRT2.1 represses lateral root initiation in response to nutritional cues. Proc Natl Acad Sci USA 102:13693–13698

Liu C, Muchhal US, Uthappa M, Kononowicz AK, Raghothama KG (1998) Tomato phosphate transporter genes are differentially regulated in plant tissues by phosphorus. Plant Physiol 116:91–99

Liu T-Y, Chang C-Y, Chiou T-J (2009) The long-distance signaling of mineral macronutrients. Curr Opin Plant Biol 12:312–319

Liu P-P, von Dahl CC, Park S-W, Klessig DF (2011a) Interconnection between methyl salicylate and lipid-based long-distance signaling during the development of systemic acquired resistance in Arabidopsis and tobacco. Plant Physiol 155:1762–1768

Liu T-Y, Aung K, Tseng C-Y, Chang T-Y, Chen Y-S, Chiou T-J (2011b) Vacuolar Ca2+/H+ transport activity is required for systemic phosphate homeostasis involving shoot-to-root signaling in Arabidopsis. Plant Physiol 156:1176–1189

López-Ráez JA, Charnikhova T, Gómez-Roldán V, Matusova R, Kohlen W, De Vos R, Verstappen F, Puech-Pages V, Bécard G, Mulder P, Bouwmeester H (2008) Tomato strigolactones are derived

from carotenoids and their biosynthesis is promoted by phosphate starvation. New Phytol 178: 863–874
Lough TJ, Lucas WJ (2006) Integrative plant biology: role of phloem long-distance macromolecular trafficking. Ann Rev Plant Biol 57:203–232
Ma Y, Miura E, Ham B-K, Cheng H-W, Lee Y-J, Lucas WJ (2010) Pumpkin eIF5A isoforms interact with components of the translational machinery in the cucurbit sieve tube system. Plant J 64:536–550
Maffei ME, Mithofer A, Boland W (2007) Before gene expression: early events in plant–insect interaction. Trends Plant Sci 12:310–316
Makandar R, Nalam V, Chaturvedi R, Jeannotte R, Sparks AA, Shah J (2010) Involvement of salicylate and jasmonate signaling pathways in *Arabidopsis* interaction with *Fusarium graminearum*. Mol Plant Microbe Interact 23:861–870
Maldonado AM, Doerner P, Dixon RA, Lamb CJ, Cameron RK (2002) A putative lipid transfer protein involved in systemic resistance signalling in *Arabidopsis*. Nature 419:399
Malinowski R, Higgins R, Luo Y, Piper L, Nazir A, Bajwa V, Clouse S, Thompson P, Stratmann J (2009) The tomato brassinosteroid receptor BRI1 increases binding of systemin to tobacco plasma membranes, but is not involved in systemin signaling. Plant Mol Biol 70:603–616
Mancuso S (1999) Hydraulic and electrical transmission of wound-induced signals in *Vitis vinifera*. Aust J Plant Physiol 26:55–61
Manosalva PM, Park S-W, Forouhar F, Tong L, Fry WE, Klessig DF (2010) Methyl esterase 1 (StMES1) is required for systemic acquired resistance in potato. Mol Plant Microbe Interact 23:1151–1163
Martin A, Adam H, Díaz-Mendoza M, Żurczak M, González-Schain ND, Suárez-López P (2009) Graft-transmissible induction of potato tuberization by the microRNA *miR172*. Development 136:2873–2881
Mathieu J, Warthmann N, Küttner F, Schmid M (2007) Export of FT protein from phloem companion cells is sufficient for floral induction in *Arabidopsis*. Curr Biol 17:1055–1060
Michaels SD, Amasino RM (1999) *FLOWERING LOCUS C* encodes a novel MADS domain protein that acts as a repressor of flowering. Plant Cell 11:949–956
Michaels SD, Himelblau E, Kim SY, Schomburg FM, Amasino RM (2005) Integration of flowering signals in winter-annual Arabidopsis. Plant Physiol 137:149–156
Miller AJ, Fan X, Shen Q, Smith SJ (2008) Amino acids and nitrate as signals for the regulation of nitrogen acquisition. J Exp Bot 59:111–119
Miller G, Schlauch K, Tam R, Cortes D, Torres MA, Shulaev V, Dangl JL, Mittler R (2009) The plant NADPH oxidase RBOHD mediates rapid systemic signaling in response to diverse stimuli. Sci Signal 2:ra45
Minchin PEH, Thorpe MR (1987) Measurement of unloading and reloading of photoassimilate within the stem of bean. J Exp Bot 38:211–220
Mishina TE, Zeier J (2007) Pathogen-associated molecular pattern recognition rather than development of tissue necrosis contributes to bacterial induction of systemic acquired resistance in Arabidopsis. Plant J 50:500–513
Misson J, Thibaud MC, Bechtold N, Raghothama K, Nussaume L (2004) Transcriptional regulation and functional properties of *Arabidopsis* Pht1;4, a high affinity transporter contributing greatly to phosphate uptake in phosphate deprived plants. Plant Mol Biol 55:727–741
Montoya T, Nomura T, Farrar K, Kaneta T, Yokota T, Bishop GJ (2002) Cloning the tomato *Curl3* gene highlights the putative dual role of the leucine-rich repeat receptor kinase tBRI1/SR160 in plant steroid hormone and peptide hormone signaling. Plant Cell 14:3163–3176
Mudge SR, Rae AL, Diatloff E, Smith FW (2002) Expression analysis suggests novel roles for members of the Pht1 family of phosphate transporters in *Arabidopsis*. Plant J 31:341–353
Mullendore DL, Windt CW, Van As H, Knoblauch M (2010) Sieve tube geometry in relation to phloem flow. Plant Cell 22:579–593

Muszynski MG, Dam T, Li B, Shirbroun DM, Hou Z, Bruggemann E, Archibald R, Ananiev EV, Danilevskaya ON (2006) *Delayed flowering1* encodes a basic leucine zipper protein that mediates floral inductive signals at the shoot apex in maize. Plant Physiol 142:1523–1536

Mutasa-Göttgens E, Hedden P (2009) Gibberellin as a factor in floral regulatory networks. J Exp Bot 60:1979–1989

Narváez-Vásquez J, Ryan C (2004) The cellular localization of prosystemin: a functional role for phloem parenchyma in systemic wound signaling. Planta 218:360–369

Nelson T, Tausta SL, Gandotra N, Liu T (2006) Laser microdissection of plant tissue: what you see is what you get. Annu Rev Plant Biol 57:181–201

Ohto M, Onai K, Furukawa Y, Aoki E, Araki T, Nakamura K (2001) Effects of sugar on vegetative development and floral transition in Arabidopsis. Plant Physiol 127:252–261

Oparka KJ, Turgeon R (1999) Sieve elements and companion cells-traffic control centers of the phloem. Plant Cell 11:739–750

Orians C (2005) Herbivores, vascular pathways, and systemic induction: facts and artifacts. J Chem Ecol 31:2231–2242

Orozco-Cardenas ML, Narvaez-Vasquez J, Ryan CA (2001) Hydrogen peroxide acts as a second messenger for the induction of defense genes in tomato plants in response to wounding, systemin, and methyl jasmonate. Plant Cell 13:179–191

Palauqui JC, Elmayan T, Pollien JM, Vaucheret H (1997) Systemic acquired silencing: transgene-specific post-transcriptional silencing is transmitted by grafting from silenced stocks to non-silenced scions. EMBO J 16:4738–4745

Pant BD, Buhtz A, Kehr J, Scheible W-R (2008) MicroRNA399 is a long-distance signal for the regulation of plant phosphate homeostasis. Plant J 53:731–738

Park S-W, Kaimoyo E, Kumar D, Mosher S, Klessig DF (2007) Methyl salicylate is a critical mobile signal for plant systemic acquired resistance. Science 318:113–116

Parker JE (2009) The quest for long-distance signals in plant systemic immunity. Sci Signal 2:pe31

Peer WA, Blakeslee JJ, Yang H, Murphy AS (2011) Seven things we think we know about auxin transport. Mol Plant 4:487–504

Peng M, Hannam C, Gu H, Bi Y-M, Rothstein SJ (2007) A mutation in *NLA*, which encodes a RING-type ubiquitin ligase, disrupts the adaptability of *Arabidopsis* to nitrogen limitation. Plant J 50:320–337

Périlleux C, Bernier G (2002) The control of flowering: do genetical and physiological approaches converge? In: O'Neill SD, Roberts JA (eds) Plant reproduction, vol 6. Sheffield Academic Press, Sheffield, pp 1–32

Pin PA, Benlloch R, Bonnet D, Wremerth-Weich E, Kraft T, Gielen JJL, Nilsson O (2010) An antagonistic pair of *FT* homologs mediates the control of flowering time in sugar beet. Science 330:1397–1400

Pnueli L, Gutfinger T, Hareven D, Ben-Naim O, Ron N, Adir N, Lifschitz E (2001) Tomato SP-interacting proteins define a conserved signaling system that regulates shoot architecture and flowering. Plant Cell 13:2687–2702

Proust H, Hoffmann B, Xie X, Yoneyama K, Schaefer DG, Yoneyama K, Nogué F, Rameau C (2011) Strigolactones regulate protonema branching and act as a quorum sensing-like signal in the moss *Physcomitrella patens*. Development 138:1531–1539

Ragni L, Nieminen K, Pacheco-Villalobos D, Sibout R, Schwechheimer C, Hardtke CS (2011) Mobile gibberellin directly stimulates *Arabidopsis* hypocotyl xylem expansion. Plant Cell 23:1322–1336

Rahayu YS, Walch-Liu P, Neumann G, Römheld V, von Wirén N, Bangerth F (2005) Root-derived cytokinins as long-distance signals for NO3- -induced stimulation of leaf growth. J Exp Bot 56:1143–1152

Ranjan A, Fiene G, Fackendahl P, Hoecker U (2011) The *Arabidopsis* repressor of light signaling SPA1 acts in the phloem to regulate seedling de-etiolation, leaf expansion and flowering time. Development 138:1851–1862

Remans T, Nacry P, Pervent M, Filleur S, Diatloff E, Mounier E, Tillard P, Forde BG, Gojon A (2006) The *Arabidopsis* NRT1.1 transporter participates in the signaling pathway triggering root colonization of nitrate-rich patches. Proc Natl Acad Sci USA 103:19206–19211

Rhodes JD, Thain JF, Wildon DC (1996) The pathway for systemic electrical signal conduction in the wounded tomato plant. Planta 200:50–57

Roberts K, Love AJ, Laval V, Laird J, Tomos AD, Hooks MA, Milner JJ (2007) Long-distance movement of *Cauliflower mosaic virus* and host defence responses in *Arabidopsis* follow a predictable pattern that is determined by the leaf orthostichy. New Phytol 175:707–717

Rocher F, Chollet JF, Jousse C, Bonnemain JL (2006) Salicylic acid, an ambimobile molecule exhibiting a high ability to accumulate in the phloem. Plant Physiol 141:1684–1693

Roldán M, Gómez-Mena C, Ruiz-García L, Salinas J, Martínez-Zapater JM (1999) Sucrose availability on the aerial part of the plant promotes morphogenesis and flowering of *Arabidopsis* in the dark. Plant J 20:581–590

Rubin G, Tohge T, Matsuda F, Saito K, W-Rd S (2009) Members of the *LBD* family of transcription factors repress anthocyanin synthesis and affect additional nitrogen responses in *Arabidopsis*. Plant Cell 21:3567–3584

Rustérucci C, Espunya MC, Díaz M, Chabannes M, Martínez MC (2007) S-nitrosoglutathione reductase affords protection against pathogens in Arabidopsis, both locally and systemically. Plant Physiol 143:1282–1292

Schachtman DP, Shin R (2007) Nutrient sensing and signaling: NPKS. Annu Rev Plant Biol 58:47–69

Scheer JM, Ryan CA (2002) The systemin receptor SR160 from *Lycopersicon* peruvianum is a member of the LRR receptor kinase family. Proc Natl Acad Sci USA 99:9585–9590

Schilmiller AL, Howe GA (2005) Systemic signaling in the wound response. Curr Opin Plant Biol 8:369–377

Searle I, He Y, Turck F, Vincent C, Fornara F, Krober S, Amasino RA, Coupland G (2006) The transcription factor FLC confers a flowering response to vernalization by repressing meristem competence and systemic signaling in *Arabidopsis*. Genes Dev 20:898–912

Seskar M, Shulaev V, Raskin I (1998) Endogenous methyl salicylate in pathogen-inoculated tobacco plants. Plant Physiol 116:387–392

Shah J (2009) Plants under attack: systemic signals in defence. Curr Opin Plant Biol 12:459–464

Shalit A, Rozman A, Goldshmidt A, Alvarez JP, Bowman JL, Eshed Y, Lifschitz E (2009) The flowering hormone florigen functions as a general systemic regulator of growth and termination. Proc Natl Acad Sci USA 106:8392–8397

Shannon S, Meeks-Wagner DR (1991) A mutation in the *Arabidopsis TFL1* gene affects inflorescence meristem development. Plant Cell 3:877–892

Sheard LB, Tan X, Mao H, Withers J, Ben-Nissan G, Hinds TR, Kobayashi Y, Hsu F-F, Sharon M, Browse J, He SY, Rizo J, Howe GA, Zheng N (2010) Jasmonate perception by inositol-phosphate-potentiated COI1-JAZ co-receptor. Nature 468:400–405

Sheldon CC, Burn JE, Perez PP, Metzger J, Edwards JA, Peacock WJ, Dennis ES (1999) The *FLF* MADS box gene: a repressor of flowering in Arabidopsis regulated by vernalization and methylation. Plant Cell 11:445–458

Sheldon CC, Rouse DT, Finnegan EJ, Peacock WJ, Dennis ES (2000) The molecular basis of vernalization: the central role of *FLOWERING LOCUS C* (*FLC*). Proc Natl Acad Sci USA 97:3753–3758

Shin H, Shin HS, Dewbre GR, Harrison MJ (2004) Phosphate transport in *Arabidopsis*: *Pht1;1* and *Pht1;4* play a major role in phosphate acquisition from both low- and high-phosphate environments. Plant J 39:629–642

Sibout R, Plantegenet S, Hardtke CS (2008) Flowering as a condition for xylem expansion in *Arabidopsis* hypocotyl and root. Curr Biol 18:458–463

Sieburth LE, Lee D-K (2010) BYPASS1: how a tiny mutant tells a big story about root-to-shoot signaling. J Integr Plant Biol 52:77–85

Sjölund RD (1997) The phloem sieve element: a river runs through it. Plant Cell 9:1137–1146

Skøt L, Sanderson R, Thomas A, Skøt K, Thorogood D, Latypova G, Asp T, Armstead I (2011) Allelic variation in the perennial ryegrass *FLOWERING LOCUS T* gene is associated with changes in flowering time across a range of populations. Plant Physiol 155:1013–1022

Spoel SH, Johnson JS, Dong X (2007) Regulation of tradeoffs between plant defenses against pathogens with different lifestyles. Proc Natl Acad Sci USA 104:18842–18847

Staswick PE (2008) JAZing up jasmonate signaling. Trends Plant Sci 13:66–71

Stenzel I, Hause B, Maucher H, Pitzschke A, Miersch O, Ziegler J, Ryan CA, Wasternack C (2003) Allene oxide cyclase dependence of the wound response and vascular bundle-specific generation of jasmonates in tomato—amplification in wound signalling. Plant J 33:577–589

Sticher L, Mauch-Mani B, Métraux JP (1997) Systemic acquired resistance. Annu Rev Phytopathol 35:235–270

Stitt M (1999) Nitrate regulation of metabolism and growth. Curr Opin Plant Biol 2:178–186

Stratmann JW (2003) Long distance run in the wound response—jasmonic acid is pulling ahead. Trends Plant Sci 8:247–250

Suárez-López P (2005) Long-range signalling in plant reproductive development. Int J Dev Biol 49:761–771

Suzuki H, Xia Y, Cameron R, Shadle G, Blount J, Lamb C, Dixon RA (2004) Signals for local and systemic responses of plants to pathogen attack. J Exp Bot 55:169–179

Svistoonoff S, Creff A, Reymond M, Sigoillot-Claude C, Ricaud L, Blanchet A, Nussaume L, Desnos T (2007) Root tip contact with low-phosphate media reprograms plant root architecture. Nat Genet 39:792–796

Takada S, Goto K (2003) TERMINAL FLOWER 2, an Arabidopsis homolog of HETEROCHROMATIN PROTEIN1, counteracts the activation of *FLOWERING LOCUS T* by CONSTANS in the vascular tissues of leaves to regulate flowering time. Plant Cell 15:2856–2865

Takei K, Sakakibara H, Taniguchi M, Sugiyama T (2001) Nitrogen-dependent accumulation of cytokinins in root and the translocation to leaf: implication of cytokinin species that induces gene expression of maize response regulator. Plant Cell Physiol 42:85–93

Takei K, Takahashi T, Sugiyama T, Yamaya T, Sakakibara H (2002) Multiple routes communicating nitrogen availability from roots to shoots: a signal transduction pathway mediated by cytokinin. J Exp Bot 53:971–977

Tamaki S, Matsuo S, Wong HL, Yokoi S, Shimamoto K (2007) Hd3a Protein is a mobile flowering signal in rice. Science 316:1033–1036

Teper-Bamnolker P, Samach A (2005) The flowering integrator FT regulates *SEPALLATA3* and *FRUITFULL* accumulation in *Arabidopsis* leaves. Plant Cell 17:2661–2675

Thibaud M-C, Arrighi JF, Bayle V, Chiarenza S, Creff A, Bustos R, Paz-Ares J, Poirier Y, Nussaume L (2010) Dissection of local and systemic transcriptional responses to phosphate starvation in Arabidopsis. Plant J 64:775–789

Thines B, Katsir L, Melotto M, Niu Y, Mandaokar A, Liu G, Nomura K, He SY, Howe GA, Browse J (2007) JAZ repressor proteins are targets of the SCFCOI1 complex during jasmonate signalling. Nature 448:661–665

Thompson MV (2006) Phloem: the long and the short of it. Trends Plant Sci 11:26–32

Thompson GA, Goggin FL (2006) Transcriptomics and functional genomics of plant defence induction by phloem-feeding insects. J Exp Bot 57:755–766

Thorpe MR, Ferrieri AP, Herth MM, Ferrieri RA (2007) 11C-imaging: methyl jasmonate moves in both phloem and xylem, promotes transport of jasmonate, and of photoassimilate even after proton transport is decoupled. Planta 226:541–551

Thorpe MR, Furch ACU, Minchin PEH, Föller J, Van Bel AJE, Hafke JB (2010) Rapid cooling triggers forisome dispersion just before phloem transport stops. Plant Cell Environ 33:259–271

Truman W, Bennett MH, Kubigsteltig I, Turnbull C, Grant M (2007) *Arabidopsis* systemic immunity uses conserved defense signaling pathways and is mediated by jasmonates. Proc Natl Acad Sci USA 104:1075–1080

Truman WM, Bennett MH, Turnbull CGN, Grant MR (2010) Arabidopsis auxin mutants are compromised in systemic acquired resistance and exhibit aberrant accumulation of various indolic compounds. Plant Physiol 152:1562–1573

Tsuji H, Taoka K-i, Shimamoto K (2011) Regulation of flowering in rice: two florigen genes, a complex gene network, and natural variation. Curr Opin Plant Biol 14:45–52

Turck F, Fornara F, Coupland G (2008) Regulation and identity of florigen: FLOWERING LOCUS T moves center stage. Ann Rev Plant Biol 59:573–594

Turgeon R, Wolf S (2009) Phloem transport: cellular pathways and molecular trafficking. Ann Rev Plant Biol 60:207–221

Turnbull CGN, Booker JP, Leyser HMO (2002) Micrografting techniques for testing long-distance signalling in *Arabidopsis*. Plant J 32:255–262

Umehara M, Hanada A, Magome H, Takeda-Kamiya N, Yamaguchi S (2010) Contribution of strigolactones to the inhibition of tiller bud outgrowth under phosphate deficiency in rice. Plant Cell Physiol 51:1118–1126

Urbanus SL, Martinelli AP, Dinh QD, Aizza LCB, Dornelas MC, Angenent GC, Immink RGH (2010) Intercellular transport of epidermis-expressed MADS domain transcription factors and their effect on plant morphology and floral transition. Plant J 63:60–72

Valverde F, Mouradov A, Soppe W, Ravenscroft D, Samach A, Coupland G (2004) Photoreceptor regulation of CONSTANS protein in photoperiodic flowering. Science 303:1003–1006

van Bel AJ (2003a) The phloem, a miracle of ingenuity. Plant Cell Environ 26:125–149

van Bel AJ (2003b) Transport phloem: low profile, high impact. Plant Physiol 131:1509–1510

van Bel AJ, Gaupels F (2004) Pathogen-induced resistance and alarm signals in the phloem. Mol Plant Pathol 5:495–504

van Bel AJE, Knoblauch M, Furch ACU, Hafke JB (2011a) Questions on phloem biology. 1. Electropotential waves, Ca2+ fluxes and cellular cascades along the propagation pathway. Plant Sci 181:210–218

van Bel AJE, Furch ACU, Hafke JB, Knoblauch M, Patrick JW (2011b) Questions on phloem biology. 2. Mass flow, molecular hopping, distribution patterns and macromolecular signalling. Plant Sci 181:325–330

van Loon LC, Bakker PAHM, Pieterse CMJ (1998) Systemic resistance induced by rhizosphere bacteria. Annu Rev Phytopathol 36:453–483

Van Norman JM, Breakfield NW, Benfey PN (2011) Intercellular communication during plant development. Plant Cell 23:855–864

Vidal EA, Araus V, Lu C, Parry G, Green PJ, Coruzzi GM, Gutiérrez RA (2010) Nitrate-responsive miR393/*AFB3* regulatory module controls root system architecture in *Arabidopsis thaliana*. Proc Natl Acad Sci USA 107:4477–4482

Vilaine F, Palauqui JC, Amselem J, Kusiak C, Lemoine R, Dinant S (2003) Towards deciphering phloem: a transcriptome analysis of the phloem of *Apium graveolens*. Plant J 36:67–81

Vlot AC, Klessig DF, Park S-W (2008a) Systemic acquired resistance: the elusive signal(s). Curr Opin Plant Biol 11:436–442

Vlot AC, Liu P-P, Cameron RK, Park S-W, Yang Y, Kumar D, Zhou F, Padukkavidana T, Gustafsson C, Pichersky E, Klessig DF (2008b) Identification of likely orthologs of tobacco salicylic acid-binding protein 2 and their role in systemic acquired resistance in *Arabidopsis thaliana*. Plant J 56:445–456

Voinnet O, Baulcombe DC (1997) Systemic signalling in gene silencing. Nature 389:553

Walling LL (2008) Avoiding effective defenses: strategies employed by phloem-feeding insects. Plant Physiol 146:859–866

Walz C, Juenger M, Schad M, Kehr J (2002) Evidence for the presence and activity of a complete antioxidant defence system in mature sieve tubes. Plant J 31:189–197

Wang Y-Y, Tsay Y-F (2011) *Arabidopsis* nitrate transporter NRT1.9 is important in phloem nitrate transport. Plant Cell 23:1945–1957

Wang D, Pajerowska-Mukhtar K, Culler AH, Dong X (2007) Salicylic acid inhibits pathogen growth in plants through repression of the auxin signaling pathway. Curr Biol 17:1784–1790

Wasternack C, Stenzel I, Hause B, Hause G, Kutter C, Maucher H, Neumerkel J, Feussner I, Miersch O (2006) The wound response in tomato—role of jasmonic acid. J Plant Physiol 163:297–306

Wigge PA, Kim MC, Jaeger KE, Busch W, Schmid M, Lohmann JU, Weigel D (2005) Integration of spatial and temporal information during floral induction in *Arabidopsis*. Science 309:1056–1059

Will T, van Bel AJ (2006) Physical and chemical interactions between aphids and plants. J Exp Bot 57:729–737

Will T, Tjallingii WF, Thonnessen A, van Bel AJ (2007) Molecular sabotage of plant defense by aphid saliva. Proc Natl Acad Sci USA 104:10536–10541

Will T, Kornemann SR, Furch ACU, Tjallingii WF, van Bel AJE (2009) Aphid watery saliva counteracts sieve-tube occlusion: a universal phenomenon? J Exp Bot 212:3305–3312

Windt CW, Vergeldt FJ, Jager PAD, Van As H (2006) MRI of long-distance water transport: a comparison of the phloem and xylem flow characteristics and dynamics in poplar, castor bean, tomato and tobacco. Plant Cell Environ 29:1715–1729

Wu J, Baldwin IT (2010) New insights into plant responses to the attack from insect herbivores. Annu Rev Genet 44:1–24

Wu J, Wang L, Baldwin I (2008) Methyl jasmonate-elicited herbivore resistance: does MeJA function as a signal without being hydrolyzed to JA? Planta 227:1161–1168

Xia Y, Suzuki H, Borevitz J, Blount J, Guo Z, Patel K, Dixon RA, Lamb C (2004) An extracellular aspartic protease functions in *Arabidopsis* disease resistance signaling. EMBO J 23:980–988

Yamaguchi A, Kobayashi Y, Goto K, Abe M, Araki T (2005) *TWIN SISTER OF FT* (*TSF*) acts as a floral pathway integrator redundantly with *FT*. Plant Cell Physiol 46:1175–1189

Yano M, Katayose Y, Ashikari M, Yamanouchi U, Monna L, Fuse T, Baba T, Yamamoto K, Umehara Y, Nagamura Y, Sasaki T (2000) *Hd1*, a major photoperiod sensitivity quantitative trait locus in rice, is closely related to the Arabidopsis flowering time gene *CONSTANS*. Plant Cell 12:2473–2483

Yoneyama K, Takeuchi Y, Sekimoto H (2007) Phosphorus deficiency in red clover promotes exudation of orobanchol, the signal for mycorrhizal symbionts and germination stimulant for root parasites. Planta 225:1031–1038

Yoo BC, Kragler F, Varkonyi-Gasic E, Haywood V, Archer-Evans S, Lee YM, Lough TJ, Lucas WJ (2004) A systemic small RNA signaling system in plants. Plant Cell 16:1979–2000

Yoo SJ, Chung KS, Jung SH, Yoo SY, Lee JS, Ahn JH (2010) *BROTHER OF FT AND TFL1* (*BFT*) has *TFL1*-like activity and functions redundantly with *TFL1* in inflorescence meristem development in Arabidopsis. Plant J 63:241–253

Zeevaart JAD (2008) Leaf-produced floral signals. Curr Opin Plant Biol 11:541–547

Zhang J, Zhou J-M (2010) Plant immunity triggered by microbial molecular signatures. Mol Plant 3:783–793

Zhao M, Ding H, Zhu J-K, Zhang F, Li W-X (2011) Involvement of miR169 in the nitrogen-starvation responses in Arabidopsis. New Phytol 190:906–915

Zhu Q-H, Helliwell CA (2011) Regulation of flowering time and floral patterning by miR172. J Exp Bot 62:487–495

Zimmermann MR, Maischak H, Mithofer A, Boland W, Felle HH (2009) System potentials, a novel electrical long-distance apoplastic signal in plants, induced by wounding. Plant Physiol 149:1593–1600

Intercellular Signaling During Floral Development

Balaji Enugutti and Kay Schneitz

Abstract Flowers are central to sexual reproduction in higher plants and during evolution floral organs have acquired diverse morphologies to aid in this process. Cells need to communicate to allow floral morphogenesis to happen. The flow of information between plant cells occurs through signaling mechanisms that involve cell surface receptors, cell wall diffusible factors, and plasmodesmata. Transcription factors and small RNAs are now known to move between floral cells to regulate cell identity and morphogenesis. A growing number of cell surface receptor-like kinases have been identified that play a role in intercellular communication in the floral meristem (FM), the specification of the male germline, and the formation of the ovule integuments. In this chapter, we highlight some of the progress that has been made toward an understanding of these types of signaling mechanisms.

1 Introduction

Flowers have fascinated mankind for thousands of years if not for their value as seed-producing entities of the plant then because of their dazzling variety and uplifting beauty. The last 25 years have witnessed significant progress in the understanding of the molecular mechanisms that govern for example floral induction or the specification of floral meristems and floral organs. Although impressive advances have been made as well, comparably little is known about intercellular communication processes that are required for proper floral organ development.

Flowers derive from lateral or axillary floral meristem (FM). In a typical plant, such as the model plant *Arabidopsis thaliana*, lateral organs are produced postembryonically at the periphery of the shoot apical meristem (SAM), at this

B. Enugutti • K. Schneitz (✉)
Entwicklungsbiologie der Pflanzen, Wissenschaftszentrum Weihenstephan, Technische Universität München, Freising, Germany
e-mail: balaji@wzw.tum.de; schneitz@wzw.tum.de,

G. Witzany and F. Baluška (eds.), *Biocommunication of Plants*,
Signaling and Communication in Plants 14,
DOI 10.1007/978-3-642-23524-5_7,

stage also known as inflorescence meristem (IM). The FM produces four types of organs: sepals, petals, stamen, and a gynoecium, usually made up of two or more fused carpels. Within the gynoecium, ovules develop from the placental tissue of the carpels and eventually produce the egg cell proper. The floral organs are organized in concentric whorls with the first whorl being occupied by sepals, the second whorl by petals, the third whorl by stamens, and the fourth whorl by carpels. A set of homeotic genes, encoding mostly MADS-box transcription factors with spatially overlapping activities, regulate the identities of individual whorls in a combinatorial fashion and together with a number of cofactors (Coen and Meyerowitz 1991; Causier et al. 2010; Melzer et al. 2010). The so-called ABC model states that A function specifies whorl 1, A and B function whorl 2, B and C function whorl 3, and C function whorl 4. In addition, genes of the A and C classes repress each other. In *Arabidopsis*, *APETALA1* (*AP1*) and *AP2* represent A-class genes, *AP3* and *PISTILLATA* (*PI*) are B-class genes, while *AGAMOUS* (*AG*) carries C function.

Aboveground meristems are characterized by clonally distinct, so-called histogenic layers (Satina et al. 1940). Cells of the outermost or L1 layer and the first subepidermal or L2 layer stereotypically divide in an anticlinal fashion, thereby maintaining the layers. Cells of the inner core or L3 divide in an essentially random fashion. The L1/L2 and the L3 are also known as tunica and corpus, respectively. Both meristems are also organized into different types of zones. The central zone (CZ) harbors the stem cells a group of infrequently dividing cells that ultimately give rise to all aboveground plant organs. Eventually, cells in the CZ become displaced to the side and into the peripheral zone (PZ) where they divide more frequently. It is the PZ from which organ primordia originate. Beneath the CZ, the rib meristem will generate interior tissues of the shoot or flower.

Plant organs are made up of cells originating from all histogenic layers. Cells of the L1 contribute to the epidermis while cells of the L2/L3 layers generate different types of internal tissues. Interestingly, the relative contributions of L2- and L3-derived cells to a given tissue can vary between the same organs of different individuals indicating that cells within a tissue coordinate their behavior (Szymkowiak and Sussex 1996). This coordination requires communication. In plants, the cell wall constitutes a natural barrier to intercellular communication. Two general types of mechanisms evolved to overcome the cell wall, a natural barrier to this process. Information transfer can occur via small, cell wall–penetrating ligands, for example peptides or phytohormones, and their receptors, or via plasmodesmata (PD), channels that traverse the cell wall and interconnect the cytoplasm of neighboring cells. As it will become clear throughout this chapter, both types of signaling mechanisms have been invoked during the evolution of flowers. In this chapter, we focus on selected examples. Several excellent reviews deal with other aspects, such as intercellular communication in other organs, auxin signaling in reproductive development, size control, gametophyte development, or fertilization (Lucas et al. 2009; Sundberg and Ostergaard 2009; Breuninger and Lenhard 2010; Chapman and Goring 2010; Chitwood and Timmermans 2010; Ma and Sundaresan 2010; Van Norman et al. 2011).

2 Plasmodesmata-Based Intercellular Communication in Flowers

PD are symplasmic channels that interconnect cells and mediate cell-to-cell trafficking of a wide array of molecules (Ehlers and Kollmann 2001; Kim 2005; Lucas et al. 2009) either in a targeted or nontargeted (by passive diffusion) manner. Primary PD form during cytokinesis and can either exist as relatively simple channels or may develop a complex array of branches. Secondary PD are generated de novo in existing cell walls, often during cell expansion. Intercellular communication via movement of molecules to neighboring cells through PD has been well documented. For example, there is developmental regulation of symplasmic trafficking through plasmodesmata in apices (Rinne and van der Schoot 1998; Gisel et al. 1999; Rinne et al. 2001) and in ovules (Werner et al. 2010), both at the temporal and spatial levels. TFs, such as KNOTTED1 or SHORTROOT (Lucas et al. 1995; Nakajima et al. 2001; Jackson 2002; Cui et al. 2007), move between cells likely through PD as does the microRNA miRNA165/166 (Carlsbecker et al. 2010). It is therefore not surprising that PD-mediated intercellular signaling is also important for floral development.

2.1 Intercellular Protein Trafficking

In a series of landmark studies, it was recently shown that the control of floral induction by day length (photoperiod) requires long-distance protein movement. This process has long been known to involve a long-range communication between the leaves and the apex. Upon perception of the photoperiodic signal, leaves produce a secondary signal, or florigen, that moves from leaves to the apex and initiates the production of flowers (Zeevaart 1976). Several labs have recently contributed to the decipherment of the molecular basis of florigen and its mode of action (Turck et al. 2008; Amasino 2010). In short, the TF CONSTANS (CO) mediates the light response and activates the TF FLOWERING LOCUS T (FT) in phloem companion cells. Subsequently, the FT protein moves to the apex via the phloem (Corbesier et al. 2007; Jaeger and Wigge 2007; Mathieu et al. 2007; Tamaki et al. 2007) where it forms a heterodimer in floral organ anlagen with the FD protein already present in this tissue and activates downstream targets, such as the floral regulator *AP1*.

Intercellular communication between histogenic layers of the IM and FM is important for proper floral specification of cells and floral morphogenesis. For example, indeterminacy of the *Antirrhinum* and *Arabidopsis* IMs appears to depend in part on noncell-autonomous function of the two related genes *CENTRORADIALIS* (*CEN*) and *TERMINAL FLOWER1* (*TFL1*), respectively (Bradley et al. 1996, 1997). With regard to floral organogenesis, the epidermis promotes and restricts organ growth (Savaldi-Goldstein et al. 2007) and plays a large influence on petal shape in several species (Perbal et al. 1996; Efremova et al. 2001; Jenik and Irish 2001; Vincent et al. 2003). Such interactions do not only take place in the outside-in direction but also occur in the opposite, inside-out, direction. For example, the L3 layer dictates FM size in tomato (Szymkowiak and Sussex 1992). In addition, floral determinacy requires the action of *AG* in the L2/L3 layers, and cells in L2 are able to confer cell identity to cells in L1 (Sieburth et al. 1998).

Many of the above-described interactions depend on intercellular movement of known or assumed TFs. Recent data indicate that the noncell-autonomous function of *TFL1* in maintaining IM indeterminacy depends on controlled movement of TFL1 protein (Conti and Bradley 2007). *LFY* and its *Antirrhinum* homolog *FLORICAULA* (*FLO*) were shown to act in a noncell-autonomous fashion in the floral meristem (Carpenter and Coen 1995; Sessions et al. 2000), and the LFY protein was found to move between cells in a nontargeted fashion (Wu et al. 2003). It was proposed that passive diffusion was the default mode for many proteins unless they are efficiently retained in the cell by various means.

Apart from LFY, other floral regulators were shown to move between histogenic layers. The *Antirrhinum* B-factors DEFICIENS (DEF) and GLOBOSA (GLO) move from the L2 to the L1 in a developmentally regulated manner (Perbal et al. 1996). By contrast, the *Arabidopsis* DEF and GLO orthologs AP3 and PI do not travel between layers (Jenik and Irish 2001; Urbanus et al. 2010), indicating that there exist species-specific differences in the control of intercellular TF movement.

The *Arabidopsis* C-class gene *AG* is responsible for the specification of reproductive organs and floral determinacy (Yanofsky et al. 1990; Lenhard et al. 2001; Lohmann et al. 2001). AG RNA and protein are expressed throughout the center of the flower (Yanofsky et al. 1990; Drews et al. 1991; Urbanus et al. 2009; Wollmann et al. 2010). Nevertheless, several lines of evidence indicate that *AG* acts in a noncell-autonomous fashion (Sieburth et al. 1998; Jenik and Irish 2000; Cartolano et al. 2009). Recent work addressed this issue carefully (Urbanus et al. 2010). Interestingly, these authors could show that translational fusions between green fluorescent protein (GFP) and AG, AP3, PI, and their cofactor SEP3 were able to move between epidermal cells of the FM but only AG:GFP could move from the L1 to subepidermal layers. In addition, epidermal expression of AG:GFP was sufficient to rescue an *ag* mutant. It was proposed that this transport is likely to work in both directions (Sieburth et al. 1998; Urbanus et al. 2010). Epidermal cells are connected via primary PD while the connections between epidermis and subepidermis are achieved through secondary PD (Ehlers and Kollmann 2001). The result indicates that movement of the tested TFs through primary PD is either nontargeted or differentially regulated from their movement through secondary PD (Urbanus et al. 2010). In particular, the movement of AG:GFP between the L1/L2 is unlikely to be due to a passive mechanism as the other fusion proteins exhibited roughly the same molecular weight. It was suggested that early intercellular movement of AG in the FM helps to rapidly establish its stable and broad expression domain required for further development (Urbanus et al. 2010).

2.2 *Movement of Small RNAs in Floral Development*

Apart from proteins, it has become clear that certain small RNA species (sRNAs) with a length between 21 and 24 nucleotides move between cells in a regulated and biologically relevant manner (Chitwood and Timmermans 2010; Van Norman et al. 2011). It is generally assumed that sRNAs move symplastically through PD. The sRNA variants differ in the way they are generated and their biological function (Chapman and Carrington 2007; Voinnet 2009). siRNAs are formed from perfectly matching dsRNAs, act in a noncell-autonomous fashion with at least 21-nucleotide species moving as siRNA duplexes (Dunoyer et al. 2010), and are involved in posttranscriptional gene silencing of viruses and transgenes. In addition, endogenous mobile siRNAs are of 24-nucleotide length, derived from transposable elements (TEs) or other methylated DNA regions, and can direct DNA methylation at target loci (Molnar et al. 2010). miRNAs and *trans*acting siRNAs (tasiRNAs) regulate gene silencing. In general, siRNAs seem to move further than miRNAs (de Felippes et al. 2011).

Defects in the biogenesis of tasiRNAs result in leaf and floral phenotypes (Peragine et al. 2004; Adenot et al. 2006), and it has become clear that miRNA 165/166 and miR390/tasiRNA tasiR-ARF affect leaf patterning (Husbands et al. 2009). For example, in *Arabidopsis*, miR390 spreads from its subSAM origin of expression to the young lateral primordia where it participates in the biogenesis of tasiRNA directed against the abaxial factors ARF3 and ARF4 (tasiR-ARF) (Chitwood et al. 2009). Production of tasiR-ARF is restricted to the adaxial cell layers from which tasiR-ARF moves toward abaxial layers generating a corresponding adaxial-abaxial gradient of tasiR-ARF. This gradient likely results in the translational repression of *ARF3* in adaxial cells and the presence of ARF3 protein in abaxial cells only (Husbands et al. 2009). It is not known if miRNA165/66, regulating abaxial patterning of lateral organs, moves in leaves or floral organs; however, this miRNA was recently shown to move from the endodermis into the vascular cylinder, thereby regulating xylem differentiation (Carlsbecker et al. 2010).

In ovules, a single megaspore mother cell (MMC) originates from a group of L2-derived cells in the nucellus. The MMC undergoes meiosis resulting in a tetrad of megaspores. As a rule, three megaspores degenerate and the sole surviving functional megaspore further develops into the female gametophyte with the egg cell proper. Recently, it was shown that an *AGO9*-dependent siRNA pathway plays an essential role in singling out the MMC in a noncell-autonomous fashion (Olmedo-Monfil et al. 2010). In *ago9*, *rdr6*, or *sgs3* mutants, several MMC-like cells develop in the nucellus, although only one continues with meiosis. Still, one or several of the other enlarged cells acquire female gametophyte identity despite the absence of meiosis, a situation resembling apospory. Interestingly, AGO9 protein could only be detected in the epidermis cells of the nucellus, was shown to preferentially associate with 24-nucleotide sRNAs, and was required for the silencing of endogenous TEs in the egg and synergids. Importantly, AGO9-dependent TE inactivation apparently restricts female gametophyte formation to a single precursor cell (the MMC) through a 24-nucleotide siRNA biosynthetic pathway. The authors suggested that inactivation of TEs in all subepidermal cells of the nucellus except the MMC somehow prevents those cells to enter gametophyte development, although how this is achieved remains to be investigated. The MMC, however, appears to be somehow isolated, and thus, the silencing signal cannot enter this cell. Indeed, the MMC is known to become symplastically isolated (Werner et al. 2010), possibly due to accumulation of high levels of callose around the MMC (Schneitz et al. 1995).

Movement of siRNAs also appears to be important for maintenance of genome stability in sperm cells. In the vegetative nucleus of pollen, TEs become reactivated resulting in the generation of a high level of siRNAs. By contrast, TEs in the sperm cells remain silent, possibly at least in part as a consequence of siRNAs moving from the vegetative cell into the sperm where they could act in the epigenetic silencing of the TEs (Slotkin et al. 2009).

3 Receptor-Like-Kinase-Mediated Intercellular Signaling in Flowers

Cell surface receptor-like kinases (RLKs) are natural mediators of information transfer between cells and are involved in many short-range intercellular signaling processes. The *Arabidopsis* genome encodes more than 600 RLK genes (Shiu and Bleecker 2001); a growing number of which are known to affect several aspects of organogenesis (Hématy and Höfte 2008; De Smet et al. 2009; Steinwand and Kieber 2010; Gish and Clark 2011).

Regulation of stem cell maintenance in SAMs and FMs is mediated through an autoregulatory feedback loop involving the signal peptide CLAVATA3 (CLV3), the leucine-rich repeat (LRR) RLK CLV1, and the homeobox transcription factor WUSCHEL (WUS) (Clark et al. 1997; Mayer et al. 1998; Fletcher et al. 1999; Brand et al. 2000; Schoof et al. 2000). WUS is an indirect positive regulator of stem cells which in turn express CLV3 that negatively regulates WUS through CLV1 and the plasma membrane–localized phosphatases POLTERGEIST (POL) and PLL1 (Yu et al. 2003; Gagne and Clark 2010). More recently, it was found that this feedback loop also involves the direct negative control of *CLV1* by WUS (Busch et al. 2010). Apart from regulating *CLV1* expression, WUS seems to foster stem cell development by influencing the hormonal control of the stem cell niche in the SAM (Leibfried et al. 2005; Gordon et al. 2009; Zhao et al. 2010). Interestingly, *WUS* also regulates chalaza formation in a nonautonomous fashion (Gross-Hardt et al. 2002; Sieber et al. 2004). The mechanism is not understood but does not involve the *CLV* genes.

Perception of the CLV3 peptide has proven to be more complex than initially appreciated. First, it was realized that a processed form of the CLV3 peptide directly binds to CLV1 and CLV2 (Kondo et al. 2006; Ogawa et al. 2008; Guo et al. 2010).

Second, it is now apparent that several receptor complexes act in parallel in the perception of CLV3 at the cell surface. One receptor complex consists of constitutive CLV1 homodimers. In addition, CLV1 can form heterodimers with the closely related and redundantly acting BAM receptors (DeYoung et al. 2006; DeYoung and Clark 2008; Bleckmann et al. 2010; Guo et al. 2010; Zhu et al. 2010). Furthermore, the receptor-like protein CLV2 (Kayes and Clark 1998; Jeong et al. 1999), which carries but a small cytoplasmic domain, forms a receptor complex with the transmembrane putative kinase CORYNE (CRN) which itself carries a transmembrane domain but only a small extracellular domain (Miwa et al. 2008; Müller et al. 2008). In addition, homo-oligomers formed by the RLK RECEPTOR-LIKE PROTEIN KINASE 2 (RPK2)/TOADSTOOL 2 (TOAD2) represent a third CLV3-transmitting receptor complex (Kinoshita et al. 2010).

RLKs are also involved in interhistogenic-layer communication in the SAM, FM, and the organs derived from those meristems. The underlying communication can go in two directions. For example, the epidermis has an important influence on subepidermal cell behavior (Reinhardt et al. 2003). The brassinosteroid receptor BRASSINOSTEROID INSENSITIVE 1 (BRI1) (Li and Chory 1997; Kinoshita et al. 2005) was demonstrated to participate in the communication between epidermis and subepidermis in the control of cell morphogenesis by providing an epidermis-derived nonautonomous signal (Savaldi-Goldstein et al. 2007). An "inside to outside" mechanism of intercell-layer communication in floral organs is suggested by the subcellular localization of the epidermally expressed RLK ARABIDOPSIS CRINKLY 4 (ACR4) (Gifford et al. 2003). ACR4 is the *Arabidopsis* homolog of maize CRINKLY 4 (CR4) (Becraft et al. 1996; Becraft et al. 2001) and involved in the regulation of epidermal cell organization in ovule integuments, sepals, and leaves (Gifford et al. 2003; Watanabe et al. 2004; Gifford et al. 2005). ACR4-dependent control of epidermis development also involves the RLK ABNORMAL LEAF SHAPE 2 (ALE2) (Tanaka et al. 2007).

The *STRUBBELIG* (*SUB*) locus encodes a LRR-RLK that was implied in intercell-layer communication in flowers as well (Chevalier et al. 2005; Yadav et al. 2008). *SUB*, also known as *SCRAMBLED* (*SCM*) (Kwak et al. 2005), regulates cell morphogenesis in FMs and ovules in a noncell-autonomous fashion. In the FM, expression of functional SUB:GFP fusion protein from the L1 was sufficient to rescue cellular defects in the L2 while nucellar expression of SUB:GFP was able to rescue integument defects to a large extent in ovules. *SUB* interacts with the RLK gene *ERECTA* (*ER*) (Torii et al. 1996) in a synergistic fashion in stem development but interestingly not in ovules (Vaddepalli et al. 2011). How SUB affects the behavior of neighboring cells is currently being investigated. With *QUIRKY* (*QKY*), *ZERZAUST* (*ZET*), and *DETORQEO* (*DOQ*), three additional components of the *SUB* signaling pathway have recently been identified genetically (Fulton et al. 2009). *QKY* was found to encode a putative membrane-anchored C2-domain protein. On the basis of related domain architecture in animal proteins such as synaptotagmins or ferlins, QKY was hypothesized to function in membrane trafficking. Additional postulated scenarios include a role of *SUB* and *QKY* in cell wall biology or the regulation of PD function.

Interestingly, kinase activity of SUB is not required for its function in vivo (Chevalier et al. 2005; Vaddepalli et al. 2011). In vitro kinase assays were negative, but critically, transgenes carrying several well-characterized mutations in the SUB kinase domain were able to rescue the *sub* mutant phenotype. Thus, SUB is a plant representative of the unusual class of atypical or "dead" kinases that is best studied in animals (Kroiher et al. 2001; Boudeau et al. 2006; Castells and Casacuberta 2007). However, it should be noted that in some instances even the signaling mechanism of biochemically active RLKs, such as ACR4 or FEI1, may not absolutely require a functional kinase domain in vivo (Gifford et al. 2005; Xu et al. 2008). Thus, it is conceivable that redundant activities exist in multiprotein receptor complexes that could substitute for the absence of kinase activity of a single receptor. In any case, it is an exciting challenge to unravel a signaling pathway mediated by an atypical RLK.

Anthers are the male reproductive tissues of plants. They constitute microsporangia within which the male germline develops. The pollen mother cells (PMCs), or microsporocytes, which will undergo meiosis, are contained in the four corners of the anther and within concentric cell somatic layers, the tapetum, the middle layer, and the endothecium subjacent to the epidermis. The PMCs and the cell layers are derived from an archesporial cell through a set of regulated stereotypic cell divisions. As a model system to study organogenesis, early anther development has met with considerable interest and it has become apparent that a number of RLKs are involved in the establishment of the different cell layers during early anther ontogenesis (Feng and Dickinson 2007; Feng and Dickinson 2010a).

Somatic cell fate in general appears to be under the control of the redundantly acting *CLV1* homologs *BAM1* and *BAM2* (Hord et al. 2006). It was suggested that *BAM1/2* restrict proliferation of sporogenous cells and/or promote differentiation of the peripheral somatic cells. Formation of the tapetum is under the control of another set of LRR-RLKs. Defects in the RLK genes *EXTRA MICROSPOROCYTES1* (*EMS1*)/*EXTRA SPOROGENOUS CELLS* (*EXS*) (Canales et al. 2002; Zhao et al. 2002) and *SOMATIC EMBRYOGENESIS RECEPTOR KINASE1* (*SERK1*) and *SERK2* (Albrecht et al. 2005; Colcombet et al. 2005) result in an overproliferation of PMCs and the absence of the tapetum. A similar phenotype is observed in mutants with a defect in *TAPETUM DETERMINANT1* (*TPD1*) predicted to encode a small and secreted protein (Yang et al. 2003). Interestingly, the function of the *EMS1/EXS* and *TPD1* genes is conserved in evolution (Nonomura et al. 2003; Zhao et al. 2008). The similar phenotypes suggest that all genes act in the same process, and genetic and biochemical data indicate that EMS1/EXS and TPD1 constitute a receptor-ligand pair (Jia et al. 2008). There is evidence that SERK1/2 can form homodimers in plant cells (Albrecht et al. 2005) but it remains to be shown if SERK1/2 are part of the EMS1/EXS receptor complex or if they act in parallel to EMS1/EXS.

The mutant phenotype of *ems1/exs*, *serk1/2*, and *tpd1* mutants suggests that this signaling pathway either regulates PMC proliferation or the specification of tapetal cells. Recent evidence, however, indicates that *EMS1/EXS* regulates cell proliferation in the tapetal cell monolayer (Feng and Dickinson 2010b). Tapetum development, and middle layer formation, is also regulated by the RLK RPK2 (Mizuno et al. 2007).

4 Conclusions

With this chapter, we have provided a brief account of some of the better understood intercellular signaling aspects of floral development. The last 20 years have witnessed a series of landmark papers and an overall impressive body of exciting work. The observation that TFs and sRNAs can move between cells has revolutionized our thinking about how plant cells communicate. After decades of hard but fruitless work, the molecular nature of the florigen is finally being unraveled. While 25 years ago some people would argue that the cell wall would make it unlikely that plants possess RLKs, we now know that RLKs do exist and in truly staggering numbers which by far exceed the number of different RLKs in humans. As one may expect, however, a number of questions remain. For example, why do some TFs and sRNAs move through PD and others don't? There is a perhaps surprising specificity in the mechanisms that regulate transport of molecules through PD. How is this achieved? With respect to RLK signaling mechanisms, a major area requiring even more research relates to the identification and analysis of their ligands, as we know only a handful of specific ligands. In addition, the downstream signaling components have been identified for only a few RLKs, and the function of only a comparably small number of RLKs is known at all. Finally, how is the information flow mediated by different RLKs integrated to direct proper cellular behavior? As already indicated, we have come a long way. These are exciting times in plant signal transduction, and no doubt research will be very rewarding for many years to come.

Acknowledgments We thank members of our lab for fruitful discussions. We apologize to colleagues whose work we could not cite due to space restrictions. Work on signaling in floral organs in the Schneitz lab is funded by grants SCHN 723/1-3, SCHN 723/3-2, and SCHN 723/6-1 from the German Research Council (DFG) and by the Free State of Bavaria.

References

Adenot X, Elmayan T, Lauressergues D, Boutet S, Bouche N, Gasciolli V, Vaucheret H (2006) DRB4-dependent *TAS3 trans*-acting siRNAs control leaf morphology through AGO7. Curr Biol 16:927–932

Albrecht C, Russinova E, Hecht V, Baaijens E, de Vries S (2005) The *Arabidopsis thaliana* SOMATIC EMBRYOGENESIS RECEPTOR-LIKE KINASES1 and 2 control male sporogenesis. Plant Cell 17:3337–3349

Amasino R (2010) Seasonal and developmental timing of flowering. Plant J 61:1001–1013

Becraft PW, Stinard PS, McCarty DR (1996) CRINCLY4: a TNFR-like receptor kinase involved in maize epidermal differentiation. Science 273:1406–1409

Becraft PW, Kang SH, Suh SG (2001) The maize CRINKLY4 receptor kinase controls a cell-autonomous differentiation response. Plant Physiol 127:486–496

Bleckmann A, Weidtkamp-Peters S, Seidel CA, Simon R (2010) Stem cell signaling in Arabidopsis requires CRN to localize CLV2 to the plasma membrane. Plant Physiol 152:166–176

Boudeau J, Miranda-Saavedra D, Barton GJ, Alessi DR (2006) Emerging roles of pseudokinases. Trends Cell Biol 16:443–452

Bradley D, Carpenter R, Copsey L, Vincent C, Rothstein S, Coen E (1996) Control of inforescence architecture in *Antirrhinum*. Nature 379:791–797

Bradley D, Ratcliffe O, Vincent C, Carpenter R, Coen E (1997) Inflorescence commitment and architecture in *Arabidopsis*. Science 275:80–83

Brand U, Fletcher JC, Hobe M, Meyerowitz EM, Simon R (2000) Dependence of stem cell fate in *Arabidopsis* on a feedback loop regulated by *CLV3* activity. Science 289:617–619

Breuninger H, Lenhard M (2010) Control of tissue and organ growth in plants. Curr Top Dev Biol 91:185–220

Busch W, Miotk A, Ariel FD, Zhao Z, Forner J, Daum G, Suzaki T, Schuster C, Schultheiss SJ, Leibfried A, Haubeiss S, Ha N, Chan RL, Lohmann JU (2010) Transcriptional control of a plant stem cell niche. Dev Cell 18:849–861

Canales C, Bhatt AM, Scott R, Dickinson H (2002) EXS, a putative LRR receptor kinase, regulates male germline cell number and tapetal identity and promotes seed development in Arabidopsis. Curr Biol 12:1718–1727

Carlsbecker A, Lee JY, Roberts CJ, Dettmer J, Lehesranta S, Zhou J, Lindgren O, Moreno-Risueno MA, Vaten A, Thitamadee S, Campilho A, Sebastian J, Bowman JL, Helariutta Y, Benfey PN (2010) Cell signalling by microRNA165/6 directs gene dose-dependent root cell fate. Nature 465:316–321

Carpenter R, Coen ES (1995) Transposon induced chimeras show that *floricaula*, a meristem identity gene, acts non-autonomously between cell layers. Development 121:19–26

Cartolano M, Efremova N, Kuckenberg M, Raman S, Schwarz-Sommer Z (2009) Enhanced *AGAMOUS* expression in the centre of the Arabidopsis flower causes ectopic expression over its outer expression boundaries. Planta 230:857–862

Castells E, Casacuberta JM (2007) Signalling through kinase-defective domains: the prevalence of atypical receptor-like kinases in plants. J Exp Bot 58:3503–3511

Causier B, Schwarz-Sommer Z, Davies B (2010) Floral organ identity: 20 years of ABCs. Semin Cell Dev Biol 21:73–79

Chapman EJ, Carrington JC (2007) Specialization and evolution of endogenous small RNA pathways. Nat Rev Genet 8:884–896

Chapman LA, Goring DR (2010) Pollen-pistil interactions regulating successful fertilization in the Brassicaceae. J Exp Bot 61:1987–1999

Chevalier D, Batoux M, Fulton L, Pfister K, Yadav RK, Schellenberg M, Schneitz K (2005) STRUBBELIG defines a receptor kinase-mediated signaling pathway regulating organ development in Arabidopsis. Proc Natl Acad Sci USA 102:9074–9079

Chitwood DH, Timmermans MC (2010) Small RNAs are on the move. Nature 467:415–419

Chitwood DH, Nogueira FT, Howell MD, Montgomery TA, Carrington JC, Timmermans MC (2009) Pattern formation via small RNA mobility. Genes Dev 23:549–554

Clark SE, Williams RW, Meyerowitz EM (1997) The *CLAVATA1* gene encodes a putative receptor kinase that controls shoot and floral meristem size in Arabidopsis. Cell 89:575–585

Coen ES, Meyerowitz EM (1991) The war of the whorls: genetic interactions controlling flower development. Nature 353:31–37

Colcombet J, Boisson-Dernier A, Ros-Palau R, Vera CE, Schroeder JI (2005) Arabidopsis SOMATIC EMBRYOGENESIS RECEPTOR KINASES1 and 2 are essential for tapetum development and microspore maturation. Plant Cell 17:3350–3361

Conti L, Bradley D (2007) TERMINAL FLOWER1 is a mobile signal controlling *Arabidopsis* architecture. Plant Cell 19:767–778

Corbesier L, Vincent C, Jang S, Fornara F, Fan Q, Searle I, Giakountis A, Farrona S, Gissot L, Turnbull C, Coupland G (2007) FT protein movement contributes to long-distance signaling in floral induction of *Arabidopsis*. Science 316:1030–1033

Cui H, Levesque MP, Vernoux T, Jung JW, Paquette AJ, Gallagher KL, Wang JY, Blilou I, Scheres B, Benfey PN (2007) An evolutionarily conserved mechanism delimiting SHR movement defines a single layer of endodermis in plants. Science 316:421–425

de Felippes FF, Ott F, Weigel D (2011) Comparative analysis of non-autonomous effects of tasiRNAs and miRNAs in *Arabidopsis thaliana*. Nucleic Acids Res 39:2880–2889

De Smet I, Voß U, Jürgens G, Beeckman T (2009) Receptor-like kinases shape the plant. Nat Cell Biol 11:1166–1173

DeYoung BJ, Clark S (2008) BAM receptors regulate stem cell specification and organ development through complex interactions with CLAVATA signaling. Genetics 180:895–904

DeYoung BJ, Bickle KL, Schrage KJ, Muskett P, Patel K, Clark SE (2006) The CLAVATA1-related BAM1, BAM2 and BAM3 receptor kinase-like proteins are required for meristem function in Arabidopsis. Plant J 45:1–16

Drews GN, Bowman JL, Meyerowitz EM (1991) Negative regulation of the Arabidopsis homeotic gene *AGAMOUS* by the *APETALA2* product. Cell 65:991–1002

Dunoyer P, Schott G, Himber C, Meyer D, Takeda A, Carrington JC, Voinnet O (2010) Small RNA duplexes function as mobile silencing signals between plant cells. Science 328:912–916

Efremova N, Perbal MC, Yephremov A, Hofmann WA, Saedler H, Schwarz-Sommer Z (2001) Epidermal control of floral organ identity by class B homeotic genes in Antirrhinum and Arabidopsis. Development 128:2661–2671

Ehlers K, Kollmann R (2001) Primary and secondary plasmodesmata: structure, origin, and functioning. Protoplasma 216:1–30

Feng X, Dickinson HG (2007) Packaging the male germline in plants. Trends Genet 23:503–510

Feng X, Dickinson HG (2010a) Cell–cell interactions during patterning of the Arabidopsis anther. Biochem Soc Trans 38:571–576

Feng X, Dickinson HG (2010b) Tapetal cell fate, lineage and proliferation in the Arabidopsis anther. Development 137:2409–2416

Fletcher JC, Brand U, Running MP, Simon R, Meyerowitz EM (1999) Signaling of cell fate decisions by *CLAVATA3* in Arabidopsis shoot meristems. Science 283:1911–1914

Fulton L, Batoux M, Vaddepalli P, Yadav RK, Busch W, Andersen SU, Jeong S, Lohmann JU, Schneitz K (2009) DETORQUEO, QUIRKY, and ZERZAUST represent novel components involved in organ development mediated by the receptor-like kinase STRUBBELIG in *Arabidopsis thaliana*. PLoS Genet 5:e1000355

Gagne JM, Clark SE (2010) The Arabidopsis stem cell factor POLTERGEIST is membrane localized and phospholipid stimulated. Plant Cell 22:729–743

Gifford ML, Dean S, Ingram GC (2003) The Arabidopsis ACR4 gene plays a role in cell layer organisation during ovule integument and sepal margin development. Development 130:4249–4258

Gifford ML, Robertson FC, Soares DC, Ingram GC (2005) ARABIDOPSIS CRINKLY4 function, internalization, and turnover are dependent on the extracellular crinkly repeat domain. Plant Cell 17:1154–1166

Gisel A, Barella S, Hempel FD, Zambryski PC (1999) Temporal and spatial regulation of symplastic trafficking during development in *Arabidopsis thaliana* apices. Development 126:1879–1889

Gish LA, Clark SE (2011) The RLK/Pelle family of kinases. Plant J 66:117–127

Gordon SP, Chickarmane VS, Ohno C, Meyerowitz EM (2009) Multiple feedback loops through cytokinin signaling control stem cell number within the Arabidopsis shoot meristem. Proc Natl Acad Sci USA 106:16529–16534

Gross-Hardt R, Lenhard M, Laux T (2002) *WUSCHEL* signaling functions in interregional communication during *Arabidopsis* ovule development. Genes Dev 16:1129–1138

Guo Y, Han L, Hymes M, Denver R, Clark SE (2010) CLAVATA2 forms a distinct CLE-binding receptor complex regulating Arabidopsis stem cell specification. Plant J 63:889–900

Hématy K, Höfte H (2008) Novel receptor kinases involved in growth regulation. Curr Opin Plant Biol 11:321–328

Hord CL, Chen C, Deyoung BJ, Clark SE, Ma H (2006) The BAM1/BAM2 receptor-like kinases are important regulators of Arabidopsis early anther development. Plant Cell 18:1667–1680
Husbands AY, Chitwood DH, Plavskin Y, Timmermans MC (2009) Signals and prepatterns: new insights into organ polarity in plants. Genes Dev 23:1986–1997
Jackson D (2002) Double labeling of KNOTTED1 mRNA and protein reveals multiple potential sites of protein trafficking in the shoot apex. Plant Physiol 129:1423–1429
Jaeger KE, Wigge PA (2007) FT protein acts as a long-range signal in Arabidopsis. Curr Biol 17:1050–1054
Jenik PD, Irish VF (2000) Regulation of cell proliferation patterns by homeotic genes during *Arabidopsis* floral development. Development 127:1267–1276
Jenik PD, Irish VF (2001) The Arabidopsis floral homeotic gene *APETALA3* differentially regulates intercellular signaling required for petal and stamen development. Development 128:13–23
Jeong S, Trotochaud AE, Clark SE (1999) The *Arabidopsis CLAVATA2* gene encodes a receptor-like protein required for the stability of the CLAVATA1 receptor-like kinase. Plant Cell 11:1925–1934
Jia G, Liu X, Owen HA, Zhao D (2008) Signaling of cell fate determination by the TPD1 small protein and EMS1 receptor kinase. Proc Natl Acad Sci USA 105:2220–2225
Kayes JM, Clark SE (1998) *CLAVATA2*, a regulator of meristem and organ development in *Arabidopsis*. Development 125:3843–3851
Kim JY (2005) Regulation of short-distance transport of RNA and protein. Curr Opin Plant Biol 8:45–52
Kinoshita T, Caño-Delgado A, Seto H, Hiranuma S, Fujioka S, Yoshida S, Chory J (2005) Binding of brassinosteroids to the extracellular domain of plant receptor kinase BRI1. Nature 433:167–171
Kinoshita A, Betsuyaku S, Osakabe Y, Mizuno S, Nagawa S, Stahl Y, Simon R, Yamaguchi-Shinozaki K, Fukuda H, Sawa S (2010) RPK2 is an essential receptor-like kinase that transmits the CLV3 signal in *Arabidopsis*. Development 137:3911–3920
Kondo T, Sawa S, Kinoshita A, Mizuno S, Kakimoto T, Fukuda H, Sakagami Y (2006) A plant peptide encoded by CLV3 identified by in situ MALDI-TOF MS analysis. Science 313:845–848
Kroiher M, Miller MA, Steele RE (2001) Deceiving appearances: signaling by "dead" and "fractured" receptor protein-tyrosine kinases. Bioessays 23:69–76
Kwak SH, Shen R, Schiefelbein J (2005) Positional signaling mediated by a receptor-like kinase in Arabidopsis. Science 307:1111–1113
Leibfried A, To JP, Busch W, Stehling S, Kehle A, Demar M, Kieber JJ, Lohmann JU (2005) WUSCHEL controls meristem function by direct regulation of cytokinin-inducible response regulators. Nature 438:1172–1175
Lenhard M, Bohnert A, Jürgens G, Laux T (2001) Termination of stem cell maintenance in Arabidopsis floral meristems by interactions between *WUSCHEL* and *AGAMOUS*. Cell 105:805–814
Li JM, Chory J (1997) A putative leucine-rich repeat receptor kinase involved in brassinosteroid signal transduction. Cell 90:929–938
Lohmann JU, Hong RL, Hobe M, Busch M, Parcy F, Simon R, Weigel D (2001) A molecular link between stem cell regulation and floral patterning in Arabidopsis. Cell 105:793–803
Lucas WJ, Bouche-Pillon S, Jackson DP, Nguyen L, Baker L, Ding B, Hake S (1995) Selective trafficking of KNOTTED1 homeodomain protein and its mRNA through plasmodesmata. Science 270:1980–1983
Lucas WJ, Ham BK, Kim JY (2009) Plasmodesmata—bridging the gap between neighboring plant cells. Trends Cell Biol 19:495–503
Ma H, Sundaresan V (2010) Development of flowering plant gametophytes. Curr Top Dev Biol 91:379–412

Mathieu J, Warthmann N, Kuttner F, Schmid M (2007) Export of FT protein from phloem companion cells is sufficient for floral induction in Arabidopsis. Curr Biol 17:1055–1060
Mayer FX, Schoof H, Haecker A, Lenhard M, Jürgens G, Laux T (1998) Role of *WUSCHEL* in regulating stem cell fate in the Arabidopsis shoot meristem. Cell 95:805–815
Melzer R, Wang YQ, Theissen G (2010) The naked and the dead: the ABCs of gymnosperm reproduction and the origin of the angiosperm flower. Semin Cell Dev Biol 21:118–128
Miwa H, Betsuyaku S, Iwamoto K, Kinoshita A, Fukuda H, Sawa S (2008) The receptor-like kinase SOL2 mediates CLE signaling in Arabidopsis. Plant Cell Physiol 49:1752–1757
Mizuno S, Osakabe Y, Maruyama K, Ito T, Osakabe K, Sato T, Shinozaki K, Yamaguchi-Shinozaki K (2007) Receptor-like protein kinase 2 (RPK 2) is a novel factor controlling anther development in *Arabidopsis thaliana*. Plant J 50:751–766
Molnar A, Melnyk CW, Bassett A, Hardcastle TJ, Dunn R, Baulcombe DC (2010) Small silencing RNAs in plants are mobile and direct epigenetic modification in recipient cells. Science 328:872–875
Müller R, Bleckmann A, Simon R (2008) The receptor kinase CORYNE of Arabidopsis transmits the stem cell-limiting signal CLAVATA3 independently of CLAVATA1. Plant Cell 20:934–946
Nakajima K, Sena G, Nawy T, Benfey PN (2001) Intercellular movement of the putative transcription factor SHR in root patterning. Nature 413:307–311
Nonomura K, Miyoshi K, Eiguchi M, Suzuki T, Miyao A, Hirochika H, Kurata N (2003) The MSP1 gene is necessary to restrict the number of cells entering into male and female sporogenesis and to initiate anther wall formation in rice. Plant Cell 15:1728–1739
Ogawa M, Shinohara H, Sakagami Y, Matsubayashi Y (2008) Arabidopsis CLV3 peptide directly binds CLV1 ectodomain. Science 319:294
Olmedo-Monfil V, Duran-Figueroa N, Arteaga-Vazquez M, Demesa-Arevalo E, Autran D, Grimanelli D, Slotkin RK, Martienssen RA, Vielle-Calzada JP (2010) Control of female gamete formation by a small RNA pathway in Arabidopsis. Nature 464:628–632
Peragine A, Yoshikawa M, Wu G, Albrecht HL, Poethig RS (2004) *SGS3* and *SGS2/SDE1/*i are required for juvenile development and the production of *trans*-acting siRNAs in Arabidopsis. Genes Dev 18:2368–2379
Perbal MC, Haughn G, Saedler H, Schwarz-Sommer Z (1996) Non-cell-autonomous function of the Antirrhinum floral homeotic proteins *DEFICIENS* and *GLOBOSA* is exerted by their polar cell-to-cell trafficking. Development 122:3433–3441
Reinhardt D, Frenz M, Mandel T, Kuhlemeier C (2003) Microsurgical and laser ablation analysis of interactions between the zones and layers of the tomato shoot apical meristem. Development 130:4073–4083
Rinne PL, van der Schoot C (1998) Symplasmic fields in the tunica of the shoot apical meristem coordinate morphogenetic events. Development 125:1477–1485
Rinne PL, Kaikuranta PM, van der Schoot C (2001) The shoot apical meristem restores its symplasmic organization during chilling-induced release from dormancy. Plant J 26:249–264
Satina S, Blakeslee AF, Avery AG (1940) Demonstration of the three germ layers in the shoot apex of Datura by means of induced polyploidy in periclinal chimeras. Am J Bot 27:895–905
Savaldi-Goldstein S, Peto C, Chory J (2007) The epidermis both drives and restricts plant shoot growth. Nature 446:199–202
Schneitz K, Hülskamp M, Pruitt RE (1995) Wild-type ovule development in *Arabidopsis thaliana*: a light microscope study of cleared whole-mount tissue. Plant J 7:731–749
Schoof H, Lenhard M, Haecker A, Mayer KFX, Jürgens G, Laux T (2000) The stem cell population of Arabidopsis shoot meristems is maintained by a regulatory loop between the *CLAVATA* and *WUSCHEL* genes. Cell 100:635–644
Sessions A, Yanofsky MF, Weigel D (2000) Cell-cell signaling and movement by the floral transcription factors LEAFY and APETALA1. Science 289:779–781
Shiu SH, Bleecker AB (2001) Receptor-like kinases from Arabidopsis form a monophyletic gene family related to animal receptor kinases. Proc Natl Acad Sci USA 98:10763–10768

Sieber P, Gheyeselinck J, Gross-Hardt R, Laux T, Grossniklaus U, Schneitz K (2004) Pattern formation during early ovule development in *Arabidopsis thaliana*. Dev Biol 273:321–334

Sieburth LE, Drews GN, Meyerowitz EM (1998) Non-autonomy of *AGAMOUS* function in flower development: use of a Cre/*loxP* method for mosaic analysis in Arabidopsis. Development 125:4303–4312

Slotkin RK, Vaughn M, Borges F, Tanurdzic M, Becker JD, Feijo JA, Martienssen RA (2009) Epigenetic reprogramming and small RNA silencing of transposable elements in pollen. Cell 136:461–472

Steinwand BJ, Kieber JJ (2010) The role of receptor-like kinases in regulating cell wall function. Plant Physiol 153:479–484

Sundberg E, Ostergaard L (2009) Distinct and dynamic auxin activities during reproductive development. Cold Spring Harb Perspect Biol 1:a001628

Szymkowiak EJ, Sussex IM (1992) The internal meristem layer (L3) determines floral meristem size and carpel number in tomato periclinal chimeras. Plant Cell 4:1089–1100

Szymkowiak EJ, Sussex IM (1996) What chimeras can tell us about plant development. Annu Rev Plant Physiol Plant Mol Biol 47:351–376

Tamaki S, Matsuo S, Wong HL, Yokoi S, Shimamoto K (2007) Hd3a protein is a mobile flowering signal in rice. Science 316:1033–1036

Tanaka H, Watanabe M, Sasabe M, Hiroe T, Tanaka T, Tsukaya H, Ikezaki M, Machida C, Machida Y (2007) Novel receptor-like kinase ALE2 controls shoot development by specifying epidermis in Arabidopsis. Development 134:1643–1652

Torii KU, Mitsukawa N, Oosumi T, Matsuura Y, Yokoyama R, Whittier RF, Komeda Y (1996) The Arabidopsis *ERECTA* gene encodes a putative receptor protein kinase with extracellular leucine-rich repeats. Plant Cell 8:735–746

Turck F, Fornara F, Coupland G (2008) Regulation and identity of florigen: FLOWERING LOCUS T moves center stage. Annu Rev Plant Biol 59:573–594

Urbanus SL, de Folter S, Shchennikova AV, Kaufmann K, Immink RG, Angenent GC (2009) *In planta* localisation patterns of MADS domain proteins during floral development in *Arabidopsis thaliana*. BMC Plant Biol 9:5

Urbanus SL, Martinelli AP, Dinh QD, Aizza LC, Dornelas MC, Angenent GC, Immink RG (2010) Intercellular transport of epidermis-expressed MADS domain transcription factors and their effect on plant morphology and floral transition. Plant J 63:60–72

Vaddepalli P, Fulton L, Batoux M, Yadav RK, Schneitz K (2011) Structure-function analysis of STRUBBELIG, an Arabidopsis atypical receptor-like kinase involved in tissue morphogenesis. PLoS One 6:e19730

Van Norman JM, Breakfield NW, Benfey PN (2011) Intercellular communication during plant development. Plant Cell 23:855–864

Vincent CA, Carpenter R, Coen ES (2003) Interactions between gene activity and cell layers during floral development. Plant J 33:765–774

Voinnet O (2009) Origin, biogenesis, and activity of plant microRNAs. Cell 136:669–687

Watanabe M, Tanaka H, Watanabe D, Machida C, Machida Y (2004) The ACR4 receptor-like kinase is required for surface formation of epidermis-related tissues in *Arabidopsis thaliana*. Plant J 39:298–308

Werner D, Gerlitz N, Stadler R (2010) A dual switch in phloem unloading during ovule development in Arabidopsis. Protoplasma 248:225–235

Wollmann H, Mica E, Todesco M, Long JA, Weigel D (2010) On reconciling the interactions between *APETALA2*, miR172 and *AGAMOUS* with the ABC model of flower development. Development 137:3633–3642

Wu X, Dinneny JR, Crawford KM, Rhee Y, Citovsky V, Zambryski PC, Weigel D (2003) Modes of intercellular transcription factor movement in the Arabidopsis apex. Development 130:3735–3745

Xu SL, Rahman A, Baskin TI, Kieber JJ (2008) Two leucine-rich repeat receptor kinases mediate signaling, linking cell wall biosynthesis and ACC synthase in Arabidopsis. Plant Cell 20:3065–3079

Yadav RK, Fulton L, Batoux M, Schneitz K (2008) The Arabidopsis receptor-like kinase STRUBBELIG mediates inter-cell-layer signaling during floral development. Dev Biol 323:261–270

Yang SL, Xie LF, Mao HZ, Puah CS, Yang WC, Jiang L, Sundaresan V, Ye D (2003) *TAPETUM DETERMINANT1* is required for cell specialization in the Arabidopsis anther. Plant Cell 15:2792–2804

Yanofsky MF, Ma H, Bowman JL, Drews GN, Feldmann KA, Meyerowitz EM (1990) The protein encoded by the Arabidopsis homeotic gene *agamous* resembles transcription factors. Nature 346:35–39

Yu LP, Miller AK, Clark SE (2003) *POLTERGEIST* encodes a protein phosphatase 2 C that regulates CLAVATA pathways controlling stem cell identity at Arabidopsis shoot and flower meristems. Curr Biol 13:179–188

Zeevaart JAD (1976) Physiology of flower formation. Annu Rev Plant Physiol 27:321–348

Zhao DZ, Wang GF, Speal B, Ma H (2002) The excess microsporocytes1 gene encodes a putative leucine-rich repeat receptor protein kinase that controls somatic and reproductive cell fates in the Arabidopsis anther. Genes Dev 16:2021–2031

Zhao X, de Palma J, Oane R, Gamuyao R, Luo M, Chaudhury A, Herve P, Xue Q, Bennett J (2008) OsTDL1A binds to the LRR domain of rice receptor kinase MSP1, and is required to limit sporocyte numbers. Plant J 54:375–387

Zhao Z, Andersen SU, Ljung K, Dolezal K, Miotk A, Schultheiss SJ, Lohmann JU (2010) Hormonal control of the shoot stem-cell niche. Nature 465:1089–1092

Zhu Y, Wang Y, Li R, Song X, Wang Q, Huang S, Jin J, Liu C, Lin J (2010) Analysis of interactions among the CLAVATA3 receptors reveals a direct interaction between CLAVATA2 and CORYNE in Arabidopsis. Plant J 61:223–233

Plant Cell Wall Signaling in the Interaction with Plant-Parasitic Nematodes

Krzysztof Wieczorek and Georg J. Seifert

Abstract Plant cell wall signaling, or more generally cell wall performance and integrity control, is thought to play crucial roles in the regulation of plant growth and development in the presence of abiotic and biotic stresses. While, analogous to the well-characterized cell wall integrity response in yeast, the hallmarks of plant cell wall signaling are stress-induced global alterations in the expression of genes related to cell wall biosynthesis and remodeling, its molecular players are only beginning to become defined at the genetic level. Biochemical, molecular, and genetic studies have implicated cell wall signaling with the response to various plant pathogens including fungi and bacteria. Here we speculate how cell wall performance and integrity control might be involved in the infection of roots by sedentary plant-parasitic nematodes. We recapitulate that analogous to various typical cell wall stress scenarios, changes in the expression of cell wall–related genes are a major characteristic of nematode infection.

1 Introduction

In every multicellular organism, the extracellular matrix (ECM) primarily acts to mechanically integrate individual cells within tissues. However, next to its mechanical function, the ECM is an important platform of communication both between cells as well as between the organism and its environment. Owing to this central

K. Wieczorek
Department of Plant Protection, University of Natural Resources and Life Sciences, Vienna, Austria
e-mail: krzysztof.wieczorek@boku.ac.at

G.J. Seifert (✉)
Department for Applied Genetics and Cell Biology, University of Natural Resources and Life Sciences, Vienna, Austria
e-mail: georg.seifert@boku.ac.at

G. Witzany and F. Baluška (eds.), *Biocommunication of Plants*,
Signaling and Communication in Plants 14,
DOI 10.1007/978-3-642-23524-5_8,

importance, all classes of eukaryotes have developed a system to gauge the mechanical performance and structural integrity of their ECM. While the ECM interconnects cells within tissues of all higher organisms, it is also the major mechanical load-bearing network in plants and fungi, organisms that grow by cell enlargement. Due to this analogous mode of growth, both kingdoms might have evolved analogous systems to control the mechanical performance and structural integrity of their cell walls. Generally, a cell wall performance and integrity control system has to sense the state of cell wall polymer structure as well as mechanical parameters such as rigidity and yield to turgor. These stimuli are passed on from the cell surface to the machinery of signal transduction. Either the activated signal transduction system directly influences the cellular machinery involved in biosynthesis or remodeling of the cell wall or in osmotic control, or it affects gene regulation at the transcriptional level to reset the structure and performance of the cell wall to the required level. Thereby the structure and mechanical properties of cell wall polymers are steadily readjusted to fulfill the demands of developmental programming and environmental fluctuations. This also means that the activation of such a signaling system could be recognized by its massive downstream effect on the transcriptional activity of cell wall–related genes.

2 A Paradigm from Yeast

Genetic and biochemical studies performed mainly in *Saccharomyces cerevisiae* have revealed the cell wall integrity (CWI) signaling pathway in yeast (Levin 2005). In brief, five cell surface receptors called Wsc1, Wsc2, Wsc3, Mid2, and Mtl1 stimulate the small G protein Rho1. Rho1 is positively and negatively regulated by guanidine nucleotide exchange factors (GEFs) and GTPase-activating proteins (GAPs), respectively, and directly affects the activity of cell wall glucan synthase (GS) protein Fks1, two regulators of polarized secretion, a mitogen-activated protein (MAP) kinase cascade, and a transcription factor. The relatively linear CWI MAP kinase pathway underlies regulation by protein phosphatases and regulates the activity of transcription factors Rlm1 and Swi4/Swi6 which, among others, regulate the expression of the alternative GS protein Fks2. Thereby Rho1 regulates GS activity either directly on a posttranslational basis or indirectly by transcriptional control. Rlm1, on the other hand, is responsible for the transcriptional regulation of at least 25 genes, many of which are involved in cell wall biosynthesis or remodeling. The key feature of the yeast CWI pathway is its convergence onto a single switch (Rho1) and subsequently a single MAP kinase pathway (Levin 2005). Thus, the activation state of the downstream MAP kinase Mpk1 that is specifically activated by cell wall stress is used as an indicator for the state of CWI signaling in various developmental and environmental situations. Despite its overall linearity, there is cross talk between CWI sensing and several other control pathways including sensors of thermal, pH, osmotic, and oxidative stress (Fuchs and Mylonakis 2009). This cross talk is crucial, for organisms that live with cell walls need to manage the function of this vital organelle throughout development and during various stresses.

3 Plant Cell Wall Signaling Is Inferred from Transcriptional Alterations of Cell Wall–Related Genes in Response to Cell Wall Stress

It was probably just this kind of cross talk between different types of stress responses that has first drawn wider attention to the presence of cell wall performance and integrity control in plants (Pilling and Hofte 2003; Somerville et al. 2004; Vorwerk et al. 2004). The first line of evidence indicated that structural alterations of cell wall polymers trigger physiological responses such as disease resistance [reviewed in Pilling and Hofte (2003); Vorwerk et al. (2004); also see Table 1 in Seifert and Blaukopf (2010) and below]. A second series of observations suggested that when subjected to artificial disturbances to their cell walls such as mutations or drug treatments, plants tend to induce the expression of genes targeted at cell wall biosynthesis and remodeling, presumably in an attempt to repair cell wall damage [see Table 1 in Seifert and Blaukopf (2010)]. For instance, it has been observed that in mutants defective in cellulose biosynthesis or in wild-type cells treated with cellulose biosynthesis inhibitors, there is an induction of pectin biosynthetic genes and cell wall remodeling genes (Manfield et al. 2004; Errakhi et al. 2008;

Table 1 Receptor-like molecules hypothetically involved in cell wall signaling in *Arabidopsis*

Family	Genes in Ath genome	Selected representatives thought to act in cell wall signaling	Key data	Reference
WAK	5 (+20 WAK-like genes)	WAK1, WAK2	Pectin binding	Kohorn et al. (2006) and Brutus et al. (2010)
CrRLKL1	17	THE1	THE1 required for secondary effects of cesA6 mutation	Hématy et al. (2007)
LRR-RLKs	216	FEI1/FEI2	Required for normal level of cellulose under sugar and salt stress conditions	Xu et al. (2008)
Lec-RLKs	46	At5g60300	RGD binding	Gouget et al. (2006)
PERK	11	PERK4	Solubilized by pectinase, involved in ABA sensitivity	Bai et al. (2009)
LRX	7	LRX1	Required for root hair development	Baumberger et al. (2001)

Bischoff et al. 2009; Duval and Beaudoin 2009) and deposition of ectopic lignin (Cano-Delgado et al. 2003; Hamann et al. 2009). Monitoring of transcript profiles during the first 36 h after seedlings were treated with the cellulose synthase inhibitor isoxaben revealed both upregulation of genes involved in lignin biosynthesis and downregulation of expansins and arabinogalactan proteins (Hamann et al. 2009). Treatments that specifically target cell wall structures such as the *in muro* interference with arabinogalactan proteins induce the transcription of numerous cell wall–related genes in addition to wound and stress response (Guan and Nothnagel 2004). Genetic suppression of L-rhamnose biosynthesis results in the transcriptional upregulation of cell wall remodeling genes (Diet et al. 2006). Likewise, in the *mur4* mutant that has arabinose-deficient cell walls (Burget and Reiter 1999), the largest group of differentially regulated genes was involved in the biosynthesis and modification of the cell wall. Interestingly, both the group of upregulated genes and the set of downregulated genes (in *mur4* mutants compared to wild type) were highly enriched in cell wall–related genes (Li et al. 2007). The feedback from defective cell wall structure to the regulation of cell wall biosynthesis and remodeling is not only apparent in mutants affecting the primary cell wall but also in secondary cellulose mutants such as *cesA4*, *cesA7*, and *cesA8*, a major functional group of differentially expressed genes that is related to cell walls. Interestingly, arabinogalactan proteins were exclusively downregulated in secondary cellulose-defective mutants compared to wild-type plants (Hernandez-Blanco et al. 2007).

These examples show that in plants like in yeast, the cell wall stress-induced transcriptional reprogramming of cell wall biosynthesis and remodeling is a telltale "fingerprint" of CWI signaling. This observation will play an important role for the second part of this article.

4 The Plant Cell Wall Signaling Pathway Is Still Poorly Characterized

Unfortunately, compared to the elegant model of the yeast CWI pathway, our genetic and biochemical insights into cell wall performance and integrity control in plants are much less coherent. However, although apparent plant orthologs for key components of the yeast CWI either are lacking or are represented by very large gene families, the *S. cerevisiae* model might provide a very useful working model for a hypothetical plant CWI pathway. Starting to fill in the empty spots in a network for plant CWI signaling, *Arabidopsis thaliana* researchers have recently identified some genetic loci and interesting proteins involved [thoroughly reviewed by Hématy and Hofte (2008); Ringli (2010); Seifert and Blaukopf (2010); Steinwand and Kieber (2010)]. Because the plant cell wall is structurally and developmentally more complex than the yeast ECM, the sensors that gauge cell wall structure and performance are likely to be more diverse. Indeed, plant genomes contain a vast number of receptor-like protein kinases (RLKs), some of which act in developmental regulation, while others are involved in pathogen perception. A subset, however, is involved in cell wall signaling (Table 1).

At present, the evidence that any of the hypothetical cell wall receptors indeed transmits information on the status of the cell wall into the cell is fragmentary in most cases. For some candidates, mutant and transgenic data indicate a genetic interaction of a receptor-like molecule with cell wall biosynthesis or structure, as is the case for THESEUS1 (THE1) (Hématy et al. 2007), the FEI1/FEI2 pair (Xu et al. 2008), or LRX1 (Baumberger et al. 2001; Diet et al. 2006). However, the extracellular ligands of the mentioned receptor-like proteins are not known. In other cases, binding to potential ligands, as in the case of interaction of some leguminous L-type lectin RLKs with the peptide RGD (Gouget et al. 2006) or association of PERK4 with pectin (Bai et al. 2009), remains to be complemented by genetic evidence for a role for cell wall signaling.

The best characterized paradigm for cell wall signaling is the wall-associated kinases (WAKs). The WAK family in *Arabidopsis* contains five linked genes coding for type I transmembrane proteins that contain a carboxy-proximal serine/threonine protein kinase domain exposed to the cytosol and an amino-proximal epidermal growth factor (EGF)–like repeat domain exposed to the extracellular space. Originally, WAKs were defined by their tight association with the cell wall *in planta* that can be solubilized with pectin-degrading enzymes (He et al. 1996; Wagner and Kohorn 2001). It has been shown that the domain upstream of the EGF-like repeat binds to Ca^{2+}-cross-linked oligogalacturonides (OG) in vitro (Decreux and Messiaen 2005). Using a chimeric receptor strategy, it was found that the WAK1 extracellular domain fused to the kinase domain of the receptor for the bacterial elicitor Ef-Tu (EFR) triggers EFR-like responses in dependence of OG (Brutus et al. 2010). One explanation for this observation is that WAK1 is a receptor of OG and thereby involved in sensing damage-associated molecular patterns (DAMPs) released while pathogens attack cell wall polymers. Could WAKs be involved in CWI sensing even beyond pathogen-induced cell wall damage? On the one hand, there is genetic evidence for a role of WAKs in cell elongation and in growth under sugar-limited conditions in whole plants (Lally et al. 2001; Wagner and Kohorn 2001; Kohorn et al. 2006). On the other hand, *WAK2* is responsible for the homogalacturonan-triggered induction of vacuolar invertase in protoplasts, a response that could be reconciled with a feedback loop sensing the state of the cell wall and triggering a compensatory response at the level of turgor control. Hence, WAKs might sense the conformation and integrity of pectic homogalacturonan in the control of innate immune response and growth. What processes might lie downstream the WAKs? Initial evidence implicates a MAP kinase cascade including MAPK3 in the pectin-triggered and *WAK2*-dependent induction of vacuolar invertase (Kohorn et al. 2009). With this finding, our view of plant cell wall signaling has moved a little bit closer to the well-elaborated model of the yeast CWI pathway. However, at present, it needs to be stated that there are no *bona fide* reporters for the specific activation of plant cell wall performance and integrity signaling. While the activation of *S. cerevisiae* Mpk1 and the induction of Rlm1-responsive reporter genes are specific indicators for CWI signaling in yeast (Levin 2005), MAPK3 activation occurs in response to a vast array of different stresses (Colcombet and Hirt 2008). Moreover, no plant cell wall stress-specific cis-responsive element has yet been identified. Therefore, stimulus-triggered transcript profiles that are enriched with differentially regulated genes dedicated to cell wall biosynthesis and remodeling remain the best indirect indicator that cell wall signaling is at work. Under this premise, we will now consider the potential involvement of cell wall signaling in the interaction between nematodes and their plant hosts in the following section.

5 Is Cell Wall Signaling Active in the Interaction Between Plant-Parasitic Nematodes and Roots?

Recent genetic observations have implicated cell wall signaling with various plant diseases (reviewed in Hématy et al. 2009). In particular, it has been found that genetic alterations of diverse cell wall polymers trigger pathogen resistance. Some examples include the powdery mildew–resistant (*pmr*) mutants *pmr4*, *pmr5*, and *pmr6* that are affecting callose synthase, and two genes required for normal pectin structure, respectively (Vogel et al. 2002; Nishimura et al. 2003; Vogel et al. 2004). On the one hand, a mutation in the *MUR3* locus that is required for normal xyloglucan structure in primary cell walls leads to resistance to *Hyaloperonospora parasitica* (Tedman-Jones et al. 2008). Defective secondary cellulose, on the other hand, triggers resistance to some bacterial and fungal pathogens (Hernandez-Blanco et al. 2007). More recently, it was described that normal cell wall polymer acetylation is required for wild-type susceptibility toward a range of fungal necrotrophic pests (Manabe et al. 2011). While these phenomena suggest that there is intense cross talk between the sensing and signaling of defective cell walls and the control of innate immune response, there is little insight in how cell wall signaling might act during plant-parasite interactions. One of the most sophisticated of these relationships is established by the sedentary plant-parasitic nematodes that manipulate root vascular cells into large hypertrophied and multinucleate nematode feeding sites (NFS), as reviewed in Gheysen and Goellner-Mitchum (2011). Specific changes in the expression of the cell wall biosynthetic, degrading, and modifying enzymes and proteins as well as highly elaborate alterations in the cell wall architecture are important prerequisites for a successful infection. These soil-born worms belong to the economically most important pests in agriculture nowadays. Two groups are of great interest: cyst nematodes (CNs; *Heterodera spp.* and *Globodera spp.*) and root-knot nematodes (RKNs; *Meloidogyne spp.*). CNs invade roots of host plants in the root elongation zone and migrate intracellularly in search of a suitable root cell (Golinowski et al. 1996; Sobczak et al. 2005). This process is facilitated by cell wall enzymes produced in the subventral glands and secreted by the nematode (Davis et al. 2008). Right after a CN larva has chosen the eligible cell, it pierces this cell with the mouth stylet, a protrusible hollow spear, and injects widely unknown effectors into the cytoplasm (Fig. 1). This event triggers the formation of the initial cell followed by induction of a so-called syncytium, a sink and feeding organ composed of hundreds of root cells that fuse together as a result of a partial cell wall dissolution. In contrast, at later stages, the outer syncytial cell wall is thickened, which enables it to withstand the high osmotic pressure within the growing feeding site. Even more pronounced alterations of the wall can be observed in syncytia. At the border to xylem elements, elaborate cell wall ingrowths are formed (Jones and Gunning 1976; Jones and Northcote1972; Golinowski et al. 1996) in order to increase the membrane surface, thus enhancing water transport (Pate and Gunning 1972; Gunning 1977; Offler et al. 2003).

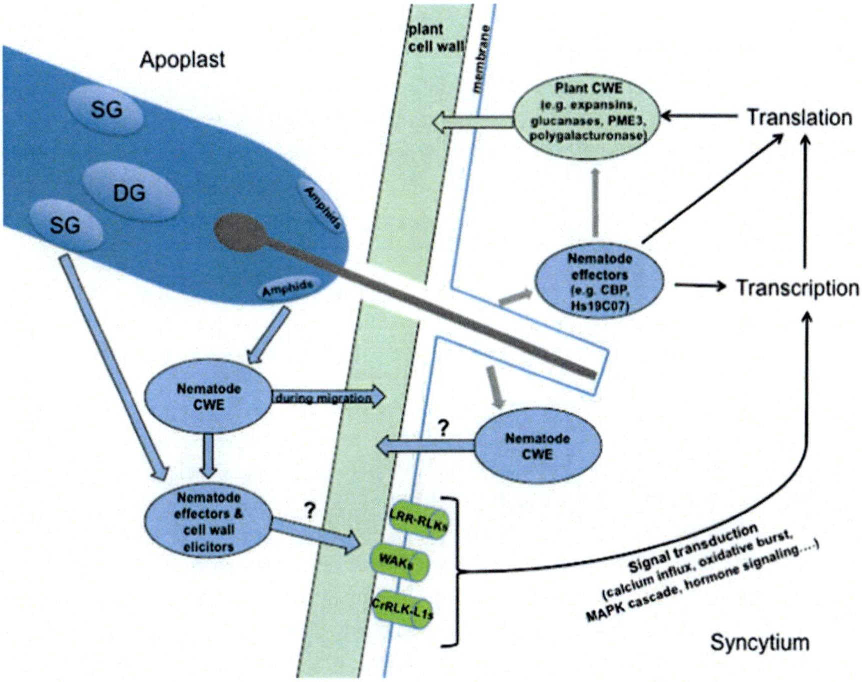

Fig. 1 Action of nematode cell wall–degrading enzymes and effector proteins in host plant tissue and in the nematode feeding site. This figure schematically illustrates the current state of knowledge and hypotheses concerning the role of cell wall and cell wall signaling during the infection of sedentary plant-parasitic nematodes, cyst, as well as root-knot nematodes. Cell wall–degrading enzymes produced in subventral gland cells are released through the stylet and facilitate the nematode migration in the root tissue. Cell wall elicitors might be recognized by a set of different plant receptors (LRR-RLKs, WAKs, CrRLK-L1s) that trigger the signal transduction followed by changes in the expression of cell wall–related genes such as expansins or glucanases. During the sedentary stage, nematodes inject effectors from dorsal gland cells into the feeding site that can interact with plant cell wall enzymes, such as nematode CBP that interacts with plant PME3, or Hs19C07 that triggers the auxin-dependent activation of plant cell wall enzymes, such as polygalacturonase

By contrast, RKNs invade roots of the host plants at the root tip and, by dissolving the middle lamella by means of secreted cell wall–degrading enzymes, are able to migrate intercellularly (Wyss 1992). They trigger the formation of several giant cells in the vascular parenchyma that are embedded in gall tissue (root knot). These cells undergo repeated mitosis without following cytokinesis and contain multiple enlarged nuclei (de Almeida-Engler et al. 1999; Goverse et al. 2000). Cell walls of giant cells are thickened, and extensive wall ingrowths are formed, but in contrast to syncytia, the dissolution of the cell wall does not take place.

During the migration phase, both groups of nematodes use cell wall–degrading and cell wall–modifying enzymes and proteins that are produced in subventral glands. At later stages, when the nematodes become sedentary and the formation of the feeding site is in progress, they inject effector molecules from dorsal glands that are thought to change the plant response. Among those (summarized by Davis et al. 2008), there are also cell wall effectors; however, only few of them are analyzed in detail so far.

Here, we summarize experimental data from nematodes and, based on this and findings obtained from other systems such as *Saccharomyces cerevisiae*, try to hypothesize about the role of the plant cell wall and cell wall signaling in the interaction between plant-parasitic nematodes and their host plants.

5.1 Does the Plant Sense the Nematode While It Migrates Within the Root Tissue?

One of the important differences between the cyst and root-knot nematodes is the way they migrate within the plant root. After entering, the CNs move intracellularly, while RKNs migrate intercellularly. To facilitate this process, both groups secrete a cocktail of enzymes and proteins produced in the subventral glands that soften or degrade the structure of the plant cell wall (Fig. 1). For both CNs and RKNs, it contains 1,4-β-glucanases that hydrolyze 1-4-beta-D-glucosidic linkages in glucans such as cellulose or lichenin (Smant et al. 1998; de Boer et al. 1999; Lilley et al. 1999; Rosso et al. 1999; Wang et al. 1999; Goellner et al. 2000; Gao et al. 2004; Chen et al. 2005; Ledger et al. 2006; Ithal et al. 2007a; Roze et al. 2008; Rehman et al. 2009). It was shown that knocking down the cellulase in *Globodera rostochiensis* (Chen et al. 2005) by soaking the larvae in dsRNA led to reduced penetration of the plant by the nematodes. In order to dissolve the middle lamella that contains mainly pectin, CNs and RKNs produce and secrete a set of enzymes that hydrolyze this polymer. In secretions from beetroot cyst nematode *H. schachtii*, an arabinogalactan endo-1,4-beta-galactosidase was found (Vanholme et al. 2006, 2009). In other CNs and RKNs, pectate lyases, endoxylanases, and polygalacturonases were found (Vanholme et al. 2006, 2009; Popeijus et al. 2000; de Boer et al. 2002; Doyle and Lambert 2002; Huang et al. 2005; Ithal et al. 2007a; Bellafiore et al. 2008; Roze et al. 2008; Mitreva-Dautova et al. 2006; Jaubert et al. 2002). Furthermore, migrating larvae produce proteins that are thought to facilitate degradation of the plant cell wall, such as cellulose-binding protein (Ding et al. 1998; Huang et al. 2003; Gao et al. 2004; Ithal et al. 2007a; Adam et al. 2008; Bellafiore et al. 2008; Hewezi et al. 2008; Jones et al. 2009) and expansins and expansin-like proteins (Ding et al. 1998; Qin et al. 2004; Kudla et al. 2005). Taking all these into account, we can speculate that plants are able to sense the migrating nematode either via detection of damage of wall polysaccharides, release of oligosaccharides, or deformation of the plasma membrane in damaged or weakened plant cells. Unfortunately, there is no information on involvement of any plant receptors recognizing the extracellular cell wall degradation products in the nematode perception of the nematode during the migration phase. Therefore, we can only speculate that potential extracellular receptors involved in CWI control might play a role in sensing the migrating nematode. Possibly the L-type lectin RKL and CrRLK-L1 might be involved in the recognition of the cell wall damage caused by endoglucanases secreted by nematodes (cellodextrin). During later stages of the feeding site development, however, most of these receptor-like kinases do not show any changes in their expression or are downregulated (Szakasits et al. 2009). Similarly, WAKs could be involved in sensing the pectin-derived degrading products, such as OG, that are associated with the action of the nematode pectin-degrading enzymes, e.g., pectate lyases. However, until now, no experimental data concerning early stages of the nematode-plant interaction support that, but in older syncytia, most of the WAKs are downregulated, or their expression remains unchanged (Szakasits et al. 2009). It is an intriguing possibility that the nematodes during migration actively trigger downregulation of these receptors facilitating the parasitism process. Concerning the signal transduction, there are only hints on specific upregulation of genes, including downstream MAP kinases and their regulators during the early developmental stages of syncytia induced by *H. schachtii* (Hofmann, unpublished results). There are few examples of resistant host plants that can be invaded by the nematode, but soon after the larva becomes surrounded by necrotized cells, neither induction of the initial cell nor further development of the nematode takes place (Bleve-Zacheo et al. 1982; Kouassi et al. 2004; Paulson and Webster 1972). Are these plants successfully countering the nematode's efforts by quickly reacting with a hypersensitive response? Until now, due to the lack of information about the genetic background of these phenomena, this question still remains unanswered and requires experimental data.

5.2 *Nematode Effectors Interfere with Cell Wall Performance in the Nematode Feeding Site*

More information is available about the sedentary stage of the parasitism process concerning the nematode proteins that might influence cell wall performance within the NFS. Recently reports about nematode effectors that are produced in dorsal glands and injected through the stylet into the syncytia of CN were published (Fig. 1). One of these effectors is the cellulose-binding protein (CBP) from the beet cyst nematode *H. schachtii* (Ding et al. 1998; Hewezi et al. 2008). Hs CBP is expressed only during the early phase of feeding site induction but not during the migratory phase. It was shown that it strongly and specifically interacts with *Arabidopsis* pectin methylesterase (PME3) and plants overexpressing CBP show increased PME3 activity. These results indicate that nematodes are able to target and influence directly plant cell wall enzymes, thereby facilitating nematode parasitism. The second recent study reported on the function of an esophageal gland cell protein from *H. schachtii*, Hs19C07, during the development of syncytia (Lee et al. 2011). This novel effector interacts with *Arabidopsis* auxin influx transporter LAX3 that is expressed in lateral root primordia where it provides a hormonal signal that triggers the expression of cell wall–modifying enzymes. This in turn allows lateral roots to emerge. Increased activity of LAX3 triggers the auxin-dependent induction of a polygalacturonase. Both genes are expressed within the syncytia as well as in the adjacent root cells that will be incorporated into the NFS. A decrease in the number of females was observed on the double *aux1lax3* and quadruple mutant *aux1lax1lax2lax3*, suggesting an important role of LAX transporters during the development of syncytia. Hs10C07 is thought to function in LAX3-mediated auxin influx into the syncytium. Moreover, it might allow the nematode to regulate the auxin flow in both NFS and adjacent root cells, thereby activating the hydrolysis of the plant cell wall and thus facilitating syncytium development. This is greatly supported by studies on the hormone distribution in the NFS which show that nematodes have evolved to manipulate the hormone network and use it for their own purposes (de Meutter et al. 2005, Goverse et al. 2000; Karczmarek et al. 2004; Grunewald et al. 2008; Grunewald et al. 2009a; Grunewald et al. 2009b).

5.3 *Expression of Cell Wall Enzymes in the Nematode Feeding Site Is Specifically Affected*

As a reaction to a nematode infection, specific changes in the expression of plant cell wall enzymes occur within the NFS. However, whether these alterations are due to the innate immune response of the host or are caused by the pathogen itself is hard to distinguish. What are the triggers that cause these massive changes? It is known that nematodes use different ways to control the genetic machinery of the plant. They secrete effectors that can activate plant enzymes, as described in the previous section (Ding et al. 1998; Hewezi et al. 2008). Alternatively, they are able to change the hormonal balance, and they thereby potentially affect host hormone-responsive genes (Goverse et al. 2000; Karczmarek et al. 2004; Grunewald et al. 2008, Grunewald et al. 2009a; Grunewald et al. 2009b; Lee et al. 2011). On the other hand, assuming the plant senses nematodes via various receptors and gauges the integrity of its own wall, it can be predicted that at least a part of the changes is based on the host response. The above-mentioned activation of certain MAP kinases and their regulators during the early stages of syncytium development could certainly play an important role in this process (Hofmann, unpublished results).

During the nematode infection, these different triggers affect a vast number of different groups of cell wall–related genes, including genes involved in the biosynthesis, degradation, and remodeling of the cell wall. In several studies mostly based on gene chip analysis, changes in the expression of CesA and different classes of CSL genes were described (Ithal et al. 2007b; Hudson 2008; Szakasits et al. 2009; Barcala et al. 2009). It seems that the cellulose biosynthesis in giant cells and syncytia differs because most of the corresponding genes in RKN feeding sites were downregulated, whereas in CN-induced syncytia, they were mostly upregulated. This is supported by the functional analysis of CesA mutants (CesA1-8), as the development of the RKN females on these plants was decreased (Hudson 2008), while on selected CesA lines, there was a significant increase in the number of CN females (Wieczorek, unpublished results).

Early studies indicating that glucanases might be involved in the formation of the NFS were done on CN *Globodera tabacum* and RKN *Meloidogyne incognita* in tobacco roots (Goellner et al. 2001). Subsequently, further evidence came from studies using various other nematode species and different plants, including *Arabidopsis* (Sukno et al. 2006; Wang et al. 2007; Karczmarek et al. 2008; Wieczorek et al. 2008; Swiecicka et al. 2009). In one of these reports, mutants of two endo-1,4-β-glucanases, *kor3* and *cel2*, showed a reduction in the number of developing females by 45% and 48%, respectively (Wieczorek et al. 2008). Furthermore, there are examples of pectin-degrading and pectin-modifying enzymes of plant origin that are differentially expressed in the NFS, such as pectin acetylesterase (Vercauteren et al. 2002), pectate lyases (Puthoff et al. 2003, Jammes et al. 2005; Wieczorek, unpublished results), and polygalacturonases (Mahalingham et al. 1999). Finally, not only expansins secreted by the nematode play a role in the parasitism process, but also plant expansins are crucial for the development of the feeding site and the parasite. Apart from several microarray studies showing different expression of expansin isoforms (Jammes et al. 2005; Bar-Or et al. 2005; Gal et al. 2006; Ithal et al. 2007a, 2007b, Puthoff et al. 2007; Tucker et al. 2007), expression of these cell wall–modifying proteins expressed in NFS has been investigated in more detail. It was shown that certain *Arabidopsis* and tomato expansin isoforms are specifically expressed in syncytia induced by *H. schachtii* (Wieczorek et al. 2006) and *G. rostochiensis* (Fudali et al. 2008). In case of *Arabidopsis*, the study revealed that EXPA3 and EXPA16 are and are not present in other parts of the root. In tomato, it was shown that EXPA5 might be involved in the cell wall relaxation that supports the hypertrophy of the feeding site.

6 Summary

CWI signaling in yeast provides a paradigm that helps to envisage analogous pathways in plants. Being a presumably far more complex signaling network, only some components of plant cell wall performance and integrity control are presently known. However, genome-wide alterations in transcript levels of enzymes related to cell wall biosynthesis and remodeling are hallmarks of activation of this elusive signaling pathway. How cell wall performance and integrity control is involved in the interaction between host plants and sedentary plant-parasitic nematodes still remains largely unknown. However, there are first hints about the changes to the cell wall caused by infecting nematodes and which cell wall enzymes they secrete to facilitate the migration within the root tissue. Furthermore, nematode effectors that are injected into the feeding site were found to interact with plant cell wall–remodeling enzymes. Many microarray studies provided information about how genes encoding plant cell wall–synthesizing, cell wall–degrading, and cell wall–remodeling enzymes as well as genes involved directly in the cell wall signaling process are affected by the nematode infection. On the one hand, there are some suggestions how nematodes could trigger these massive changes in gene expression. On the other hand, knowledge about possible receptors and further signal transduction pathways is still very limited. Thus, there are still gaps in knowledge about the signaling role of the cell wall in the parasitic process of sedentary plant-parasitic nematodes, and therefore, more experimental work is needed to shed more light on this fascinating topic.

Acknowledgment The authors receive financial support of the Austrian Science Fund (FWF) grants P19778, P21782, and P21067.

References

Adam MAM, Phillips MS, Jones JT, Blok VC (2008) Characterisation of the cellulose-binding protein Mj-cbp-1 of the root-knot nematode, *Meloidogyne javanica*. Physiol Mol Plant Pathol 72:21–28

Bai L, Zhang G, Zhou Y, Zhang Z, Wang W, Du Y, Wu Z, Song CP (2009) Plasma membrane-associated proline-rich extensin-like receptor kinase 4, a novel regulator of Ca signalling, is required for abscisic acid responses in *Arabidopsis thaliana*. Plant J 60:314–327

Barcala M, García A, Cabrera J, Casson S, Lindsey K, Favery B, García-Casado G, Solano R, Fenoll C, Escobar C (2009) Early transcriptomic events in microdissected Arabidopsis nematode-induced giant cells. Plant J 61:698–712

Bar-Or C, Kapulnik Y, Koltai H (2005) A broad characterization of the transcriptional profile of the compatible tomato response to the plant parasitic root knot nematode *Meloidogyne javanica*. Eur J Plant Pathol 111:181–192

Baumberger N, Ringli C, Keller B (2001) The chimeric leucine-rich repeat/extensin cell wall protein LRX1 is required for root hair morphogenesis in *Arabidopsis thaliana*. Genes Dev 15:1128–1139

Bellafiore S, Shen ZX, Rosso MN, Abad P, Shih P, Briggs SP (2008) Direct identification of the *Meloidogyne incognita* secretome reveals proteins with host cell reprogramming potential. PLoS Pathog 4:e1000192

Bischoff V, Cookson SJ, Wu S, Scheible WR (2009) Thaxtomin A affects CESA-complex density, expression of cell wall genes, cell wall composition, and causes ectopic lignification in Arabidopsis thaliana seedlings. J Exp Bot 60:955–965

Bleve-Zacheo T, Zacheo G, Melillo MT, Lamberti F (1982) Ultrastructural aspects of the hypersensitive reaction in tomato root cells resistant to *Meloidogyne incognita*. Nematol Mediter 10:81–90

Brutus A, Sicilia F, Macone A, Cervone F, De Lorenzo G (2010) A domain swap approach reveals a role of the plant wall-associated kinase 1 (WAK1) as a receptor of oligogalacturonides. Proc Natl Acad Sci USA 107:9452–9457

Burget EG, Reiter WD (1999) The *mur4* mutant of Arabidopsis is partially defective in the de novo synthesis of uridine diphospho L-arabinose. Plant Physiol 121:383–389

Cano-Delgado A, Penfield S, Smith C, Catley M, Bevan M (2003) Reduced cellulose synthesis invokes lignification and defense responses in *Arabidopsis thaliana*. Plant J 34:351–362

Chen Q, Rehman S, Smant G, Jones JT (2005) Functional analysis of pathogenicity proteins of the potato cyst nematode *Globodera rostochiensis* using RNAi. Mol Plant Microbe Interact 18:621–625

Colcombet J, Hirt H (2008) Arabidopsis MAPKs: a complex signalling network involved in multiple biological processes. Biochem J 413:217–226

Davis EL, Hussey RS, Mitchum MG, Baum TJ (2008) Parasitism proteins in nematode–plant interactions. Curr Opin Plant Biol 11:360–366

de Almeida EJ, De Vleesschauwer V, Burssens S, Celenza JL, Inzé D, Van Montagu M, Engler G, Gheysen G (1999) Molecular markers and cell cycle inhibitors show the importance of cell cycle progression in nematode-induced galls and syncytia. Plant Cell 11:793–807

de Boer JM, Yan YT, Wang XH, Smant G, Hussey RS, Davis EL, Baum TJ (1999) Developmental expression of secretory beta-1,4-endoglucanases in the subventral esophageal glands of *Heterodera glycines*. Mol Plant Microbe Interact 12:663–669

de Boer JM, McDermott JP, Davis EL, Hussey RS, Popeijus H, Smant G, Baum TJ (2002) Cloning of a putative pectate lyase gene expressed in the subventral esophageal glands of *Heterodera glycines*. J Nematology 34:9–11

de Meutter J, Tytgat T, Prinsen E, Gheysen G, van Onckelen H (2005) Production of auxin related compounds by the plant parasitic nematodes *Heterodera schachtii* and *Meloidogyne incognita*. Commun Agric Appl Biol Sci 70:51–60

Decreux A, Messiaen J (2005) Wall-associated kinase WAK1 interacts with cell wall pectins in a calcium-induced conformation. Plant Cell Physiol 46:268–278

Diet A, Link B, Seifert GJ, Schellenberg B, Pauly M, Wangar U, Reiter WD, Ringli C (2006) The Arabidopsis root hair cell wall formation mutant *lrx1* is suppressed by mutations in the *RHM1* gene encoding a UDP-L-rhamnose synthase. Plant Cell 18:1630–1641

Ding X, Shields J, Allen R, Hussey RS (1998) A secretory cellulose-binding protein cDNA cloned from the root-knot nematode (*Meloidogyne incognita*). Mol Plant Microbe Interact 11:952–959

Doyle EA, Lambert KN (2002) Cloning and characterization of an esophageal-gland-specific pectate lyase from the root-knot nematode *Meloidogyne javanica*. Mol Plant Microbe Interact 15:549–556

Duval I, Beaudoin N (2009) Transcriptional profiling in response to inhibition of cellulose synthesis by thaxtomin A and isoxaben in *Arabidopsis thaliana* suspension cells. Plant Cell Rep 28:811–830

Errakhi R, Dauphin A, Meimoun P, Lehner A, Reboutier D, Vatsa P, Briand J, Madiona K, Rona JP, Barakate M, Wendehenne D, Beaulieu C, Bouteau F (2008) An early Ca^{2+} influx is a prerequisite to thaxtomin A-induced cell death in *Arabidopsis thaliana* cells. J Exp Bot 59:4259–4270

Fuchs BB, Mylonakis E (2009) Our paths might cross: the role of the fungal cell wall integrity pathway in stress response and cross talk with other stress response pathways. Eukaryot Cell 8:1616–1625

Fudali S, Janakowski S, Sobczak M, Griesser M, Grundler FMW, Golinowski W (2008) Two tomato α-expansins show distinct spatial and temporal expression patterns during development of nematode-induced syncytia. Physiol Plant 132:370–383

Gal TZ, Aussenberg ER, Burdman S, Kapulnik Y, Koltai H (2006) Expression of a plant expansin is involved in the establishment of root knot nematode parasitism in tomato. Planta 224:155–162

Gao B, Allen R, Davis EL, Baum TJ, Hussey RS (2004) Molecular characterisation and developmental expression of a cellulose-binding protein gene in the soybean cyst nematode *Heterodera glycines*. Int J Parasitol 34:1377–1383

Gheysen G, Goellner-Mitchum M (2011) How nematodes manipulate plant development pathways for infection. Curr Opin Plant Biol 14:1–7

Goellner M, Smant G, De Boer JM, Baum TJ, Davis EL (2000) Isolation of beta-1, 4-endoglucanase genes from *Globodera tabacum* and their expression during parasitism. J Nematol 32:154–165

Goellner M, Wang XH, Davis EL (2001) Endo-beta-1,4-glucanase expression in compatible plant–nematode interactions. Plant Cell 13:2241–2255

Golinowski W, Grundler FMW, Sobczak M (1996) Changes in the structure of Arabidopsis thaliana during female development of the plant-parasitic nematode Heterodera schachtii. Protoplasma 194:103–116

Gouget A, Senchou V, Govers F, Sanson A, Barre A, Rouge P, Pont-Lezica R, Canut H (2006) Lectin receptor kinases participate in protein–protein interactions to mediate plasma membrane-cell wall adhesions in Arabidopsis. Plant Physiol 140:81–90

Goverse A, Engler JD, Verhees J, van der Krol S, Helder J, Gheysen G (2000) Cell cycle activation by plant parasitic nematodes. Plant Mol Biol 43:747–761

Grunewald W, Karimi M, Wieczorek K, Van de Cappelle E, Wischnitzki E, Grundler F, Inzé D, Beeckman T, Gheysen G (2008) A Role for AtWRKY23 in feeding site establishment of plant-parasitic nematodes. Plant Physiol 148:358–368

Grunewald W, Cannoot B, Friml J, Gheysen G (2009a) Parasitic nematodes modulate PIN-mediated auxin transport to facilitate infection. PLoS Pathog 5:e1000266

Grunewald W, van Noorden G, van Isterdael G, Beeckman T, Gheysen G, Mathesius U (2009b) Manipulation of auxin transport in plant roots during *Rhizobium* symbiosis and nematode parasitism. Plant Cell 21:2553–2562

Guan Y, Nothnagel EA (2004) Binding of arabinogalactan proteins by Yariv phenylglycoside triggers wound-like responses in Arabidopsis cell cultures. Plant Physiol 135:1346–1366

Gunning BES (1977) Transfer cells and their roles in transport of solutes in plants. Sci Prog 64:539–568

Hamann T, Bennett M, Mansfield J, Somerville C (2009) Identification of cell-wall stress as a hexose-dependent and osmosensitive regulator of plant responses. Plant J 57:1015–1026

He ZH, Fujiki M, Kohorn BD (1996) A cell wall-associated, receptor-like protein kinase. J Biol Chem 271:19789–19793

Hématy K, Hofte H (2008) Novel receptor kinases involved in growth regulation. Curr Opin Plant Biol 11:321–328

Hématy K, Sado PE, Van Tuinen A, Rochange S, Desnos T, Balzergue S, Pelletier S, Renou JP, Höfte H (2007) A receptor-like kinase mediates the response of Arabidopsis cells to the inhibition of cellulose synthesis. Curr Biol 17:922–931

Hématy K, Cherk C, Somerville S (2009) Host–pathogen warfare at the plant cell wall. Curr Opin Plant Biol 12:406–413

Hernandez-Blanco C, Feng DX, Hu J, Sanchez-Vallet A, Deslandes L, Llorente F, Berrocal-Lobo M, Keller H, Barlet X, Sanchez-Rodriguez C, Anderson LK, Somerville S, Marco Y, Molina A (2007) Impairment of cellulose synthases required for Arabidopsis secondary cell wall formation enhances disease resistance. Plant Cell 19:890–903

Hewezi T, Howe P, Maier TR, Hussey RS, Mitchum MG, Davis EL, Baum TJ (2008) Cellulose binding protein from the parasitic nematode *Heterodera schachtii* interacts with *Arabidopsis* pectin methylesterase: co-operative cell wall modification during parasitism. Plant Cell 20:3080–3093

Huang G, Gao B, Maier T, Allen R, Davis EL, Baum TJ, Hussey RS (2003) A profile of putative parasitism genes expressed in the esophageal gland cells of the root-knot nematode *Meloidogyne incognita*. Mol Plant Microbe Interact 16:376–381

Huang G, Dong R, Allen R, Davis EL, Baum TJ, Hussey RS (2005) Developmental expression and molecular analysis of two *Meloidogyne incognita* pectate lyase genes. Int J Parasitol 35:685–692

Hudson LC (2008) Analysis of cell wall synthesis genes in feeding cells formed by root-knot nematodes. PhD thesis. Plant pathology, North Carolina State University, Raleigh

Ithal N, Recknor J, Nettleton D, Hearne L, Maier T, Baum TJ, Mitchum MG (2007a) Parallel genome-wide expression profiling of host and pathogen during soybean cyst nematode infection of soybean. Mol Plant Microbe Interact 20:293–305

Ithal N, Recknor J, Nettleton D, Maier T, Baum TJ, Mitchum MG (2007b) Developmental transcript profiling of cyst nematode feeding cells in soybean roots. Mol Plant Microbe Interact 20:510–525

Jammes F, Lecomte P, de Almeida-Engler J, Bitton F, Martin-Magniette ML, Renou JP, Abad P, Favery B (2005) Genome-wide expression profiling of the host response to root-knot nematode infection in Arabidopsis. Plant J 44:447–458

Jaubert S, Laffaire JB, Abad P, Rosso MN (2002) A polygalacturonase of animal origin isolated from the root-knot nematode *Meloidogyne incognita.* FEBS Lett 522:109–112

Jones MGK, Gunning BES (1976) Transfer cells and nematode induced giant cells in Helianthemum. Protoplasma 87:273–279

Jones MGK, Northcote DH (1972) Nematode-induced syncytium—a multinucleate transfer cell. J Cell Sci 10:789–809

Jones JT, Kumar A, Pylypenko LA, Thirugnanasambandum A, Castelli L (2009) Identification and functional characterization of effectors in expressed sequence tags from various life cycle stages of the potato cyst nematode *Globodera pallida.* Mol Plant Pathol 10:815–828

Karczmarek A, Overmars H, Helder J, Goverse A (2004) Feeding cell development by cyst and root-knot nematodes involves a similar early, local and transient activation of a specific auxin-inducible promoter element. Mol Plant Pathol 5:343–346

Karczmarek A, Fudali S, Lichocka M, Sobczak M, Kurek W, Janakowski S, Roosien J, Golinowski W, Bakker J, Goverse A, Helder J (2008) Expression of two functionally distinct plant endo-β-1,4-glucanases is essential for compatible interaction between potato cyst nematode and its hosts. Mol Plant Microbe Interact 21:791–798

Kohorn BD, Kobayashi M, Johansen S, Riese J, Huang LF, Koch K, Fu S, Dotson A, Byers N (2006) An Arabidopsis cell wall-associated kinase required for invertase activity and cell growth. Plant J 46:307–316

Kohorn BD, Johansen S, Shishido A, Todorova T, Martinez R, Defeo E, Obregon P (2009) Pectin activation of MAP kinase and gene expression is WAK2 dependent. Plant J 60:974–982

Kouassi AB, Kerlan MC, Sobczak M, Dantec JP, Rouaux C, Ellissèche D, Mugniéry D (2004) Resistance to the root-knot nematode *Meloidogyne fallax* in *Solanum sparsipilum*: analysis of the mechanisms. Nematology 6:389–400

Kudla U, Qin L, Milac A, Kielak A, Maissen C (2005) Origin, distribution and 3D-modeling of Gr-EXPB1, an expansin from the potato cyst nematode *Globodera rostochiensis.* FEBS Lett 579:2451–2457

Lally D, Ingmire P, Tong HY, He ZH (2001) Antisense expression of a cell wall-associated protein kinase, WAK4, inhibits cell elongation and alters morphology. Plant Cell 13:1317–1331

Ledger TN, Jaubert S, Bosselut N, Abad P, Rosso MN (2006) Characterization of a new beta-1, 4-endoglucanase gene from the root-knot nematode *Meloidogyne incognita* and evolutionary scheme for phytonematode family 5 glycosyl hydrolases. Gene 382:121–128

Lee C, Chronis D, Kenning C, Peret B, Hewezi T, Davis EL, Baum TJ, Hussey R, Bennett M, Goellner Mitchum M (2011) The novel cyst nematode effector protein 19 C07 interacts with the Arabidopsis auxin influx transporter LAX3 to control feeding site development. Plant Physiol 155:866–880

Levin DE (2005) Cell wall integrity signaling in Saccharomyces cerevisiae. Microbiol Mol Biol Rev 69:262–291

Li Y, Smith C, Corke F, Zheng L, Merali Z, Ryden P, Derbyshire P, Waldron K, Bevan MW (2007) Signaling from an altered cell wall to the nucleus mediates sugar-responsive growth and development in *Arabidopsis thaliana.* Plant Cell 19:2500–2515

Lilley CJ, Urwin PE, Atkinson HJ (1999) Characterization of plant nematode genes: identifying targets for a transgenic defence. Parasitology 118:63–72

Mahalingam R, Wang G, Knap HT (1999) Polygalacturonase and polygalacturonase inhibitor protein: gene isolation and transcription in *Glycine max-Heterodera glycines* interactions. Mol Plant Microbe Interact 12:490–498

Manabe Y, Nafisi M, Verhertbruggen Y, Orfila C, Gille S, Rautengarten C, Cherk C, Marcus SE, Somerville S, Pauly M, Knox JP, Sakuragi Y, Scheller HV (2011) Loss-of-function mutation of REDUCED WALL ACETYLATION2 in Arabidopsis leads to reduced cell wall acetylation and increased resistance to *Botrytis cinerea*. Plant Physiol 155:1068–1078

Manfield IW, Orfila C, McCartney L, Harholt J, Bernal AJ, Scheller HV, Gilmartin PM, Mikkelsen JD, Paul Knox J, Willats WG (2004) Novel cell wall architecture of isoxaben-habituated Arabidopsis suspension-cultured cells: global transcript profiling and cellular analysis. Plant J 40:260–275

Mitreva-Dautova M, Roze E, Overmars H, de Graaff L, Schots A, Helder J, Goverse A, Bakker J, Smant G (2006) Asymbiont-independent endo-1,4-b-xylanase from the plant-parasitic nematode *Meloidogyne incognita*. Mol Plant Microbe Interact 19:521–529

Nishimura MT, Stein M, Hou BH, Vogel JP, Edwards H, Somerville SC (2003) Loss of a callose synthase results in salicylic acid-dependent disease resistance. Science 301:969–972

Offler CE, McCurdy DW, Patrick JW, Talbot MJ (2003) Transfer cells: cells specialized for special purpose. Ann Rev Plant Biol 54:431–454

Pate JS, Gunning BES (1972) Transfer cells. Ann Rev Plant Physiol 23:174–196

Paulson RE, Webster JM (1972) Ultrastructure of the hypersensitive reaction in roots of tomato, *Lycopersicon esculentum* L. to infection by the root-knot nematode, *Meloidogyne incognita*. Physiol Plant Pathol 2:227–234

Pilling E, Hofte H (2003) Feedback from the wall. Curr Opin Plant Biol 6:1–6

Popeijus H, Overmars H, Jones J, Blok V, Goverse A, Helder J, Schots A, Bakker J, Smant G (2000) Enzymology: degradation of plant cell walls by a nematode. Nature 406:36–37

Puthoff DP, Nettleton D, Rodermel SR, Baum TJ (2003) Arabidopsis gene expression changes during cyst nematode parasitism revealed by statistical analyses of microarray expression profiles. Plant J 33:911–921

Puthoff DP, Ehrenfried ML, Vinyard BT, Tucker ML (2007) GeneChip profiling of transcriptional responses to soybean cyst nematode, *Heterodera glycines*, colonization of soybean roots. J Exp Bot 58:3407–3418

Qin L, Kudla U, Roze EHA, Goverse A, Popeijus H (2004) Plant degradation: a nematode expansin acting on plants. Nature 427:30

Rehman S, Butterbach P, Popeijus H, Overmars H, Eric LD, Jones JT, Goverse A, Bakker J, Smant G (2009) Identification and characterization of the most abundant cellulase in stylet secretions from *Globodera rostochiensis*. Phytopathology 99:194–202

Ringli C (2010) Monitoring the outside: cell wall-sensing mechanisms. Plant Physiol 153:1445–1452

Rosso MN, Favery B, Piotte C, Arthaud L, De Boer JM, Hussey RS, Bakker J, Baum TJ, Abad P (1999) Isolation of a cDNA encoding a beta-1,4-endoglucanase in the root-knot nematode *Meloidogyne incognita* and expression analysis during plant parasitism. MPMI 12:585–591

Roze E, Hanse B, Mitreva M, Vanholme B, Bakker J, Smant G (2008) Mining the secretome of the root-knot nematode *Meloidogyne chitwoodi* for candidate parasitism genes. Mol Plant Pathol 9:1–10

Seifert GJ, Blaukopf C (2010) Irritable walls: the plant extracellular matrix and signaling. Plant Physiol 153:467–478

Smant G, Stokkermans JPWG, Yan Y, de Boer JM, Baum TJ et al (1998) Endogenous cellulases in animals: isolation of b-1,4-endoglucanase genes from two species of plant-parasitic cyst nematodes. Proc Natl Acad Sci USA 95:4906–4911

Sobczak M, Avrova A, Jupowicz J, Phillips MS, Ernst K, Kumar A (2005) Characterization of susceptibility and resistance responses to potato cyst nematode (Globodera spp.) infection of tomato lines in the absence and presence of the broad-spectrum nematode resistance Hero gene. MPMI 18:158–168

Somerville C, Bauer S, Brininstool G, Facette M, Hamann T, Milne J, Osborne E, Paredez A, Persson S, Raab T, Vorwerk S, Youngs H (2004) Toward a systems approach to understanding plant cell walls. Science 306:2206–2211

Steinwand BJ, Kieber JJ (2010) The role of receptor-like kinases in regulating cell wall function. Plant Physiol 153:479–484

Sukno S, Shimerling O, McCuiston J, Tsabary G, Shani Z, Shoseyov O, Davis EL (2006) Expression and regulation of the *Arabidopsis thaliana Cel1* endo 1,4 β glucanase gene during compatible plant-nematode interactions. J Nematol 38:354–361

Swiecicka M, Filipecki M, Lont D, Van Vliet J, Qin L, Goverse A, Bakker J, Helder J (2009) Dynamics in the tomato root transcriptome on infection with the potato cyst nematode *Globodera rostochiensis*. Mol Plant Pathol 10:487–500

Szakasits D, Heinen P, Wieczorek K, Hofmann J, Wagner F, Kreil DP, Sykacek P, Grundler FMW, Bohlmann H (2009) The transcriptome of syncytia induced by the cyst nematode *Heterodera schachtii* in *A. thaliana* roots. Plant J 57:771–784

Tedman-Jones JD, Lei R, Jay F, Fabro G, Li X, Reiter WD, Brearley C, Jones JD (2008) Characterization of Arabidopsis mur3 mutations that result in constitutive activation of defence in petioles, but not leaves. Plant J 56:691–703

Tucker ML, Burke A, Murphy CA, Thai VK, Ehrenfried ML (2007) Gene expression profiles for cell wall-modifying proteins associated with soybean cyst nematode infection, petiole abscission, root tips, flowers, apical buds, and leaves. J Exp Bot 58:3395–3406

Vanholme B, Mitreva M, Van Criekinge W, Logghe M, Bird D, McCarter J, Gheysen G (2006) Detection of putative secreted proteins in the plant parasitic nematode *Heterodera schachtii*. Parasitol Res 98:414–424

Vanholme B, Haegeman A, Jacob J, Cannoot B, Gheysen G (2009) Arabinogalactan endo-1, 4-b-galactosidase: a putative plant cell wall degrading enzyme of plant-parasitic nematodes. Nematology 11:739–747

Vercauteren I, de Almeida Engler J, de Groodt R, Gheysen G (2002) An Arabidopsis thaliana pectin acetylesterase gene is upregulated in nematode feeding sites induced by root-knot and cyst nematodes. MPMI 15, 404–407

Vogel JP, Raab TK, Schiff C, Somerville SC (2002) *PMR6*, a pectate lyase-like gene required for powdery mildew susceptibility in Arabidopsis. Plant Cell 14:2095–2106

Vogel JP, Raab TK, Somerville CR, Somerville SC (2004) Mutations in *PMR5* result in powdery mildew resistance and altered cell wall composition. Plant J 40:968–978

Vorwerk S, Somerville S, Somerville C (2004) The role of plant cell wall polysaccharide composition in disease resistance. Trends Plant Sci 9:203–209

Wagner TA, Kohorn BD (2001) Wall-associated kinases are expressed throughout plant development and are required for cell expansion. Plant Cell 13:303–318

Wang XH, Meyers D, Yan YT, Baum T, Smant G, Hussey R, Davis E (1999) In planta localization of a beta-1,4-endoglucanase secreted by *Heterodera glycines*. MPMI 12:64–67

Wang X, Li H, Hu Y, Fu P, Xu J (2007) Molecular cloning and analysis of a new venom allergen-like protein gene from the root-knot nematode *Meloidogyne incognita*. Exp Parasitol 117:133–140

Wieczorek K, Golecki B, Gerdes L, Heinen P, Szakasits D, Durachko DM, Cosgrove DJ, Kreil DP, Puzio PS, Bohlmann H, Grundler FMW (2006) Expansins are involved in the formation of nematode-induced syncytia in roots of *Arabidopsis thaliana*. Plant J 48:98–112

Wieczorek K, Hofmann J, Blöchl A, Szakasits D, Bohlmann H, Grundler FMW (2008) Arabidopsis endo-1,4-β-glucanases are involved in the formation of root syncytia induced by *Heterodera schachtii*. Plant J 53:336–351

Wyss U (1992) Observations on the feeding behaviour of *Heterodera schachtii* throughout development, including events during moulting. Fundam Appl Nematol 15:75–89

Xu SL, Rahman A, Baskin TI, Kieber JJ (2008) Two leucine-rich repeat receptor kinases mediate signaling, linking cell wall biosynthesis and ACC synthase in Arabidopsis. Plant Cell 20:3065–3079

Plant Root Interactions

Angela Hodge

Abstract The modular structure of plant root systems enables a high degree of flexibility (or plasticity) in responding to prevailing conditions in the soil, including resource distribution. However, more recently, it has been suggested that root-root interactions are more sophisticated than simply being driven by resource availability alone. Some evidence suggests that plant roots may be able to recognise their own roots from those of other plants even when the other plant is a genetically identical individual, while other studies suggest plants may be able to identify related individuals (kin) from non-related individuals and modify their competitive interactions as a result. The results of these studies together with their limitations will be reviewed here.

1 Introduction

The key functions of roots are to acquire nutrients and water and provide anchorage for the plant. In some cases, roots have adapted specifically for these purposes such as buttress roots that aid in tree stability (Crook et al. 1997) and ephemeral 'rain' roots produced by some desert plants (e.g. *Agave deserti*) specifically to take advantage of light rainfall events but can then be shed when drought conditions resume (Hunt et al. 1987; Nobel et al. 1990). Being sessile organisms, plants have developed a number of mechanisms to enable them to respond to environmental resources and compete with other organisms including other plants (reviewed by Sultan 2009). The vast majority of studies have focused upon resource acquisition by roots and the responses to the heterogeneous soil environment (reviewed by Hodge 2004, 2009a, b). More recently, however, the ability of plant roots to recognise both their own roots ('self' recognition) and those from related

A. Hodge (✉)
Department of Biology, University of York, York, UK
e-mail: angela.hodge@york.ac.uk

G. Witzany and F. Baluška (eds.), *Biocommunication of Plants*,
Signaling and Communication in Plants 14,
DOI 10.1007/978-3-642-23524-5_9,

individuals ('kin' recognition) and the resulting implications for plant competition has been the subject of much research effort, and it is this evidence that will be reviewed here.

2 Root Responses to Own Versus Different Species Roots

At least some plant species can modify their root growth, or that of others, when grown together (Krannitz and Caldwell 1995; Mahall and Callaway 1991). This modification can occur without direct contact. For example, studies by Mahall and Callaway in the early 1990s (Mahall and Callaway 1991, 1992) on the desert shrubs *Larrea tridentata* and *Ambrosia dumosa* demonstrated that *Larrea* roots inhibited the elongation of either *Ambrosia* or other *Larrea* roots without physical contact being required. Plants were grown in rectangular root observation chambers filled with sand which was kept continually moist and periodically (c. every 8–10 days) flushed with one-eighth strength nutrient solution. These conditions were selected to reduce the likelihood of either nutrient or water depletion being a factor in the experiment given it is well established that roots can respond to both (reviewed by Hodge 2004, 2009a). The addition of activated carbon, a strong absorber of organic compounds (but see also Sect. 3.2), resulted in a reduction of the inhibitory effect by the *Larrea* roots suggesting an unidentified diffusible substance was responsible for the observed inhibition (Mahall and Callaway 1992). The release of such a non-specific inhibitory compound by *Larrea* is surprising given that it may also be expected to reduce the elongation of the *Larrea* roots that had released it and so inhibit its own elongation rate, yet this does not appear to occur. Contact with other *Larrea* roots on the same plant rarely occurred, so any impact on related roots also could not be determined. This implies that the typical root length density of an individual plant species may be important: if a plant normally has a low root length density, then a generalist inhibition mechanism may be advantageous. At a high root length density, however, such a mechanism would not be advantageous as it would result in inhibition of its own roots.

In contrast, *Ambrosia* roots inhibited elongation of other *Ambrosia* roots only, and even then, only following physical contact. Unlike the *Larrea* roots, the inhibitory effect of *Ambrosia* roots was not affected by the addition of activated carbon. Moreover, contact among *Ambrosia* roots on the same plant did not result in inhibition. This, Mahall and Callaway (1992) suggested, indicated a self-non-self recognition response by the *Ambrosia* plants. This recognition response was later found to occur only among *Ambrosia* plants from the same population, and so Mahall and Callaway (1996) argued against a genetic mechanism behind the observed 'self-recognition' response. The study by Gruntman and Novoplansky (2004) on *Buchloe dactyloides* (buffalo grass) also suggested against a genetic mechanism. In this case, plants were able to differentiate between self and non-self neighbours and produced fewer and shorter roots when grown in the presence of roots from the same individual. However, when cuttings from the same node were separated from each other by growing in different pots, they became increasingly alienated from each other and responded to each other's presence as if genetically and physiologically different clones, suggesting, with separation, the original recognition mechanism was lost.

In the sagebrush steppe of the USA, the native tussock grass *Pseudoroegneria spicata* has frequently been observed to be a less effective competitor than the introduced tussock grass *Agropyron desertorum* with the shrub *Artemisia tridentata* (Caldwell et al. 1991, 1985; Eissenstat and Caldwell 1988). Thus, it might be expected that when both these grasses are grown together, *Agropyron* would be a superior competitor than *Pseudoroegneria*. However, although *Agropyron* responded to fertilisation by developing more tillers and biomass than *Pseudoroegneria*, there was no evidence based on relative performance of *Agropyron* in monoculture and mixture that it was a superior competitor when both were grown together. More striking, however, was the observation that *Pseudoroegneria* appeared to be able to detect if it was grown with conspecifics (i.e. the same species) or with *Agropyron* individuals and responded by morphological changes. Some of these altered morphological changes were not affected by fertilisation while others (such as root:shoot ratio and altered density and number of tillers) were. However, Huber-Sannwald et al. (1996) concluded that these changes were unlikely a result of resource competition because there was no difference in overall biomass. In a further study, *Pseudoroegneria* was found to alter its root growth following physical contact with *Agropyron* roots but not after contact with other *Pseudoroegneria* roots (Krannitz and Caldwell 1995) suggesting *Pseudoroegneria* can distinguish between related and unrelated species.

3 Response to Self Versus Non-self Roots

3.1 Avoiding Other Plants' Roots

Spatial distributions of roots from a number of plant species have been shown to be modified by the presence of other plants (Fitter 1986; D'Antonio and Mahall 1991; Caldwell et al. 1996; de Kroon et al. 2003). Spatial segregation (or avoidance) of other plant roots may be a means to reduce competition for resources, which has led some workers (e.g. Schenk et al. 1999) to propose that root segregation will be higher when grown with related than unrelated individuals. Intuitively, this would make sense as it would reduce competition among related individuals; however, there is evidence both in support and some evidence against this suggestion. For example, in a monospecific stand of silver fir (*Abies alba*) trees, the roots of individuals (identified by molecular techniques) were found to overlap (Brunner et al. 2004). Similarly, in a monospecific beech (*Fagus sylvatica*) forest, no evidence for root segregation was found (Lang et al. 2010). In both cases, root competition among related individuals probably did occur.

In contrast, Holzapfel and Alpert (2003) found that connected clones of *Fragaria chiloensis* (wild strawberry) with a high degree of physiological integration segregated their roots and so avoided competition with each other. Adult connected plant pairs accumulated as much biomass as singly grown plants and more biomass than disconnected clones grown together. *Pisum sativum* (pea) plants also altered their root allocation pattern and segregated roots when given the opportunity to do so (i.e. when the plant had its roots split between two pots, termed a 'fence-sitter'; Gersani et al. 1998). The fence-sitter allocated its roots in proportion to the level of nutrients in each of the two pots. When an increasing number of competitor pea plants were added to one of the pots, the fence-sitter plant segregated its root biomass away from the pot containing the competitor plants and into the pot it alone occupied. This strategy enabled the fence-sitter to maintain its total root weight (although root weight differed *between* the pots) and, more importantly, its overall fitness (determined as fruit dry weight). When the pot to which the competitors were added contained twice the level of nutrients of the other and only one competitor pea plant, the fence-sitter allocated its roots equally between the pots. Thus, the fence-sitter was able to detect the higher nutrient level and, despite the presence of one competitor, allocated its roots accordingly. This strategy again enabled the fence-sitter to maintain its fitness while that of the competitor plants declined (Gersani et al. 1998). In this case, the segregation of roots may have been due to the decline in nutrient availability in one of the pots rather than a detection of the presence of competitor plants *per se*. However, this does not explain the results of the study by Holzapfel and Alpert (2003) where the strawberry clones placed less root mass between the two plants when connected compared to unconnected clones.

Cahill et al. (2010) have recently demonstrated that *Abutilon theophrasti* plants modify their root placement depending on if another *A. theophrasti* plant is present *and* the distribution of resources. If grown alone, roots followed a broad foraging strategy that was largely noncommittal to tracking resource distribution. However, in the presence of a competitor, a more restricted foraging strategy was adopted that was, in turn, modified by nutrient distribution. When resources were uniformly distributed, root distribution was more restricted and spatial segregation occurred. When nutrients were concentrated in a patch between the plants, root distribution overlapped and was not segregated. Thus, the observed response by plants is a result of integrating various pieces of information on their environment and is therefore likely context dependant (see also Hodge 2004, 2009b).

3.2 *Overproduction of Roots in Response to Another Plant's Roots*

Other studies also using the fence-sitter approach but with two plants as fence-sitters sharing two pots have come to very different conclusions from that by

Gersani et al. (1998) discussed above (Sect. 3.1). Instead of the plants segregating their roots away from each other (i.e. effectively into one pot each) as may be expected, an overproduction of roots and an overall decline in fitness compared to controls (i.e. two plants each in their own pot) has been reported. This response has been referred to as a 'Tragedy of the Commons' scenario (Maina et al. 2002; Gersani et al. 2001) as the overproduction of roots comes at the cost of a decline in overall fitness, and so at the expense to all, including the next generation. Such a Tragedy of the Commons response has been reported for a number of plant species including *Glycine max* (soybean), *Phaseolus vulgaris* (bean) and *P. sativum* (O'Brien et al. 2005; Maina et al. 2002; Gersani et al. 2001). Given that these plants are all agriculturally important species, and have been bred to ensure similarity, it is perhaps surprising that the individual plants can distinguish between their own and other, genetically identical, roots of the other plant, but as previously mentioned (Sect. 2), there is little evidence to support a genetic mechanism behind the observed self-recognition response (Gruntman and Novoplansky 2004). However, such an overproduction of roots is also counter to the earlier literature where root segregation was primarily reported among conspecific neighbours (reviewed by Schenk et al. 1999).

There are several other problems with the two fence-sitter approach and the comparison to 'control' plants grown in their own pot, however, as have been highlighted by several researchers (see Schenk 2006; Hess and de Kroon 2007; Semchenko et al. 2007; but see also O'Brien and Brown 2008) including variation in pot volume and, often, total nutrients available. In addition, Laird and Aarssen (2005) suggested that taking average values of both fence-sitter roots (due to problems trying to separate intermingled root systems) mathematically biased the results towards a Tragedy of the Commons response, even when this may not have occurred at the individual plant level. Such a bias was refuted by O'Brien and Brown (2008) who argued that the results of some of this earlier work (e.g. Gersani et al. 2001; O'Brien et al. 2005) was the same regardless if an individual plant from the two fence-sitters was selected or if the average values were taken. However, the issue of variation in absolute soil volume among the comparisons made is less easy to counter. Schenk (2006) reanalysed Maina et al. (2002) and Gersani et al. (2001) data to show that the plants appeared to respond to absolute soil volume rather than the presence of a competitor, although the reanalysis was based on relatively few data points. So can the response seen in these Tragedy of the Commons studies really be simply due to variation in physical volume? Surprisingly, few studies have investigated how plants respond to variation in soil volume *per se*. Some of the early work on plants' response to elevated atmospheric CO_2 concentration did highlight potential confounding issues of pot volume, size and shape influencing the responses subsequently reported (see Berntson et al. 1993; McConnaughay et al. 1993; Thomas and Strain 1991). In another study, McConnaughay and Bazzaz (1991) found that increased pot volume enhanced vegetative growth even when nutrient supply was kept equal. However, while reproductive biomass also varied with space available, the response was more complex and depended on the plant species present. This does suggest that physical space influences both plant growth and reproductive biomass; thus, extra care is required when interpreting the results of studies were this is not held constant among comparisons.

Moreover, other studies have found no adverse impact upon reproductive biomass (thus no Tragedy of the Commons response) despite using similar experimental conditions and, in some cases, the same plant species (see Murphy and Dudley 2007; O'Brien et al. 2005). Semchenko et al. (2007) also did not observe a Tragedy of the Commons response in their study on oats (*Avena sativa*), although actual reproductive biomass was not measured directly but, instead, total plant biomass was followed. Following activated carbon addition, when the oat root systems were allowed to interact, the activated carbon did not alter plant mass or root-shoot allocation. However, when partitions (either as plastic or mesh barriers) were present to separate the root systems of the two plants, plant performance increased in the presence of the activated carbon. Physical obstacles may restrict root growth and the addition of activated carbon or potassium permanganate (a strong oxidizer of organic compounds) can alleviate this restriction (Falik et al. 2005) suggesting a role for root-released compounds in regulating root growth. However, while activated carbon addition has been widely used in root interactions investigations (see Mahall and Callaway 1992; Semchenko et al. 2007; Kulmatiski and Beard 2006), the results need to be treated with caution and appropriate controls included, as activated carbon can result in a number of other modifications to the substrate and/or plant. These include altering the nutrient availability (Lau et al. 2008; Weisshuhn and Prati 2009), modifying the pH or water retention of the substrate (Inderjit and Callaway 2003; Kabouw et al. 2010), influencing plant germination (Kabouw et al. 2010) and altering establishment of mutualistic symbioses (Wurst et al. 2010; Wurst and van Beersum 2009). However, these effects are also likely context dependant as while nodulation was reported to be reduced in one study (Wurst and van Beersum 2009), it was unaffected in another (Wurst et al. 2010). The source of the activated carbon applied may also be important in determining the response obtained (Kabouw et al. 2010).

4 Kin Recognition Responses

Dudley and File (2007) sparked renewed interest and indeed criticism (Klemens 2008; but see also Dudley and File 2008; Bhatt et al. 2011) in root-root interactions when they reported that *Cakile edentula* (sea rocket) plants that shared the same mother allocated *less* biomass to their roots when grown together compared to when grown with plants that had different mothers. Dudley and File (2007) referred to this phenomenon as 'kin recognition'. However, reproductive biomass (a measure of plant fitness) did not differ among the groups. Thus, there was no evidence for the arguably more important phenomenon of 'kin selection' (see also Callaway and Mahall 2007). Dudley and File (2007) were by no means the first to study the response of plants when grown with kin, although no census emerges from these previous studies with both positive and negative effects when grown with kin compared to non-kin reported (Allard and Adams 1969; Antonovics and Ellstrand 1984; Cheplick and Kane 2004; Donohue 2003; Willson et al. 1987). However, compared to the large number of studies upon kin selection in animals, this area still remains a less well-investigated aspect of plant ecology.

In another study, Murphy and Dudley (2009) examined kin recognition in a North American species of *Impatiens*. *Impatiens* cf. *pallida* seedlings were grown with kin (sharing the same mother) or 'strangers' (unrelated conspecifics from four different families from the same field population). Kin recognition occurred but only in the presence of another plant's roots (i.e. solitary plants grown on their own did not differ between kin and strangers for any plant trait measured). However, in this case, the response was mainly in aboveground structures. Plant height, branch number and elongation increased in response to kin, while allocation to leaves relative to both stems and roots increased in response to strangers (non-kin). Thus, for *C. edentula*, the kin recognition response was mainly belowground (Dudley and File 2007; Bhatt et al. 2011), while for *I. pallida*, the response was mainly aboveground (Murphy and Dudley 2009) albeit with the presence of the other plants roots being essential to elicit the aboveground response. Murphy and Dudley (2009) suggested that it was due to the differing ecologies experienced by these two species: *I. pallida* grows in wooded areas thus competition for light is important, while *C. edentula* is found in relatively nutrient poor beach soils thus is more limited by resource acquisition. However, given the very limited data available on so few species, this obviously has to be more rigorously investigated before such a suggestion can be confirmed. Moreover, Karban and Shiojiri (2009) recently demonstrated that sagebrush (*Artemisia tridentata*) plants that received volatile cues from genetically identical cuttings accumulated less damage than plants that had received cues from non-self cuttings. However, in contrast to the findings of Murphy and Dudley (2009), in this case, the response was not mediated through roots as the individuals were grown in separate pots.

In a further study, this time on *Arabidopsis thaliana*, root 'exudates' (although more correctly root-released compounds as root secretions were also present) were reported to be responsible for recognising kin versus non-kin (Biedrzycki et al. 2010). *Arabidopsis* was exposed to liquid media containing root-released compounds from siblings, strangers (i.e. non-kin) or themselves. When exposed to non-kin media, lateral root numbers were higher than when grown on media that had contained either kin or their own roots except following addition of sodium orthovanadate, a root secretion inhibitor. However, chemical analysis of the growth media was not made; thus, which compounds were responsible for the observed effect on lateral root numbers is unknown. It is also possible that nutrient availability also differed among the treatments, which is well established as having an impact upon *Arabidopsis* root architecture (Malamy 2005; Hodge et al. 2009). The addition of the root secretion inhibitor may also have had unknown side effects, but again, this was not tested for. In contrast, Masclaux et al. (2010), also using *Arabidopsis thaliana*, found no evidence of positive interactions and reduced competition when kin compared to non-kin are grown together but instead concluded that the outcome of the interaction depended upon the competitive ability of the various accessions screened. Moreover, whole-genome microarray analysis showed no genes differentially expressed when grown with kin compared to non-kin. Milla et al. (2009) also found no evidence for a positive kin recognition response in *Lupinus angustifolius*, but instead, genetic relatedness actually resulted in both decreased individual and group fitness with reductions in both flowering and vegetative biomass. The lack of consensus from the various kin recognition studies therefore may argue against kin recognition as a widespread evolutionary mechanism among plant species, even though it may occur under some conditions. There are also important issues as to what the different authors of various studies actually mean by 'kin', i.e. the exact degree of relatedness of the plants used.

5 Mycorrhizal Considerations

Many of the studies reported in the previous sections have not considered the impact mycorrhizal (literally meaning 'fungus-root') associations may have upon the various root-root recognition responses. This is surprising given that mycorrhizal symbiosis is ubiquitous in the natural environment. There are seven different types of mycorrhizal association depending on the plant species and fungi involved; the two most important are the arbuscular mycorrhizal (AM) and ectomycorrhizal (ECM) association. The AM is the most common and ancient type and can form on c. two thirds of all land plant species. The ECM is less widespread and almost all the plants involved are woody perennials, thus the association is of great economic value in forestry systems (Smith and Read 2008; Fitter and Moyersoen 1996). Mycorrhizal symbiosis is known to confer a number of benefits to the host plant including enhanced nutrient acquisition and protection against pathogens (Smith and Read 2008; Newsham et al. 1995; Read and Perez-Moreno 2003). However, while there is considerable evidence that mycorrhizal colonisation modifies both qualitatively and quantitatively the compounds released by roots (see reviews by Jones et al. 2004, 2009), whether mycorrhizal colonisation also alters the compounds suggested to be involved in root-root recognition is currently unknown. Given the near ubiquity of the mycorrhizal symbiosis, this area certainly warrants further research. Most of the data on how mycorrhizal symbiosis may be affected by plant-plant interactions comes from plant invasion studies (see review by Pringle et al. 2009). The results of some studies suggest that the fungal symbionts may have a role in protecting their host plant from allelopathic substances released from invasive plant roots (Barto et al. 2010), while the results of other studies suggest that the mycorrhizal fungi themselves may be negatively impacted (Roberts and Anderson 2001; Callaway et al. 2008; Wolfe et al. 2008), but different fungal symbionts may be affected in different ways (Zhang et al. 2010). However, many of these studies have followed the impact of allelopathic substances on AM spore germination and subsequent colonisation of hosts rather than on already established mycorrhizal networks as would already be in place under natural conditions. If these released substances also affect such established networks remains to be determined. Song et al. (2010) recently reported that healthy tomato (*Lycopersicon esculentum*) plants connected via a common mycelial network (CMN) of arbuscular mycorrhizal hyphae to tomato plants infected with the pathogen *Alternaria solani* showed elevated defensive enzyme levels and defence-related gene expression. This, Song et al. (2010) suggested, demonstrated that communication had occurred via the CMN. Tomato plants grown near the infected plants but not connected to the CMN did not show such an enhanced defence expression. However, many of the defensive enzymes measured are also activated following mycorrhizal colonisation; thus, further research is required to fully understand the elevated defence response observed. Moreover, the mechanism by which such a signal could be transferred via the fungal hyphae seems hard to explain based on current understanding of these fungi—although here also a large knowledge gap exists.

6 Conclusions

While there is evidence that certain plant species may be able to identify their own roots from those of other plants (even when related), there is also evidence to counter these claims. In many cases, the studies have been conducted under highly artificial conditions which may lead to disproportionate importance being attached to the observed response. The use of appropriate controls is also essential. Further, the actual extent of relatedness among individuals needs to be taken into consideration when comparing the results of different studies. Although (often unidentified) root-released compounds are often suggested to be responsible for the interactions observed in the natural soil environment, any root-released compound will be subject to possible degradation by the microbial community and/or binding to soil particles which may reduce its effectiveness. This is why a more realistic medium needs to be used in studies. This is not to say that these compounds may still operate given there are many examples where molecules released from plant roots are known to act as signals (such as in the mycorrhizal and *Rhizobium*-legume nitrogen-fixing symbioses). Different plant species may show a continuum of responses which can be expressed under different conditions. We have to ensure that by selecting experimental conditions to favour a plant response, we do not inflate its importance, i.e. just because a plant *can* respond in a certain way does not necessary mean it will in the natural environment. Thus, while plants have been shown to display a wide range of quite sophisticated responses, i.e. the ability to: (1) avoid obstacles (Falik et al. 2005), (2) recognise kin from non-kin (Dudley and File 2007), and (3) detect nutrient patches and modify its root system growth (Hodge 2009a, b) and so forth, in the soil environment, the plant and its root system will be subject to a whole range of these signals, and it may simply be the strongest of these signals (be it nutrient availability or the presence of other plant roots) that is the one that the plant ultimately responds to, otherwise it may simply be overwhelmed. In other words, the context in which the response is observed is ultimately as important as the response itself.

Acknowledgement I thank Alastair Fitter for his insightful comments on an earlier draft of this text.

References

Allard RW, Adams J (1969) Population studies in predominantly self-pollinating species. XIII Intergenotypic competition and population structure in barley and wheat. Am Naturalist 103:621–645

Antonovics J, Ellstrand NC (1984) Experimental studies of the evolutionary significance of sexual reproduction. I. A test of the frequency-dependent selection hypothesis. Evolution 38:103–115

Barto K, Friese C, Cipollini D (2010) Arbuscular mycorrhizal fungi protect a native plant from allelopathic effects of an invader. J Chem Ecol 36:351–360

Berntson GM, McConnaughay KDM, Bazzaz FA (1993) Elevated CO_2 alters deployment of roots in 'small' growth containers. Oecologia 94:558–564

Bhatt MV, Khandelwal A, Dudley SA (2011) Kin recognition, not competitive interactions, predicts root allocation in young *Cakile edentula* seedling pairs. New Phytologist 189:1135–1142

Biedrzycki ML, Jilany TA, Dudley SA, Bais HP (2010) Root exudates mediate kin recognition in plants. Commun Integrat Biol 3:28–35

Brunner I, Ruf M, Lüscher P, Sperisen C (2004) Molecular markers reveal extensive intraspecific below-ground overlap of silver fir fine roots. Mol Ecol 13:3595–3600

Cahill JF, McNickle GG, Haag JJ, Lamb EG, Nyanumba SM, St. Clair CC (2010) Plants integrate information about nutrients and neighbours. Science 328:1657

Caldwell MM, Eissenstat DM, Richards JH, Allen MF (1985) Competition for phosphorus: differential uptake from dual-isotope-labelled soil interspaces between shrub and grass. Science 229:384–386

Caldwell MM, Manwaring JH, Jackson RB (1991) Exploitation of phosphate from fertile soil microsites by three Great Basin perennials when in competition. Funct Ecol 5:757–764

Caldwell MM, Manwaring JH, Durham SL (1996) Species interactions at the level of fine roots in the field: influence of soil nutrient heterogeneity and plant size. Oecologia 106:440–447

Callaway RM, Mahall BE (2007) Family roots. Nature 448:145–147

Callaway RM, Cipollini D, Barto K, Thelen GC, Hallett SG, Prati D, Stinson K, Klironomos J (2008) Novel weapons: invasive plant suppresses fungal mutualists in America but not in its native Europe. Ecology 89:1043–1055

Cheplick GP, Kane KH (2004) Genetic relatedness and competition in *Triplasis purpurea* (Poaceae): resource partitioning or kin selection? Int J Plant Sci 165:623–630

Crook MJ, Ennos AR, Banks JR (1997) The function of buttress roots: a comparative study of the anchorage systems of buttressed (*Aglaia* and *Nephelium ramboutan* species) and non-buttressed (*Mallotus wrayi*) tropical trees. J Exp Bot 48:1703–1716

D'Antonio CM, Mahall BE (1991) Root profiles and competition between the invasive, exotic perennial, *Carpobrotus edulis* and two native shrub species in California coastal scrub. Am J Bot 78:885–894

de Kroon H, Mommer L, Nishiwaki A (2003) Root competition: towards a mechanistic understanding. In: de Kroon H, Visser EJW (eds) Root ecology, vol 168. Springer, Berlin, pp 215–234

Donohue K (2003) The influence of neighbour relatedness on multilevel selection in the Great Lakes sea rocket. Am Naturalist 162:77–92
Dudley SA, File AL (2007) Kin recognition in an annual plant. Biol Lett 3:435–438
Dudley SA, File AL (2008) Yes, kin recognition in plants. Biol Lett 4:69–70
Eissenstat DM, Caldwell MM (1988) Competitive ability is linked to rates of water extraction: a field study of two aridland tussock grasses. Oecologia 75:1–7
Falik O, Reides P, Gersani M, Novoplansky A (2005) Root navigation by self inhibition. Plant Cell Environ 28:562–569
Fitter AH (1986) Spatial and temporal patterns of root activity in a species-rich alluvial grassland. Oecologia 69:594–599
Fitter AH, Moyersoen B (1996) Evolutionary trends in root-microbe symbiosis. Philos Trans R Soc Lond B Biol Sci 351:1367–1375
Gersani M, Abramsky Z, Falik O (1998) Density-dependent habitat selection in plants. Evol Ecol 12:223–234
Gersani M, Brown JS, O'Brien EE, Maina GM, Abramsky Z (2001) Tragedy of the commons as a result of root competition. J Ecol 89:660–669
Gruntman M, Novoplansky A (2004) Physiologically mediated self/non-self discrimination in roots. Proc Natl Acad Sci USA 101:3863–3867
Hess L, de Kroon H (2007) Effects of rooting volume and nutrient availability as an alternative explanation for root self/non-self discrimination. J Ecol 95:241–251
Hodge A (2004) The plastic plant: root responses to heterogeneous supplies of nutrients. New Phytologist 162:9–24
Hodge A (2009a) Roots: the acquisition of water and nutrients from the heterogeneous soil environment. In: Lüttge U, Beyschlag W, Büdel B, Francis D (eds) Progress in botany, vol 71. Springer, Berlin, pp 307–337
Hodge A (2009b) Root decisions. Plant Cell Environ 32:628–640
Hodge A, Berta G, Doussan C, Merchan F, Crespi M (2009) Plant root growth, architecture and function. Plant Soil 321:153–187
Holzapfel C, Alpert P (2003) Root cooperation in a clonal plant: connected strawberries segregate roots. Oecologia 134:72–77
Huber-Sannwald E, Pyke DA, Caldwell MM (1996) Morphological plasticity following species-specific recognition and competition in two perennial grasses. Am J Bot 83:919–931
Hunt ER, Zakir NJD, Nobel PS (1987) Water cost and water revenues for established and rain-induced roots of *Agave deserti*. Funct Ecol 1:125–129
Inderjit, Callaway RM (2003) Experimental designs for the study of allelopathy. Plant Soil 256:1–11
Jones DL, Hodge A, Kuzyakov Y (2004) Plant and mycorrhizal regulation of rhizodeposition. New Phytologist 163:459–480
Jones DL, Nguyen C, Finlay RD (2009) Carbon flow in the rhizosphere: carbon trading at the soil–root interface. Plant Soil 321:5–33
Kabouw P, Nab M, van Dam NM (2010) Activated carbon addition affects substrate pH and germination of six plant species. Soil Biol Biochem 42:1165–1167
Karban R, Shiojiri K (2009) Self-recognition affects plant communication and defense. Ecol Lett 12:502–506
Klemens JA (2008) Kin recognition in plants? Biol Lett 4:67–68
Krannitz PG, Caldwell MM (1995) Root growth responses of three Great Basin perennials to intra- and interspecific contact with other roots. Flora 190:161–167
Kulmatiski A, Beard KH (2006) Activated carbon as a restoration tool: potential for control of invasive plants in abandoned agricultural fields. Restoration Ecol 14:251–257
Laird RA, Aarssen LW (2005) Size inequality and the tragedy of the commons phenomenon in plant competition. Plant Ecol 179:127–131
Lang C, Dolynska A, Finkeldey R, Polle A (2010) Are beech (*Fagus sylvatica*) roots territorial? Forest Ecol Manage 260:1212–1217

Lau JA, Puliafico KP, Kopshever JA, Steltzer H, Jarvis EP, Schwarzländer M, Strauss SY, Hufbauer RA (2008) Inference of allelopathy is complicated by effects of activated carbon on plant growth. New Phytologist 178:412–423

Mahall BE, Callaway RM (1991) Root communication among desert shrubs. Proc Natl Acad Sci USA 88:874–876

Mahall BE, Callaway RM (1992) Root communication mechanisms and intracommunity distributions of two Mojave desert shrubs. Ecology 73:2145–2151

Mahall BE, Callaway RM (1996) Effects of regional origin and genotype on intraspecific root communication in the desert shrub *Ambrosia dumosa* (Asteraceae). Am J Bot 83:93–98

Maina GM, Brown JS, Gersani M (2002) Intra-plant versus inter-plant root competition in beans: avoidance, resource matching or tragedy of the commons. Plant Ecol 160:235–247

Malamy JE (2005) Intrinsic and environmental response pathways that regulate root system architecture. Plant Cell Environ 28:67–77

Masclaux F, Hammond RL, Meunier J, Gouhier-Darimont C, Keller L, Reymond P (2010) Competitive ability not kinship affects growth of *Arabidopsis thaliana* accessions. New Phytologist 185:322–331

McConnaughay KDM, Bazzaz FA (1991) Is physical space a soil resource? Ecology 72:94–103

McConnaughay KDM, Berntson GM, Bazzaz FA (1993) Limitations to CO_2-induced growth enhancement in pot studies. Oecologia 94:550–557

Milla R, Forero DM, Escudero A, Iriondo JM (2009) Growing with siblings: a common ground for cooperation or for fiercer competition among plants? Proc R Soc Lond Ser B Biol Sci 276:2531–2540

Murphy GP, Dudley SA (2007) Above- and below-ground cues elicit independent responses. J Ecol 95:261–272

Murphy GP, Dudley SA (2009) Kin recognition: competition and cooperation in *Impatiens* (Balsaminaceae). Am J Bot 96:1990–1996

Newsham KK, Fitter AH, Watkinson AR (1995) Multi-functionality and biodiversity in arbuscular mycorrhizas. Trends Ecol Evol 10:407–411

Nobel PS, Schulte PJ, North GB (1990) Water influx characteristics and hydraulic conductivity for roots of *Agave deserti* Engelm. J Exp Bot 41:409–415

O'Brien EE, Brown JS (2008) Games roots play: effects of soil volume and nutrients. J Ecol 96:438–446

O'Brien EE, Gersani M, Brown JS (2005) Root proliferation and seed yield in response to spatial heterogeneity of below-ground competition. New Phytologist 168:401–412

Pringle A, Bever JD, Gardes A, Parrent JL, Rillig MC, Klironomos JN (2009) Mycorrhizal symbioses and plant invasions. Annu Rev Ecol Evol Syst 40:699–715

Read DJ, Perez-Moreno J (2003) Mycorrhizas and nutrient cycling in ecosystems—a journey towards relevance? New Phytologist 157:475–492

Roberts KJ, Anderson RC (2001) Effect of garlic mustard [*Alliaria petiolata* (Beib. Cavara & Grande)] extracts on plants and arbuscular mycorrhizal (AM) fungi. Am Midland Naturalist 146:146–152

Schenk HJ (2006) Root competition: beyond resource depletion. J Ecol 94:725–739

Schenk HJ, Callaway RM, Mahall BE (1999) Spatial root segregation: are plants territorial? Adv Ecol Res 28:145–180

Semchenko M, Hutchings MJ, John EA (2007) Challenging the tragedy of the commons in root competition: confounding effects of neighbour presence and substrate volume. J Ecol 95:252–260

Smith SE, Read DJ (2008) Mycorrhizal symbiosis. Academic Press Ltd, London

Song YY, Zeng RS, Xu JF, Li J, Shen X, Yihdego WG (2010) Interplant communication of tomato plants through underground common mycorrhizal networks. PloS One 5:e13324

Sultan SE (2009) Plant developmental responses to the environment: eco-devo insights. Curr Opin Plant Biol 13:1–6

Thomas RB, Strain BR (1991) Root restriction as a factor in photosynthetic acclimation of cotton seedlings grown in elevated carbon dioxide. Plant Physiol 96:627–634

Weisshuhn K, Prati D (2009) Activated carbon may have undesired side effects for testing allelopathy in invasive plants. Basic Appl Ecol 10:500–507

Willson MF, Hoppes WG, Goldman DA, Thomas PA, Katusic-Malmborg PL, Bothwell JL (1987) Sibling competition in plants: an experimental study. Am Naturalist 129:304–311

Wolfe BE, Rodgers VL, Stinson KA, Pringle A (2008) The invasive plant *Alliaria petiolata* (garlic mustard) inhibits ectomycorrhizal fungi in its introduced range. J Ecol 96:777–783

Wurst S, van Beersum S (2009) The impact of soil organism composition and activated carbon on grass–legume competition. Plant Soil 314:1–9

Wurst S, Vender V, Rillig MC (2010) Testing for allelopathic effects in plant competition: does activated carbon disrupt plant symbioses? Plant Ecol 211:19–26

Zhang Q, Yang R, Tang J, Yang H, Hu S, Chen X (2010) Positive feedback between mycorrhizal fungi and plants influences plant invasion success and resistance to invasion. PloS One 5: e12380

Control of Programmed Cell Death During Plant Reproductive Development

Yadira Olvera-Carrillo, Yuliya Salanenka, and Moritz K. Nowack

Abstract Programmed cell death (PCD) is an actively controlled, genetically encoded self-destruct mechanism of the cell. While many forms of PCD have been described and molecularly dissected in animals, we know to date only little about the control of PCD processes in plants. Nevertheless, plant PCD is a crucial component of a plant's reaction to its biotic and abiotic environment and a central theme during plant development. In this chapter, we review the communication events triggering and executing, or preventing, PCD during plant reproductive development. These comprise intracellular communication, as well as signaling between cells and tissues, and the intricate communication between genetically distinct individuals that are necessary for successful plant reproduction.

1 Introduction

Sexual reproduction is one of the key events in the life of most organisms. It involves communication through signaling at multiple levels, from intracellular signaling between organelles, over cross talk between cells and tissues, up to the complex communication between genetically distinct individuals.

In the seed-bearing plants (spermatophytes) that form the vast majority of recent species of land plants, the seed has become the central organ of sexual reproduction. The plant seed essentially is a desiccation-tolerant capsule formed within the parent plant that contains the next plant generation in form of an embryo. During its development, the seed provides room and shelter for the growing embryo and sustains it with maternally produced nutrients. The mature seed protects the

Y. Olvera-Carrillo • Y. Salanenka • M. Nowack (✉)
Department of Plant Systems Biology, VIB, Ghent, Belgium

Department Plant Biotechnology and Bioinformatics, Ghent University, Ghent, Belgium
e-mail: moritz.nowack@psb.ugent.be

G. Witzany and F. Baluška (eds.), *Biocommunication of Plants*,
Signaling and Communication in Plants 14,
DOI 10.1007/978-3-642-23524-5_10,

desiccated, dormant embryo from abiotic and biotic harm, serves as dispersal unit, and provides a stockpile of nutrients that allow the germinating embryo to establish itself as a seedling in its new habitat (Bewley and Black 1994).

Land plants have alternating sexual and asexual generations: the typically haploid gametophytes and the typically diploid sporophytes. During land plant evolution, the gametophytic generation became gradually reduced and consists in most recent flowering plants (angiosperms) of merely three cells in the male gametophyte (pollen) and seven cells in the female gametophyte (embryo sac, see Fig. 1). The modern gametophytes develop inside and sustained by the mother sporophyte that forms the tangible body of a plant. The tasks of the gametophytes consist in producing the actual gametes (sperm cells in the pollen and egg cell and central cell in the female gametophyte) and facilitate their fusion during fertilization. In angiosperms, fertilization occurs in a unique mode called "double fertilization," during which two male gametes fuse with two female gametes in one fertilization event. Thus, two very different fertilization products are generated, the diploid embryo representing the next plant generation and the triploid endosperm, an accessory tissue that serves mainly to nourish the embryo. Embryo and endosperm are surrounded by the maternally derived seed coat, and the trinity of these three genetically distinct organisms forms the entity of the plant seed (Fig. 1).

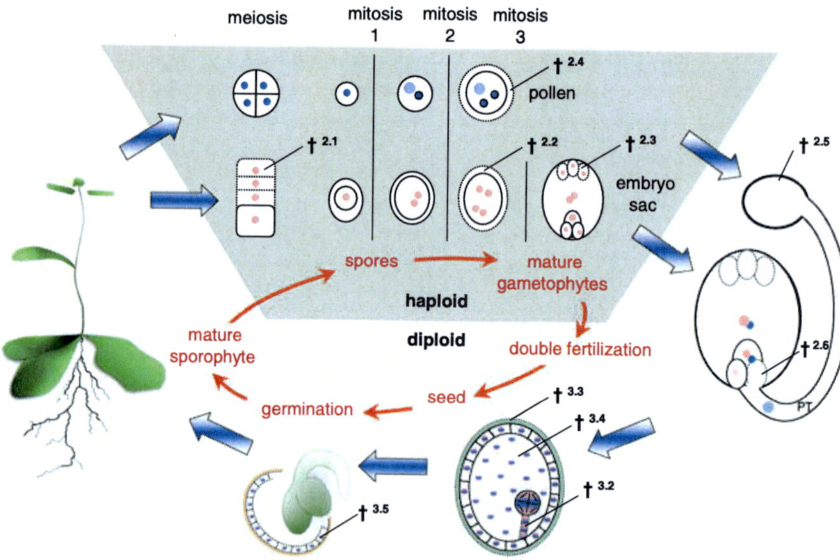

Fig. 1 PCD during plant reproductive development. Cases of PCD are ubiquitous during plant reproductive development and crucial for its success. Examples described in this chapter (the numbers in the figure correspond to the numbering of the section headings). 2.1 Cell death of the nonfunctional megaspores. 2.2 Cell death of the nucellus. 2.3 Cell death of the antipodal cells. 2.4 Cell death of the tapetum layer. 2.5 Pollen cell death during the SI response. 2.6 Synergid cell death during fertilization. 3.2 Embryonic suspensor cell death. 3.3 Cell death during seed coat formation. 3.4 Central endosperm cell death. 3.5 Aleurone cell death during germination

During plant sexual reproduction, multiple and repeated cross-communication events have to be accomplished for a successful fertilization and subsequent development of the seed. Intriguingly, the "placental habit" of plants, i.e., the development of the offspring inside and at the expense of the parent organism (Harper et al. 1970), requires additional levels of communication between the different generations of parent sporophyte, male and female gametophytes, and the next generation of sporophyte.

But not only growth and proliferation of seed tissues is important for successful reproductive development—of equal scope are developmentally controlled instances of programmed cell death (PCD). Triggers from inside or outside the cell can elicit PCD, leading to downstream signaling events that result in a cascade of hydrolytic activity, organizing the shutdown of the cellular metabolism and the ordered succession of events eventually leading to cellular death. In contrast to animals, in plants, the different forms of PCD are still poorly defined (van Doorn et al. 2011). For the time being, we will thus use the term "PCD" to cover all forms of actively controlled, genetically encoded cellular events leading to the eventual death of a cell.

While we currently only begin to understand plant PCD control at the molecular level, it is evident that PCD is of utmost importance for plant life. For instance, cell death is part of the plant's defense system against biotrophic pathogens and viruses. These pathogens can elicit the so-called hypersensitive response (HR) that compromises an oxidative burst followed by PCD of a restricted number of cells at the infection site (Ma and Berkowitz 2007; Mur et al. 2008; Hayward et al. 2009). Also, abiotic stresses such as heat, drought, or irradiation can lead to PCD (Gadjev et al. 2008; Taylor et al. 2009).

Furthermore, PCD is part of the normal developmental program of many plant tissues. Well-studied plant-PCD model systems include leaf and floral organ senescence (Lim et al. 2007; van Doorn and Woltering 2008; Guiboileau et al. 2010), the self-incompatibility response (Chen et al. 2010; Poulter et al. 2010; Tantikanjana et al. 2010), or tracheary element differentiation (Dahiya 2003; Turner et al. 2007; Ohashi-Ito and Fukuda 2010).

While in animals the molecular control of different forms of cell death have been elucidated in great detail from PCD initiation (or prevention) down to PCD execution, cell death research in plants is still in its infancy. Though a number of PCD regulatory elements have been determined in the various plant-PCD model systems, we are still far from a comprehensive picture of PCD regulation. Over the last years, it became evident that major animal cell death regulators are not conserved in plant genomes, including the pro- and antiapoptotic Bcl-2 family members Bax and Bcl-X, or caspases, the central proteolytic executers of apoptosis (Williams and Dickman 2008; Cacas 2010).

Intriguingly, however, there seem to be a number of parallels between plant and animal PCD: In several plant PCD systems, loss of mitochondrial integrity and cytochrome c release have been described, although the functional importance of these processes have not been unambiguously demonstrated (Rogers 2005; Reape et al. 2008; Qi et al. 2011). In animals, Bax translocates to the mitochondria upon cell death initiation and contributes to the formation of channels in the outer mitochondrial membrane (Antonsson et al. 1997). If expressed in plants, mammalian Bax causes cell death that exhibits hallmarks of PCD (Kawai-Yamada et al. 2001; Baek et al. 2004; Suomeng et al. 2008). Moreover, specific Bax inhibitors are conserved in animal and plant genomes and reciprocally function in both taxa (Kawanabe et al. 2006; Watanabe and Lam 2006, 2009; Henke et al. 2011). Also, caspase-like activity has been measured during many forms of plant PCD, and synthetic as well as virally expressed caspase inhibitors have been shown to counteract PCD progression in plants (Bosch et al. 2008; Woltering 2010). The search for plant proteases functioning analogous to animal caspases has resulted in the identification of several candidate proteases involved in PCD regulation, e.g., vacuolar processing enzymes (VPEs), phytaspases, saspases, and metacaspases (Bozhkov et al. 2010; Woltering 2010; Hara-Nishimura and Hatsugai 2011; Tsiatsiani et al. 2011; Vartapetian et al. 2011). It is tempting to speculate that in animals and plants, a common core machinery of PCD regulation and execution exists that is independently targeted by different regulators in both taxa.

In this chapter, we focus on the current knowledge about communication events that regulate PCD during plant reproductive development from meiosis to seed germination (see Fig. 1). In short, we will cover the cell death of the nonfunctional megaspores, nucellar and antipodal cell deaths, pollen PCD during incompatible pollen-pistil interactions, and synergid cell death during fertilization. Further, we will cover the cell death events during seed development: PCD of the extraembryonic suspensor, central endosperm cell death, cell death during seed coat formation, and finally aleurone cell death during germination (see Fig. 1 for an overview).

Seeing how ubiquitously cell death occurs in reproductive development, it comes as no surprise that a finely tuned control of PCD both in space and time is critical for successful plant reproduction. Within the context of this book, we will present developmental PCD as a paradigm for biocommunication within and between cells, between different tissues and organs, and between different individual organisms.

2 Cell Death During Gametophyte Development

2.1 *Megaspore Cell Death After Meiosis*

In most angiosperms, the female gametophyte (FG) or embryo sac is a haploid organism generated by a single meiotic product, the functional megaspore (FM). After meiosis is initiated in the megaspore mother cells, four megaspores are generated, but only one of them differentiates into the FM. The three other megaspores degenerate, undergoing developmental PCD (Yang et al. 2010), a feature shared by many oogamous eukaryotes. In some plants, for instance, *Arabidopsis* and rice, the proximal (chalazal) megaspore survives, while in others, the distal (micropylar) megaspore lives on (Rodkiewicz 1970).

The specification and survival of the FM appears to be position dependent: In *switch/dyad* mutants in *Arabidopsis*, meiosis is altered to a mitotic division, generating two unreduced diploid megaspores. Yet, only the chalazal megaspore expresses FM marker genes (Ravi et al. 2008).

As of now, molecular mechanisms that determine the survival of the FM and the cell death of the three remaining megaspores remain elusive. In *Tillandsia* and other plant taxa, callose depositions have been described, that are formed around the megaspore mother cell wall before meiosis. Together with the occlusion of plasmodesmata, callose is thought to act as a block for uncontrolled nutrient and signal fluxes between the mother plant and the megaspore mother cell (Papini et al. 2011). After meiosis, the callose depositions surround all four megaspores but are asymmetrically removed in the cell wall of the FM facing the parental sporophyte tissue. In species in which the chalazal megaspore survives, the callose disappears in the chalazal cell wall, while in species that retain the micropylar megaspore, the micropylar cell wall gets freed from callose (Rodkiewicz 1970). These callose depositions have been implicated with the control of PCD, as they completely surround only the dying megaspores, and are absent from plant taxa in which all four megaspores survive (Madrid and Friedman 2010). It has been speculated that the callose blocks PCD inhibiting signals that thus only reach the surviving functional megaspore (Papini et al. 2011). Alternatively, it is possible that the mere lack of nutrients caused by the isolation from the maternal tissue suffices to starve the remaining megaspores and thus trigger cell death (Ingram 2010). Detailed molecular analysis in model species such as *Arabidopsis* or maize using genetics and single cell-omics approaches will be required to answer these open questions.

2.2 *Nucellar Cell Death*

The plant megaspores are produced in a megasporangium termed "nucellus." Nucellar cells are symplastically linked to the chalazal pole of the megaspore mother cell, suggesting they are a nutritive tissue providing support to the developing female gametophyte after meiosis (Ingram 2010).

In *Arabidopsis*, the single layer of nucellar cells degenerate by the time the ovule reaches maturity, and this step appears to depend on the correct development of the female gametophyte because in the *Sporocyteless* (*Spl*) mutant, the nucellus remains viable (Yang et al. 1999). In the case of other plants, such as *Ricinus*, cereals, and cucurbits, there is a proliferation of the nucellar cells before fertilization, and degradation of this tissue occurs early after fertilization to support the developing embryo sac, triggered by a mechanism not yet fully understood (Dominguez et al. 2001; Greenwood et al. 2005). In *Ginkgo biloba*, nucellar cell death is linked with archegonium chamber formation to lead the motile spermatozoids to their fertilization targets, and the role of an early uptake of Ca^{2+} in mitochondria from nucellar cells was suggested in the pathway of events leading to PCD (Li et al. 2007a). The nucellus of *Pinus densiflora* undergoes PCD in response to pollen tube penetration. This is thought to sustain pollen tube growth by means of vesicular transport of degraded material which is taken up by the pollen tube via endocytosis (Hiratsuka and Terasaka 2011). The involvement of reactive oxygen species (ROS) and nitric oxide (NO) as signaling components in the pathway to PCD of nucellar cells and activation of caspase-like activities has been shown in *Sechium edule* (Lombardi et al. 2007a, 2010).

2.3 *Antipodal Cell Death During Embryo Sac Maturation*

The development of a fully receptive female gametophyte (FG) depends on the coordinated development and communication of sporophytic and gametophytic tissues, through symplastic movement of RNA and proteins, as well as apoplastic signaling cascades involving receptor kinases, as well as auxin movement and perception (Shi and Yang 2011). Generally, FG development comprises three rounds of free nuclear divisions, followed by cellularization to form a seven-celled, eight-nucleate embryo sac (Fig. 1). The mature FG usually consists of two gametes, the egg cell and the homodiploid central cell, and five accessory cells, two synergids and three antipodals (Ma and Sundaresan 2010).

Among these cells, antipodals are the most variable in terms of the number of cells, function, and lifespan. In maize, wheat, and other grasses, antipodals undergo mitoses, and they are thus referred as proliferative-type antipodals (Holloway and Friedman 2008). A nutritive function for long-lived antipodals has been suggested: They contain many well-developed mitochondria, dictyosomes, and endoplasmic reticulum, while ephemeral antipodals show fewer organelles (Sprunck and Gross-Hardt 2011). In *Arabidopsis* and many other species, the antipodals are composed of three cells that undergo PCD before embryo sac maturation and their longevity is at least in part controlled by the central cell. The *syco-1* mutant shows extended antipodal lifespan, caused by a defect in a mitochondrial localized cysteinyl-tRNA synthetase expressed in the central cell (Kagi et al. 2010). The remarkable developmental plasticity of antipodals is suggested to be an adaptive (or derived) character, which is also seen in the *lachesis* and *clotho* mutants, where the antipodal cells can express attributes characteristic of central cells and escape PCD. Both mutants are defective in putative core spliceosomal components (PRP4 and SNU114, respectively), suggesting a close link between pre-mRNA splicing factors and cell specification in the female gametophyte (Gross-Hardt et al. 2007; Moll et al. 2008). What is more, the overexpression of auxin biosynthesis genes shows that antipodals can adopt egg cell fate (Pagnussat et al. 2009). The function of ephemeral antipodal cells remains elusive, but the recent knowledge acquired from these mutants suggests a backup role in case of gametic failure (Kagi et al. 2010). There is still scarce information, and further experiments are needed to dissect the multiple roles that antipodals may have depending on the developmental program they follow.

2.4 Cell Death in the Tapetum Layer

The male gametophyte consists of mature pollen grains produced within the anthers. The anthers are commonly two-lobed structures containing two microsporangia (locules) each. Within these, the so-called tapetal cells surround a central region of sporogenous tissue (Goldberg et al. 1993). The tapetum is thought to be a nutritive tissue supporting gametophyte development, comparable to the nucellus in the megasporangium.

There are several events during anther development involving programmed PCD, and all are intimately related to male fertility. Best studied among these is tapetal breakdown, which occurs after microspore release from the tetrad. Similar to the nucellus, also the tapetum undergoes developmentally controlled death that is important to provide the pollen grains with a robust outer pollen shell (Riggs 2004). The use of mutants with defects in specific cell types shows that sporophytic tissues are vital for the proper development of viable pollen grains. Over the last years, significant progress has been achieved in understanding the process of pollen development, especially through the use of male sterile mutants in *Arabidopsis* as a model dicot and rice as a model crop monocot (Wilson and Zhang 2009). Some of the identified genes required for normal tapetal function and viable pollen production include the rice *UNDEVELOPED TAPETUM 1* (*UDT1*, (Jung et al. 2005)), its *Arabidopsis* ortholog *DYSFUNCTIONAL TAPETUM 1* (*DYT1* (Zhang et al. 2006)), and *ABORTED MICROSPORE* (*AMS* (Sorensen et al. 2003; Xu et al. 2010)), which encode basic helix-loop-helix (bHLH) transcription factors, while MALE STERILITY 1 (MS1) is a plant homeodomain transcription factor (Ito et al. 2007). The *ms1* mutant shows delayed tapetal breakdown and a switch from PCD to necrotic cell death (Vizcay-Barrena and Wilson 2006). The rice ortholog of MS1 is *PERSISTENT TAPETAL CELL 1* (*PTC1*), and it is expressed when the wild-type tapetal cells initiate PCD. Unlike the *ms1* mutant, the *ptc1* mutant displays a phenotype of uncontrolled tapetal proliferation (Li et al. 2011a). The rice mutant defective in TAPETAL DEGENERATION RETARDATION (OsTDR) also shows delayed tapetal PCD and failure of pollen wall deposition, resulting in microspore abortion. Also, it has been shown that OsTDR plays an important role in the composition of aliphatic sporopollenin, the main component of the outer pollen wall (Li et al. 2006; Zhang et al. 2008). A previously unknown pathway for regulating PCD during tapetum degeneration in rice was characterized with APOPTOSIS INHIBITOR5 (API5), a putative homolog of antiapoptotic protein API5 in animals. Rice API5 is a nuclear protein that interacts with two DEAD-box ATP-dependent RNA helicases, API5-INTERACTING PROTEIN1 (AIP1), and AIP2. They are homologs of yeast proteins involved in transcription elongation and pre-mRNA splicing (Li et al. 2011b). These results substantiate the importance of RNA processing for the correct development of male as well as female reproductive organs and are thus promising targets for future research.

2.5 Pollen Cell Death During Incompatible Pollen-Pistil Interactions

Pollen-pistil interactions are early key regulators of pollination and fertilization in flowering plants, and many components of the signaling cascades triggered by the communication between pollen and stigmatic tissues have been identified (Bosch and Franklin-Tong 2008; Hiscock and Allen 2008; Higashiyama 2010). Self-incompatibility (SI) is an adaptation to prevent inbreeding and has evolved independently several times in plants, since at least three distinct SI systems have been described at the molecular level (Takayama and Isogai 2005). SI responses differ in these systems; the response can be either the abortion of the pollen tube (PT) in the transmitting tract of the style, known as gametophytic SI, or inhibition of the PT germination triggered by the diploid parent via its stigmatic tissues, thus referred to as sporophytic SI.

Once landed on the stigma, the pollen behavior is controlled by the multiallelic S-locus, and the combination of different haplotypes allows discriminating between self (incompatible) and nonself (compatible) pollen. In the Brassicaceae-type SI, the pollen ligand is a small cysteine-rich protein (SP11/SCR) present in the pollen coat. Its receptor is a kinase in the stigmatic papilla cells which, once activated, induces incompatibility signaling (Ivanov et al. 2010). The SI determinants in the Solanaceae, Plantaginaceae, and Rosaceae are S-ribonucleases (S-RNases), allelic products of the pistil which encode secreted glycoproteins expressed in the stigma and the transmitting tract of the style. In these plant families, the male determinants are at least three types of divergent S-locus F-box proteins (SLF/SLB), which recognize and detoxify a specific subset of nonself S-RNAses inside the pollen tubes via the ubiquitin 26S proteasome (Chen et al. 2010; Kubo et al. 2010). Finally, in the Papaveraceae-type SI, the pistil determinant (PrsS) is a small secreted protein that interacts with the pollen determinant (PrpS, a highly polymorphic transmembrane protein) and induces a Ca^{2+}-dependent signaling cascade in incompatible pollen (Bosch and Franklin-Tong 2008). This cascade triggers depolymerization of the actin cytoskeleton, the phosphorylation of inorganic pyrophosphatases, and activation of a protein kinase (MAPK). These events eventually culminate in PCD through the stimulation of caspase-3-like activity (DEVDase) and DNA fragmentation in incompatible pollen (Li et al. 2007b; Bosch et al. 2008; Wheeler et al. 2009). Recently, the signaling role of ROS and NO in SI responses in *Papaver* was demonstrated to be upstream of the formation of actin remodeling and caspase-like activities (Wilkins et al. 2011).

3 Cell Death Decisions During Fertilization and Seed Development

Multiple signaling events have to occur successfully after pollination before the pollen tube can finally join the female gametophyte (Lausser and Dresselhaus 2010; Marton and Dresselhaus 2010; Okuda and Higashiyama 2010). Double fertilization results in two individual fertilization products: the diploid zygote that will develop into the embryo and the triploid nutritive tissue of the endosperm. The fate of the two fertilization products could not be more different—while the embryo lives on to form the next sporophytic plant generation, the endosperm's function and life is restricted to seed development and ends with germination (Berger 2003; Berger et al. 2006; Sabelli and Larkins 2009).

3.1 Synergid Cell Death During Fertilization

Successful fertilization critically depends on a precise and fine-tuned reciprocal communication between the pollen tube and the female gametophytic tissues inside the ovule. The two female gametophytic synergid cells have been shown to play a central role in attracting the pollen tube over the last distance to the micropyle in the ovule and facilitating fertilization (Dresselhaus 2006; Marton and Dresselhaus 2010). During the fertilization process, one of the synergids degenerates after the pollen tube enters the micropyle (Sandaklie-Nikolova et al. 2007). The pollen tube enters the synergid cell, arrests its tip growth, and releases its two sperm cells into the degenerating synergid, upon which they fuse with the egg cell and the central cell, respectively (Hamamura et al. 2011).

Synergid cell death has been put forward as a case of developmentally controlled cell death triggered by the approaching pollen tube, as in many species synergid cell death is fertilization dependent (Russell 1992; Christensen et al. 1998; Faure et al. 2002). To date, few molecular details are known that regulate synergid cell death (Sandaklie-Nikolova et al. 2007).

Three female gametophytic mutants defective in synergid cell death have been described, *sirene/feronia, nortia*, and *gfa2* (Christensen et al. 2002; Rotman et al. 2008; Kessler et al. 2010). *FERONIA* encodes a receptor-like kinase (RLK) localized at the synergid plasma membrane at the site of pollen tube reception, in the so-called filiform apparatus (Kessler et al. 2010). In *feronia* mutants, the pollen tube enters the female gametophyte but fails to penetrate one of the synergids. It continues to grow without discharging the sperm cells, forming a coiled structure inside the female gametophyte. None of the synergid cells degenerate (Huck et al. 2003; Rotman et al. 2003). Thus, the mere physical contact with the pollen tube is not sufficient to trigger synergid cell death, i.e., proper pollen tube reception, and synergid cell death requires a *FERONIA*-dependent signaling process.

Recently, *nortia* has been described as another female gametophytic mutant showing a similar phenotype as *feronia*. *NORTIA* is expressed in synergid cells of unfertilized ovules. Upon pollen tube arrival, *NORTIA* accumulates at the filiform apparatus. Interestingly, *FERONIA* is required for this polarized localization, suggesting that *FERONIA* and *NORTIA* function in one pathway to control pollen tube reception. *NORTIA* is allelic to *AtMLO7*, a member of a family of transmembrane proteins known to be required for powdery mildew resistance. The *FERONIA* pathway is also required for successful pathogenicity of powdery mildew, suggesting a common mechanism functional in hyphal penetration of epidermal cells by a fungal pathogen and in pollen tube penetration of synergid cells (Kessler et al. 2010).

GFA2 encodes a homolog of the yeast Mdj1p, a mitochondrial chaperone required for survival at elevated temperatures and for inheritance of intact mitochondrial DNA in yeast (Duchniewicz et al. 1999). In Arabidopsis *gfa2* mutants, the pollen tube is correctly attracted to the micropyle, but pollen tube penetration and synergid cell death do not occur (Christensen et al. 2002). As *Arabidopsis GFA2* localizes to the mitochondria and partly complements the yeast Mdj1p phenotype, it is tempting to speculate that its function is conserved and that mitochondrial dysfunction leads to a failure to execute the synergid cell death program.

Alternatively, the death of the receiving synergid could be caused by its mere physical disruption by the pollen tube after signaling events that lead to successful penetration (Higashiyama 2002). However, in many plant species, synergid cell death has been described to occur well before physical contact with the pollen tube, arguing for a pollen derived, long-range signal triggering synergid cell death (Sandaklie-Nikolova et al. 2007) and papers cited therein. Live cell imaging of the fertilization process in *Arabidopsis* has shown that cell death of the synergid cells is only initiated after contact of the pollen tube with the synergid cell but before pollen tube penetration and discharge. These data suggest that, at least in *Arabidopsis*, short-range communication rather than mere physical disruption causes synergid cell death (Sandaklie-Nikolova et al. 2007).

However, more recent high-resolution life cell imaging of the fertilization process has shown that synergid nuclei remain intact until mere minutes before sperm cell discharge and breakdown in the minutes following discharge (Hamamura et al. 2011). Though the question of synergid cell death was not discussed in this study, a nuclear destruction occurring within minutes around pollen tube discharge makes a physical destruction of the synergid cell by the penetrating and rupturing pollen conceivable. High-resolution life cell imaging in combination with unambiguous cell death markers will have to finally resolve this issue.

3.2 *Embryonic Suspensor Cell Death*

In most angiosperms, the zygote divides unequally into a smaller apical and a larger basal cell. While the apical cell develops in the embryo proper, the basal cell undergoes a limited number of cell divisions, forming the embryo suspensor.

It serves to push the young embryo proper into the endosperm lumen and anchor it at its micropylar position and is thought to contribute to the early embryonic nutrient uptake (Kawashima and Goldberg 2010). In some plants, like *Arabidopsis*, the fully developed suspensor consists only of a single row of seven cells, while in other species, massive structures containing hundreds of cells develop (Kawashima and Goldberg 2010).

The suspensor is a short-lived organ, undergoing developmentally controlled PCD during seed development. Although detailed molecular mechanisms have been elucidated about the auxin-based cell-to-cell signaling processes that determine the fates of embryo proper and suspensor (Larsson et al. 2008; Moller and Weijers 2009), little is known about how the suspensor cell death is initiated and executed, and how cell death is prevented in the adjacent cells of the embryo proper (Kawashima and Goldberg 2010).

A major model system for research on PCD in embryonic suspensor tissues has been developed by Peter Bozhkov and his colleagues, exploiting somatic embryogenesis of Norway spruce (*Picea abies*). During the cell death of spruce suspensors, caspase-like VEIDase activity was detected and inhibition of this proteolytic activity led to a failure of embryo-suspensor differentiation (Bozhkov et al. 2004). Additionally, a type-II metacaspase activity has been implicated with the control of suspensor degeneration via nuclear envelope disassembly and chromatin degradation (Bozhkov et al. 2005). A Tudor staphylococcal nuclease (TSN) has been found to be a natural substrate of type-II metacaspase in spruce (Sundstrom et al. 2009). Interestingly, during apoptosis in humans, the human TSN homolog is cleaved by caspase 3, leading to a breakdown of its ribonuclease activity and ability to activate mRNA splicing. Both processes are essential for cell viability, and reduction of TSN activity in *Arabidopsis* caused ectopic cell death in pollen, ovules, and developing seeds, leading to a strong reduction in fertility. Intriguingly, TSN is a target of both animal caspases as well as of the unrelated plant metacaspases, suggesting that TSN degradation has independently evolved in these taxa to initiate PCD (Sundstrom et al. 2009).

Also in angiosperms, PCD in embryonic suspensors have been described, for instance, in maize and *Phaseolus* (Giuliani et al. 2002; Lombardi et al. 2007b). Recently, a first molecular component has been reported from suspensor cell death in *Arabidopsis thaliana*. KISS OF DEATH (KOD) encodes a 25-amino-acid peptide that is specifically expressed in suspensor cells before degeneration, as well as after biotic and abiotic stresses. Loss of KOD function leads to a decreased rate of suspensor degeneration and heat-shock induced PCD in root hair cells (Blanvillain et al. 2011). Ectopic KOD expression in tobacco leaves and *Arabidopsis* seedlings lead to induction of PCD and induced caspase-3-like DEVDase activity. Furthermore, KOD expression caused loss of mitochondrial membrane potential, an early step described from other plant PCD processes. Conversely, coexpression of KOD with the antiapoptotic AtBI-1 or the caspase inhibitor p35 was shown to strongly reduce the cell death rate of transiently transfected onion cells (Blanvillain et al. 2011).

3.3 Cell Death During Seed Coat Formation

The seed coat consists of the maternally derived integuments around the developing embryo and endosperm. Throughout its development, the seed integuments fulfill many functions, from transferring nutrients to the developing offspring to eventual seed dispersal, dormancy control, and hydration during seed germination. To achieve these functions, the integumental tissues undergo a differentiated developmental program. Notably, the final differentiation step of all cells in the seed coat is cell death, and all functions carried out by the mature seed coat are accomplished by dead tissues.

The different layers of the seed coat are derived from the ovule integuments. Fertilization of the two female gametes triggers growth of the integuments and differentiation into the different seed coat tissues (Beeckman et al. 2000). In *Arabidopsis*, the five integument layers (two outer integuments, oi1 and oi2; and three inner integuments, ii1, ii2, and ii3) follow four different developmental pathways. Though molecular components have been identified that control growth and developmental differentiation of the seed coat (Haughn and Chaudhury 2005), very little is known about the cell death program executed at specific time points for the individual integument layers.

After fertilization, the first two layers to undergo cell death are ii1 and ii2. In contrast to the other seed coat tissues, these two layers do not go through any obvious morphological differentiation before entering PCD. While the molecular control of PCD of the other integument layers is still completely unknown, there is experimental evidence of a participation of vacuolar processing enzymes (VPEs) in cell death regulation of the two inner integument layers (Nakaune et al. 2005). VPEs are cysteine proteases that reside in the lumen of lytic vacuoles that take up the major part of most mature plant cells. Lytic vacuoles contain a great variety of hydrolytic enzymes that recycle cellular material that is sequestered into the vacuole. Vacuolar proteins are synthesized at the endoplasmic reticulum and then transported to the vacuole, where they are processed to mature forms by VPEs (Yamada et al. 2005). Furthermore, VPEs have been implicated as central players in vacuolar cell death, an emerging major PCD *modus* in plants (Hatsugai et al. 2006; Hara-Nishimura and Hatsugai 2011). This PCD variant culminates in the rupture of the vacuolar membrane, the tonoplast, and subsequent release of vacuolar hydrolytic enzymes such as proteases and nucleases into the cytosol. The rupture of the tonoplast leads to a disintegration of various organelles like plastids, mitochondria, and the nucleus. Vacuolar cell death has been described to be VPE dependent during a virus-induced hypersensitive response (HR). VPE appears to act early during the HR as its levels come on early and decline before visible lesions are formed (Hatsugai et al. 2004). Notably, VPEs possess caspase-1-like activity, though they are neither genetically nor structurally related to animal caspases. Thus, though the cell death players might not be conserved between animals and plants, a common, potentially ancient, core cell death mechanism might exist in both taxa. Next to their role in HR cell death, one of four *Arabidopsis* VPEs, δVPE, has been shown to control developmental cell death in the two inner integument layers during seed development. δVPE is specifically expressed in these two layers before onset of cell death, and mutant plants deficient of δVPE show a delayed cell death and collapse of these inner integument cells (Nakaune et al. 2005).

3.4 Central Endosperm Cell Death

During mid-seed development, drastic changes are preparing: While, initially, the rapidly proliferating free nuclear endosperm dominated the developing seed, its growth now comes to a halt. Its free nuclear divisions stop, and a wave of cellularization sweeps across the endosperm. At this time point, the embryo starts to expand massively, consuming—depending on the plant taxon—a minor or a major part of the endosperm and the nutrients that it accumulated during early seed development. There are two major types of endosperm fate: In plant taxa with persistent endosperm (e.g., cereals), the bulk of the endosperm is maintained. In plants with an ephemeral endosperm (e.g., legumes and *Arabidopsis*), all or nearly all endosperm is consumed by the growing embryo. Whichever the mode, in most mature plant seeds, only two living tissues remain: The dormant plant embryo and the so-called aleurone layer, a typically single-celled layer that lines the inside of the dead seed coat and represents the last living remains of the endosperm (Berger 2003; Costa et al. 2004; Olsen 2004; Sabelli and Larkins 2009; Nowack et al. 2010).

During mid-seed development, two forms of cell death terminate the life of the endosperm bulk. A first type of PCD, a consumptive form of cell death, is executed in the endosperm adjacent to the expanding embryo, termed embryo surrounding region (ESR). During this PCD process, endosperm cells in the ESR undergo complete autolysis, freeing nutrients that fuel embryo growth and making space for the expanding embryo (Ingram 2010). In taxa with ephemeral endosperm, the growing embryo incorporates the bulk of the endosperm, and only the aleurone layer is preserved in mature seeds. In taxa with persistent endosperm (e.g., grasses), the ESR cell death is rather restricted, and most of the endosperm is preserved (in cereals) as the starchy endosperm. This invasive growth of the embryo in a nutritive tissue is reminiscent of the embryo invasion into the nutrient-rich female gametophyte in nonflowering seed plants (gymnosperms, for review see (Vuosku et al. 2009)). It is tempting to speculate that the molecular mechanisms of cell death and autolysis of female gametophyte and endosperm are evolutionary conserved.

Still, very little is known about the PCD mechanisms in the ESR. So far, only one gene has been described to exert a function in this context in *Arabidopsis*, *ZHOUPI/RETARDED GROWTH OF EMBRYO1* (*ZOU/RGE1*, (Kondou et al. 2008; Yang et al. 2008)). This gene encodes a helix-loop-helix transcription factor exclusively expressed in the endosperm. In *zou/rge1* mutants, the ESR cell death is reduced, and endosperm persists in the mature seeds. Still, some ESR markers such as SUC5 are still expressed in *zou/rge1*, while others (e.g., ALE1) are not detectable in this mutant endosperm. Further research of targets of ZOU/RGE1 using transcriptomics and proteomics will be necessary to discover the functional genetic network downstream of this transcriptional activator.

A second type of PCD happens in the bulk of the starchy endosperm of cereals, during which cells are killed but the cellular corpses remain unprocessed (Young and Gallie 2000b). Ingram (2010) speculates that the cell death regulation in ESR and starchy endosperm must be quite distinct, as they serve different purposes and end in different results, autolysis in the ESR versus maintenance of the cellular corpses in the starchy endosperm. The latter form of PCD is thought to optimize nutrient storage and to facilitate the embryo's rapid access to the storage compounds at germination (Sabelli and Larkins 2009; Sreenivasulu et al. 2010).

While cell death of maternal tissues in cereals is marked by expression of a plethora of hydrolytic enzymes such as amylases, lipases, proteases, and cell wall degrading cellulases and glucanases, cell death in the starchy endosperm is characterized by the transcription of genes from more selective degradation pathways. These include the ubiquitin pathway, target recognition by F-box proteins, and protein degradation by the proteasome complex (Sreenivasulu et al. 2006). Furthermore, a caspase-6-like VEIDase activity has been localized in potential autophagosomes in barley starchy endosperm undergoing PCD (Boren et al. 2006). Still, so far there only exists circumstantial evidence for the activity of lytic enzymes, and individual cell death effectors have so far eluded detection (Sabelli and Larkins 2009). However, concrete evidence exists for the participation of phytohormone signaling in cereal endosperm cell death. Elevated ethylene levels have been associated with PCD in maize central endosperm (Young et al. 1997). Furthermore, the ethylene biosynthetic machinery and signal transduction components are upregulated before PCD in barley central endosperm (Sreenivasulu et al. 2006). In contrast to ethylene, ABA appears to inhibit PCD in central endosperm cells by negatively regulating ethylene biosynthesis. In maize *vp1* and *vp9* mutants, which are deficient in ABA perception and biosynthesis, respectively, elevated ethylene levels were coinciding with premature onset of DNA fragmentation and cell death (Young and Gallie 2000a, 2000b).

Next to its function of gathering and passing on nutrients to the developing embryo, the endosperm has also been put forward as a major sensor of genomic imbalance, as produced by interspecies hybridization or polyploidization. In cases of less severe imbalance, endosperm growth and development is altered, but when maternal and paternal genomes differ too drastically, endosperm failure can lead to seed abortion (Birchler 1993; Scott et al. 1998; Ishikawa et al. 2011). Thus, the endosperm serves as an effective postzygotic barrier that inhibits interspecies hybridization and allows speciation events, for instance via polyploidization (Costa et al. 2004; Kinoshita 2007). It is an appealing hypothesis that PCD mechanisms in the developing seed might have been recruited for the rapid execution of seed abortion in incompatible crosses.

4 Cell Death During Germination

4.1 Survival and Cell Death Decisions During Aleurone Development

After maturation, there remain only two living entities in most seeds: The embryo and the aleurone, a typically single-celled outer endosperm layer that escapes endosperm cell death during mid-seed development. Both embryo and aleurone remain in a dormant state as long as the seed does not encounter favorable conditions for germination. As soon as those conditions are given, however, embryo and aleurone restart their metabolism, and the process of germination begins. Seed imbibition, i.e., the exposure of the dry seed to a moist environment, will trigger germination as long as no specific factors imposing seed dormancy are present in the seed (Holdsworth et al. 2008). During germination, embryo and aleurone follow a very different fate—while the embryo starts its new life as a seedling, the aleurone terminally differentiates and dies. In *Arabidopsis*, the aleurone layer has been implicated in the control of seed dormancy and germination and contributes with its storage compounds to seedling establishment (Penfield et al. 2005; Bethke et al. 2007). In cereals, the aleurone additionally has an important function as a secretory tissue, producing hydrolases (e.g., alpha-amylases) that mobilize the reserve compounds in the starchy endosperm.

It has been shown that cereal aleurone cells undergo PCD regulated by plant hormones (Bethke et al. 1999, 2007; Beligni et al. 2002), but also, in dicots such as *Arabidopsis*, phytohormones trigger terminal aleurone development (Bethke et al. 2007). Both in *Arabidopsis* and cereal aleurones, gibberellic acid (GA) serves as a key signal molecule leading to extensive vacuolation resulting from fusion of protein storage vacuoles (PSVs). In cereals, this is followed by loss of plasma membrane integrity and turgor loss and subsequent cytoplasm shrinkage (Bethke et al. 1999). GA is not synthesized endogenously in aleurone cells: Upon imbibition, the embryo starts GA production and uses it as signaling molecule to communicate with the aleurone, where GA then exerts its effects (Yamaguchi et al. 2001; Ogawa et al. 2003; Mitchum et al. 2006). On the other hand, embryonic GA-biosynthesis depends on nitric oxide (NO) produced in the aleurone upon imbibition (Bethke et al. 2004, 2007). In contrast to GA and NO, abscisic acid (ABA) maintains seed dormancy and prevents cereal aleurone vacuolation and cell death. ABA is produced and accumulated in the aleurone layer during late maturation (Bethke et al. 1999, 2002; Fath et al. 2000). Thus, the hormonal cross talk between the embryo and aleurone is the basis for terminal differentiation and cell death of the aleurone.

Despite the extensive evidence on the hormonal control of aleurone PCD including GA, ABA, and NO, our knowledge on the actual sequence of the cell death execution events is still rather hypothetical. A model summarizing the current knowledge about the signaling between embryo and aleurone is shown in Fig. 2.

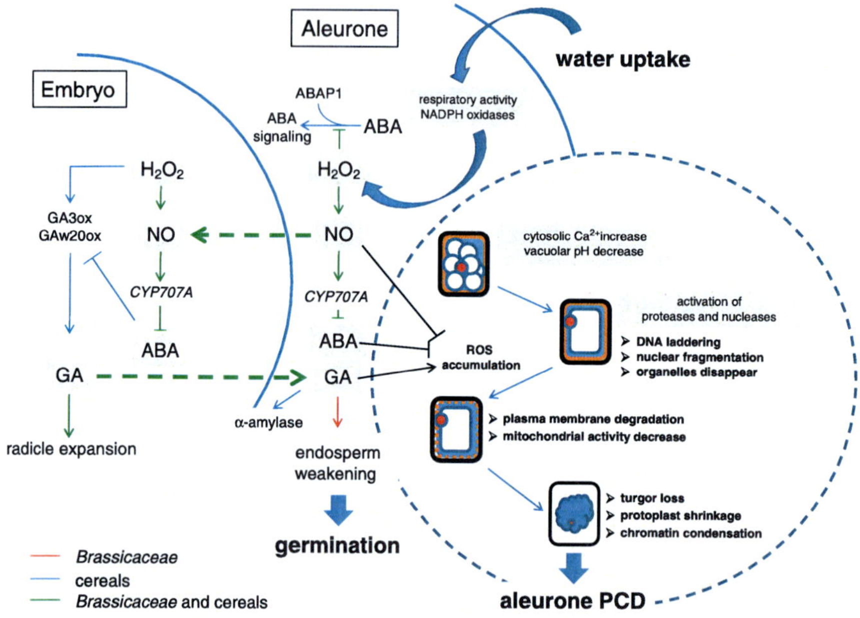

Fig. 2 Model of signaling pathways during seed germination leading to terminal aleurone differentiation and cell death. Upon the imbibition, the embryo and the aleurone become metabolically active, followed by ROS accumulation and NO synthesis in aleurone. H_2O_2 and NO mediate ABA catabolism and stimulate GA synthesis in the embryo via regulation of *CYP707A*, as well as via initiation of *GA3ox* and *GAw20ox* gene expression. GA synthesized in the embryo signals to the aleurone and triggers α-amylase synthesis in cereals and endosperm weakening in Brassicaceae. In parallel, GA induces the vacuolation of aleurone cells and triggers the PCD process in cereals. For further details see text

Although the irreversible initiation of cereal aleurone death program starts at early stage of imbibition, the cellular execution of PCD occurs only after the tissue has accomplished its functions and often proceeds as a postgermination event. In the cereal aleurone, PCD follows the vacuolation process and is as well tightly regulated by GA and ABA: While GA induces vacuolation and onset of aleurone PCD, ABA delays vacuolation and cell death execution (Kuo et al. 1996; Wang et al. 1996; Bethke et al. 1999, 2007). Vacuolation serves to mobilize the reserve compounds and enzymes stored in the aleurone. The abundant PSVs are first acidified and then coalesce into large lytic vacuoles (Swanson and Jones 1996; Fath et al. 2000). Extensive vacuolation is followed by loss of plasma membrane integrity, accompanied by turgor loss and cellular collapse (Bethke et al. 1999).

The signal transduction cascade of GA-triggered PCD in the cereal aleurone is not completely uncovered yet, but several putative PCD regulators have been identified: Upon the vacuolation of aleurone cells, a rapid increase in cytosolic Ca^{2+} occurs. In the presence of syntide-2, a synthetic substrate for Ca^{2+}, the aleurone cell vacuolation process was arrested and cell's life extended (Ritchie and Gilroy 1998), suggesting that a certain Ca^{2+} level is essential to induce the aleurone's developmental program leading to PCD. Furthermore, aleurone PCD can be blocked by the phosphatase inhibitor okadaic acid; thus, posttranslational modification by phosphatases appears to play a role in the early PCD signaling steps (Kuo et al. 1996). Finally, a cyclic guanosine monophosphate (cGMP, a second messenger mediating NO responses in mammalian systems) could also be a part of the signaling cascade leading to PCD. The inhibition of cGMP by LY83583 reduced GAMYB and α-amylase gene expression and inactivated intracellular nucleases preventing DNA degradation in barley aleurone cells (Bethke et al. 1999; Fath et al. 1999).

Next to Ca^{2+}, also ROS play an important role in the PCD regulation of aleurone cells. ROS can act in two ways: On the one hand, high levels of ROS directly damage proteins, nucleic acids, and membrane systems. On the other hand, ROS (especially the long-lived H_2O_2) are known to act as signaling molecules, causing the expression of genes involved in PCD. The finding that the levels of ROS scavengers such as superoxide dismutase (SOD), ascorbate peroxidase (APX), or catalase (CAT) (Fath et al. 2001, 2002) and haem oxygenase (Wu et al. 2011) are reduced in GA-treated aleurone cells supports the hypothesis that GA reduces ability of aleurone cells to detoxify ROS. This in turn could lead to oxidative damage and culminate in a rapid cell death (Bethke and Jones 2001; Palma and Kermode 2003). Additionally, ROS could alter the expression of the GA-dependent genes via direct regulation of gene transcription. Notably, it has been shown that GAMYB, a R2R3-type MYB transcription factor which is involved in GA-dependent gene regulation in cereal aleurone, requires reducing conditions for DNA binding (Williams and Grotewold 1997; Heine et al. 2004). It is worth to mention that GAMYB has been implicated in both PCD of tapetal cells (Aya et al. 2009) and of cereal aleurone cells (Guo and Ho 2008).

ABA regulates PCD in aleurone cells in a tight coordination with GA, inhibiting its activity and delaying PCD. In the presence of ABA, the cells of isolated aleurone layers or protoplasts arrested their vacuolation process and could be kept alive up to several months (Bethke et al. 1999, 2002, 2007; Fath et al. 2000). In contrast to GA, ABA treatment leads to upregulation of ROS scavengers (Bethke and Jones 2000; Fath et al. 2001, 2002). The inhibition of the aleurone cell vacuolation by ABA is associated with HVA22, an ABA-responsive protein accumulating in barley aleurone at late maturation (Guo and Ho 2008). Activated by high levels of ABA, HVA22 negatively regulates the vesicle trafficking and PSV fusion and thus inhibits GA-induced PCD (Guo and Ho 2008). ABA also decreases cytosolic Ca^{2+} concentration and increases intracellular pH and MAP kinase activity (Gilroy and Jones 1992; Heimovaara-Dijkstra et al. 1994; Knetsch et al. 1996) and restricts DNA fragmentation (Wang et al. 1996).

In contrast to the hormonal regulation of vacuolation and PCD onset, we know surprisingly little about the actual execution of PCD in cereal aleurone cells. There is evidence for hydrolytic activities of nucleases and proteases, but the actual enzymes remain largely elusive to date.

As a typical hallmark of many PCD variants, internucleosomal DNA degradation resulting in so-called DNA ladders was found in cells undergoing PCD in aleurone layers of barley, wheat, and maize (Wang et al. 1998; Dominguez et al. 2004). Ca^{2+}/Mg^{2+}-dependant nucleases were identified in nuclei of dying cells of wheat aleurone and strongly induced by GA application (Dominguez et al. 2004). DNA fragmentation appears at first in the aleurone cells adjacent to the embryo and extends later to the distal part of barley caryopsis, revealing a precise spatial and temporal regulation of DNA degradation as decisive step during PCD execution.

Also, proteases known as principal players in plant PCD were found in aleurone cells entering PCD (Beers et al. 2000). Accumulation of two aspartic proteases and three cysteine proteases upon GA application has been reported from barley aleurone cells. Although caspase-like activities were detected in the vacuole of barley aleurone cells, these were not dependent on GA (Fath et al. 2000). Upregulation of the transcript numbers of several cysteine proteases was found specifically in the micropylar endosperm cap during *Lepidium* seed germination (Morris et al. 2011). Also in *Arabidopsis*, cysteine proteases were implicated in the final stage of cellular collapse of aleurone cells during germination, and *CEP1* promoter activity was found in the remnants of the *Arabidopsis* aleurone layer after germination (Helm et al. 2008).

5 Conclusions

PCD is a central theme during plant reproductive development, and precise control of PCD execution, or its prevention, are intimately linked with successful plant reproduction. Despite its importance and ubiquitous occurrence throughout plant reproductive development, we still know only very little about the molecular communication events that control PCD in the diverse reproductive organs. So far, only some isolated PCD players have been identified; the signaling network as a whole remains largely unknown. The future challenge will thus consist not only in identifying more individual components of the PCD control machinery but also in applying systems biology approaches to gain an insight in the regulatory networks that take a cell's decision on the matter of life or death.

References

Antonsson B, Conti F, Ciavatta A, Montessuit S, Lewis S, Martinou I, Bernasconi L, Bernard A, Mermod JJ, Mazzei G, Maundrell K, Gambale F, Sadoul R, Martinou JC (1997) Inhibition of Bax channel-forming activity by Bcl-2. Science 277:370–372

Aya K, Ueguchi-Tanaka M, Kondo M, Hamada K, Yano K, Nishimura M, Matsuoka M (2009) Gibberellin modulates anther development in rice via the transcriptional regulation of GAMYB. Plant Cell 21:1453–1472

Baek D, Nam J, Koo YD, Kim DH, Lee J, Jeong JC, Kwak SS, Chung WS, Lim CO, Bahk JD, Hong JC, Lee SY, Kawai-Yamada M, Uchimiya H, Yun DJ (2004) Bax-induced cell death of

Arabidopsis is meditated through reactive oxygen-dependent and -independent processes. Plant Mol Biol 56:15–27
Beeckman T, De Rycke R, Viane R, Inzé D (2000) Histological study of seed coat development in *Arabidopsis thaliana*. J Plant Res 113:139–148
Beers EP, Woffenden BJ, Zhao C (2000) Plant proteolytic enzymes: possible roles during programmed cell death. Plant Mol Biol 44:399–415
Beligni MV, Fath A, Bethke PC, Lamattina L, Jones RL (2002) Nitric oxide acts as an antioxidant and delays programmed cell death in barley aleurone layers. Plant Physiol 129:1642–1650
Berger F (2003) Endosperm: the crossroad of seed development. Curr Opin Plant Biol 6:42–50
Berger F, Grini PE, Schnittger A (2006) Endosperm: an integrator of seed growth and development. Curr Opin Plant Biol 9:664–670
Bethke PC, Jones RL (2000) Vacuoles and prevacuolar compartments. Curr Opin Plant Biol 3:469–475
Bethke PC, Jones RL (2001) Cell death of barley aleurone protoplasts is mediated by reactive oxygen species. Plant J 25:19–29
Bethke PC, Lonsdale JE, Fath A, Jones RL (1999) Hormonally regulated programmed cell death in barley aleurone cells. Plant Cell 11:1033–1046
Bethke PC, Fath A, Spiegel YN, Hwang Y, Jones RL (2002) Abscisic acid, gibberellin and cell viability in cereal aleurone. Euphytica 126:3–11
Bethke PC, Badger MR, Jones RL (2004) Apoplastic synthesis of nitric oxide by plant tissues. Plant Cell 16:332–341
Bethke PC, Libourel IG, Aoyama N, Chung YY, Still DW, Jones RL (2007) The Arabidopsis aleurone layer responds to nitric oxide, gibberellin, and abscisic acid and is sufficient and necessary for seed dormancy. Plant Physiol 143:1173–1188
Bewley JD, Black M (1994) Seeds: physiology of development and germination. Plenum Publishing, New York
Birchler JA (1993) Dosage analysis of maize endosperm development. Annu Rev Genet 27:181–204
Blanvillain R, Young B, Cai YM, Hecht V, Varoquaux F, Delorme V, Lancelin JM, Delseny M, Gallois P (2011) The Arabidopsis peptide kiss of death is an inducer of programmed cell death. EMBO J 30:1173–1183
Boren M, Hoglund AS, Bozhkov P, Jansson C (2006) Developmental regulation of a VEIDase caspase-like proteolytic activity in barley caryopsis. J Exp Bot 57:3747–3753
Bosch M, Franklin-Tong VE (2008) Self-incompatibility in papaver: signalling to trigger PCD in incompatible pollen. J Exp Bot 59:481–490
Bosch M, Poulter NS, Vatovec S, Franklin-Tong VE (2008) Initiation of programmed cell death in self-incompatibility: role for cytoskeleton modifications and several caspase-like activities. Mol Plant 1:879–887
Bozhkov PV, Filonova LH, Suarez MF, Helmersson A, Smertenko AP, Zhivotovsky B, von Arnold S (2004) VEIDase is a principal caspase-like activity involved in plant programmed cell death and essential for embryonic pattern formation. Cell Death Differ 11:175–182
Bozhkov PV, Suarez MF, Filonova LH, Daniel G, Zamyatnin AA Jr, Rodriguez-Nieto S, Zhivotovsky B, Smertenko A (2005) Cysteine protease mcII-Pa executes programmed cell death during plant embryogenesis. Proc Natl Acad Sci USA 102:14463–14468
Bozhkov PV, Smertenko AP, Zhivotovsky B (2010) Aspasing out metacaspases and caspases: proteases of many trades. Sci Signal 3:pe48
Cacas JL (2010) Devil inside: does plant programmed cell death involve the endomembrane system? Plant Cell Environ 33:1453–1473
Chen G, Zhang B, Zhao Z, Sui Z, Zhang H, Xue Y (2010) 'A life or death decision' for pollen tubes in S-RNase-based self-incompatibility. J Exp Bot 61:2027–2037
Christensen CA, Subramanian S, Drews GN (1998) Identification of gametophytic mutations affecting female gametophyte development in Arabidopsis. Dev Biol 202:136–151

Christensen CA, Gorsich SW, Brown RH, Jones LG, Brown J, Shaw JM, Drews GN (2002) Mitochondrial GFA2 is required for synergid cell death in Arabidopsis. Plant Cell 14:2215–2232

Costa LM, Gutierrez-Marcos JF, Dickinson HG (2004) More than a yolk: the short life and complex times of the plant endosperm. Trends Plant Sci 9:507–514

Dahiya P (2003) Role of death in providing lifeline to plants. Trends Plant Sci 8:462–465

Dominguez F, Moreno J, Cejudo FJ (2001) The nucellus degenerates by a process of programmed cell death during the early stages of wheat grain development. Planta 213:352–360

Dominguez F, Moreno J, Cejudo FJ (2004) A gibberellin-induced nuclease is localized in the nucleus of wheat aleurone cells undergoing programmed cell death. J Biol Chem 279:11530–11536

Dresselhaus T (2006) Cell–cell communication during double fertilization. Curr Opin Plant Biol 9:41–47

Duchniewicz M, Germaniuk A, Westermann B, Neupert W, Schwarz E, Marszalek J (1999) Dual role of the mitochondrial chaperone Mdj1p in inheritance of mitochondrial DNA in yeast. Mol Cell Biol 19:8201–8210

Fath A, Bethke PC, Jones RL (1999) Barley aleurone cell death is not apoptotic: characterization of nuclease activities and DNA degradation. Plant J 20:305–315

Fath A, Bethke P, Lonsdale J, Meza-Romero R, Jones R (2000) Programmed cell death in cereal aleurone. Plant Mol Biol 44:255–266

Fath A, Bethke PC, Jones RL (2001) Enzymes that scavenge reactive oxygen species are down-regulated prior to gibberellic acid-induced programmed cell death in barley aleurone. Plant Physiol 126:156–166

Fath A, Bethke P, Beligni V, Jones R (2002) Active oxygen and cell death in cereal aleurone cells. J Exp Bot 53:1273–1282

Faure JE, Rotman N, Fortune P, Dumas C (2002) Fertilization in *Arabidopsis thaliana* wild type: developmental stages and time course. Plant J 30:481–488

Gadjev I, Stone JM, Gechev TS (2008) Programmed cell death in plants: new insights into redox regulation and the role of hydrogen peroxide. Int Rev Cell Mol Biol 270:87–144

Gilroy S, Jones RL (1992) Gibberellic acid and abscisic acid coordinately regulate cytoplasmic calcium and secretory activity in barley aleurone protoplasts. Proc Natl Acad Sci USA 89:3591–3595

Giuliani C, Consonni G, Gavazzi G, Colombo M, Dolfini S (2002) Programmed cell death during embryogenesis in maize. Ann Bot (Lond) 90:287–292

Goldberg RB, Beals TP, Sanders PM (1993) Anther development: basic principles and practical applications. Plant Cell 5:1217–1229

Greenwood JS, Helm M, Gietl C (2005) Ricinosomes and endosperm transfer cell structure in programmed cell death of the nucellus during ricinus seed development. Proc Natl Acad Sci USA 102:2238–2243

Gross-Hardt R, Kagi C, Baumann N, Moore JM, Baskar R, Gagliano WB, Jurgens G, Grossniklaus U (2007) LACHESIS restricts gametic cell fate in the female gametophyte of Arabidopsis. PLoS Biol 5:e47

Guiboileau A, Sormani R, Meyer C, Masclaux-Daubresse C (2010) Senescence and death of plant organs: nutrient recycling and developmental regulation. C R Biol 333:382–391

Guo WJ, Ho TH (2008) An abscisic acid-induced protein, HVA22, inhibits gibberellin-mediated programmed cell death in cereal aleurone cells. Plant Physiol 147:1710–1722

Hamamura Y, Saito C, Awai C, Kurihara D, Miyawaki A, Nakagawa T, Kanaoka MM, Sasaki N, Nakano A, Berger F, Higashiyama T (2011) Live-cell imaging reveals the dynamics of two sperm cells during double fertilization in *Arabidopsis thaliana*. Curr Biol 21:497–502

Hara-Nishimura I, Hatsugai N (2011) The role of vacuole in plant cell death. Cell Death Differ 18:1298–1304

Harper JL, Lovell PH, Moore KG (1970) The shapes and sizes of seeds. Ann Rev Ecol Syst 1:327–356

Hatsugai N, Kuroyanagi M, Yamada K, Meshi T, Tsuda S, Kondo M, Nishimura M, Hara-Nishimura I (2004) A plant vacuolar protease, VPE, mediates virus-induced hypersensitive cell death. Science 305:855–858
Hatsugai N, Kuroyanagi M, Nishimura M, Hara-Nishimura I (2006) A cellular suicide strategy of plants: vacuole-mediated cell death. Apoptosis 11:905–911
Haughn G, Chaudhury A (2005) Genetic analysis of seed coat development in Arabidopsis. Trends Plant Sci 10:472–477
Hayward AP, Tsao J, Dinesh-Kumar SP (2009) Autophagy and plant innate immunity: defense through degradation. Semin Cell Dev Biol 20:1041–1047
Heimovaara-Dijkstra S, Vanduijn B, Libbenga KR, Heidekamp F, Wang M (1994) Abscisic acid-induced membrane potential changes in barley aleurone protoplasts: a possible relevance for the regulation of Rab gene expression. Plant Cell Physiol 35:743–750
Heine GF, Hernandez JM, Grotewold E (2004) Two cysteines in plant R2R3 MYB domains participate in REDOX-dependent DNA binding. J Biol Chem 279:37878–37885
Helm M, Schmid M, Hierl G, Terneus K, Tan L, Lottspeich F, Kieliszewski MJ, Gietl C (2008) KDEL-tailed cysteine endopeptidases involved in programmed cell death, intercalation of new cells, and dismantling of extensin scaffolds. Am J Bot 95:1049–1062
Henke N, Lisak DA, Schneider L, Habicht J, Pergande M, Methner A (2011) The ancient cell death suppressor BAX inhibitor-1. Cell Calcium. 2011 Sep;50(3):251–60
Higashiyama T (2002) The synergid cell: attractor and acceptor of the pollen tube for double fertilization. J Plant Res 115:149–160
Higashiyama T (2010) Peptide signaling in pollen–pistil interactions. Plant Cell Physiol 51:177–89
Hiratsuka R, Terasaka O (2011) Pollen tube reuses intracellular components of nucellar cells undergoing programmed cell death in *pinus densiflora*. Protoplasma 248:339–351
Hiscock SJ, Allen AM (2008) Diverse cell signalling pathways regulate pollen-stigma interactions: the search for consensus. New Phytol 179:286–317
Holdsworth MJ, Bentsink L, Soppe WJ (2008) Molecular networks regulating Arabidopsis seed maturation, after-ripening, dormancy and germination. New Phytol 179:33–54
Holloway SJ, Friedman WE (2008) Embryological features of *tofieldia glutinosa* and their bearing on the early diversification of monocotyledonous plants. Ann Bot 102:167–182
Huck N, Moore JM, Federer M, Grossniklaus U (2003) The Arabidopsis mutant feronia disrupts the female gametophytic control of pollen tube reception. Development 130:2149–2159
Ingram GC (2010) Family life at close quarters: communication and constraint in angiosperm seed development. Protoplasma 247:195–214
Ishikawa R, Ohnishi T, Kinoshita Y, Eiguchi M, Kurata N, Kinoshita T (2011) Rice interspecies hybrids show precocious or delayed developmental transitions in the endosperm without change to the rate of syncytial nuclear division. Plant J 65:798–806
Ito T, Nagata N, Yoshiba Y, Ohme-Takagi M, Ma H, Shinozaki K (2007) Arabidopsis MALE STERILITY1 encodes a PHD-type transcription factor and regulates pollen and tapetum development. Plant Cell 19:3549–3562
Ivanov R, Fobis-Loisy I, Gaude T (2010) When no means no: guide to brassicaceae self-incompatibility. Trends Plant Sci 15:387–394
Jung KH, Han MJ, Lee YS, Kim YW, Hwang I, Kim MJ, Kim YK, Nahm BH, An G (2005) Rice undeveloped Tapetum1 is a major regulator of early tapetum development. Plant Cell 17:2705–2722
Kagi C, Baumann N, Nielsen N, Stierhof YD, Gross-Hardt R (2010) The gametic central cell of Arabidopsis determines the lifespan of adjacent accessory cells. Proc Natl Acad Sci USA 107:22350–22355
Kawai-Yamada M, Jin L, Yoshinaga K, Hirata A, Uchimiya H (2001) Mammalian Bax-induced plant cell death can be down-regulated by overexpression of Arabidopsis Bax inhibitor-1 (AtBI-1). Proc Natl Acad Sci USA 98:12295–12300

Kawanabe T, Ariizumi T, Kawai-Yamada M, Uchimiya H, Toriyama K (2006) Abolition of the tapetum suicide program ruins microsporogenesis. Plant Cell Physiol 47:784–787

Kawashima T, Goldberg RB (2010) The suspensor: not just suspending the embryo. Trends Plant Sci 15:23–30

Kessler SA, Shimosato-Asano H, Keinath NF, Wuest SE, Ingram G, Panstruga R, Grossniklaus U (2010) Conserved molecular components for pollen tube reception and fungal invasion. Science 330:968–971

Kinoshita T (2007) Reproductive barrier and genomic imprinting in the endosperm of flowering plants. Genes Genet Syst 82:177–186

Knetsch M, Wang M, Snaar-Jagalska BE, Heimovaara-Dijkstra S (1996) Abscisic acid induces mitogen-activated protein kinase activation in barley aleurone protoplasts. Plant Cell 8:1061–1067

Kondou Y, Nakazawa M, Kawashima M, Ichikawa T, Yoshizumi T, Suzuki K, Ishikawa A, Koshi T, Matsui R, Muto S, Matsui M (2008) RETARDED GROWTH OF EMBRYO1, a new basic helix-loop-helix protein, expresses in endosperm to control embryo growth. Plant Physiol 147:1924–1935

Kubo K, Entani T, Takara A, Wang N, Fields AM, Hua Z, Toyoda M, Kawashima S, Ando T, Isogai A, Kao TH, Takayama S (2010) Collaborative non-self recognition system in S-RNase-based self-incompatibility. Science 330:796–799

Kuo A, Cappelluti S, Cervantes-Cervantes M, Rodriguez M, Bush DS (1996) Okadaic acid, a protein phosphatase inhibitor, blocks calcium changes, gene expression, and cell death induced by gibberellin in wheat aleurone cells. Plant Cell 8:259–269

Larsson E, Sitbon F, von Arnold S (2008) Polar auxin transport controls suspensor fate. Plant Signal Behav 3:469–470

Lausser A, Dresselhaus T (2010) Sporophytic control of pollen tube growth and guidance in grasses. Biochem Soc Trans 38:631–634

Li N, Zhang DS, Liu HS, Yin CS, Li XX, Liang WQ, Yuan Z, Xu B, Chu HW, Wang J, Wen TQ, Huang H, Luo D, Ma H, Zhang DB (2006) The rice tapetum degeneration retardation gene is required for tapetum degradation and anther development. Plant Cell 18:2999–3014

Li DH, Yang X, Cui KM (2007a) Formation of archegonium chamber is associated with nucellar-cell programmed cell death in *ginkgo biloba*. Protoplasma 231:173–181

Li S, Samaj J, Franklin-Tong VE (2007b) A mitogen-activated protein kinase signals to programmed cell death induced by self-incompatibility in papaver pollen. Plant Physiol 145:236–245

Li H, Yuan Z, Vizcay-Barrena G, Yang C, Liang W, Zong J, Wilson ZA, Zhang D (2011a) PERSISTENT TAPETAL CELL1 encodes a PHD-finger protein that is required for tapetal cell death and pollen development in rice. Plant Physiol 156:615–630

Li X, Gao X, Wei Y, Deng L, Ouyang Y, Chen G, Zhang Q, Wu C (2011b) Rice APOPTOSIS INHIBITOR5 coupled with Two DEAD-Box adenosine 5′-triphosphate-dependent RNA helicases regulates tapetum degeneration. Plant Cell 23:1416–1434

Lim PO, Kim HJ, Nam HG (2007) Leaf senescence. Annu Rev Plant Biol 58:115–136

Lombardi L, Casani S, Ceccarelli N, Galleschi L, Picciarelli P, Lorenzi R (2007a) Programmed cell death of the nucellus during *sechium edule* Sw. Seed development is associated with activation of caspase-like proteases. J Exp Bot 58:2949–2958

Lombardi L, Ceccarelli N, Picciarelli P, Lorenzi R (2007b) DNA degradation during programmed cell death in phaseolus coccineus suspensor. Plant Physiol Biochem 45:221–227

Lombardi L, Ceccarelli N, Picciarelli P, Sorce C, Lorenzi R (2010) Nitric oxide and hydrogen peroxide involvement during programmed cell death of *sechium edule* nucellus. Physiol Plant 140:89–102

Ma W, Berkowitz GA (2007) The grateful dead: calcium and cell death in plant innate immunity. Cell Microbiol 9:2571–2585

Ma H, Sundaresan V (2010) Development of flowering plant gametophytes. Curr Top Dev Biol 91:379–412

Madrid EN, Friedman WE (2010) Female gametophyte and early seed development in peperomia (piperaceae). Am J Bot 97:1–14
Marton ML, Dresselhaus T (2010) Female gametophyte-controlled pollen tube guidance. Biochem Soc Trans 38:627–630
Mitchum MG, Yamaguchi S, Hanada A, Kuwahara A, Yoshioka Y, Kato T, Tabata S, Kamiya Y, Sun TP (2006) Distinct and overlapping roles of two gibberellin 3-oxidases in Arabidopsis development. Plant J 45:804–818
Moll C, von Lyncker L, Zimmermann S, Kagi C, Baumann N, Twell D, Grossniklaus U, Gross-Hardt R (2008) CLO/GFA1 and ATO are novel regulators of gametic cell fate in plants. Plant J 56:913–921
Moller B, Weijers D (2009) Auxin control of embryo patterning. Cold Spring Harb Perspect Biol 1:a001545
Morris K, Linkies A, Muller K, Oracz K, Wang X, Lynn JR, Leubner-Metzger G, Finch-Savage WE (2011) Regulation of seed germination in the close Arabidopsis relative *lepidium sativum*: a global tissue-specific transcript analysis. Plant Physiol 155:1851–1870
Mur LA, Kenton P, Lloyd AJ, Ougham H, Prats E (2008) The hypersensitive response; the centenary is upon us but how much do we know? J Exp Bot 59:501–520
Nakaune S, Yamada K, Kondo M, Kato T, Tabata S, Nishimura M, Hara-Nishimura I (2005) A vacuolar processing enzyme, deltaVPE, is involved in seed coat formation at the early stage of seed development. Plant Cell 17:876–887
Nowack MK, Ungru A, Bjerkan KN, Grini PE, Schnittger A (2010) Reproductive cross-talk: seed development in flowering plants. Biochem Soc Trans 38:604–612
Ogawa M, Hanada A, Yamauchi Y, Kuwahara A, Kamiya Y, Yamaguchi S (2003) Gibberellin biosynthesis and response during Arabidopsis seed germination. Plant Cell 15:1591–1604
Ohashi-Ito K, Fukuda H (2010) Transcriptional regulation of vascular cell fates. Curr Opin Plant Biol 13:670–676
Okuda S, Higashiyama T (2010) Pollen tube guidance by attractant molecules: LUREs. Cell Struct Funct 35:45–52
Olsen OA (2004) Nuclear endosperm development in cereals and *Arabidopsis thaliana*. Plant Cell 16 Suppl:S214–S227
Pagnussat GC, Alandete-Saez M, Bowman JL, Sundaresan V (2009) Auxin-dependent patterning and gamete specification in the Arabidopsis female gametophyte. Science 324:1684–1689
Palma K, Kermode AR (2003) Metabolism of hydrogen peroxide during reserve mobilization and programmed cell death of barley (*hordeum vulgare* L.) aleurone layer cells. Free Radic Biol Med 35:1261–1270
Papini A, Mosti S, Milocani E, Tani G, Di Falco P, Brighigna L Megasporogenesis and programmed cell death in Tillandsia (Bromeliaceae). Protoplasma, 2011 Oct; 248(4):651–662
Penfield S, Graham S, Graham IA (2005) Storage reserve mobilization in germinating oilseeds: Arabidopsis as a model system. Biochem Soc Trans 33:380–383
Poulter NS, Wheeler MJ, Bosch M, Franklin-Tong VE (2010) Self-incompatibility in papaver: identification of the pollen S-determinant PrpS. Biochem Soc Trans 38:588–592
Qi Y, Wang H, Zou Y, Liu C, Liu Y, Wang Y, Zhang W (2011) Over-expression of mitochondrial heat shock protein 70 suppresses programmed cell death in rice. FEBS Lett 585:231–239
Ravi M, Marimuthu MP, Siddiqi I (2008) Gamete formation without meiosis in Arabidopsis. Nature 451:1121–1124
Reape TJ, Molony EM, McCabe PF (2008) Programmed cell death in plants: distinguishing between different modes. J Exp Bot 59:435–444
Riggs DR (2004) Programmed cell death events during reproductive development. In: Gray J (ed) Programmed cell death in plants. Blackwell Publishing, Oxford, pp 71–105
Ritchie S, Gilroy S (1998) Calcium-dependent protein phosphorylation may mediate the gibberellic acid response in barley aleurone. Plant Physiol 116:765–776
Rodkiewicz B (1970) Callose in cell wall during megasporogenesis in angiosperms. Planta 93:39–47

Rogers HJ (2005) Cell death and organ development in plants. Curr Top Dev Biol 71:225–261
Rotman N, Rozier F, Boavida L, Dumas C, Berger F, Faure JE (2003) Female control of male gamete delivery during fertilization in Arabidopsis thaliana. Curr Biol 13:432–436
Rotman N, Gourgues M, Guitton AE, Faure JE, Berger F (2008) A dialogue between the SIRENE pathway in synergids and the fertilization independent seed pathway in the central cell controls male gamete release during double fertilization in Arabidopsis. Mol Plant 1:659–666
Russell SD (1992) Double fertilization. Int Rev Cytol 140:357–388
Sabelli PA, Larkins BA (2009) The development of endosperm in grasses. Plant Physiol 149:14–26
Sandaklie-Nikolova L, Palanivelu R, King EJ, Copenhaver GP, Drews GN (2007) Synergid cell death in Arabidopsis is triggered following direct interaction with the pollen tube. Plant Physiol 144:1753–1762
Scott RJ, Spielman M, Bailey J, Dickinson HG (1998) Parent-of-origin effects on seed development in *Arabidopsis thaliana*. Development 125:3329–3341
Shi DQ, Yang WC (2011) Ovule development in Arabidopsis: progress and challenge. Curr Opin Plant Biol 14:74–80
Sorensen AM, Krober S, Unte US, Huijser P, Dekker K, Saedler H (2003) The Arabidopsis ABORTED MICROSPORES (AMS) gene encodes a MYC class transcription factor. Plant J 33:413–423
Sprunck S, Gross-Hardt R (2011) Nuclear behavior, cell polarity, and cell specification in the female gametophyte. Sex Plant Reprod 24:123–136
Sreenivasulu N, Radchuk V, Strickert M, Miersch O, Weschke W, Wobus U (2006) Gene expression patterns reveal tissue-specific signaling networks controlling programmed cell death and ABA-regulated maturation in developing barley seeds. Plant J 47:310–327
Sreenivasulu N, Borisjuk L, Junker BH, Mock HP, Rolletschek H, Seiffert U, Weschke W, Wobus U (2010) Barley grain development toward an integrative view. Int Rev Cell Mol Biol 281:49–89
Sundstrom JF, Vaculova A, Smertenko AP, Savenkov EI, Golovko A, Minina E, Tiwari BS, Rodriguez-Nieto S, Zamyatnin AA Jr, Valineva T, Saarikettu J, Frilander MJ, Suarez MF, Zavialov A, Stahl U, Hussey PJ, Silvennoinen O, Sundberg E, Zhivotovsky B, Bozhkov PV (2009) Tudor staphylococcal nuclease is an evolutionarily conserved component of the programmed cell death degradome. Nat Cell Biol 11:1347–1354
Suomeng D, Zhengguang Z, Xiaobo Z, Yuanchao W (2008) Mammalian pro-apoptotic bax gene enhances tobacco resistance to pathogens. Plant Cell Rep 27:1559–1569
Swanson SJ, Jones RL (1996) Gibberellic acid induces vacuolar acidification in barley aleurone. Plant Cell 8:2211–2221
Takayama S, Isogai A (2005) Self-incompatibility in plants. Annu Rev Plant Biol 56:467–489
Tantikanjana T, Nasrallah ME, Nasrallah JB (2010) Complex networks of self-incompatibility signaling in the brassicaceae. Curr Opin Plant Biol 13:520–526
Taylor NL, Tan YF, Jacoby RP, Millar AH (2009) Abiotic environmental stress induced changes in the *Arabidopsis thaliana* chloroplast, mitochondria and peroxisome proteomes. J Proteomics 72:367–378
Tsiatsiani L, Van Breusegem F, Gallois P, Zavialov A, Lam E, Bozhkov PV (2011) Metacaspases. Cell Death Differ 18:1279–1288
Turner S, Gallois P, Brown D (2007) Tracheary element differentiation. Annu Rev Plant Biol 58:407–433
van Doorn WG, Woltering EJ (2008) Physiology and molecular biology of petal senescence. J Exp Bot 59:453–480
van Doorn WG, Beers EP, Dangl JL, Franklin-Tong VE, Gallois P, Hara-Nishimura I, Jones AM, Kawai-Yamada M, Lam E, Mundy J, Mur LA, Petersen M, Smertenko A, Taliansky M, Van Breusegem F, Wolpert T, Woltering E, Zhivotovsky B, Bozhkov PV (2011) Morphological classification of plant cell deaths. Cell Death Differ 18:1241–1246

Vartapetian AB, Tuzhikov AI, Chichkova NV, Taliansky M, Wolpert TJ (2011) A plant alternative to animal caspases: subtilisin-like proteases. Cell Death Differ 18:1289–1297
Vizcay-Barrena G, Wilson ZA (2006) Altered tapetal PCD and pollen wall development in the Arabidopsis ms1 mutant. J Exp Bot 57:2709–2717
Vuosku J, Sutela S, Tillman-Sutela E, Kauppi A, Jokela A, Sarjala T, Haggman H (2009) Pine embryogenesis: many licences to kill for a new life. Plant Signal Behav 4:928–932
Wang M, Oppedijk BJ, Lu X, Van Duijn B, Schilperoort RA (1996) Apoptosis in barley aleurone during germination and its inhibition by abscisic acid. Plant Mol Biol 32:1125–1134
Wang M, Oppedijk BJ, Caspers MPM, Lamers GEM, Boot MJ, Geerlings DNG, Bakhuizen B, Meijer AH, van Duijn B (1998) Spatial and temporal regulation of DNA fragmentation in the aleurone of germinating barley. J Exp Bot 49:1293–1301
Watanabe N, Lam E (2006) Arabidopsis Bax inhibitor-1 functions as an attenuator of biotic and abiotic types of cell death. Plant J 45:884–894
Watanabe N, Lam E (2009) Bax inhibitor-1, a conserved cell death suppressor, is a key molecular switch downstream from a variety of biotic and abiotic stress signals in plants. Int J Mol Sci 10:3149–3167
Wheeler MJ, de Graaf BH, Hadjiosif N, Perry RM, Poulter NS, Osman K, Vatovec S, Harper A, Franklin FC, Franklin-Tong VE (2009) Identification of the pollen self-incompatibility determinant in *papaver rhoeas*. Nature 459:992–995
Wilkins KA, Bancroft J, Bosch M, Ings J, Smirnoff N, Franklin-Tong VE (2011) Reactive oxygen species and nitric oxide mediate actin reorganization and programmed cell death in the self-incompatibility response of papaver. Plant Physiol 156:404–416
Williams B, Dickman M (2008) Plant programmed cell death: can't live with it; can't live without it. Mol Plant Pathol 9:531–544
Williams CE, Grotewold E (1997) Differences between plant and animal Myb domains are fundamental for DNA binding activity, and chimeric Myb domains have novel DNA binding specificities. J Biol Chem 272:563–571
Wilson ZA, Zhang DB (2009) From Arabidopsis to rice: pathways in pollen development. J Exp Bot 60:1479–1492
Woltering EJ (2010) Death proteases: alive and kicking. Trends Plant Sci 15:185–188
Wu M, Huang J, Xu S, Ling T, Xie Y, Shen W (2011) Haem oxygenase delays programmed cell death in wheat aleurone layers by modulation of hydrogen peroxide metabolism. J Exp Bot 62:235–248
Xu J, Yang C, Yuan Z, Zhang D, Gondwe MY, Ding Z, Liang W, Wilson ZA (2010) The ABORTED MICROSPORES regulatory network is required for postmeiotic male reproductive development in *Arabidopsis thaliana*. Plant Cell 22:91–107
Yamada K, Shimada T, Nishimura M, Hara-Nishimura I (2005) A VPE family supporting various vacuolar functions in plants. Physiol Plant 123:369–375
Yamaguchi S, Kamiya Y, Sun T (2001) Distinct cell-specific expression patterns of early and late gibberellin biosynthetic genes during Arabidopsis seed germination. Plant J 28:443–453
Yang WC, Ye D, Xu J, Sundaresan V (1999) The SPOROCYTELESS gene of Arabidopsis is required for initiation of sporogenesis and encodes a novel nuclear protein. Genes Dev 13:2108–2117
Yang S, Johnston N, Talideh E, Mitchell S, Jeffree C, Goodrich J, Ingram G (2008) The endosperm-specific ZHOUPI gene of *Arabidopsis thaliana* regulates endosperm breakdown and embryonic epidermal development. Development 135:3501–3509
Yang WC, Shi DQ, Chen YH (2010) Female gametophyte development in flowering plants. Annu Rev Plant Biol 61:89–108
Young TE, Gallie DR (2000a) Programmed cell death during endosperm development. Plant Mol Biol 44:283–301
Young TE, Gallie DR (2000b) Regulation of programmed cell death in maize endosperm by abscisic acid. Plant Mol Biol 42:397–414

Young TE, Gallie DR, DeMason DA (1997) Ethylene-mediated programmed cell death during maize endosperm development of wild-type and shrunken2 genotypes. Plant Physiol 115:737–751

Zhang W, Sun Y, Timofejeva L, Chen C, Grossniklaus U, Ma H (2006) Regulation of *Arabidopsis* tapetum development and function by DYSFUNCTIONAL TAPETUM1 (DYT1) encoding a putative bHLH transcription factor. Development 133:3085–3095

Zhang DS, Liang WQ, Yuan Z, Li N, Shi J, Wang J, Liu YM, Yu WJ, Zhang DB (2008) Tapetum degeneration retardation is critical for aliphatic metabolism and gene regulation during rice pollen development. Mol Plant 1:599–610

Induction and Suppression of Herbivore-Induced Indirect Defenses

Juan M. Alba, Silke Allmann, Joris J. Glas, Bernardus C.J. Schimmel, Eleni A. Spyropoulou, Marije Stoops, Carlos Villarroel, and Merijn R. Kant

Abstract Plants release volatiles into the air. Upon herbivory, the amounts they release from the vegetative tissues increases dramatically. Although the physiological necessity for this increased emission is not fully understood, it has interesting consequences, the most important one being that foraging predators and host-searching parasitoids use these signals to track down plants with prey. This process is referred to as "indirect defense" since these responses can augment the plant's own "direct" defenses, such as structural barriers and toxins, when they result in decreased herbivory via increased predation. Here we will describe how plants organize indirect defenses and how herbivores have adapted to interfere with these processes.

1 Introduction to Plant Defenses

At first glance, plants are easy food for hungry herbivores (Bede et al. 2002; Merkx-Jacques et al. 2008). While carnivores often have to chase their prey and put up a struggle before they can eat, herbivores seem to have it much easier since plants are sessile. However, also herbivores do not get their meals for free since plants do fight back.

J.M. Alba • J.J. Glas • B.C.J. Schimmel • M. Stoops • M.R. Kant (✉)
IBED, Department of Population Biology (University of Amsterdam), Amsterdam, The Netherlands
e-mail: m.kant@uva.nl

S. Allmann • E.A. Spyropoulou
SILS, Department of Plant Physiology, University of Amsterdam, Amsterdam, The Netherlands

C. Villarroel
IBED, Department of Population Biology (University of Amsterdam), Amsterdam, The Netherlands

SILS, Department of Plant Physiology, University of Amsterdam, Amsterdam, The Netherlands

G. Witzany and F. Baluška (eds.), *Biocommunication of Plants*,
Signaling and Communication in Plants 14,
DOI 10.1007/978-3-642-23524-5_11, © Springer-Verlag Berlin Heidelberg 2012

Plants do not have easy lives which are reflected by their anatomy and physiology. Their cell membranes are shielded by walls, and their surface is covered with a protective cuticula and other structures to prevent dehydration as well as penetration by pathogens or feeding by herbivores (Eigenbrode and Espelie 1995; Hematy et al. 2009). Moreover, plants accumulate diverse substances that interfere with herbivore digestive physiology. Some herbivores withstand such constitutive plant defenses. If so, the plant will increase some of these defenses to hinder the feeding herbivore more while switching to other measures such as selective tissue death, i.e., the hypersensitive response (HR) (Hematy et al. 2009; Walling 2000), and resource allocation (Anten and Pierik 2010) collectively referred to as "direct defenses" while also increasing the release of volatiles. These herbivore-induced plant volatiles (HIPVs) can be used by foraging predators or host-searching parasitoids to track down plants with prey, thereby augmenting the direct defenses, and hence are referred to as "indirect defenses" (Schoonhoven et al. 1998). The coordination of herbivore-induced defenses runs via plant signaling molecules, mostly hormones (Pieterse et al. 2009). Herbivores, in turn, sometimes have adapted to resist or suppress these induced changes (Alba et al. 2011).

2 How Are Herbivores Recognized by Plants?

Plants can respond quickly upon imminent danger as they have adapted to recognize many of their enemies and boost their defense physiology even before herbivores take their first "bite." Herbivores can betray their presence to plants by touch, e.g., by the damage they cause with their footsteps when wandering on the leaf surface (Hall et al. 2004) or upon egg deposition. Oviposition often causes small wounds to plant tissues to which plants can respond. However, also the fluids secreted by adult female herbivores which serve to attach eggs to the leaf surface can contain substances that elicit plant defenses upon recognition by the plant (Hilker and Meiners 2010). Ovipositing pea weevil females (*Bruchus pisorum* L.) secrete so-called bruchins, i.e., mono- and bis-(3-hydroxypropanoate) esters of long-chain α,ω-diols, which stimulate cell division and neoplasm formation in plants (Doss et al. 2000; Hilker and Meiners 2010). Moreover, benzyl cyanides from the oviposition fluids of mated female cabbage white butterflies (*Pieris brassicae*) can elicit transcriptional changes of several defense-related genes and changes in leaf-surface morphology, the latter stimulating the egg parasitoid *Trichogramma brassicae* to stay around longer (Fatouros et al. 2008).

Herbivore feeding causes mechanical damage. The degree of damage, however, can vary greatly depending on how a herbivore takes up food. Herbivorous arthropods are either chewers or piercing-and-sucking stylet feeders (Labandeira 1997). Homopterans like aphids and whiteflies have long stylets and primarily feed from vascular fluids, while smaller herbivores like mites and nematodes have shorter stylets they use to feed from epidermal or mesophyll cells. Plants discriminate between real folivory and randomly occurring mechanical damage on the basis of the temporal frequency pattern of the damage (Mithöfer et al. 2005) as well as of chemical elicitors, mostly from herbivore saliva, introduced into the wound during feeding. These elicitors are key substances for the coordinated accumulation of phytohormones and the subsequent release of HIPVs to establish indirect defenses (Wu and Baldwin 2010). Since the cocktail of herbivore-secreted elicitors can be quite species specific, plants can use this information to tailor defense responses to the attacker (Schmelz et al. 2009).

Elicitors that come into contact with the plant during regurgitation (Peiffer and Felton 2009) often are relatively small nonproteinous substances. Fatty acid conjugates (FACs) are a well-studied group of defense elicitors (Bonaventure et al. 2011) from insect regurgitant which are formed in the insect gut via conjugation of a plant-derived fatty acid, i.e., predominantly linolenic acid, 17-hydroxylinolenic acid, and the corresponding linolenic acid derivatives, to an insect-derived amino acid, i.e., predominantly L-glutamine or L-glutamate. The first chemically described FAC was *N*-(17-hydroxylinolenoyl)-L-glutamine. It was isolated and identified from the oral secretions (OS) of *Spodoptera exigua* larvae and was named volicitin since it induced the emission of several terpenoids in *Zea mays* in a similar fashion as the caterpillar's OS (Alborn et al. 1997). Following the discovery of volicitin, defense-inducing FACs have been found in the OS of many lepidopteran species and, recently, also in the OS of two species of crickets and of fruit flies (Hilker and Meiners 2010). In addition to FACs, also "caeliferins," which are sulfated fatty acids, from the OS of the grasshopper *Schistocerca americana* induce, like volicitin, the release of herbivore-specific terpenes from maize seedlings (Alborn et al. 2007; Hilker and Meiners 2010; Wu and Baldwin 2010). A third group of OS-derived elicitors are proteolytic peptides, called inceptins, which were isolated from *S. frugiperda* OS after feeding on cowpea. Inceptins are formed in the insect midgut by degradation of the plant chloroplastic ATP synthase γ-subunit and stimulate the accumulation of the phytohormones jasmonic acid (JA), ethylene (Et), and salicylic acid (SA) and induce the emission of distinct HIPVs (Hilker and Meiners 2010; Schmelz et al. 2006; Wu and Baldwin 2010).

Elicitors derived from herbivore secretions can also be proteinous. Pure β-glucosidase, an enzyme which catalyzes the hydrolysis of glycosidic linkages in glycosides, was found to induce HIPVs very similar to those induced by the OS of Pieris brassicae when applied to mechanically wounded Brussels sprouts. In both case HIPVs were sufficient to attract the egg parasitoid of *P. brassicae*, the wasp *Cotesia glomerata*, to experimentally treated plants (Mattiacci et al. 1995). A different type of enzyme activity has been detected in the OS of *Manduca sexta*. When *M. sexta* caterpillars fed on *Nicotiana attenuata*, the *Z*/*E* ratio of C6 volatiles dramatically changed, and this change in the volatile bouquet tripled the foraging behavior of the generalist predator *Geocoris* spp. Interestingly, the shift from *Z*-isomers to *E*-isomers was independent from plant enzymes and solely due to an unidentified isomerase enzyme in the insect's OS (Allmann and Baldwin 2010).

3 How Do Plants Arrange Their Defenses?

After plants have detected feeding herbivores or attacking pathogens, they often undergo rapid physiological changes to reinforce constitutive defenses, and these changes are referred to as "induced direct defenses." Defenses are costly and require resources otherwise used for growth and reproduction (Walters and Heil 2007). Therefore, plants have evolved a complex, largely hormonal, signaling network to arrange defense and resource allocation and set the physiological priorities (Pieterse et al. 2009).

Whereas plant resistance against immobile pathogens often is characterized by an HR, defense against herbivores is associated more with a decrease in tissue palatability (Anten and Pierik 2010). Central in the organization of antiherbivore defenses is the plant hormone jasmonic acid (JA) and its active derivative JA-isoleucine (JA-Ileu) which rapidly accumulates during herbivory. The mode of action of JA has been studied in detail using JA biosynthesis- or perception-impaired mutant plants which are often preferred by herbivores in choice tests while allowing for higher herbivore fitness (Howe and Jander 2008). Accumulation of JA-dependent defense proteins and metabolites is often coregulated by Et in a synergistic manner. In contrast, SA antagonizes the action of JA (Pieterse et al. 2009). SA is well known for its signaling role in defenses induced by biotrophic pathogens, but many stylet-feeding herbivores, like mites, whiteflies, and aphids, induce a cocktail of JA- and SA-related responses (Kant et al. 2008). Although it is not clear to which extent this mixed response is required for the plant to establish the appropriate defenses, the "decoy hypothesis" suggests that in some cases, the herbivore could benefit from a SA-mediated suppression of the JA defenses (Zarate et al. 2007). Finally, also the hormones auxin and abscisic acid (ABA) influence the properties of the signaling network mostly via antagonizing the action of JA and SA (Pieterse et al. 2009). The dynamics of this complex regulatory network, in which hormonal synergisms and antagonisms determine the final output of the defense response, depend largely on the type of herbivore as well as on the physiological status of the plant.

4 How Do Herbivores Deal with Plant Defenses?

Plants produce numerous secondary metabolites that can interfere with a herbivore's physiology, and hence, herbivores need to select the most suitable host for themselves and their progeny on the basis of visual, tactile, and chemical cues (Bernays 1999). A well-studied group of JA-dependent induced plant defense compounds is the proteinase inhibitors (PIs). PIs inhibit digestive proteases in the gut of the herbivore and will slow down herbivore development (Hartl et al. 2011) since they hinder the uptake of (essential) amino acids (Zhu-Salzman et al. 2008). In addition, plants can produce a wide variety of toxins, like alkaloids and glucosinolates, which besides interfering with herbivore survival and development directly, also can restrain herbivore compensatory feeding in response to PIs. For example, in *Nicotiana attenuata*, the alkaloid nicotine prevents the generalist herbivore *Spodoptera exigua* from simply eating more to compensate for its inefficient digestion of plant material due to induced proteinase inhibitor activity (Steppuhn and Baldwin 2007).

The most straightforward solution for a herbivore to deal with plant defenses is to avoid contact with putative harmful host plants or to depart shortly after arrival (Alba et al. 2009; Bleeker et al. 2011). However, herbivores can also select plant tissues or parts with low levels of toxins. For example, the cotton bollworm *Helicoverpa armigera* eats preferably from *Arabidopsis thaliana* leaf tissues where the concentration of glucosinolates is low (Shroff et al. 2008). Finally, in cases when defenses cannot be avoided, natural selection can cause the rapid emergence of resistances in populations, and it was shown that many insects and mites develop resistances against a broad range of substances, e.g., via adjusted detoxification physiology or toxin insensitivity (Shuler 1996; Feyereisen 1999; Li et al. 2004; Despres et al. 2007; Van Leeuwen et al. 2008).

Plants can face a variety of attackers simultaneously or sequentially, while some herbivores are generalists that feed on many different species or specialists which have a narrower host range. Specialist herbivores often have evolved effective resistances to cope with the physical and chemical defenses of their host possibly as a consequence of coevolution (Schoonhoven et al. 1998), and some specialists have adapted to use host-specific defenses as cues to identify their host. Such "counterproductive" defenses may persist when their positive impact on plant fitness via deterring generalists outweighs the negative effect of attracting specialists (Poelman et al. 2008). Furthermore, herbivores may adapt to use plant defenses for their own defense against parasitoids or predators. For example, wild tobacco *Nicotiana attenuata* stops producing costly nicotine when it is attacked by the nicotine-tolerant specialist herbivore *Manduca sexta*. Simultaneously, the plant increases the emission of volatiles and thereby possibly prevents the caterpillar to become an unsuitable host for parasitoids which are attracted by the volatiles (Kahl et al. 2000).

5 Where Are Plant Volatiles Produced?

Plants have evolved specialized structures for the production and storage of secondary metabolites. Plant volatiles are usually lipophilic substances with high vapor pressures and can be released from flowers, fruits, and vegetative tissue into the atmosphere but also from the roots into the rhizosphere. In the flower petals, the biosynthesis of plant volatiles takes place in specialized or nonspecialized epidermal cells, and their emission is in the vast majority tightly correlated with attraction of pollinators (Pichersky et al. 2006). Also, the roots contain secretory cells that release volatiles which play a role in the direct defense against microbial pathogens as well as in indirect defense, e.g., via the attraction of entomopathogenic nematodes (Rasmann et al. 2005; Wenke et al. 2010). Other common anatomical structures where plant volatiles are stored and released include secretory cavities present in the skin of many fruits and special ducts, such as those found on evergreens, in which resins are stored in a mixture with volatile chemicals to keep the resin fluid but which can evaporate during exposure to air upon mechanical damage such that the resin hardens and seals the wound (Maffei 2010). However, especially well studied are the glandular trichomes which can be found on vegetative plant tissues of many plant species and which are the source of many HIPVs. Glandular trichomes are classified in different types according to their shape and structure, and they can be divided into peltate and capitate trichomes: The peltate trichomes consist of one basal cell, one stalk cell, and many secretory cells (typically 4–18) while the capitate trichomes comprise a basal cell, a single or multicellular stalk, and a head consisting of one or two cells (Werker 2000; Maffei 2010). Alternatively, trichomes can be categorized as one of seven types as found in the family of the Solanaceae. For example, the type VI glandular trichomes of cultivated tomato consist of a stalk and a four-celled head. These four cells are small and have a large wall-less subcellular cavity on top in which secondary metabolites are stored (Simmons and Gurr 2005). Cutin is often deposited in the wall of the lowest stalk cell of glandular trichomes in order to prevent the synthesized products to flow back into the plant (Fahn 1988). Hence, trichome constituents, which can be autotoxic, are stored safely away from the other plant tissues in the subcuticular space. Finally, volatiles can be released when the head is ruptured by herbivore movement or be transported, actively or passively, out of the trichome into the air upon upregulation of their biosynthesis during indirect defenses (Gershenzon et al. 1992; Pichersky et al. 2006).

6 Which Induced Volatiles Do Plants Produce?

Flower volatiles and HIPVs establish interactions with the biotic environment of the plant. Since volatile blends contain information on the state of the plant, i.e., it has fertile flowers or is damaged by herbivores, they can be considered signals that establish biocommunication. It appeared that often the qualitative and quantitative composition of the scent bouquet rather than the characteristics of its individual components determine its communicative function (Bruce et al. 2005; Riffell et al. 2009; Van Wijk et al. 2011). The majority of organic plant volatiles are either terpenoids, fatty acid derivatives, or aromates like benzenoids or phenylpropanoids and, despite the complex interactions these volatiles play a role in, are derived from a very limited number of biochemical pathways (Dudareva et al. 2006).

Green leaf volatiles (GLVs) constitute a class of volatile C6 aldehydes, alcohols, and their esters and are released within seconds after wounding or herbivore attack. Since emission of GLVs is almost completely restricted to the wounded tissue and is incredibly fast, it is thought to result from *de novo* GLV formation when substrates and enzymes are mixed during wounding (Arimura et al. 2009). Like

JA, GLVs are derived from the octadecanoid pathway which starts when one or more lipases form linolenic acid from plasma membrane phospholipids. Linolenic acid is then oxygenated in the plastid by 13-lipoxygenase (LOX) to form C13-hydroperoxy linolenic acid (13-HP) (Wasternack 2007). Cleavage of 13-HP by fatty acid 13-hydroperoxide lyase (HPL) renders the basic volatile C6 aldehydes which can be processed into alcohols by alcohol dehydrogenases and subsequently into their corresponding acetate by acyltransferases (Arimura et al. 2009) and can be isomerized (Allmann and Baldwin 2010). Although 13-HP serves as a precursor for both GLVs and JA, there is most likely no metabolic competition since the biosynthetic enzymes of both pathways seem to have different subcellular locations (Arimura et al. 2009), while the lipoxygenases might be structurally different (Bonaventure and Baldwin 2010).

While GLVs are released rapidly after wounding, the emission of terpenes takes longer to increase significantly and typically peaks during the next photophase after wounding (Allmann and Baldwin 2010). Despite their immense variety, terpenes are in principle all assemblies of basic C5 isoprene units, and different classes of terpenes are produced mostly in the cytosol or plastids but also in the mitochondria (reviewed in Dudareva et al. 2004, 2006; Tholl 2006).

The first step in the biosynthesis of terpenes comprises the formation of the C5 "building blocks": isopentenyl diphosphate (IPP) and its isomer dimethylallyl diphosphate (DMAPP). These isoprenoids can be produced via the cytosolic mevalonate (MVE) pathway from acetyl-CoA, or via the plastidial 2-methylerythritol 4-phosphate (MEP) pathway from pyruvate and glyceraldehyde-3-phosphate (Dudareva et al. 2004, 2006; Tholl 2006). The second step is orchestrated by distinct prenyltransferases, which catalyze chain elongation of a single DMAPP by successive head-to-tail condensations of one or more IPP units to generate linear (Z)- or (E)-C10, C15, and C20 isoprenyl diphosphate molecules (Dudareva et al. 2006; Sallaud et al. 2009; Schilmiller et al. 2009). Finally, these isoprenyl diphosphates then serve as precursors for an array of primary and secondary plant substances such as sterols, carotenoids, chlorophyll, gibberellins, abscisic acid (ABA), and brassinosteroids but also the volatile terpenoids via the action of a large family of terpene synthases. These enzymes first remove the diphosphate group from the precursor after which the highly unstable intermediates can undergo secondary transformations which include reduction or removal of carboxyl groups, addition of hydroxyl groups, and the formation of esters and ethers, leading to a variety of volatile terpenoids, predominantly monoterpenes (C10) and sesquiterpenes (C15) (Dudareva et al. 2004, 2006; Tholl 2006; Sallaud et al. 2009; Schilmiller et al. 2009; Lee et al. 2010).

Since some terpenoids are only emitted by petal tissue, while others are *de novo* produced in glandular trichomes upon herbivory, there is spatial and temporal regulation at the level of transcription of terpene biosynthetic genes and via modification of precursor molecules to control substrate flux and availability (Tholl 2006). Hence, the transcription of many of the terpene biosynthetic genes in plants is also under control of herbivore- or pathogen-induced phytohormone signaling (Ozawa et al. 2000; Ament et al. 2004; Kant et al. 2004; Ament et al. 2006; Dudareva et al. 2006; Van Schie et al. 2007; Ament et al. 2010; Lee et al. 2010).

Volatile aromatics, such as the SA-derived volatile methyl-SA (MeSA), indole, and benzenoids, are derived from chorismate (Colquhoun et al. 2010), which is a precursor for an array of primary and secondary plant metabolites such as several amino acids, anthocyanins, flavonoids, and auxins. Their biosynthesis pathways are mostly regulated at the level of gene expression of key biosynthetic enzymes and depend on substrate availability (Tzin and Galili 2010). For example, the emission of MeSA is dependent on SA availability, JA signaling, and SA-methyl transferase (SAMT) activity (Ament et al. 2004; Dudareva et al. 2006; Pichersky et al. 2006).

7 How Do Induced Plant Volatiles Contribute to Plant Defenses?

HIPVs mediate indirect plant defenses, i.e., they attract foraging natural enemies of herbivores, and this is a widely observed phenomenon (Sabelis et al. 2001; Kant et al. 2009). Dicke and Sabelis (1988) were the first to show, by means of a Y-tube olfactometer assay, that the blind predatory mite *Phytoseiulus persimilis* indeed uses HIPVs for finding plants infested with its prey, the spider mite *Tetranychus urticae*. Turlings et al. (1990) showed that also host-searching parasitoids use HIPVs and, by using beet armyworm *S. exigua*-infested *Zea mays* plants, showed that females of the parasitic wasp *Cotesia marginiventris* can learn to associate HIPVs with the presence of a suitable host. Subsequently, De Moraes et al. (1998) showed that the parasitic wasp *Cardiochiles nigriceps* could discriminate between the HIPVs induced by hosts and nonhosts. Thaler (1999) showed that treatment of tomato *Solanum lycopersicum* plants with synthetic JA was sufficient to increase the parasitism of *S. exigua* larvae by the endoparasitic wasp *Hyposoter exigua* in the field. Further field experiments by Kessler and Baldwin (2001) revealed that also synthetic analogues of HIPVs can reduce herbivory in nature. Mimicking naturally herbivore-induced emissions from *N. attenuata* with synthetic volatiles, i.e., (Z)-3-hexen-1-ol, linalool, or cis-α-bergamotene, increased predation rates of *M. sexta* eggs by the generalist predator *Geocoris pallens*. Moreover, the compound linalool alone decreased oviposition rates of the herbivore *M. quinquemaculata* as did the natural HIPV blend from infested plants. In a later study, Kessler et al. (2004) planted *N. attenuata* plants, that were genetically silenced for genes involved in JA signaling, in the plant's native habitat and observed that these plants were vulnerable to their normal herbivore species but also attracted novel species. These results showed that also under natural conditions, JA signaling is essential for establishing direct and indirect defenses properly.

Apparently, HIPVs contribute to defenses in two ways: They are direct defenses when they repel herbivores, and they facilitate indirect defenses when attracting predators or parasitoids to infested plants (Sabelis et al. 2001). De Moraes et al. (2001) reported that HIPVs also repel nocturnal herbivores. They showed that the HIPVs of tobacco *Nicotiana tabacum* infested with *Heliothis virescens* caterpillars were repellent to conspecific female moths searching for a place to oviposit at night. However, it is unclear to which extend infested plants really benefit from repelling ovipositing moths since single larvae can defoliate complete plants. Hence, this behavior is more likely advantageous to the moth that avoids its offspring having to deal with competitors and preinduced direct and indirect defenses (De Moraes et al. 2001), although not all herbivores are repelled by HIPVs (Dicke and van Loon 2000). Taken together, HIPVs contain freely available information on the well-being of plants and that this information can be used by enemies and allies (Sabelis et al. 2001).

The emission of HIPVs and the establishment of indirect defenses are not limited to aboveground plant parts. Rasmann et al. (2005) showed that maize-root-feeding *Diabrotica virgifera virgifera* larvae induce the belowground release of (*E*)-β-caryophyllene which attracts the entomopathogenic soilborne nematode *Heterorhabditis megidis*. In field experiments, a fivefold higher nematode infection rate of *D. v. virgifera* larvae and a twofold reduction of the emergence of adult beetles were observed on (*E*)-β-caryophyllene-emitting maize variety compared to a maize variety that cannot emit this volatile. Moreover, a rescue of this deficient variety with a (*E*)-β-caryophyllene synthase transgene restored the indirect defense response (Degenhardt et al. 2009).

There are only few studies in which evidence is presented that indirect defense via HIPVs can benefit a plant's fitness. Van Loon et al. (2000) showed that *Arabidopsis thaliana* plants on which parasitized *Pieris rapae* caterpillars had fed produced significantly more seeds compared to plants attacked by unparasitized caterpillars. Similar studies came from maize plants infested with *Spodoptera littoralis* caterpillars parasitized by *Cotesia marginiventris* or *Campoletis sonorensis*. Parasitized larvae ate less from their host plants than larvae not parasitized and, consequently, these host plants suffered from less feeding damage and produced about 30% more seeds compared to control plants (Hoballah and Turlings 2001). Taken together, it is generally assumed that a reduction in plant damage will be beneficial for plant fitness.

8 How Do Herbivores Manipulate Induced Plant Defenses?

Direct and indirect plant defenses put selection pressure on herbivores as is evident from the diverse strategies described by which herbivores avoid defenses and develop resistances. However, there is also evidence that herbivores have adapted to manipulate direct and indirect plant defenses (Alba et al. 2011) such as to suppress induced plant defenses. The mechanisms by which herbivores suppress plant defenses are not well understood but often may come down to manipulation of hormonal signaling as is the case for galling insects (Tooker et al. 2008). However, the suppression of plant defenses can already be initiated by insect eggs. Bruessow et al. (2010) described that a nonprotein elicitor released from the eggs of the cabbage butterfly *P. brassicae* induces local accumulation of SA surrounding the oviposition site, thereby preinhibiting JA-dependent defenses induced by subsequent feeding of the future larvae. However, a positive effect on the weight gain of these larvae was only observed for the first instars of *S. litoralis* but not of *P. brassicae*. Hence, it is unclear if the observed induction of SA really serves to inhibit JA responses or whether this SA is involved in the development of a local HR possibly to defend the oviposition site against opportunistic pathogens. Moreover, SA accumulation could coincide with the production of volatile MeSA which may attract egg predators (Ament et al. 2010) or parasitoids and may repel other herbivores like the cabbage moth *Mamestra brassicae* (Ulland et al. 2008).

Eichenseer et al. (1999) showed that the saliva from *Helicoverpa zea* contains the enzyme glucose oxidase (GOX) which has multiple functions, i.e., to protect the larvae against pathogens on the one hand (Musser et al. 2005b) while suppressing the induced JA-dependent nicotine accumulation of *Nicotiana tabacum* on the other (Musser et al. 2005a) and the expression of genes involved in volatile production (Bede et al. 2006). GOX is widely present in the saliva of Lepidoptera (Eichenseer et al. 2010) while GOX activity is highest when feeding (Eichenseer et al. 1999). GOX is an oxidoreductase that catalyzes the oxidation of glucose-producing hydrogen peroxide (H_2O_2) and gluconic acid. Why GOX interferes with defense-related gene expression is not clear, but the accumulation of H_2O_2 causes a change in the redox stage of the plant tissue, which possibly interferes with the expression of downstream defense genes (Bede et al. 2006; Musser et al. 2006) and can induce SA accumulation (Diezel et al. 2009). In addition, Weech et al. (2008) showed that the saliva of *S. exigua* contains an unknown effector that alters JA-dependent plant defenses in a similar way as GOX but downstream from JA accumulation. Hence, although the JA/SA antagonism may play a role, the metabolic or genetic targets of GOX-mediated suppression of JA responses are unclear. Finally, it was found that GOX activity was on average higher in the saliva of generalist herbivore species than in the saliva of specialists (Eichenseer et al. 2010), suggesting that GOX activity may be correlated with a herbivore's host range.

Not only chewing herbivores were found to suppress plant defenses. Zarate et al. (2007) reported that the phloem-feeding whitefly *Bemisia tabaci* suppresses JA defenses via inducing SA defenses in *A. thaliana*. Zhang et al. (2009) reported that the whitefly *B. tabaci* feeding on lima bean *Phaseolus lunatus* suppressed spider mite *T. urticae* induced JA-dependent HIPV production and reduced the attractiveness of the plant to the mite's natural enemy *Phytoseiulus persimilis* while not affecting the plant's SA accumulation. Hence, it is unclear to which extend the JA/SA antagonism is responsible for defense suppression by whiteflies.

Suppression of induced plant defenses has also been observed in tomato *S. lycopersicum* when attacked by spider mites. The spider mite *T. urticae* harbors different genotypes of which most induce JA defenses while some suppress these, and it was possible to select for such distinct genotypes from natural mite populations (Kant et al. 2008). Suppression of defenses by these genotypes affects both SA and JA responses but is not absolute, i.e., the induction is lowered. Spider mites induce a cocktail of JA and SA defenses in tomato (Kant et al. 2004) as well as of JA-and SA-dependent HIPVs (Ament et al. 2004). Suppressor genotypes induce only low levels of JA-marker gene expression, do not induce significant increase in

PI activity, and do not induce a significant emission of JA-related plant volatiles. Importantly, the fitness of mite genotypes that induce these responses normally increases when sharing the feeding site with suppressor mites, suggesting that the absence of induction really is suppression (Kant et al. 2008). While the generalist mites *T. kanzawai* (Matsushima et al. 2006) and *T. urticae* (Takabayashi et al. 2000) clearly harbor genetic variation for induction and suppression of direct and indirect defenses (Kant et al. 2008), the latter trait may have come to fixation more in a tomato specialist, the spider mite *T. evansi*. Sarmento et al. (2011) showed that this mite species does not induce significant expression of SA and JA marker genes while downregulating the plant's PI activity levels to below housekeeping levels such that its fitness increases. Moreover, like the suppressor genotype of *T. urticae*, also the accumulation of JA, JA-Ileu as well of SA is suppressed by *T. evansi* albeit not below housekeeping levels (Alba et al. unpublished data). Surprisingly, although the emission of the well-known JA-dependent tomato volatiles (Ament et al. 2004) is suppressed by *T. evansi*, its natural enemies, the predatory mites *P. longipes* and *P. macropilis*, still respond to the odors of infested plants. This shows that suppression of a subset of well-known HIPVs does not necessarily disrupt indirect defenses.

It is not immediately evident why herbivores would be under selection to suppress induced plant defenses, assuming that resistance to defenses is the alternative trait (Kant et al. 2008). Defense suppression has the obvious disadvantage that competing herbivores may also benefit from it (Kant et al. 2008; Sarmento et al. 2011). Moreover, herbivores that suppress defenses may lose the traits that make them resistant to induced plant defenses since these traits are not under selective pressure any longer. Possibly, defense suppression can emerge coincidentally and persist when it allows herbivores to expand their host range (Kant et al. 2008) in cases when suppression targets conserved elements in, for example, the upstream hormonal signaling pathways of different plant species. At first glance, selection for suppression of indirect defenses may be easier to imagine than resistance to predation since the latter e.g., via regulated sequestration of induced defense products, may be a complex trait. However, since the metabolic regulatory networks of direct and indirect defenses are highly entangled (Walling 2000; Kant et al. 2009; Wu and Baldwin 2010), the physiological possibilities for a plant to uncouple direct defenses from HIPV production could be very limited, and hence, herbivores that suppress only induced indirect defenses may be rare.

References

Alba JM, Montserrat M, Fernández-Muñoz R (2009) Resistance to the two-spotted spider mite (*Tetranychus urticae*) by acylsucroses of wild tomato (*Solanum pimpinellifolium*) trichomes studied in a recombinant inbred line population. Exp Appl Acarol 47:35–47

Alba JM, Glas JJ, Schimmel BCJ, Kant MR (2011) Avoidance and suppression of plant defenses by herbivores and pathogens. J Plant Int 6:221–227

Alborn HT, Turlings TCJ, Jones TH, Stenhagen G, Loughrin JH, Tumlinson JH (1997) An elicitor of plant volatiles from beet armyworm oral secretion. Science 276:945–949

Alborn HT, Hansen TV, Jones TH, Bennett DC, Tumlinson JH, Schmelz EA, Teal PEA (2007) Disulfooxy fatty acids from the American bird grasshopper *Schistocerca americana*, elicitors of plant volatiles. Proc Natl Acad Sci USA 104:12976–12981

Allmann S, Baldwin IT (2010) Insects betray themselves in nature to predators by rapid isomerization of green leaf volatiles. Science 329:1075–1078

Ament K, Kant MR, Sabelis MW, Haring MA, Schuurink RC (2004) Jasmonic acid is a key regulator of spider mite-induced volatile terpenoid and methyl salicylate emission in tomato. Plant Physiol 135:2025–2037

Ament K, Van Schie CC, Bouwmeester HJ, Haring MA, Schuurink RC (2006) Induction of a leaf specific geranylgeranyl pyrophosphate synthase and emission of (E, E)-4,8,12-trimethyltrideca-1,3,7,11-tetraene in tomato are dependent on both jasmonic acid and salicylic acid signaling pathways. Planta 224:1197–1208

Ament K, Krasikov V, Allmann S, Rep M, Takken FLW, Schuurink RC (2010) Methyl salicylate production in tomato affects biotic interactions. Plant J 62:124–134

Anten NPR, Pierik R (2010) Moving resources away from the herbivore: regulation and adaptive significance. New Phytol 188:644–645

Arimura G, Matsui K, Takabayashi J (2009) Chemical and molecular ecology of herbivore-induced plant volatiles: proximate factors and their ultimate functions. Plant Cell Physiol 50:911–923

Bede JC, McNeil JN, Tobe SS (2002) The role of neuropeptides in caterpillar nutritional ecology. Peptides 28:185–196

Bede JC, Musser RO, Felton GW, Korth KL (2006) Caterpillar herbivory and salivary enzymes decrease transcript levels of *Medicago truncatula* genes encoding early enzymes in terpenoids biosynthesis. Plant Mol Biol 60:519–531

Bernays (1999) When host choice is a problem for a generalist herbivore: experiments with the whitefly, *Bemisia tabaci*. Ecol Entomol 24:260–267

Bleeker PM, Diergaarde PJ, Ament K, Schütz S, Johne B, Dijkink J, Hiemstra H, de Gelder R, de Both MTJ, Sabelis MW, Haring MA, Schuurink RC (2011) Tomato-produced 7-epizingiberene and R-curcumene act as repellents to whiteflies. Phytochemistry 72:68–73

Bonaventure G, Baldwin IT (2010) Transduction of wound and herbivory signals in plastids. Commun Integr Biol 34:313–317

Bonaventure G, VanDoorn A, Baldwin IT (2011) Herbivore-associated elicitors: FAC signaling and metabolism. http://www.cell.com/trends/plant-sci- 639 ence/abstract/S1360-1385%2811%2900021-5. Trends Plant Sci 16:294–299

Bruce TJA, Wadhams LJ, Woodcock CM (2005) Insect host location: a volatile situation. Trends Plant Sci 10:269–274

Bruessow F, Gouhier-Darimont C, Buchala A, Metraux JP, Reymond P (2010) Insect eggs suppress plant defence against chewing herbivores. Plant J 62:876–885

Colquhoun TA, Schimmel BCJ, Kim JY, Reinhardt D, Cline K, Clark DG (2010) A *Petunia chorismate* mutase specialized for the production of floral volatiles. Plant J 61:145–155

De Moraes CM, Lewis WJ, Paré PW, Alborn HT, Tumlinson JH (1998) Herbivore-infested plants selectively attract parasitoids. Nature 393:570–573

De Moraes CM, Mescher M, Tumlinson JH (2001) Caterpillar-induced nocturnal plant volatiles repel conspecific females. Nature 410:577–580

Degenhardt J, Hiltpold I, Kollner TG, Frey M, Gierl A, Gershenzon J, Hibbard BE, Ellersieck MR, Turlings TCJ (2009) Restoring a maize root signal that attracts insect-killing nematodes to control a major pest. Proc Natl Acad Sci USA 106:13213–13218

Despres L, David JP, Gallet C (2007) The evolutionary ecology of insect resistance to plant chemicals. Trends Ecol Evol 22:298–307

Dicke M, Sabelis MW (1988) How plants obtain predatory mites as bodyguards. Neth J Zool 38:148–165

Dicke M, van Loon JJA (2000) Multitrophic effects of herbivore-induced plant volatiles in an evolutionary context. Ent Exp Appl 97:237–249

Diezel C, von Dahl CC, Gaquerel E, Baldwin IT (2009) Different lepidopteran elicitors account for cross-talk in herbivory-induced phytohormone signaling. Plant Physiol 150:1576–1586

Doss RP, Oliver JE, Proebsting WM, Potter SW, Kuy S, Clement SL, Williamson RT, Carney JR, DeVilbiss ED (2000) Bruchins: insect-derived plant regulators that stimulate neoplasm formation. Proc Natl Acad Sci USA 97:6218–6223

Dudareva N, Pichersky E, Gershenzon (2004) Biochemistry of plant volatiles. Plant Physiol 135:1893–1902

Dudareva N, Negre F, Nagegowda DA, Orlova I (2006) Plant volatiles: recent advances and future perspectives. Crit Rev Plant Sci 25:417–440

Eichenseer H, Mathews MC, Bi JL, Murphy JB, Felton GW (1999) Salivary glucose oxidase: multifunctional roles for *Helicoverpa zea*? Arch Insect Biochem Physiol 42:99–109

Eichenseer H, Mathews MC, Powell JS, Felton GW (2010) Survey of a salivary effector in caterpillars: glucose oxidase variation and correlation with host range. J Chem Ecol 36:885–897

Eigenbrode SD, Espelie KE (1995) Effects of plant epicuticular lipids on insect herbivores. Annu Rev Entomol 49:171–194

Fahn A (1988) Secretory-tissues in vascular plants. New Phytol 108:229–257

Fatouros NE, Broekgaarden C, Bukovinszkine'Kiss G, van Loon JJA, Mumm R, Huigens ME, Dicke M, Hilker M (2008) Male-derived butterfly anti-aphrodisiac mediates induced indirect plant defense. Proc Natl Acad Sci USA 105:10033–10038

Feyereisen R (1999) Insect p450 enzymes. Annu Rev Entomol 44:507–533

Gershenzon J, McCaskill D, Rajaonarivony JI, Mihaliak C, Karp F, Croteau R (1992) Isolation of secretory cells from plant glandular trichomes and their use in biosynthetic studies of monoterpenes and other gland products. Anal Biochem 200:130–138

Hall D, MacGregor K, Nijsse J, Bown A (2004) Footsteps from insect larvae damage leaf surfaces and initiate rapid responses. Eur J Plant Pathol 110:441–447

Hartl M, Giri AP, Kaur H, Baldwin IT (2011) Serine protease inhibitors specifically defend *Solanum nigrum* against generalist herbivores but do not influence plant growth and development. Plant Cell 22:4158–4175

Hematy K, Cherk C, Somerville S (2009) Host–pathogen warfare at the plant cell wall. Curr Opin Plant Biol 12:406–413

Hilker M, Meiners T (2010) How do plants "notice" attack by herbivorous arthropods? Biol Rev 85:267–280

Hoballah MEF, Turlings TCJ (2001) Experimental evidence that plants under caterpillar attack may benefit from attracting parasitoids. Evol Ecol Res 3:553–565

Howe GA, Jander G (2008) Plant immunity to insect herbivores. Annu Rev Plant Biol 59:41–66

Kahl J, Siemens DH, Aerts RJ, Gabler R, Kuhnemann F, Preston CA, Baldwin IT (2000) Herbivore-induced ethylene suppresses a direct defense but not a putative indirect defense against an adapted herbivore. Planta 210:336–342

Kant MR, Ament K, Sabelis MW, Haring MA, Schuurink RC (2004) Differential timing of spider mite-induced direct and indirect defenses in tomato plants. Plant Physiol 135:483–495

Kant MR, Sabelis MW, Haring MA, Schuurink RC (2008) Intraspecific variation in a generalist herbivore accounts for induction and impact of host-plant defenses. Proc Royal Soc B Biol Sci 275:443–452

Kant MR, Bleeker PM, Van Wijk M, Schuurink RC, Haring MA (2009) Plant volatiles in defence. Adv Bot Res 51:613–666

Kessler A, Baldwin IT (2001) Defensive function of herbivore-induced plant volatile emissions in nature. Science 291:2141–2144

Kessler A, Halitschke R, Baldwin IT (2004) Silencing the jasmonate cascade: induced plant defenses and insect populations. Science 305:665–668

Labandeira CC (1997) Insect mouthparts: ascertaining the paleobiology of insect feeding strategies. Annu Rev Ecol Sys 28:153–193

Lee S, Badieyan S, Bevan DR, Herde M, Gatz C, Tholl D (2010) Herbivore-induced and floral homoterpene volatiles are biosynthesized by a single P450 enzyme (CYP82G1) in Arabidopsis. Proc Natl Acad Sci USA 107:21205–21210

Li XC, Baudry J, Berenbaum MR, Schuler MA (2004) Structural and functional divergence of insect CYP6B proteins: from specialist to generalist cytochrome P450. Proc Natl Acad Sci USA 101:2939–2944

Maffei ME (2010) Sites of synthesis, biochemistry and functional role of plant volatiles. South African J Bot 76:612–631

Matsushima R, Ozawa R, Uefune M, Gotoh T, Takabayashi J (2006) Intraspecific variation in the kanzawa spider mite differentially affects induced defensive response in lima bean plants. J Chem Ecol 32:2501–2512

Mattiacci L, Dicke M, Posthumus MA (1995) Beta-glucosidase—an elicitor of herbivore-induced plant odor that attracts host-searching parasitic wasps. Proc Natl Acad Sci USA 92:2036–2040

Merkx-Jacques M, Despland E, Bede JC (2008) Nutrient utilization by caterpillars of the generalist beet armyworm, *Spodoptera exigua*. Physiol Entomol 33:51–61

Mithöfer A, Wanner G, Boland W (2005) Effects of feeding *Spodoptera littoralis* on lima bean leaves. II. Continuous mechanical wounding resembling insect feeding is sufficient to elicit herbivory-related volatile emission. Plant Physiol 137:1160–1168

Musser RO, Cipollini DF, Hum-Musser SM, Williams SA, Brown JK, Felton GW (2005a) Evidence that the caterpillar salivary enzyme glucose oxidase provides herbivore offense in solanaceous plants. Arch Insect Biochem Physiol 58:128–137

Musser RO, Kwon HS, Williams SA, White CJ, Romano MA, Holt SM, Bradbury S, Brown JK, Felton GW (2005b) Evidence that caterpillar labial saliva suppresses infectivity of potential bacterial pathogens. Arch Insect Biochem Physiol 58:138–144

Musser RO, Farmer E, Peiffer M, Williams SA, Felton GW (2006) Ablation of caterpillar labial salivary glands: technique for determining the role of saliva in insect–plant interactions. J Chem Ecol 32:981–992

Ozawa R, Arimura G, Takabayashi J, Shimoda T, Nishioka T (2000) Involvement of jasmonate- and salicylate-related signaling pathways for the production of specific herbivore-induced volatiles in plants. Plant Cell Physiol 41:391–398

Peiffer M, Felton GW (2009) Do caterpillars secrete “oral secretions”? J Chem Ecol 35:326–335

Pichersky E, Noel JP, Dudareva N (2006) Biosynthesis of plant volatiles: nature’s diversity and ingenuity. Science 311:808–811

Pieterse CMJ, Leon-Reyes A, Van der Ent S, Van Wees SCM (2009) Networking by small-molecule hormones in plant immunity. Nat Chem Biol 5:308–316

Poelman EH, van Loon JJA, Dicke M (2008) Consequences of variation in plant defense for biodiversity at higher trophic levels. Trends Plant Sci 13:534–541

Rasmann S, Köllner TG, Degenhardt J, Hiltpold I, Toepfer S, Kuhlmann U, Gershenzon J, Turlings TCJ (2005) Recruitment of entomopathogenic nematodes by insect-damaged maize roots. Nature 434:732–737

Riffell JA, Lei H, Christensen TA, Hildebrand JG (2009) Characterization and coding of behaviorally significant odor mixtures. Curr Biol 19:335–340

Sabelis MW, Janssen A, Kant MR (2001) The enemy of my enemy is my ally. Science 291:2104–2105

Sallaud C, Rontein D, Onillon S, Jabès F, Duffé P, Giacalone C, Thoraval S, Escoffier C, Herbette G, Leonhardt N, Causse M, Tissier A (2009) A novel pathway for sesquiterpene biosynthesis from Z, Z-farnesyl pyrophosphate in the wild tomato *Solanum habrochaites*. Plant Cell 21:301–317

Sarmento RA, Lemos F, Bleeker PM, Schuurink RC, Pallini A, Oliveira MGA, Lima GR, Kant MR, Sabelis MW, Janssen A (2011) A herbivore that manipulates plant defence. Ecol Lett 14:229–236

Schilmiller AL, Schauvinhold I, Larson M, Xu R, Charbonneau AL, Schmidt A, Wilkerson C, Last RL, Pichersky E (2009) Monoterpenes in the glandular trichomes of tomato are synthesized from a neryl diphosphate precursor rather than geranyl diphosphate. Proc Natl Acad Sci USA 106:10865–10870

Schmelz EA, Carroll MJ, LeClere S, Phipps SM, Meredith J, Chourey PS, Alborn HT, Teal PEA (2006) Fragments of ATP synthase mediate plant perception of insect attack. Proc Natl Acad Sci USA 103:8894–8899

Schmelz EA, Engelberth J, Alborn HT, Tumlinson JH, Teal PEA (2009) Phytohormone-based activity mapping of insect herbivore-produced elicitors. Proc Natl Acad Sci USA 106:653–657

Schoonhoven LM, Jermy T, van Loon JJA (1998) Insect–plant biology: from physiology to evolution. Chapman & Hall, London

Shroff R, Vergara F, Muck A, Svatos A, Gershenzon J (2008) Nonuniform distribution of glucosinolates in *Arabidopsis thaliana* leaves has important consequences for plant defense. Proc Natl Acad Sci USA 105:6196–6201

Shuler MA (1996) The role of cytochrome P450 monooxygenases in plant–insect interactions. Plant Phys 112:1411–1419

Simmons AT, Gurr GM (2005) Trichomes of lycopersicon species and their hybrids: effects on pests and natural enemies. Agric For Entomol 7:265–276

Steppuhn A, Baldwin IT (2007) Resistance management in a native plant: nicotine prevents herbivores from compensating for plant proteinase inhibitors. Ecol Lett 10:499–511

Takabayashi J, Shimoda T, Dicke M, Ashihara W, Takafuji A (2000) Induced response of tomato plants to injury by green and red strains of *Tetranychus urticae*. Exp Appl Acarol 24:377–383

Thaler J (1999) Jasmonate-inducible plant defences cause increased parasitism of herbivores. Nature 399:686–688

Tholl D (2006) Terpene synthases and the regulation, diversity and biological roles of terpene metabolism. Curr Opin Plant Biol 9:297–304

Tooker JF, Rohr JR, Abrahamson WG, De Moraes CM (2008) Gall insects can avoid and alter indirect plant defenses. New Phytol 178:657–671

Turlings TCJ, Tumlinson JH, Lewis WJ (1990) Exploitation of herbivore-induced plant odors by host-seeking parasitic wasps. Science 250:1251–1253

Tzin V, Galili G (2010) New insights into the shikimate and aromatic amino acids biosynthesis pathways in plants. Mol Plant 3:956–972

Ulland S, Ian E, Mozuraitis R, Borg-Karlson AK, Meadow R, Mustaparta H (2008) Methyl salicylate, identified as primary odorant of a specific receptor neuron type, inhibits oviposition by the moth *Mamestra brassicae* L. (Lepidoptera, noctunidae). Chem Senses 33:35–46

Van Leeuwen T, Vanholme B, Van Pottelberge S, Van Nieuwenhuyse P, Nauen R, Tirry L, Denholm I (2008) Mitochondrial heteroplasmy and the evolution of insecticide resistance: non-mendelian inheritance in action. Proc Natl Acad Sci USA 105:5980–5985

Van Loon JJA, de Boer JG, Dicke M (2000) Parasitoid–plant mutualism: parasitoid attack of herbivore increases plant reproduction. Ent Exp Appl 97:219–227

Van Schie CC, Haring MA, Schuurink RC (2007) Tomato linalool synthase is induced in trichomes by jasmonic acid. Plant Mol Biol 64:251–263

Van Wijk M, De Bruijn PJA, Sabelis MW (2011) Complex odor from plants under attack: herbivore's enemies react to the whole, not its parts. http://www.plosone.org/article/info%3Adoi%2F10.1371%2Fjournal.pone.0021742. PLoS One 6:e21742

Walling LL (2000) The myriad plant responses to herbivores. J Plant Growth Regul 19:195–216

Walters D, Heil M (2007) Costs and trade-offs associated with induced resistance. Physiol Mol Plant Pathol 71:3–17

Wasternack C (2007) Jasmonates: an update on biosynthesis, signal transduction and action in plant stress response, growth and development. Ann Bot 100:681–697

Weech MH, Chapleau M, Pan L, Ide C, Bede JC (2008) Caterpillar saliva interferes with induced *Arabidopsis thaliana* defense responses via the systemic acquired resistance pathway. J Exp Botany 59:2437–2448

Wenke K, Kai M, Piechulla B (2010) Belowground volatiles facilitate interactions between plant roots and soil organisms. Planta 231:499–506

Werker E (2000) Plant trichomes. In: Advances in botanical research incorporating advances in plant pathology, vol 31. Academic, San Diego, pp 6–12

Wu JQ, Baldwin IT (2010) New insights into plant responses to the attack from insect herbivores. Annu Rev Genet 44:1–24

Zarate SI, Kempema LA, Walling LL (2007) Silverleaf whitefly induces salicylic acid defenses and suppresses effectual jasmonic acid defenses. Plant Physiol 143:866–875

Zhang PJ, Zheng SJ, van Loon JJA, Boland W, David A, Mumm R, Dicke M (2009) Whiteflies interfere with indirect plant defense against spider mites in lima bean. Proc Natl Acad Sci USA 106:21202–21207

Zhu-Salzman K, Luthe DS, Felton GW (2008) Arthropod-inducible proteins: broad spectrum defenses against multiple herbivores. Plant Physiol 146:852–858

Nematode Communication with Plants is Surprisingly Complex and Multidimensional

David Mc.K. Bird and Peter M. DiGennaro

Abstract Over a century of nematology research has focused on plant parasites that establish intimate symbioses with their host plants, yet the molecular basis of this interaction remains largely unknown. Central to the lifecycle of these obligate parasites is their ability to manipulate host tissue into specialized and dedicated feeding sites. This process is predicated on the ability of the nematodes to interject signaling cues to exploit the developmental plasticity of the host. Recent evidence, including the availability of significant amounts of parasitic genome data, points to diverse interactions that underpin a complex communication network. In this chapter, we examine the hierarchy of these interactions and propose a framework for placing the interactions in a formal context of parasitic symbioses.

1 Introduction

Nematodes are a large and speciose phylum of unsegmented roundworms (Bird and Bird 1991; Blaxter et al. 1998). They typically are microscopic (although many gut parasites of mammals are substantially larger) and, at hatch, share a remarkably uniform body plan. All nematodes develop through four larval stages (L1-L4; also known as juveniles: J1-J4) to the reproductive adult. Postembryonic development is remarkably plastic and has permitted nematodes to acquire the adaptations necessary for the phylum to occupy essentially every ecological niche (Borgonie et al. 2011) including being parasites of every other multicellular organism (Blaxter and Bird 1997). Here, we focus on those species that parasitize plants (plant-parasitic nematodes: PPN).

Collectively, PPN exploit all plant tissues and occupy niches in plant organs above and belowground. Their impact on humans is largely reflective of the

D.Mc.K. Bird (✉) • P.M. DiGennaro
Department of Plant Pathology, NC State University, Raleigh, NC, USA
e-mail: bird@ncsu.edu

G. Witzany and F. Baluška (eds.), *Biocommunication of Plants*,
Signaling and Communication in Plants 14,
DOI 10.1007/978-3-642-23524-5_12,

importance of the particular crop in question. For example, in China alone, the reduction of rice yield attributable to nematode infection exceeds USD 22 billion annually (McCarter 2009), making the foliar nematode parasite of rice, *Aphelenchoides oryzae*, an organism of significant global importance. However, despite the importance of nematodes that infect aerial tissues (including leaves, stems, and seeds), much of the research effort on PPN has been focused on a handful of species that function as obligate, sedentary parasites of roots, namely, the cyst nematodes (CN: *Globodera* and *Heterodera* spp.) and root-knot nematodes (RKN: *Meloidogyne* spp.). Because these cosmopolitan species are responsible for substantial yield losses on many crops worldwide (McCarter 2009), this emphasis is understandable. But beyond this, CN and RKN establish very intimate symbioses with their hosts in which the plant's innate developmental processes are manipulating by the nematodes to elicit specialized and dedicated feeding sites. Understanding how PPN successfully subvert their host's biology is a major goal of many nematologists, and a substantial body of (mostly descriptive) literature has accumulated over the last century. Yet despite this, the molecular basis for sedentary plant parasitism remains largely unknown. It is our contention that the absence of a conceptual framework for the parasitic interaction has contributed to this lack of progress. In this chapter, we attempt to redress this deficiency by proposing specific and testable models that we believe can be generalized to understanding metazoan-plant symbioses *per se*. Within that context, our specific focus is on plant responses to nematode effectors. We refer interested readers to an early iteration of these models (Bird 1996).

2 "Effectors"

According to the American Phytopathological Society (www.apsnet.org/edcenter/illglossary), the definition of an effector is "a pathogen molecule, usually a protein, which is translocated into host cells where it may act to directly manipulate host innate immunity." However, examination of the recent literature (e.g. Abad and Williamson 2010) reveals that PPN researchers typically expand this definition to encompass the manipulation of host processes beyond merely "host innate immunity." We concur and prefer a broader definition. Here, we propose that for the context of understanding the basis for parasitism by PPN, an "effector" be defined as "*a pathogen derived molecule(s) able to be perceived by the host to directly or indirectly act in an essential but not necessarily sufficient manner to elicit a host response germane to the pathogenic phenotype*."

Note that this definition differs from that of the APS in several subtle but key aspects. First, the nature of the molecule is not specified, and indeed, as we argue below, there is indirect evidence that PPN produce a range of nonprotein signaling molecules. Although most effectors might be expected to be secreted molecules (released from the worm, at least), this is not a strict requirement. It needs to be stressed that the corollary argument also need not be true. In other words, the fact that a molecule is secreted by the nematode into the host is not a sufficient requirement to declare that molecule to be an effector. The cellulases secreted into the host apoplast by migrating RKN and CN are good examples (Bird et al. 2009; Abad and Williamson 2010). Although these enzymes are presumably part of the parasite's armory, RNAi knockdown experiments have shown that they are not essential (Chen et al. 2005), nor do these molecules elicit observable host responses. Thus, by our definition, they are not effectors.

A second point on which our definition differs from that of the APS concerns the site of the interaction. Unlike the APS model, which dictates translocation of the effector into the symplast, our model permits apoplastic location. As we detail below, this is consistent with the presence of large families of receptors which span the host cell membrane and are poised to perceive events in the apoplast. It also is consistent with what is known about the *in planta* ecology of CN and RKN: These nematodes reside in the apoplast. Using antibodies raised to nematode proteins, it has been demonstrated that CN-derived molecules can enter the host cytoplasm (Wang et al. 2010), but to the best of our knowledge, no such demonstration has been made for RKN, although RKN-derived proteins can be unambiguously located to the apoplast (Jaubert et al. 2005).

Finally, our definition does not dictate the molecular target of the interaction; indeed, it not only allows for elicitors that are involved in processes beyond "host innate immunity" but, in fact, requires the induction of responses broadly contributing to parasitism ("germane to the pathogenic phenotype"). In the case of RKN and CN, relevant phenotypes would include, among others, the number of feeding sites formed and fecundity of individual nematode females.

Our intent in redefining "effector" is not to disparage APS but rather is to provide the context to better describe the mechanisms underpinning the PPN-host interaction. We are mindful that what we propose must both reflect the constraints provided by the biology of the interacting systems and also permit the full diversity of observed responses to be accounted. For example, proposed effectors must be able to exert influence both locally (e.g., at the feeding site) and remotely (e.g., in the shoot) because local and global responses are observed (Loveys and Bird 1973). Similarly, the complement of possible nematode effectors must, by necessity, be restricted by the range of functions able to be executed by the endogenous pathways; a better understanding of these host constraints will inevitably inform our understanding of PPN biology. We propose that PPN produce effectors for communication with their host at three levels (Table 1). Primary effectors interact directly. Secondary effectors interact to modify some aspect of host regulatory or physiological machinery. Tertiary effectors interact in a complex manner such that the existence of the elicitor can only be deduced from observation of the "pathogenic phenotype."

We believe that a full understanding of the "*pathogen derived molecule(s) able to be perceived by the host to directly or indirectly act in an essential but not necessarily sufficient manner to elicit a host response germane to the pathogenic phenotype*" will provide a comprehensive understanding of the nematode-plant interaction, and the goal of this chapter is to present the argument for this model and to provide the supporting data.

Table 1 Multidimensional signaling events between plant parasitic nematode and their hosts

Level	Interaction	Example
1°		CLE, CEP, and cytokinin
2°		Chorismate mutase
3°		ENOD40 and ccs52

3 Plant Parasitic Nematodes

Here, we present a very brief introduction into the biology, life cycle, and relative phylogenetic relationships within Nematoda of our major protagonists: root-knot, soybean cyst, and lesion nematodes. These topics are expanded in comprehensive reviews (Berg and Taylor 2009; Bird et al. 2009; Perry et al. 2009; Zunke 1990). Valuable and powerful insights into the communication that underpins the nematode-plant interaction can be gained by simply observing the life cycles. For example, it is clear that many developmental decisions in the nematode (including initiation of feeding, sex determination, and hatch status of their eggs) are based on perception of events in the plant. Genetic resources have been developed for *M. hapla* (Opperman et al. 2008), and genome data are available for all three of our exemplars. It is most extensive for RKN, with complete genomes obtained for *M. incognita* (Abad et al. 2008) and *M. hapla* VW9 (Opperman et al. 2008), plus deep skims from *M. hapla* VW8 and *M. hapla* LM (unpublished data). A robust assembly has recently been obtained for the lesion nematode, *Pratylenchus coffeae* (unpublished), and GenBank contains a draft sequence of *H. glycines*. Collectively, these resources represent a powerful tool kit to dissect PPN biology.

3.1 *Root-Knot Nematodes (RKN)*

The genus *Meloidogyne* probably infects all species of seed plants as well as lower plants such as ferns (Fig. 1). Reduced yield of infected crops equates to an annual economic impact that may approach USD 60 billion worldwide annually.

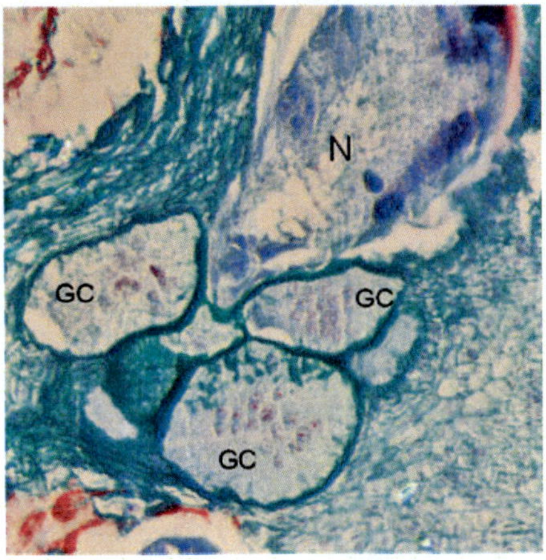

Fig. 1 RKN-induced feeding site on a primitive host. Mature root-knot nematode female (*N*) feeding from multiple multinucleate giant cells (*GC*) induced on the roots of an unidentified fern (Source: Image courtesy of Drs. Darlene DeMason and Manuel Mundo-Ocampo, University of California-Riverside)

Each RKN female has the potential to lay over 1,000 eggs which hatch in the soil as developmentally arrested second-stage larvae (L2 or J2), which typically reinfect the same plant. Infection usually occurs at the root tip by mechanical (and possible enzymatic) mediated penetration. As the nematode migrates intercellularly (i.e., apoplastically) into the vasculature, copious amounts of protein, including cell wall–degrading enzymes, are visibly secreted from the feeding stylet (Davis and Mitchum 2005). Within the vascular cylinder, the L2 "selects" up to ten vascular parenchyma cells which undergo developmental reprogramming into a unique cell type termed a *giant cell* (GC). One model (Bird 1996) postulates that GC is a novel chimera of (1) a xylem cell arrested in an early stage of differentiation, with (2) a transfer cell. Transfer cells normally form in response to a metabolic sink, which presumably is provided by the feeding nematode. Consistent with this, initiation of feeding occurs in pari passu with the appearance of GC. At the same time, the L2 commits to a sedentary lifestyle via loss of the somatic musculature. GC undergoes multiple rounds of karyokinesis without cytokinesis, and consequently, GC contains many polyploidy nuclei as well as thickened cell walls and an increased number of organelles. Depending on the RKN species or isolate, tissue surrounding GC undergoes variable degrees of hyperplasia, creating the noticeable galls (knots) characteristic of RKN infection.

3.2 Cyst Nematodes (CN)

Although there are many superficial similarities between the lifecycles and host-parasite interaction of RKN and CN, it is important to note the distinctions separating their biology. Like RKN, CN (*Heterodera* and *Globodera* spp.) are devastating sedentary obligate parasites of many crop plants, albeit with a much restricted host range. The term "cyst" is derived from the tanned body of the adult female that retains the eggs until host signals are perceived that elicit hatching as developmentally arrested L2. CN also migrate through host tissue in the apoplast toward vasculature to initiate permanent and dedicated feeding sites, simply termed *syncytia*. Importantly, the ontogeny of CN-induced syncytia differs markedly from RKN GC. Instead of multiple rounds of nuclear division without cell division, the syncytia arise from the coalescence of numerous adjacent cells. Also distinct from RKN, CN have been shown to have direct access to host cytoplasm into which proteins may be injected to alter host development and from which the worms presumably feed (Wang et al. 2010).

3.3 Lesion Nematodes (LN)

As obligate migratory endoparasites, the biology of lesion nematode (LN: *Pratylenchus* spp.) presents a contrast to our primary sedentary, endoparasitic protagonists. Recent LN genome data and comparisons with current PPN genomes may provide insight into the requirements for plant parasitic life and more specifically may point to loci involved in the formation of nematode-induced plant structures including GC, syncytia, and galls. Because LN remain vermiform and motile throughout their larval and adult stages, such loci might be absent from LN genomes. LN penetrate host roots behind the root tips and migrate to the cortex where the nematode uses its style to puncture host cells, into which it will enter and directly ingest cytoplasm. LN move and feed destructively, producing a lesion from the decaying and necrotizing tissue of spent cells (hence the common name). Severe root lesions result in secondary aboveground symptoms including stunting, chlorosis in leaves, and significant yield loss in crop plants.

3.4 Phylogenetic Relationships

Consistent with the distinctions in the life cycle and parasitic biology between CN and RKN, phylogenetic analyses (Holterman et al. 2008) show an ancient divergence for the ancestors of these sedentary obligate parasites. Four major clades were proposed for the order Tylenchida. *Heterodera* and *Rotylenchulus* spp. (a sedentary semiendoparasite) grouped together in clade A, whereas RKN mapped to clade B along with *Pratylenchus* spp. and *Nacobbus* spp.; the latter also induce GC. The similarities between CN and RKN larvae and their mature feeding sites presumably have arisen independently in each clade. This is an important point and cautions against strictly modeling the host-parasite interactions of one PPN genus with that of another. The constraints provided by host biology must limit the mechanistic options for the formation of feeding sites and likely serve as a driving force for convergent evolution.

4 Modes of Communication

As introduced above (Table 1), we propose that PPN have evolved diverse strategies to intervene in plant regulatory and developmental processes. To help focus discussion, we characterize these strategies as being 1°, 2°, or 3°, depending on the nature of the communication and in particular, the nematode-encoded effectors.

4.1 Primary Communication in Which Nematode Effectors Interact Directly with Host Machinery to Elicit the Pathogenic Phenotype

The concept that nematode-encoded functions play a role in the parasitic interaction is not new. Indeed, Linford (1937) hypothesized that RKN secretions play fundamental roles in the formation of GC. However, it was not until recently that genes expressed in pharyngeal glands of RKN and CN have been isolated as encoding candidate elicitors (Gao et al. 2001; Huang et al. 2003). Based largely on the location of expression plus the presence of signal sequences, it has been argued that these genes encode proteins secreted by the nematode into the host, but definitive experiments confirming these models are largely lacking. Nonetheless, modes of action have been proposed based on inferred function. The most tantalizing example comes from the discovery that PPN encode plant peptide hormone mimics.

The idea that feeding site formation might productively be considered in the context of plant developmental biology was first codified by Bird (1996), who proposed (in a very general sense) that mimics of plant peptide hormones (*pph*) might be secreted by nematodes. The first corroborative evidence of *pph* mimics came from a computational screen (Olsen and Skriver 2003), revealing that a gene (SYV-46) previously cloned from soybean cyst nematode (SCN) likely encoded a clavata-like element (CLE) ligand. CLE is a family of secreted plant peptide hormone ligands (typically 12 amino acids in their active, processed form) responsible for regulating many developmental events, the canonical function being meristem maintenance. Previously identified as a protein secreted from the stylet of SCN (Gao et al. 2001), SYV-46 was shown to bind CLV2 (a *bone fide* CLE receptor subunit in plants) and also to complement *clv3-1* mutants in *Arabidopsis* (Wang et al. 2005). Collectively, these data point to SCN encoding a genuine CLE, but the functional role of this protein as produced by the worm in the host-parasite interaction remains untested. It appears that *syv-46* has undergone a recent gene duplication event, as SCN contains two CLE encoding genes, differing in just three bases (all outside the active domain). In potato cyst nematode (PCN: *Globodera rostochiensis*), the CLE mimic family is even more expansive and diverse (Lu et al. 2009).

Whether or not the RKN genome encodes CLEs is controversial. Like *H. glycines* SYV-46, a gene from *M. incognita* called 16D10 was initially isolated from pharyngeal glands (Huang et al. 2003) and was later noticed to exhibit sequence similarity with *Arabidopsis* CLE (Huang et al. 2006). Transgenic overexpression of this nematode gene in which the protein was targeted to the cytoplasm in *Arabidopsis* gave a root developmental response and, through yeast two-hybrid assays, was found to be a ligand for scarecrow-like (SCL) proteins (Huang et al. 2006). SCL proteins are transcription regulators and members of the GRAS family that play central roles in rhizobial nodulation and meristem specification (Hirsch et al. 2009). Interestingly, these processes have multiple molecular and developmental similarities to GC induction (Bird 2004; Weerasinghe et al. 2005). The surprising result that 16D10 interacts with a nuclear protein rather than a transmembrane receptor in the apoplast led Mitchum et al. (2008) to conclude that 16D10 does not encode a CLE. We are unable to reconcile the findings of Huang et al. (2006) with our preliminary data and argue that 16D10 and its homologues in other RKN species encode *bona fide* CLE. In fact, we hypothesize that RKN genomes encode multiple families of *pph* mimics (including 16D10) that act through established signaling pathways in the host apoplast, consistent with a primary mode of communication between parasite and host.

4.2 PPN-Encoded CLE

We mined the completed genomes of *M. incognita* (Abad et al. 2008) and *M. hapla* (Opperman et al. 2008) with the double-affine Smith-Waterman algorithm and revealed five and eight candidate CLE loci, respectively (unpublished data). These loci not only exhibit sequence similarity to active plant peptides but also encode a secretion signal sequence and contain a predicted cleavage site directly upstream of the active domain. Intriguingly in nematodes, these two domains (signal sequence and active peptide) are not separated by an additional "pro" domain common to native plant peptide hormones. For native CLE, cleavage of the "pro" domain from the active peptide occurs in the apoplast (Ni et al. 2011) and presumably serves as an additional regulatory function against unwanted activity of these potent ligands. Absence of the "pro" domain from RKN-encoded mimics is consistent with the secretion active peptide hormones directly into the host apoplast where they presumably interact with transmembrane receptor-like kinases (RLK). Consistent with this is the finding that *Lotus japonicus* plants carrying mutations in the orthologue of CLV1 (a known CLE receptor in *Arabidopsis*) exhibit hyper infection by RKN (Lohar and Bird 2003).

Although direct evidence for RKN secretions into the host symplast is lacking, the evidence for SCN being able to secrete proteins into the host cytoplasm is strong (Wang et al. 2010). But the proposed behavior of SCN-encoded CLE is quite complex. Unlike RKN-encoded CLE, SCN CLE mimics contain a "pro" domain, and this domain has been implicated in the transportation of *H. glycines* CLE (HgCLE) mimics from host cytoplasm to the apoplast where it may act through endogenous pathways. However, GFP-tagged antibodies raised against HgCLE appear to indicate that these peptides localize in the cytoplasm of the syncytium, while in the same report, transient overexpression of protein fusions and subsequent bioassays indicates an apoplastic mode of action (Wang et al. 2010). The fusion proteins used were constructs of variable domains from plant CLEs with nematode active domains and vice versa. Based on root developmental phenotypes, the variable domains between HgCLE and native CLEs are reported as interchangeable, despite the contradicting ascribed functions of extracellular transport and apoplastic regulation, respectively (Wang et al. 2010).

Understanding native plant hormone action may illuminate the endogenous mechanisms exploited by RKN. CLEs are the most well-studied family of *pph*, and the signaling pathway is a paradigm for all *pph*. Functional analyses of plant CLE have split the family into two classes, "A" and "B." A-type CLEs, which includes CLV3, act to promote cell differentiation at meristems by antagonizing the general transcription factor *WUS*, which aborts root growth. B-type CLEs do not promote cell differentiation but rather inhibit cell differentiation in *Zinnia elegans* xylem elements. The two ascribed functions for these classes are not necessarily in opposition; rather, they are described as being agonistic with A-type CLE potentiating the activity of B-type CLE (Whitford et al. 2008). This degree of communication is able to balance the development of a complex vascular system through the regulation of proliferation and specification. Based on sequence similarity, both types of CLEs are found in RKN (unpublished), possibly indicating the developmental reprogramming potential required to initiate feeding sites and galls. Further, specific residues within the active domain of B-type CLE have been shown through alanine scanning experiments to be critical to peptide function (Ito et al. 2006). These residues (amino acids 1, 3, 6, 8, 9, and 12) are highly conserved within the global sequence similarity between native and *M. hapla* CLE, further pointing to nematode-encoded CLE as being analogues of native CLE.

4.3 RKN-Encoded CEP

Typically being small, genes encoding *pph* ligands necessarily have low information content compared to the entire genome. Consequently, *pph* tend to be recalcitrant to traditional, genome-wide computationally screens. To circumvent this, Ohyama et al. (2008) developed an algorithm to screen the *Arabidopsis* genome for novel *pph* families based on several assumptions (1) *pph* are encoded by multiple paralogous genes encoding relatively small products (70–110 amino acids) that (2) lack clear potential for secondary structure, such as cysteine-mediated disulphide linkages. The Ohyama algorithm expects genes to encode a secretion (signal) signal but permits a high degree of sequence diversity. The peptide domain exists as a conserved domain at the carboxyl terminus. Using this algorithm, a novel family of *pph* was identified, collectively known as CEPs (c-terminally encoded peptides). CEPs are expressed in lateral root primorida and are postulated to be *pph* based on the presence of a signal sequence and mass spectrometry data revealing the active c-terminal domain *in planta*. The overexpression phenotype of lateral root inhibition can be rescued by the application of exogenous CEP peptide, congruent with CEP being a *pph*. In the original report, Ohyama et al. (2008) classified five genes encoding CEP. Consistent with the role of regulating lateral root development, CEPs are widely distributed across vascular plants but appear absent from mosses or unicellular green algae.

Screening RKN genomes reveals 8 and 9 CEP genes in *M. incognita* and *M. hapla*, respectively. Like their plant analogues, RKN CEPs encode a signal sequence at the amino-terminus and a single CEP motif at the carboxyl terminus. As is the case with plant CLE, plant CEPs contain a "pro" domain between the signal sequence and the active carboxyl terminus, likely representing a measure of tertiary control over ligand activity. Akin to RKN CLE, RKN CEPs lack this "pro" domain, possibly allowing for the direct introduction of an active peptide into the host apoplast. Extensive experimentation is underway to fully understand the role of RKN CLE and CEP in the nematode-host interaction.

4.4 CLE and CEP Loci

It is widely accepted that horizontal gene transfer (HGT) from soil-borne bacteria has permitted PPN to acquire many functions, including an arsenal of cell wall–degrading enzymes (Bird et al. 2009). It is appealing to speculate that PPN may have acquired *pph* genes from their host in a similar manner, but the evidence necessary for such an inference (i.e., phylogenetic incongruence between species and gene trees) is lacking, perhaps due to the restricted phylogenetic signal available from the short sequences. For the same reason, phylogenetic reconstruction of the RKN CEP fails to reveal clear homology. However, cladograms derived from merged nematode and plant CEP reveal patterns of similarity, which likely reflects analogy (i.e., equivalent function). A reasonable hypothesis (Sikora et al. 2005; Mitchum et al. 2008) is that these nematode mimics have arisen de novo (i.e., convergently) rather than by HGT. Examination of the *M. hapla* genome reveals the CEP genes to be grouped into two tightly linked clusters within otherwise gene pauce regions. Comparison of the CEP loci between sequenced *M. hapla* isolates (VW8, VW9 and LM) indicates that these regions are hypervariable. Collectively, we hypothesize that these regions may be under high diversifying pressure and are exhibiting rapid evolution. Perhaps CEP function in the RKN-plant interaction is currently expanding its role. In contrast, CLE seem to be more evolutionarily ancient, based both on phylogenetic analyses and upon their distribution at discrete loci within the RKN and CN genomes.

4.5 Cytokinin

Because of their role in modulating cell cycle and cell division, cytokinins have long been postulated to play a role in plant parasitism, most likely via the execution of programs downstream of the actual nematode-plant interaction. During the 1960s, a number of studies on whole plants revealed elevated cytokinin levels in RKN-infected plants (e.g., Krupasagar and Barker 1969), although experiments involving the direct application of cytokinin failed to show an increase in RKN infection (Dropkin et al. 1969). However, application of exogenous cytokinin to a tomato cultivar carrying a gene (*Mi*) that conditions resistance to RKN resulted in loss of resistance (Dropkin et al. 1969). These studies implicated cytokinin as an important regulator of the host-parasite interaction, yet the mechanism underlying this affect was not apparent nor was the source of cytokinin. Remarkably, using bioassays, RKN was shown to produce biologically active cytokinin (Bird and Loveys 1980; de Meutter et al. 2003), although the role of such activity in the parasitic interaction remains questionable.

To better understand the temporal relationship between cytokinin levels and the formation of feeding sites, Lohar et al. (2004) used the *ARR5* promoter driving reporter constructs in transgenic plants. Although a response was not evident upon RKN infection or during apoplastic migration, a strong *ARR5* response was observed once the L2 reached the vascular bundle, the site of GC induction. Further, it was apparent that the cytokinin response occurs before the L2 reach the differentiation zone, although the spatial mapping of *ARR5* expression did not have the resolution required to determine if the cytokinin response occurs in those vascular parenchyma cells destined to become GC (Lohar et al. 2004), but this seems likely. Supporting the hypothesis that cytokinin is required at the initiation of GC, the use of cell cycle inhibitors revealed an initial transient requirement for cycle activation during GC formation (de Almeida Engler et al. 1999). Further, in an elegant experiment exploiting the temperature sensitivity of the *Mi* gene, Dropkin et al. (1969) demonstrated that the ability of *Mi* to confer resistance to RKN is restricted to the initial period of GC induction.

The evidence supporting the transient requirement for cytokinin in the induction of RKN feeding sites may have broader impacts on our understanding of the temporal aspects of the host-parasite interaction. Recently, microarray experiments have revealed a number of cytokinin-related genes that are differentially regulated in SCN-infected roots. Placing these genes into appropriate regulatory cascades will likely be very informative as to the precise role of cytokinins in the nematode-plant interaction. And it needs to be established if RKN truly produces cytokinin in a manner germane to the parasitic interaction. A very simple model has been proposed (Bird 1992) in which parasitism by RKN is attributed to the synergistic effects of cytokinin and cellulases secreted by RKN L2.

5 Secondary Communication in Which Nematode-Derived Effectors Modulate Innate Plant Regulatory Pathways

In this scenario, the nematode indirectly influences host biology by modulating host biochemistry, and we present several scenarios. The first involves chorismate mutase (CM), which is a plant enzyme central to the shikimate pathway. CM executes a claisen rearrangement on chorismate to yield prephenate, thus directing the shikimate pathway toward the biosynthesis of tyrosine and phenylalanine and away from tryptophan, the precursors of salicylic acid (SA) and auxins, respectively. Because of the obvious roles that may be played by SA and auxin, CM is a tantalizing candidate for being a 2° effector. Other mechanisms by which auxin is modulated by the parasite are similarly interesting.

5.1 *Chorismate Mutase*

RKN has been postulated to encode a secreted form of CM (Lambert et al. 1999). As noted, based on the role of CM in the biosynthesis of plant developmental and defense regulator precursors, a role in either initiating GC and/or suppression of host defense response seems tantalizing; controlling upstream pathways involved in the production of crucial host regulatory molecules is an appealing target for an exploitive parasite. Two lines of evidence point to the RKN enzyme as being a true CM. Complementation experiments demonstrate that RKN CM can rescue CM-deficient *E. coli* (Lambert et al. 1999). However, it is important to note that this experiment leaves other possible functions and substrates untested. The second line of evidence (and perhaps the strongest) comes from the aborted lateral root phenotype exhibited by transgenic soybean hairy roots overexpressing the RKN CM gene. This phenotype can be rescued by applying auxin, consistent with an auxin-deficient plant.

However, nematodes other than RKN also appear to encode CM. For example, examination of the *P. coffeae* genome (unpublished data) reveals a CM gene, yet this migratory nematode does not initiate feeding sites nor suppress host defense responses, which are the postulated roles of RKN-produced CM. To complicate the story, pathogenic organisms other than nematodes, including the human bacterial pathogen *Mycobacterium tuberculosis*, also secrete a functional CM (Sasso et al. 2005; Kim et al. 2006). In this case, the role of this enzyme in the pathogenic interaction remains unclear as the host (human) lacks the shikimate pathway. This observation might point to another role for CM that is unrelated to pathogenicity.

5.2 *Modulation of Local Auxin Concentrations by Endoparasitic Nematodes*

Although an active nematode-derived CM would have a counterintuitive effect on host auxin levels (driving the shikimate pathway away from auxin precursors), the ultimate response to changes in secondary metabolism cannot be predicted with certainty using available data. Irrelevant to the possible action of an enzyme resembling CM encoded within PPN genomes is the evidence for a local increase in auxin in GC and syncytia. As one of the earliest responses to nematode infection, the question of how such a change is achieved remains. An alternative hypothesis to CM is that manipulation of polar auxin transport (required for normal plant development and growth) resulting in the observed changes in auxin levels may be due to a local host defense response toward the invading nematode (Jones et al. 2007). Recently, corroborative evidence has shown that PPN infection induces rearrangements in PIN and AUX/LAX proteins (auxin transporters) possibly by nematode effectors interfering with auxin transport regulators (Grunewald et al. 2009).

A concept integral to our definition of effector, immaterial to the level of interaction, is the requirement for host perception. Following this, an alternative explanation to the observed manipulation of host auxin hormones upon nematode infection might be a change in auxin sensitivity and perception in the host. Consistent with this hypothesis is the rapid, nematode-mediated auxin-independent induction of the general transcription factor WRKY23, the promoter of which contains four auxin regulatory elements (Grunewald et al. 2008). The relatively rapid increase in expression and auxin-less induction of WRKY23 in feeding site formation possibly indicates a hijacking of plant gene expression by a nematode effector. Corroborating evidence comes from the detection of low molecular weight compounds in CN secretion which were shown to stimulate tobacco protoplast proliferation, in the presence of auxin and cytokinin, a possible indication of increased auxin sensitivity (Goverse et al. 1999).

6 Tertiary Communication: Perception of the Nematode Is Deduced from an Observable Plant Phenotype

Although the development of tools for forward and reverse genetics in *M. hapla* (Opperman et al. 2008) provides a strategy to investigate the host-parasite interaction without preconceived ideas of mechanism, much of what is known about how the plant recognizes the nematode must be gleaned from studying the plant response, which is likely to be removed by several steps from the primary interaction.

6.1 Ethylene

Long recognized as arising from infection by PPN, ethylene production in plants was thought to be a secondary response due to biotic stress (Glazer et al. 1983, 1985). However, functional analysis of ethylene production and signaling mutants in *Arabidopsis* revealed that this hormone is essential for the proper formation of feeding cells by CN. Ethylene overproducing mutants resulted in hypersusceptibility, correlating ethylene levels with the initiation of syncytia in roots (Wubben et al. 2001; Goverse et al. 2000). Conversely, ethylene insensitive mutants demonstrated a significant reduction in *H. Schachtii* development (Wubben et al. 2001). Supporting the requirement for de novo ethylene production in syncytial development, transient transcript increases have been demonstrated for the rate-limiting enzyme of ethylene production (ACC synthase) during syncytial development (Yamagami et al. 2003). Further, due to an increase in cell wall ingrowths from syncytia, ethylene was postulated to have a primary role in cell wall modification, increasing solute exchange between feeding cells and neighboring vascular tissue (Goverse et al. 2000). Intriguingly, in experiments designed to assess the role of ethylene in feeding cell development and further examination of the infection process, Wubben et al. (2001) revealed that the hypersusceptibility of ethylene overproducing mutants may be a result of enhanced host attraction to *H. Shachtii* L2, prior to root penetration.

Despite the obvious role of ethylene in CN parasitism, the metabolite's role in RKN infection is less clear. Although transgenic expression of the *Arabidopsis etr1-1* allele in *Lotus japonicas* conferred ethylene resistance and hypernodulation with rhizobial infection, nematode infection of transgenic lines was indistinguishable from wild type (Lohar and Bird 2003).

6.2 Pathways Shared with Rhizobia

The molecular signaling similarities between the beneficial symbioses of legumes and rhizobia and the parasitic symbiosis of RKN infection reflect the constraints host biology places on communication. Both symbionts induce structures that resemble meristems, presumably reflecting an overlap of regulatory pathways. Indeed, temporal and spatial expression of *PHAN* and *KNOX* (two genes central to meristematic maintenance) in *Medicago* is similar in nodules and nematode feeding sites (Koltai and Bird 2000; Koltai et al. 2001), consistent with endosymbionts utilizing and reprogramming normal developmental regulatory systems. Physiological similarities are also apparent between nodules and feeding sites, particularly the presence of "giant" polyploidy cells. Differentiation of nodules is regulated by *ccs52*, which is responsible for the division arrest and transformation of mitotic cycles to endocycles, producing multinucleate cells (Cebolla et al. 1999). Further, the small (12–13 amino acids) deduced open reading frame protein of *ENOD40* is a primary initiator of nodule formation and stimulates cortical division. Expression of *ccs52* and *ENOD40* in giant cells recapitulates the parallels between nodules, nematode feeding sites, and meristems (Koltai et al. 2001). ENOD40 has also been shown to be induced by cytokinin and is present in nonlegumes. Collectively, this suggests that the role of ENOD40 in feeding site initiation has broader implications beyond meristem and nodule induction (Foucher and Kondorosi 2000), possibly a result of a tertiary effect of nematode parasitism.

7 Conclusions

Plant parasitic nematodes have coevolved with their host plants to a very high degree, and this is most strikingly seen both in the signaling molecules deployed by the nematode (such as hormone mimics) and in the fundamental nature of the pathways the nematode manipulates. In designing targets for nematode control, there are two key points that come from understanding this level of interaction. The first is that individually, these molecules will most likely not be essential to nematode viability and thus poor targets for control. Second, and in contrast, they are paramount to plant developmental regulation; we presume that evolution has not equipped host plants with such self-debilitating defense responses.

References

Abad P, Williamson VM (2010) Plant nematode interaction: a sophisticated dialogue. Adv Bot Res 53:147–192

Abad P, Gouzy J, Aury J-M, Castagnone-Sereno P, Danchin EGJ, Deleury E, Perfus-Barbeoch L et al (2008) Genome sequence of the metazoan plant-parasitic nematode *Meloidogyne incognita*. Nat Biotechnol 8:909–915

Berg RH, Taylor CG (2009) Cell biology of plant-nematode interactions, plant cell monographs, vol 15. Springer, Berlin, p 273

Bird DM (1992) Mechanisms of the *Meloidogyne*–host interaction. In: Gommers FJ, Maas PWTH (eds) Nematology from molecule to ecosystem. ESN press, Wageningen

Bird DM (1996) Manipulation of host gene expression by root-knot nematodes. J Parasitol 82:881–888

Bird DM (2004) Signaling between nematodes and plants. Curr Opin Plant Biol 7:372–376

Bird AF, Bird J (1991) The structure of nematodes. Academic, San Diego

Bird AF, Loveys BR (1980) The involvement of cytokinins in a host–parasite relationship between the tomato (*Lycopersiconesculentum*) and a nematode (*Meloidogynejavanica*). Parasitology 80:497–505

Bird DM, Williamson VM, Abad P, McCarter J, Danchin EGJ, Castagnone-Sereno P, Opperman CH (2009) The genomes of root-knot nematodes. Annu Rev Phytopathol 47:333–351

Blaxter, M, Bird, D (1997) Parasitic nematodes. In: Riddle, D.L et al (eds) C. elegans II. Cold Spring Harbor Laboratory Press, p 851–878

Blaxter ML, DeLey P, Garey J, Liu LX, Scheldeman P, Vierstraete A, Vanfletern J, Mackey LY, Dorris M, Frisse LM, Vida JT, Thomas WK (1998) A molecular evolutionary framework for the phylum Nematoda. Nature 392:71–75

Borgonie G, Garcia-Moyano A, Litthauer D, Bert W, Bester A, van Heerden E, Moller C, Erasmus M, Onstott TC (2011) Nematoda from the terrestrial deep subsurface of South Africa. Nature 474:79–82

Cebolla A, VinardellJ M, Kiss E, Oláh B, Roudier F, Kondorosi A, Kondorosi E (1999) The mitotic inhibitor ccs52 is required for endoreduplication and ploidy-dependent cell enlargement in plants. EMBO J 18:4476–4484

Chen Q, Rehman S, Smant G, Jones JT (2005) Functional analysis of pathogenicity proteins of the potato cyst nematode *Globoderarostochiensis* using RNAi. Mol Plant–Microbe Interact 18:621–625

Davis EL, Mitchum MG (2005) Nematodes. Sophisticated parasites of legumes. Plant Physiology 137:1182–1188

de Almeida EJ, Vleesschauwer VD, Burssens S, Celenza JL Jr, Inzé D, Van Montagu M, Engler G, Gheysen G (1999) Molecular markers and cell cycle inhibitors show the importance of cell cycle progression in nematode-induced galls and synctia. Plant Cell 11:793–807

deMeutter J, Tytgat T, Witters E, Gheysen G, van Onckelen H, Gheysen G (2003) Identification of cytokinins produced by the plant parasitic nematodes *Heteroderaschachtii* and *Meloidogyne incognita*. Mol Plant Pathol 4:271–277

Dropkin VH, Helgeson JP, Upper CD (1969) The hypersensitivity reaction of tomato resistant to *Meloidogyne incognita*: reversal by cytokinins. J Nematol 1:55–61

Foucher F, Kondorosi E (2000) Cell cycle regulation in the course of nodule organogenesis in *Medicago*. Plant Mol Biol 43:773–786

Gao B, Allen R, Maier T, Davis EL, Baum TJ, Hussey RS (2001) Identification of putative parasitism genes expressed in the esophageal gland cells of the soybean cyst nematode *Heteroderaglycines*. Mol Plant–Microbe Interact 14:1247–1254

Glazer I, Orion D, Apelbaum A (1983) Interrelationships between ethylene production, gall formation, and root-knot nematode development in tomato plants infected with *Meloidogyne javanica*. J Nematol 15:539–544

Glazer I, Apelbaum A, Orion D (1985) Effect of inhibitors and stimulators of ethylene production on gall development in *Meloidogynejavanica*-infected tomato roots. J Nematol 17:145–149

Goverse A, Rouppe van der Voort J, Roupe van der Voort C, Kavelaars A, Smant G, Schots A, Bakker J, Helder J (1999) Naturally-induced secretions of the potato cyst nematode co-stimulate the proliferation of both tobacco leaf protoplasts and human peripheral blood mononuclear cells. Mol Plant–Microbe Interact 12:872–881

Goverse A, de Almeida EJ, Verhees J, van der Krol S, Helder J, Gheysen G (2000) Cell cycle activation by plant-parasitic nematodes. Plant Mol Biol 43:747–761

Grunewald W, Karimi M, Wieczorek K, Van de Capelle E, Grundler F, Beeckman T, Inze D, Gheysen G (2008) A role for AtWRKY23 in feeding site establishment of plant-parasitic nematodes. Plant Physiol 148:358–368

Grunewald W, Cannoot B, Friml J, Gheysen G (2009) Parasitic nematodes modulate PIN-mediated auxin transport to facilitate infection. PLoS Pathog 5:e1000266

Hirsch S, Kim J, Muñoz A, Heckmann AB, Downie JA, Oldroyd GED (2009) GRAS proteins form a DNA binding complex to induce gene expression during nodulation signaling in *Medicago truncatula*. Plant Cell 21:545–557

Holterman M, Holovachov O, van den Elsen S, van Megen H, Bongers T, Bakker J, Helder J (2008) Small subunit ribosomal DNA-based phylogeny of basal Chromadoria (Nematoda) suggests that transitions from marine to terrestrial habitats (and vice versa) require relatively simple adaptations. Mol Phylogenet Evol 48:758–763

Huang G, Gao B, Maier T, Allen R, Davis EL, Baum TJ, Hussey RS (2003) A profile of putative parasitism genes expressed in the esophageal gland cells of the root-knot nematode *Meloidogyne incognita*. Mol Plant–Microbe Interact 16:376–381

Huang G, Dong R, Allen R, Davis EL, Baum TJ, Hussey RS (2006) A root-knot nematode secretory peptide functions as a ligand for a plant transcription factor. Mol Plant–Microbe Interact 19:463–470

Ito Y, Nakanomyo I, Motose H, Iwamoto K, Sawa S, Dohmae N, Fukuda H (2006) Dodeca-CLE peptides as suppressors of plant stem cell differentiation. Science 313:842–845

Jaubert S, Milac AL, Petrescu AJ, de Almeida-Engler J, Abad P, Rosso M-N (2005) In planta secretion of a calreticulin by migratory and sedentary stages of root-knot nematode. Mol Plant–Microbe Interact 18:1277–1284

Jones JT, Furlanetto C, Phillips MS (2007) The role of flavonoids produced in response to cyst nematode infection of *Arabidopsis thaliana*. Nematology 9:671–677

Kim S-K, Reddy SK, Nelson BC, Vasquez GB, Davis A, Howard AJ, Patterson S, Gilliland GL, Ladner JE, Reddy PT (2006) Biochemical and structural characterization of the secreted chorismatemutase (Rv1885c) from *Mycobacterium tuberculosis* H37Rv: An *aroq enzyme not regulated by the aromatic amino acids. J Bacteriol 24:8638–8648

Koltai H, Bird DM (2000) Epistatic repression of PHANTASTICA and class 1 KNOTTED genes is uncoupled in tomato. Plant J 22:455–459

Koltai H, Dhandaydham M, Opperman C, Tomas J, Bird D (2001) Overlapping plant signal transduction pathways induced by a parasitic nematode and a rhizobial endosymbiont. Mol Plant–Microbe Interact 14:1168–1177

Krupasagar V, Barker KR (1969) Increased cytokinin concentrations in tobacco infected with the root-knot nematode *Meloidogyne incognita*. Phytopathology 56:885

Lambert KN, Allen KD, Sussex IM (1999) Cloning and characterization of an esophageal-gland-specific chorismatemutase from the phytoparasitic nematode *Meloidogyne javanica*. Mol Plant–Microbe Interact 12:328–336

Linford MB (1937) Stimulated activity of natural enemies of nematodes. Science 85:123–124

Lohar DP, Bird DM (2003) *Lotus japonicus*: a new model to study root-parasitic nematodes. Plant Cell Physiol 44:1176–1184

Lohar DP, Schaff JE, Laskey JG, Kieber JJ, Bilyeu KD, Bird DM (2004) Cytokinins play opposite roles in lateral root formation, and nematode and rhizobial symbioses. Plant J 38:203–214

Loveys RR, Bird AF (1973) The influence of nematodes on photosynthesis in tomato plants. Physiol Plant Pathol 3:525–529

Lu S-W, Chen S, Wang J, Yu H, Chronis D, Mitchum MG, Wang X (2009) Structural and functional diversity of CLAVATA3/ESR (CLE)-like genes from the potato cyst nematode *Globoderarostochiensis*. Mol Plant–Microbe Interact 22:1128–1142

McCarter J (2009). Molecular approaches toward resistance to plant-parasitic nematodes. Cell Biology of Plant Nematode Parasitism:239–267

Mitchum MG, Wang X, Davis EL (2008) Diverse and conserved roles of CLE peptides. Curr Opin Plant Biol 11:75–81

Ni J, Guo Y, Jin H, Hartsell J, Clark SE (2011) Characterization of a CLE processing activity. Plant Mol Biol 75:67–75

Ohyama K, Ogawa M, Matsubayashi Y (2008) Identification of a biologically active, small, secreted peptide in Arabidopsis by in silico gene screening, followed by LC–MS-based structure analysis. Plant J 55:152–160

Olsen AN, Skriver K (2003) Ligand mimicry? Plant-parasitic nematode polypeptide with similarity to CLAVATA3. Trends Plant Sci 8:55–57

Opperman CH, Bird DM, Williamson VM, Rokhsar DS, Burke M, Cohn J, Cromer J et al (2008) Sequence and genetic map of Meloidogynehapla: a compact nematode genome for plant parasitism. Proc Natl Acad Sci USA 105:14802–14807

Perry, R, Moens, M, Starr, J (2009) Root-knot Nematodes. CABI, Wallingford, UK

Sasso S, Ramakrishnan C, Gamper M, Hilvert D, Kast P (2005) Characterization of the secreted chorismatemutase from the pathogen *Mycobacterium tuberculosis*. FEBS J 272:375–389

Sikora S, Strongin A, Godzik A (2005) Convergent evolution as a mechanism for pathogenic adaptation. Trends Microbiol 13:522–527

Wang X, Mitchum MG, Gao B, Li C, Diab H, Baum TJ, Hussey RS, Davis EL (2005) A parasitism gene from a plant-parasitic nematode with function similar to *CLAVATA3/ESR (CLE)* of *Arabidopsis thaliana*. Mol Plant Pathol 6:187–191

Wang J, Joshi S, Korkin D, Mitchum MG (2010) Variable domain I of nematode CLEs directs post-translational targeting of CLE peptides to the extracellular space. Plant Signal Behav 5:1633–1635

Weerasinghe RR, Bird DM, Allen NS (2005) Root-knot nematodes and bacterial Nod factors elicit common signal transduction events in *Lotus japonicus*. Proc Natl Acad Sci USA 102:3147–3152

Whitford R, Fernandez A, De Groodt R, Ortega E, Hilson P (2008) Plant CLE peptides from two distinct functional classes synergistically induce division of vascular cells. Proc Natl Acad Sci USA 105:18625–18630

Wubben MJE, Su H, Rodermel SR, Baum TJ (2001) Susceptibility to the sugar beet cyst nematode is modulated by ethylene signal transduction in *Arabidopsis thaliana*. Mol Plant–Microbe Interact 14:1206–1212

Yamagami, T, Atsunari T, Kayoko Y, William FH, Leslie AH, Athanasios T (2003) Biochemical diversity among the 1-amino-cyclopropane-1-carboxylate synthase isozymes encoded by the Arabidopsis gene family. Journal of Biological Chemistry 49:49102–49112

Zunke U (1990) Observations on the invasion and endoparasitic behavior of the root lesion nematode *Pratylenchuspenetrans*. J Nematol 22:309–320

Virus Operation Control Centers

Ulrich Melcher

Abstract The nucleic acids of plant viruses are the control centers that coordinate all activities associated with virus survival and propagation within cells, in whole plants and between organisms. Within cells, the viruses use a diversity of signaling mechanisms to assure the orderly production at specific subcellular locations of viral mRNAs, viral proteins, viral genomic nucleic acids, and viral particles and the export of infectious entities to neighboring cells. Within cells, viruses also signal their presence to the host cell machinery, establishing the conditions of coexistence of virus and plant in successful infections. At the plant level, the control centers direct the movement of infectious entities from one cell to another, into the vascular system, and into tissues remote from the site of initial infection. At the same time, the control centers condition the plant to be hospitable to virus reproduction and survival. They also cause the plant to issue signals to potential vectors guiding them to the plant to acquire the virus and encouraging their departure to further plants, in effect spreading the virus among multiple plants. The signals used in these processes include small molecules (hormones and volatiles), macromolecules with binding sites for other molecules (some being enzymatic), macromolecular structure conformations, genomic organizations, and others. Often, different viruses accomplish the same activity in completely different ways, although some common strategies are employed.

1 Introduction

Biology can be thought of as a large network of molecular interactions mediated by signals. Signal molecules are produced by and/or released from transmitting molecules and travel to interact with receiver molecules. The transmitting molecule

U. Melcher (✉)
Department of Biochemistry and Molecular Biology, Oklahoma State University, Stillwater, OK, USA
e-mail: u-melcher-4@alumni.uchicago.edu

G. Witzany and F. Baluška (eds.), *Biocommunication of Plants*,
Signaling and Communication in Plants 14,
DOI 10.1007/978-3-642-23524-5_13,

and/or the receiving molecule may be altered by the process, enabling recognition or failure of recognition by still other molecules. This semiochemical framework was developed by Witzany (2008) and illustrated by examples involving bacteria. It is the purpose of this chapter to extend the conceptual framework of biochemical signaling in life processes to viruses and, in particular, viruses associated with plants.

Witzany (2008) distinguished three levels of communicative interactions for bacteria: interactions within one bacterial cell, interactions between multiple cells of the same bacterial species, and interactions with organisms of other species. Viruses also exhibit these levels of interactions: activities within an infected cell, spread of the infection to neighboring cells, and transmission of the virus from one host plant to another. However, since plant viruses are subcellular entities of eukaryotic organisms, we need to explore also the semiochemical interactions of viral molecules with their host environments.

Although most viruses have physical entities that we associate with the name virus (virions), these are only storage forms of the viral genetic material. In a signaling and regulation context, the viral genetic material is the center of signaling. It determines what informational mRNAs are generated, which proteins that it encodes are made, when genome complementary strands are synthesized, when genome sense strands are synthesized, when and if virions are formed, and how the viral information is disseminated to other locations (organelles, cells, other hosts). Thus, I designate the viral genomic nucleic acid as the virus operations control center (VOCC). I ask the reader's patience since this designation leads to descriptions that are anthropomorphic and use the active voice: "the VOCC decides," "the VOCC evaluates," etc. VOCCs have developed over years of evolution selecting for fitness the ability to produce progeny capable of carrying on the genetic line. It is important to understand that viruses and VOCCs have not evolved to cause disease in plants. Plant disease as a result of virus infection is an ancillary consequence of interactions of some viruses with some plants (Malmstrom et al. 2011).

Two general categories of signals will be woven through the discussion: small molecules and macromolecules. Small molecules such as plant hormones have the ability to spread within a plant and sometimes from one plant to surrounding organisms. Macromolecules such as RNA and proteins play major roles in virus signaling. Throughout, we will be concerned with the semantics of signals as well as their nature. Do they have meanings that are context dependent? To what extent are they shared among other viruses? Are different kinds of signals used in the same context to produce the same result?

2 Intracellular Viral Communication

Virions of plant viruses are diverse in the types of nucleic acid they contain (ssDNA, dsDNA, dsRNA, negative sense (−) ssRNA, and positive sense (+) ssRNA), but all form mRNA as part of their replication strategy. Many of the links in VOCC communications are based on RNA molecules. RNA can form complex and dynamic three-dimensional structures. Intrastrand pairing of bases produces simple helical stem-loop hairpins. Various ways of stacking helices and creating turns in helical orientations lead to a wide variety of possible structures, including complex ones in which loop residues pair with residues in other parts of the chain. These structures can be semi-stable and switch conformations upon proper provocation (Wakeman et al. 2007), an ideal property for a signal transducer. RNA conformations serve as receivers for proteins including replicases, capsid proteins, translation factors, and proteins that guide the RNA to different subcellular locations.

Plant virus genomes, because of their small size, encode a limited diversity of proteins: coat proteins for formation of viral particles, movement proteins for infection spread, enzymes needed for replication of their genomes, and a few additional types not encoded by all viruses. Proteins may have baggage tags directing those proteins to specific subcellular locations. Proteins can interact with other proteins or with RNA structures through binding domains or binding surfaces. The ability of some proteins to bind simultaneously to more than one other molecule makes them excellent transmitters and receivers.

2.1 *mRNA Production*

Signaling in mRNA production is complex because of the diversity of genomic forms of plant viruses. In the case of ssDNA viruses, a host DNA polymerase must recognize the ssDNA and create a circular nucleosome-coated DNA in the nucleus of the cell. These and the analogous minichromosomes of dsDNA viruses need then to be recognized by host transcription factors that bind at appropriate sites including promoters, upstream activating sequences, and enhancers to guide the host DNA-dependent RNA polymerase to the transcription initiation point or points. When multiple promoters exist, the VOCC needs additional mechanisms to manipulate the timing and volume of transcripts produced from the various promoters (Shung and Sunter 2009). Promoters and other signal receivers are generally context independent, working equivalently in a wide variety of host plants.

In the (−) RNA viruses *Cytorhabdoviridae* and *Nucleorhabdoviridae*, genome organization is a form of signal. These genomes have the gene for an RNA-binding N-protein at the first transcribed 3′ end and the replicase L protein at the last transcribed 5′ end. The mechanism of (+) RNA synthesis means that synthesis of abundantly required proteins precedes the synthesis of the replicase such that replication cannot begin until everything is prepared for making new virions.

For viruses with (+) RNA genomes, since synthesis is in the cytoplasm rather than in the nucleus, the mRNAs do not receive caps from the nuclear-located capping enzymes. Given the importance of caps for translation initiation, some viruses encode capping enzymes. Strategies used by others will be mentioned below. Viruses with dsRNA genomes include helicases, capping enzymes, and replicases in their particles allowing transcription of mRNAs once the virion enters the cell (Roy 2008).

2.2 Replication

Replication requires separate treatment for different classes of viruses. VOCCs established by members of the *Geminiviridae* must signal the quiescent cell that it enters to resume the DNA synthesis phase of the cell cycle (Ascencio-Ibanez et al. 2008). The virus signals, Rep proteins, interact with cellular proteins that regulate cell cycle progression (Kong et al. 2000). *Beet curly top virus* (BCTV) with a nonfunctional C4 open reading frame (ORF) is unable to establish a systemic infection although it can replicate in protoplasts (Teng et al. 2010). The C4 protein induces synthesis of RKP, a cell cycle regulation protein (Lai et al. 2009). Protein expression of a C4 transgene from a *Cauliflower mosaic virus* (CaMV) 35S (Park et al. 2010) or an inducible (Mills-Lujan and Deom 2010) promoter leads to cell proliferation and developmental abnormalities. When expressed in resting yeast cells, the Rep proteins are able to stimulate them to undergo multiple rounds of DNA synthesis without mitosis (Kittelmann et al. 2009), indicating that this particular signaling route is conserved among fungi and plants.

2.3 Translation

The general language of protein synthesis results in initiation (reviewed in (Miller et al. 2011)) at the 5′ most AUG in proper context (Joshi et al. 1997; Lukaszewicz et al. 2000) and not sequestered in secondary structure. The 5′ cap structure is the site of binding by the eIF4 initiation complex. Not all viral mRNAs have 5′ cap receivers for eIF4. To compensate, some produce Vpg proteins that are covalently linked to the 5′ end and serve the same purpose as the cap. However, their presence also leads to the sequestration of eIF4E thus favoring viral translation over host translation (refs in Culver and Padmanabhan 2007). Viral RNAs with 3′ polyA tails bind the eIF4 complex (Le Gall et al. 2011) via polyA binding proteins (PABAs). Some other viral RNAs have tRNA-like structures at their 3′ ends that stimulate translation initiation by unknown mechanisms (Miller et al. 2011). Others lacking polyA ends have special RNA structures (CITES) in their 3′ nontranslated regions that serve as cap-independent translation elements and bind initiation factors. CITES are hypothesized to be important signal receivers in the VOCC since their disruption during RNA replication will prevent translation of the RNA being replicated (Miller et al. 2011).

Some VOCCs increase the frequency of translation initiation via enhancer sites that serve as receivers for initiation factors or small ribosomal subunits (Miller et al. 2011). VOCCs can also signal initiation of translation at internal RNA sites through folded RNA structures called internal ribosome entry site (IRES) (reviewed in (Miller et al. 2011)). These RNA structures are diverse in structure and in the initiation factor requirements for their function.

Should the first AUG not be in a strong context, the ribosome may scan further and initiate at a later AUG or a CUG or AUA in optimal context (Miller et al. 2011), a phenomenon called leaky scanning that results, for example, in synthesis of either longer or shorter versions of the same protein, such as the P95 and P105 proteins of *Cowpea mosaic virus* (CPMV) (Holness et al. 1989), or the synthesis of two unrelated proteins from consecutive reading frames, such as, for example, in tymoviruses, poleroviruses, and tombusviruses, a phenomenon called overprinting. VOCCs with such strategies likely have been selected in evolution as a way to achieve economy of genome length.

A further common occurrence in VOCCs is the production of different levels of proteins. In read-through translation, the protein in greater demand is encoded N-terminal of the other protein (often the RNA-dependent RNA polymerase) separated from it by either a stop codon or an RNA sequence that induces a shift in reading frame (Giedroc and Cornish 2009). Read-through of an in-frame stop codon is accomplished by recruiting selected tRNAs (selection depends on the virus) to respond to the termination codon, suppressing termination. The RNA sequence signals required for read-through have been investigated (Harrell et al. 2002). The results suggest that a handful of different signals in the sequence immediately preceding the stop codon have the ability to be used as the words that communicate "bypass termination."

2.4 Protein Processing

Theoretically, the advantage of the above multiple ways of controlling whether translation initiates at particular places provides the VOCCs excellent flexibility in managing the volume of particular proteins produced and the timing of their production relative to other proteins and other events in replication. However, other viruses have developed a polyprotein strategy to produce individual distinct proteins. The sole mRNA is translated without pause into a single long polyprotein. The polyprotein itself has recognition sites for proteases (part of the polyprotein), the cleavage of which results in the release of individual polypeptides. Relative amounts of proteins have to be varied in this system by differential degradation of those that are relatively overproduced.

Proteins can accept posttranslational modifications that will alter their ability to be recognized by other receivers or alter their ability to recognize other molecules. Phosphorylation, acetylation, and ubiquitination of viral replicase proteins have been documented (Nagy and Pogany 2011). Ubiquitination should lead to degradation and thus assists in generating the optimal ratio of proteins in infection by members of the *Potyviridae*. Phosphorylation of residues in the C-terminal end of the movement protein (MP) of *Tobacco mosaic virus* (TMV) and its relatives has been studied (Karpova et al. 1999).

2.5 Encapsidation

The VOCC chooses between using an RNA for replication and/or translation and packing it into virions, thus removing it from circulation. The signal involved is an RNA conformation (determined by sequence or composition) that is recognized by the capsid protein as a packaging signal. However, encapsidation signals vary from virus to virus and may sometimes require that the RNA being packaged also be replicating (Shin et al. 2010). Transencapsidation of one virus by the capsid of another is known to occur and is the basis for the phenomenon of satellite RNAs. These RNAs propagate themselves by mimicking the host virus' replication and encapsidation signals. Population genetic studies of virus evolution suggest that encapsidation usually wins out in competitions such that most RNA molecules do not contribute to the evolutionarily effective population size. Population genetic techniques only suggest that these effective sizes are surprisingly small.

2.6 Uncoating

Uncoating of virions has been investigated in a few systems revealing entirely different signals. Uncoating of rigid rod virions, like those of TMV, likely occurs when virions attach to the endoplasmic reticulum (Christensen et al. 2009). For TMV, the capsid subunit in its binding to RNA has a decided preference for G-containing sequences. The 5′ end of the genomic RNA of tobamoviruses is typically devoid of Gs, meaning that the capsid subunits bound at the 5′ end of the particle are only loosely bound. When virions enter a naïve cell, chemical equilibrium drives dissociation of subunits from the 5′ end, freeing the RNA for binding by translation initiation factors that bind to the 5′ cap structure present in these RNAs. Subsequent binding of the 40S ribosome, scanning for the first AUG, and consequent translation release the remainder of the RNA from its package. It is the absence of a semiochemical, the free coat protein subunit, which leads to the response. Presence of coat protein subunits leads to association with the RNA and its packaging.

The balance between the assembly of viral particles and the availability of genomes for other functions is also an issue with the ssDNA viruses whose genomes are transported, after synthesis, from the nucleus to the cytoplasm where some of them need to be transported to the neighboring cell. However, the cytoplasm is also the location of the capsid protein that could encapsidate the genome before it makes it out of the cell. The *Geminiviridae* VOCC appears to control the amount of encapsidation-competent capsid by recruiting a host acetyltransferase NSI (nuclear shuttle interactor) to bind to its nuclear shuttle protein (NSP, required for movement) and having it acetylate the CP subunits presumably reducing their affinity for the genome (Carvalho et al. 2006). Some view this signaling as a plant defense mechanism (Santos et al. 2010).

The second type of uncoating has been studied with icosahedral virions. Virions self-assemble from RNA and capsid subunits. The interactions between subunits include salt bridges and are often mediated by divalent cations. Alteration of the divalent cation concentration of the surroundings can lead to conformational changes in the subunits and therefore also in the whole virion. The virion becomes more open, and RNA can be released from it.

3 Intracellular Virus-Host Communication

Events in viral replication, translation, protein processing, assembly, and disassembly happen inside plant cells. Inevitably, communicative interactions between viral processes and cellular molecules occur. Viral fitness requires a certain level of host fitness and survival. It is therefore not unreasonable to expect that the virus will signal its presence to the host to allow the host to limit virus replication to preserve host survival.

3.1 Sequestration

The events discussed in Sect. 2 occur at specific subcellular locations. Thus, the VOCC must direct the complexes carrying out these functions to the correct location. Different virus groups have adapted diverse strategies to accomplish the localization of replication complexes and of movement and coat protein complexes.

3.1.1 Replication Complexes and Inclusion Bodies

Most, if not all, VOCCs organize electron microscopically identifiable replication complexes attached on the cytoplasmic side of membrane systems. The complexes consist of several virus-encoded proteins and some host proteins. Thus, the assembly of the complex requires coordinated binding between multiple interaction pairs. Different viruses use diverse endomembranes. For example, the formation of mitochondrially located replication complexes in *Carnation Italian ringspot virus* (CIRSV) is the result of the action of the p36 viral nonstructural protein (Hwang et al. 2008). Even for a single virus, there is flexibility as evidenced by viruses such as *Cymbidium ringspot* virus (CyRSV), which, in plant cells, replicate on peroxisomes, but in yeast cells use the endoplasmic reticulum since they do not have peroxisomes (Rubino et al. 2008). In this case, the pragmatic purpose of the signal is the same, but it results in putting replication in a different context.

Inclusion bodies form in cells as a result of the activity of multiple kinds of viruses. In the case of the *Potyviridae* members, several different types of inclusions form. These may represent garbage heaps for those parts of the polyprotein that are not required in as large amounts as the coat protein. In other cases (see Sect. 5.1), evidence indicates that inclusion bodies are programmed by the VOCC to receive signals that further the dissemination of the virus.

Within cells, virus replication complexes and MPs have been observed attached and moving along microtubules and actin microfilaments (Harries and Ding 2011). However, the significance of these associations is under dispute. In addition, the subcellular associations seem to be highly virus specific with *Turnip vein-clearing virus* exhibiting different cytoskeletal affiliations from its close relative, TMV.

3.1.2 Movement and Coat Proteins

The primary fate of CPs is assembly into virions. However, since some CP appear to have additional destinations, other roles are possible. *Grapevine rupestris stem pitting–associated virus* CP has a signal that brings it into the plant cell nucleus (Meng and Li 2010). Similarly, *Cocksfoot mottle virus* (CfMV) CP has nuclear localization signals (Olspert et al. 2010). The role of these localizations is unclear.

Subcellular locations of MPs depend on what else is present in the cell. They are thus controlled by protein interactions that determine their location (triple gene block). MPs link diverse molecules to one another. For example, the MP of *Barley yellow dwarf virus* (BYDV) has an N-terminal region responsible for its attachment to the nuclear membrane and a C-terminal region that binds RNA (Vogler et al. 2008). Thus, it serves to bridge the nucleus and the RNA, bringing them together. Correct targeting of MP to their intracellular organelle destination is important as illustrated by the impairment of cell-to-cell movement of AltMV by removal of the plastid-targeting signal of its TGP3 MP (Lim et al. 2010). Other MPs are targeted to the ER membrane (Verchot-Lubicz et al. 2010; Martinez-Gil et al. 2010).

3.2 *Intracellular Virus-Host Interactions*

Plants have a variety of mechanisms to prevent disease caused by virus infection (Palukaitis and Carr 2008). Resistance to virus infection can be via pathways that are constitutive in the plant or can be induced (signaled) as a consequence of infection (Carr et al. 2010).

3.2.1 DsRNA-Dependent Kinase

Replication of RNA viruses almost inevitably produces dsRNA. DsRNA can signal the host that a virus is present. The RNA activates a kinase that phosphorylates eIF2-alpha, greatly reducing protein synthesis. With some viruses, plants in which expression of this kinase has been silenced or knocked out have much more severe symptoms of local infection (Bilgin et al. 2003). However, studies with other viruses provide different pictures (Culver and Padmanabhan 2007).

3.2.2 Hypersensitive Response

The presence of certain protein motifs produced by translation of viral genomes is a signal to the plant cell that a virus (Liu and Coaker 2008) infects it. NBS-LRR proteins (Tameling and Joosten 2007; Drutskaya et al. 2011; Bhattacharjee et al. 2009) are a major class of plant receiver molecules that accomplish initial recognition of virus infection (Cournoyer and Dinesh-Kumar 2011). Signal reception can lead to a type of induced resistance known as the hypersensitive response (Carr et al. 2010). A series of protein interactions and enzyme reactions can lead to programmed cell death. Often, cell death occurs early enough that virus spread is limited to a small region, a local lesion. The trigger has been best studied in the TMV system where a part of the 183 kDa protein interacts with a toll-like receptor in the host cell. The signaling pathway can include components of the RNA silencing pathway (see Bhattacharjee et al. 2009), and salicylic acid (SA) has a role in the hypersensitive response (Venugopal et al. 2009).

For example, amino acid residue 461 of the *Cucumber mosaic virus* (CMV) 1a protein is critical for the hypersensitive response of tobacco to the virus (Salanki et al. 2007). Interaction of the viral elicitor with the plant receptor results in activation of a cascade of reactions leading to cell death (Takabatake et al. 2007). For TCV, the elicitor is the N-terminal region of the CP (Ren et al. 2000). At least one host gene, HRT, is required for the TCV-specific induction of HR through a series of other proteins including CRT (Kang et al. 2008). Light signaling is also involved since blue-light photoreceptors, cryptochrome 2 and phototropin 2, mediate HRT stability (Jeong et al. 2010). Basal resistance pathways may also be signaled by TCV CP through its interaction with TCV-interacting protein (TIP) (Jeong et al. 2008).

Plant mitochondria are thought to play an important role in detecting and responding to pathogen infections, including those of viruses (Amirsadeghi et al. 2007). Molecules evolved during incompatible (gene-for-gene) interactions target mitochondrial components resulting in a cascade of reactive oxygen species being produced.

In the interaction of TMV with N-gene-containing tobacco, enzymes of spermine synthesis are induced and the polyamines play a role in the programmed cell death induced by this resistance interaction (Yamakawa et al. 1998; Yoda et al. 2003). Spermine action is mediated in part by stimulation of gene induction cascades and is active in hypersensitive responses to other viral pathogens such as CMV (Mitsuya et al. 2009). CMV-induced genes overlap heavily spermine-induced genes, and prevention of spermine synthesis increases the multiplication of CMV in *A. thaliana* (Mitsuya et al. 2009). How the N-gene increases spermine synthesis is not known. TMV interaction with the N-gene product has also been asserted to alter the epigenetic marking of LRR-containing genes and actin genes, marking that leads to increased rates of genetic changes (Kathiria et al. 2010; Boyko et al. 2007). However, such transgenerational effects are far from being universally accepted (Daxinger and Whitelaw 2010). Regulation by calcium levels may provide signals for induction of plant defenses (Lu et al. 2010).

Major interaction pathways lead to what phytopathologists recognize as symptoms of disease (Culver and Padmanabhan 2007). The pathways are only now beginning to be investigated and appear to be nonuniversal communication routes, in that related viruses in the same host may produce quite different effects and the same virus will produce quite different symptoms in two different plants (Culver and Padmanabhan 2007). Some effects of infection may be attributed to the usurpation of plant nutrients through viral protein and particle synthesis (Dordas 2008).

4 Intercellular Communication

The spread of virus infection from a single initially infected cell is a critical component of viral fitness. This spread involves movement of the infection from the initially infected cell to neighboring cells through plasmodesmata (Fernandéz-Calvino et al. 2011b) until cells adjacent to vascular elements are reached. The infectious agent is then loaded into the vascular system, usually phloem, and is downloaded into cells surrounding phloem in sink tissue (Pallás et al. 2011). Cell-to-cell movement and phloem loading and unloading require cross talk between virus and host.

4.1 Infection Spread

The key proteins required for intercellular movement are called movement proteins (Lucas et al. 2009). Structurally and by sequence similarity, there are a small number of classes of these proteins. Again, different languages are used to accomplish the syntactic movement. The most studied of these is the 30 K superfamily (Melcher 2000). In this superfamily, it has been shown that the C-terminal tail of the MP is the virus-specific part of the molecule (Lee et al. 2005). Outside of this recognition, the grammar works regardless of the virus. The ToMV MP interacts with a transcription factor KELP and by binding redirects its subcellular location so that it is unable to foster movement of the infection (Sasaki et al. 2009).

Another family is the triple gene block family of three cooperating polypeptides that together move infection from cell to cell. Competing models for how such movement is accomplished have been reconciled with one another (Verchot-Lubicz et al. 2010). TGB3 is a bridge protein signal, binding both TGBP2 and a plasmodesmatal location (Tilsner et al. 2010). TGP1 (Wright et al. 2010) of PMTV localizes to nuclei and microtubules.

By as yet not understood mechanisms, virus infection is usually absent from meristems. Thus, one of the most important functions of the VOCCs is to orchestrate the delivery of viral genomes to newly developed cells. Such movement can be thought of as a triple sending of a message, first from home to the post office, second through the post office system, and finally from that system to the recipient. In the plant's case, the post office system is the vasculature. Infectious entities move from cell to cell, establishing infection in each new cell as they go, using all the signaling mechanisms already discussed. Eventually they are dumped into the phloem and go along to the developing leaves where they enter a cell in the leaf and spread from cell to cell spreading infection.

Cell-to-cell movement is thought to occur through plasmodesmata or through plasmodesmata-like tunnels created in response to virus infection (Benitez-Alfonso et al. 2010). Plasmodesmata, not yet well understood at the molecular level (Faulkner and Maule 2011), besides providing the route for viruses to traffic, also are the way that noncell-autonomous proteins, certain mRNAs, miRNAs, and siRNAs are transported from cell to cell and into phloem. This sharing of a pathway has the inevitable consequence that virus infection interferes with one of the major routes of intercellular signaling with the plant (Culver and Padmanabhan 2007). Late in infection of a cell, after movement has happened, preferential degradation of the TMV MP by proteasomes (Reichel and Beachy 2000) closes the gates. Callose deposition at plasmodesmata impedes viral transport (Zavaliev et al. 2011).

4.2 *Intercellular Virus Communication: Small Molecule Signals*

Virus infection can induce changes in levels of secondary signals, signals the plant uses to communicate intercellularly such as auxin (IAA), abscisic acid, gibberellic acid, cytokinin, brassinosteroids (Robert-Seilaniantz et al. 2007), salicylic acid, ethylene, and jasmonic acid (Roberts et al. 2007). Plants overproducing caffeine or being treated with caffeine have molecular signatures indicative of induction of plant defense responses (Kim and Sano 2008). Other compounds from other plants can serve the same function, for example, 3-acetonyl-3-hydroxyindole from *Strobilanthes cusia* (Li et al. 2008). Some small molecules are suspected to be signals, but mechanisms for their action have not been elucidated. For example, interference with phytic acid production results in an increased sensitivity of plants to virus infection (Murphy et al. 2008).

Plant cells are not competent to respond to the plant hormone IAA due to the sequestration of auxin-stimulatable transcription factors in a complex with Aux/IA proteins. In TMV infection, the latter proteins are transferred from the nucleus to the cytoplasm, presumably for proteasomal degradation, making the tissue auxin responsive. Symptoms resembling the results of auxin treatment are seen. As many as 30% of genes upregulated in TMV infection have auxin-responsive elements in their upstream regions (Padmanabhan et al. 2005). Alteration of gibberellic acid levels is likely a consequence of rice dwarf virus infection of rice. The virus-encoded P2 protein interacts directly with an enzyme in the hormone's biosynthetic pathway (Zhu et al. 2005). A three-nucleotide difference in the genomes of CMV satellites D and Dm is sufficient to turn on the ethylene synthesis pathway and produce substantial disease (Irian et al. 2007).

It has long been known that infection of a plant with one kind of pathogen can make that plant resistant to infection by another pathogen, even one of a completely different nature. For example, virus infection can induce resistance against fungal attack. This systemic acquired resistance (SAR) (Vlot et al. 2009; Vasyukova and Ozeretskovskaya 2007; Kiraly et al. 2007; Hammerschmidt 2009) is mediated by semiochemicals such as salicylic acid, jasmonic acid, and methyl salicylate (Park et al. 2007). The plant gene EDS5 (enhanced disease susceptibility 5) is induced by virus infection, and its transcription is required for SA production.

A study analyzing changes in hormone levels early during viral infection with PVY demonstrated that JA and its precursor were the only significant hormonal molecules to change (Kovac et al. 2009). The pathways producing these signals are complex. How the VOCC communicates its presence to the plant inducing the production of these chemicals is not clear. Methyl jasmonate is a volatile related compound that can be transmitted by air to another plant to induce systemic resistance in that plant. Activation of the SA pathway during infection by some viral strains of clover yellow vein virus can lead to severe disease by inducing systemic cell death (Atsumi et al. 2009).

4.3 *Intercellular Virus Communication: Macromolecular Signals*

Plants have developed, possibly in response to the presence of viruses, a universal inducible mechanism to destroy foreign RNA, particularly RNA present in high concentration and capable of forming ds RNA (Mlotshwa et al. 2008; Fernandéz-Calvino et al. 2011a). Although initiated in infected cells, this RNA silencing interacts with plant architecture to spread to other parts of the plant. Viruses have evolved a diversity of mechanisms to suppress such silencing (Burgyan 2011).

4.3.1 RNA Silencing

As discussed above, viral RNAs have secondary structures and are replicated via complexes containing both positive and negative sense strands. As a result, plants have large viral dsRNAs during active virus replication. The dsRNA is recognized and processed by a set of proteins that result in the production of small RNAs that complex with other proteins to cleave RNA molecules containing complements to the small RNA sequence (Burgyan 2011).

Once a signal is produced in one location, it spreads throughout the plant (Hyun et al. 2011) making the younger parts of a plant resistant to infection and pathogenic symptoms, a phenomenon denoted as recovery. Silencing signals spread through both symplastic (local) and vascular (long-distance) pathways. The MPs of plant viruses have been implicated in the intercellular spread of the silencing signals (Vogler et al. 2008). Long-distance signaling in plants has been conceptually divided into four phases: induction, signal movement, perception of the signal, and response (Kehr and Buhtz 2008; Champigny and Cameron 2009). Genes that are part of the siRNA pathway are required at both ends of a signal pathway for functioning of the signal (Molnar et al. 2010; Kalantidis et al. 2008; Brosnan et al. 2007). Since siRNAs are less than 30 nt long, chances are appreciable that some siRNAs produced from viral dsRNA by the hosts RNA silencing machinery will have sufficient identity to some part of some host mRNAs to induce their silencing. Alteration in host gene expression results, as exemplified by expression of a hairpin RNA gene based on viroid sequence leading to viroid-like symptoms without production of viroids (Wang et al. 2004).

Plants use the small miRNAs to regulate gene expression during development. It is likely that many symptoms of virus infection are due to interference with miRNA action (Zhang et al. 2007). Indeed, some plant viral genomes have been identified to have miRNA-like sequences. This field awaits further exploration. Nevertheless, it is clear at this point that some virus signals interact with plant developmental pathways and that plants use these pathways to coexist with viruses (Chung et al. 2008). Interaction between virus infection and specific host miRNA levels has been noted (Lang et al. 2011), but the routes of signal transduction are not known.

Small RNAs are not the only RNAs that move through the plant vasculature. A variety of mRNAs do so also, and their composition can be altered by virus infection (Ruiz-Medrano et al. 2007). They can thus act as secondary messages of virus infection.

4.3.2 Silencing Suppression

Viruses have developed multiple strategies for overcoming silencing. Most involve the elaboration of proteins that recognize and inactivate elements of the silencing pathway (Siddiqui et al. 2008). Over 50 such virus-encoded silencing suppressors have been discovered (Burgyan 2011). A few recent examples are mentioned here. The pns10 silencing suppressor protein of the *Rice dwarf phytoreovirus* (Zhou et al. 2010) intercepts the ds small RNA preventing the signal from arriving at the RISC complex (Ren et al. 2010). *Banana bunchy top virus* proteins B3 and B4 are active as silencing suppressors at different steps of the silencing pathway (Niu et al. 2009). The P50 MP of ACLSV inhibits the long-distance spread of silencing signals, whether viral or nonviral (Yaegashi et al. 2008; Yaegashi et al. 2007). Poleroviruses encode a protein, P0, which interacts with components of ubiquitin ligases (Pazhouhandeh et al. 2006), presumably targeting host-silencing proteins for degradation. The TRV-encoded 16-kDa protein acts as a silencing suppressor downstream of dsRNA formation (Martinez-Priego et al. 2008). Because the siRNA pathway includes steps that are common to the processing of miRNAs, which are important for proper development of plants, suppressors of RNA silencing produced by virus infection from viral genes also interfere with miRNA maturation and thus result in some phenotypic appearances interpretable as disease (Culver and Padmanabhan 2007).

5 Interorganismal Signaling

Although the use of volatile signals by plants to induce responses to viruses in parts of the plant remote to the site of initial infection can spill over to neighboring plants inducing resistance responses in those plants, it is likely that this is an incidental consequence of selection for intraplant communication (Heil 2001). True interorganismal communication does however occur between plants and vectors of viruses. Two kinds of interactions between viruses, cross protection and synergy, are also known and need brief description.

5.1 Vectors

To make the leap from one host to another, many VOCCs utilize mobile vectors. Arthropods are the most frequently investigated vectors, but nematodes, fungi, and mammals also serve. VOCCs can cause the plant to become more attractive to vectors (de Vos and Jander 2010). Attractiveness can be through visual or volatile clues. For example, BYDV-infected wheat plants produce volatile compounds that attract the aphid vector of the virus (Jimenez-Martinez et al. 2004). The blend of volatile compounds surrounding *Potato leafroll virus*–infected potato plants not only attracts *Myzus persicae* aphids but also induces them to stay longer on the plants (Ngumbi et al. 2007) than on control noninfected plants. On the other hand, CMV infection of a cucurbit, although elevating volatile levels and increasing attractiveness to aphids, also increases the rate at which aphids placed on these leaves emigrate, suggesting that the aphids find the infected leaves unpalatable (Mauck et al. 2010). How virus infection leads to alteration of volatile interorganismal signals or palatability to aphids is not known yet (de Vos and Jander 2010). Palatability is also of importance to mammals, although in a different way. Infection of *Kennedya rubicunda* plants with Kennedya yellow mosaic virus makes the plants less palatable to herbivores so that virus-infected plants survive longer than noninfected plants (Gibbs 1980).

Transmission of viruses by insects (Blanc and Drucker 2011) takes different forms. In one form, the insect is viruliferous for only brief periods of time, suggesting that the virus is loosely bound to mouthparts, so that transmission will happen on the next probing. This kind of transmission would benefit from a plant that attracts the vector to probe but after probing repels the insect (Mauck et al. 2010). In semi-persistent transmission, the virus is bound to a site within the stylet, such that the aphid can remain viruliferous until a shedding of the exoskeleton occurs. In persistent transmission, the virus needs to traverse several barriers in the insect, the gut wall, the basal lamina at the gut, the basal lamina at the salivary gland, and the salivary gland duct side membrane, a process that takes considerable time, reducing the urgency of causing the insect to move to another plant. In some types of persistent transmission, the virus replicates also in the insect so that, in effect, the virus has two hosts.

5.2 Cross Protection and Synergy

Cross protection refers to the immunity of a plant to superinfection with a second virus closely related to the first virus and is an example of one virus communicating with a virus of a different strain. Uncoating of viral particles (Sect. 2.6) is thought to be part of the basis of cross protection. Some cases of inhibition of replication of an unrelated virus have also appeared, but the signaling mechanism involved has not been elucidated (Yang et al. 2010).

Inter-virus communication can also occur between seemingly unrelated viruses in a phenomenon that is called synergy. In synergy, disease symptoms are considerably exacerbated relative to infection of the plant with either virus alone. Investigation of levels of viral molecules in synergistic situations reveals frequently that one virus will increase the level of replication of another virus. Mechanisms of this communication between two viruses have been only scantily investigated. The signals that different viruses produce elicit different responses from the plant. The interaction of these different signaling pathways is what accounts for the phenomenon of synergism (Garcia-Marcos et al. 2009).

6 Summary

Viruses are unique in the biological world. As a conceptual entity, they are so entwined with life that we cannot distinguish easily viral from cellular interactions within the host. For our purposes, we must consider as viral any interaction that occurs that would not occur in the absence of virus or that would occur differently in its presence.

6.1 Review of Methods

Much molecular communication of viruses with each other, with the plant, and with plant visitors occurs in the midst of a large network of activities (Culver and Padmanabhan 2007). Approaches to identifying signaling pathways include examining molecular changes resulting from virus infection, at transcript, protein, and metabolite levels. Transcriptomes have been searched by comparing EST databases of infected vs. noninfected tissue (Freitas-Astua et al. 2007; Eybishtz et al. 2009), by subtractive hybridization of cDNA libraries (Alfenas-Zerbini et al. 2009), by microarray hybridization (Whitham et al. 2003; Whitham et al. 2006; Espinoza et al. 2007; Catoni et al. 2009; Babu et al. 2008a, b; Ascencio-Ibanez et al. 2008), and by serial analysis of gene expression (SAGE) (Irian et al. 2007). Proteomic comparisons can also produce leads on signaling pathways induced by virus infection (Yang et al. 2011). Metabolomics leads to recognition that ROS are important during virus infection (Quecini et al. 2007). In pea plants, infection with PPV leads to changes in chloroplast structure and metabolites that further lead to the production of ROS defenses (Diaz-Vivancos et al. 2008).

Genetic approaches to identifying signals sent during virus infection include selective partial suppression of transcripts via virus-induced gene silencing (VIGS) (Wu et al. 2008; Sarowar et al. 2009; Cheng et al. 2010; Chen et al. 2009), complete knockout of selected genes by mutagenesis (Xia et al. 2008; Lim et al. 2010), overexpression of a plant (Vannini et al. 2006; Quilis et al. 2008) or virus (Geri et al. 1999) transgene, and two-hybrid interaction studies (Piroux et al. 2007; Carvalho et al. 2008).

6.2 Semiophoric Aspects

In multiple virus-related signaling roles, we have seen many small molecules, including hormones and volatile compounds. The usual kinds of interactions exhibited by proteins (activators, repressors, receptor proteins) also typify virus communications. Perhaps unusually characteristic of virus communication is the use of RNA molecules, both as signals themselves and as the carrier of structural elements that are signals and signal receptors. Also, more strongly important for virus communication than for others is the importance of gene order in transcription and translation.

Syntactically, there are numerous examples of multiple means of achieving the same meanings, probably due to the diversity of origins of viruses. In contrast, a single means often has different meanings depending on circumstances. There are a few instances of conserved signaling themes, such as the interaction of viral signals with NBS-LRR receptor proteins.

6.3 Applications of VOCC Signaling Knowledge

Understanding of signaling has practical applications. VIGS, in which a plant gene sequence is placed in a viral vector so that it induces silencing of the targeted gene, (Catinot et al. 2008) has been widely used to explore the functions of many plant genes. Applications include crop improvement. For example, genetically engineering tobacco to produce a variant of the antimicrobial cationic peptide polyphemusin had enhanced resistance to infection by TMV (Bhargava et al. 2007). It has been proposed to use SA application to plants as a way to increase their stress resistance (Shang et al. 2011; Chandra et al. 2007). Alternatively, compounds that induce SAR can be used to protect plants from stress by pathogens (Mandal et al. 2008). Genes for plant receptors of viral elicitors can be engineered so as to produce a color change in the intact plant, thus giving evidence of the presence of the viral pathogen (Mazarei et al. 2008).

References

Alfenas-Zerbini P, Maia IG, Favaro RD, Cascardo JCM, Brommonschenkel SH, Zerbini FM (2009) Genome-wide analysis of differentially expressed genes during the early stages of tomato infection by a potyvirus. Mol Plant–Microbe Interact 22:352–361

Amirsadeghi S, Robson CA, Vanlerberghe GC (2007) The role of the mitochondrion in plant responses to biotic stress. Physiol Plant 129:253–266

Ascencio-Ibanez JT, Sozzani R, Lee TJ, Chu TM, Wolfinger RD, Cella R, Hanley-Bowdoin L (2008) Global analysis of Arabidopsis gene expression uncovers a complex array of changes impacting pathogen response and cell cycle during geminivirus infection. Plant Physiol 148:436–454

Atsumi G, Kagaya U, Kitazawa H, Nakahara KS, Uyeda I (2009) Activation of the salicylic acid signaling pathway enhances Clover yellow vein virus virulence in susceptible pea cultivars. Mol Plant–Microbe Interact 22:166–175

Babu M, Gagarinova AG, Brandle JE, Wang AM (2008a) Association of the transcriptional response of soybean plants with soybean mosaic virus systemic infection. J Gen Virol 89:1069–1080

Babu M, Griffiths JS, Huang TS, Wang A (2008b) Altered gene expression changes in Arabidopsis leaf tissues and protoplasts in response to Plum pox virus infection. BMC Genomics 9:325. doi:10.1186/1471-2164-9-325

Benitez-Alfonso Y, Faulkner C, Ritzenthaler C, Maule AJ (2010) Plasmodesmata: gateways to local and systemic virus infection. Mol Plant–Microbe Interact 23:1403–1412

Bhargava A, Osusky M, Forward BS, Hancock RE, Kay WW, Misra S (2007) Expression of a polyphemusin variant in transgenic tobacco confers resistance against plant pathogenic bacteria, fungi and a virus. Plant Cell Tissue Organ Cult 88:301–312

Bhattacharjee S, Zamora A, Azhar MT, Sacco MA, Lambert LH, Moffett P (2009) Virus resistance induced by NB-LRR proteins involves Argonaute4-dependent translational control. Plant J 58:940–951

Bilgin DD, Liu Y, Schiff M, Dinesh-Kumar SP (2003) P58(IPK), a plant ortholog of double-stranded RNA-dependent protein kinase PKR inhibitor, functions in viral pathogenesis. Dev Cell 4:651–661

Blanc S, Drucker M (2011) Functions of virus and host factors during vector-mediated transmission. In: Caranta C, Aranda MA, Tepfer M, López-Moya JJ (eds) Recent advances in plant virology. Caister Academic Press, Norfolk, pp 103–120

Boyko A, Kathiria P, Zemp FJ, Yao YL, Pogribny I, Kovalchuk I (2007) Transgenerational changes in the genome stability and methylation in pathogen-infected plants (virus-induced plant genome instability). Nucleic Acids Res 35:1714–1725

Brosnan CA, Mitter N, Christie M, Smith NA, Waterhouse PM, Carroll BJ (2007) Nuclear gene silencing directs reception of long-distance mRNA silencing in Arabidopsis. Proc Natl Acad Sci USA 104:14741–14746

Burgyan J (2011) Mechanism of action of viral suppresssors of RNA silencing. In: Caranta C, Aranda MA, Tepfer M, López-Moya JJ (eds) Recent advances in plant virology. Caister Academic Press, Norfolk, pp 103–120

Carr JP, Lewsey MG, Palukaitis P (2010) Signaling in induced resistance. Natural and engineered resistance to plant viruses, Pt II. In: Loebenstein G, Carr J (eds) Advances in virus research, vol 76. Academic Press, Amsterdam, pp 57–121

Carvalho MF, Turgeon R, Lazarowitz SG (2006) The geminivirus nuclear shuttle protein NSP inhibits the activity of AtNSI, a vascular-expressed Arabidopsis acetyltransferase regulated with the sink-to-source transition. Plant Physiol 140:1317–1330

Carvalho CM, Santos AA, Pires SR, Rocha CS, Saraiva DI, Machado JPB, Mattos EC, Fietto LG, Fontes EPB (2008) Regulated nuclear trafficking of rpL10A mediated by NIK1 represents a defense strategy of plant cells against virus. PLoS Pathog 4:12

Catinot J, Buchala A, Abou-Mansour E, Metraux JP (2008) Salicylic acid production in response to biotic and abiotic stress depends on isochorismate in *Nicotiana benthamiana*. FEBS Lett 582:473–478

Catoni M, Miozzi L, Fiorilli V, Lanfranco L, Accotto GP (2009) Comparative analysis of expression profiles in shoots and roots of tomato systemically infected by tomato spotted wilt virus reveals organ-specific transcriptional responses. Mol Plant–Microbe Interact 22:1504–1513

Champigny MJ, Cameron RK (2009). Action at a distance: long-distance signals in induced resistance. In: Plant Innate Immunity, Advances in Botanical Research 51:123–171

Chandra A, Anand A, Dubey A (2007) Effect of salicylic acid on morphological and biochemical attributes in cowpea. J Environ Biol 28:193–196

Chen YY, Lin YM, Chao TC, Wang JF, Liu AC, Ho FI, Cheng CP (2009) Virus-induced gene silencing reveals the involvement of ethylene-, salicylic acid- and mitogen-activated protein kinase-related defense pathways in the resistance of tomato to bacterial wilt. Physiol Plant 136:324–335

Cheng SF, Huang YP, Wu ZR, Hu CC, Hsu YH, Tsai CH (2010) Identification of differentially expressed genes induced by Bamboo mosaic virus infection in Nicotiana benthamiana by cDNA-amplified fragment length polymorphism. BMC Plant Biol 10:286

Christensen N, Tilsner J, Bell K, Hammann P, Parton R, Lacomme C, Oparka K (2009) The 5′ cap of Tobacco Mosaic Virus (TMV) is required for virion attachment to the actin/endoplasmic reticulum network during early infection. Traffic 10:536–551

Chung KM, Igari K, Uchida N, Tasaka M (2008) New perspectives on plant defense responses through modulation of developmental pathways. Mol Cells 26:107–112

Cournoyer P, Dinesh-Kumar SP (2011) NB-LRR immune receptors in plant virus defence. In: Caranta C, Aranda MA, Tepfer M, López-Moya JJ (eds) Recent advances in plant virology. Caister Academic Press, Norfolk, pp 149–176

Culver JN, Padmanabhan MS (2007) Virus-induced disease: altering host physiology one interaction at a time. Annu Rev Phytopathol 45:221–243

Daxinger L, Whitelaw E (2010) Transgenerational epigenetic inheritance: more questions than answers. Genome Res 20:1623–1628

de Vos M, Jander G (2010) Volatile communication in plant–aphid interactions. Curr Opin Plant Biol 13:366–371

Diaz-Vivancos P, Clemente-Moreno MJ, Rubio M, Olmos E, Garcia JA, Martinez-Gomez P, Hernandez JA (2008) Alteration in the chloroplastic metabolism leads to ROS accumulation in pea plants in response to plum pox virus. J Exp Bot 59:2147–2160

Dordas C (2008) Role of nutrients in controlling plant diseases in sustainable agriculture. A review. Agron Sustain Dev 28:33–46

Drutskaya MS, Belousov PV, Nedospasov SA (2011) Innate mechanisms of viral recognition. Mol Biol 45:5–15

Espinoza C, Medina C, Somerville S, Arce-Johnson P (2007) Senescence-associated genes induced during compatible viral interactions with grapevine and Arabidopsis. J Exp Bot 58:3197–3212

Eybishtz A, Peretz Y, Sade D, Akad F, Czosnek H (2009) Silencing of a single gene in tomato plants resistant to Tomato yellow leaf curl virus renders them susceptible to the virus. Plant Mol Biol 71:157–171

Faulkner C, Maule A (2011) Opportunities and successes in the search for plasmodesmal proteins. Protoplasma 248:27–38

Fernandéz-Calvino L, Donaire L, Llave C (2011a) RNA silencing and the interplay between plants and viruses. In: Caranta C, Aranda MA, Tepfer M, López-Moya JJ (eds) Recent advances in plant virology. Caister Academic Press, Norfolk, pp 121–136

Fernandéz-Calvino L, Faulkner C, Maule A (2011b) Plasmodesmata as active conduits for virus cell-to-cell movement. In: Aranda MA, Tepfer M, López-Moya JJ, Caranta C (eds) Recent advances in plant virology. Caister Academic Press, Norfolk, pp 47–74

Freitas-Astua J, Bastianel M, Locali-Fabris EC, Novelli VM, Silva-Pinhati AC, Basilio-Palmieri AC, Targon M, Machado MA (2007) Differentially expressed stress-related genes in the compatible citrus-Citrus leprosis virus interaction. Genet Mol Biol 30:980–990

Garcia-Marcos A, Pacheco R, Martianez J, Gonzalez-Jara P, Diaz-Ruiz JR, Tenllado F (2009) Transcriptional changes and oxidative stress associated with the synergistic interaction between Potato virus X and Potato virus Y and their relationship with symptom expression. Mol Plant–Microbe Interact 22:1431–1444

Geri C, Cecchini E, Giannakou ME, Covey SN, Milner JJ (1999) Altered patterns of gene expression in Arabidopsis elicited by Cauliflower mosaic virus (CaMV) infection and by a CaMV gene VI transgene. Mol Plant–Microbe Interact 12:377–384

Gibbs A (1980) A plant virus that partially protects its wild legume host against herbivores. Intervirology 13:42–47

Giedroc DP, Cornish PV (2009) Frameshifting RNA pseudoknots: structure and mechanism. Virus Res 139(2):193–208. doi:10.1016/j.virusres.2008.06.008

Hammerschmidt R (2009). Systemic acquired resistance. In: Plant Innate Immunity, Advances in Botanical Research 51:173–222

Harrell L, Melcher U, Atkins JF (2002) Predominance of six different hexanucleotide recoding signals 3′ of read-through stop codons. Nucleic Acids Res 30:2011–2017

Harries P, Ding BA (2011) Cellular factors in plant virus movement: at the leading edge of macromolecular trafficking in plants. Virology 411:237–243

Heil M (2001) Plant communication. In: Encyclopedia of life sciences. Wiley, Chichester

Holness CL, Lomonossoff GP, Evans D, Maule AJ (1989) Identification of the initiation codons for translation of cowpea mosaic-virus middle component RNA using site-directed mutagenesis of an infectious cDNA clone. Virology 172:311–320

Hwang YT, McCartney AW, Gidda SK, Mullen RT (2008) Localization of the Carnation Italian ringspot virus replication protein p36 to the mitochondrial outer membrane is mediated by an internal targeting signal and the TOM complex. BMC Cell Biol 9:54

Hyun TK, Uddin MN, Rim Y, Kim JY (2011) Cell-to-cell trafficking of RNA and RNA silencing through plasmodesmata. Protoplasma 248:101–116

Irian S, Xu P, Dai XB, Zhao PX, Roossinck MJ (2007) Regulation of a virus-induced lethal disease in tomato revealed by LongSAGE analysis. Mol Plant–Microbe Interact 20:1477–1488

Jeong RD, Chandra-Shekara AC, Kachroo A, Klessig DF, Kachroo P (2008) HRT-mediated hypersensitive response and resistance to Turnip crinkle virus in Arabidopsis does not require the function of TIP, the presumed guardee protein. Mol Plant–Microbe Interact 21:1316–1324

Jeong RD, Kachroo A, Kachroo P (2010) Blue light photoreceptors are required for the stability and function of a resistance protein mediating viral defense in Arabidopsis. Plant Signal Behav 5:1504–1509

Jimenez-Martinez ES, Bosque-Perez NA, Berger PH, Zemetra R, Ding HJ, Eigenbrode SD (2004) Volatile cues influence the response of Rhopalosiphum padi (Homoptera: Aphididae) to Barley yellow dwarf virus-infected transgenic and untransformed wheat. Environ Entomol 33:1207–1216

Joshi CP, Zhou H, Huang X, Chiang VL (1997) Context sequences of translation initiation codon in plants. Plant Mol Biol 35:993–1001

Kalantidis K, Schumacher HT, Alexiadis T, Helm JM (2008) RNA silencing movement in plants. Biol Cell 100:13–26

Kang HG, Kuhl JC, Kachroo P, Klessig DF (2008) CRT1, an Arabidopsis ATPase that interacts with diverse resistance proteins and modulates disease resistance to turnip crinkle virus. Cell Host Microbe 3:48–57

Karpova OV, Rodionova NP, Ivanov KI, Kozlovsky SV, Dorokhov YL, Atabekov JG (1999) Phosphorylation of Tobacco mosaic virus movement protein abolishes its translation repressing ability. Virology 261:20–24

Kathiria P, Sidler C, Golubov A, Kalischuk M, Kawchuk LM, Kovalchuk I (2010) Tobacco Mosaic Virus infection results in an increase in recombination frequency and resistance to viral, bacterial, and fungal pathogens in the progeny of infected tobacco plants. Plant Physiol 153:1859–1870

Kehr J, Buhtz A (2008) Long distance transport and movement of RNA through the phloem. J Exp Bot 59:85–92

Kim YS, Sano H (2008) Pathogen resistance of transgenic tobacco plants producing caffeine. Phytochemistry 69:882–888

Kiraly L, Barnaz B, Kiralyz Z (2007) Plant resistance to pathogen infection: Forms and mechanisms of innate and acquired resistance. J Phytopathol 155:385–396

Kittelmann K, Rau P, Gronenborn B, Jeske H (2009) Plant geminivirus Rep protein induces rereplication in fission yeast. J Virol 83:6769–6778

Kong LJ, Orozco BM, Roe JL, Nagar S, Ou S, Feiler HS, Durfee T, Miller AB, Gruissem W, Robertson D, Hanley-Bowdoin L (2000) A geminivirus replication protein interacts with the retinoblastoma protein through a novel domain to determine symptoms and tissue specificity of infection in plants. EMBO J 19:3485–3495

Kovac M, Muller A, Jarh DM, Milavec M, Duchting P, Ravnikar M (2009) Multiple hormone analysis indicates involvement of jasmonate signalling in the early defence of potato to potato virus Y-NTN. Biol Plant 53:195–199

Lai JB, Chen H, Teng KL, Zhao QZ, Zhang ZH, Li Y, Liang LM, Xia R, Wu YR, Guo HS, Xie Q (2009) RKP, a RING finger E3 ligase induced by BSCTV C4 protein, affects geminivirus infection by regulation of the plant cell cycle. Plant J 57:905–917

Lang QL, Jin CZ, Lai LY, Feng JL, Chen SN, Chen JS (2011) Tobacco microRNAs prediction and their expression infected with Cucumber mosaic virus and Potato virus X. Mol Biol Rep 38:1523–1531

Le Gall O, Aranda MA, Caranta C (2011) Plant resistance to viruses mediated by translation initiation factors. In: Caranta C, Aranda MA, Tepfer M, López-Moya JJ (eds) Recent advances in plant virology. Caister Academic Press, Norfolk, pp 177–194

Lee JY, Taoka K, Yoo BC, Ben-Nissan G, Kim DJ, Lucas WJ (2005) Plasmodesmal-associated protein kinase in tobacco and Arabidopsis recognizes a subset of non-cell-autonomous proteins. Plant Cell 17:2817–2831

Li YM, Zhang ZK, Jia YT, Shen YM, He HM, Fang RX, Chen XY, Hao XJ (2008) 3-Acetonyl-3-hydroxyoxindole: A new inducer of systemic acquired resistance in plants. Plant Biotechnol J 6:301–308

Lim HS, Vaira AM, Bae H, Bragg JN, Ruzin SE, Bauchan GR, Dienelt MM, Owens RA, Hammond J (2010) Mutation of a chloroplast-targeting signal in Alternanthera mosaic virus TGB3 impairs cell-to-cell movement and eliminates long-distance virus movement. J Gen Virol 91:2102–2115

Liu J, Coaker G (2008) Nuclear trafficking during plant innate immunity. Mol Plant 1:411–422

Lu H, Zhao XM, Wang WX, Yin H, Xu JG, Bai XF, Du YG (2010) Inhibition effect on Tobacco mosaic virus and regulation effect on calreticulin of oligochitosan in tobacco by induced Ca^{2+} influx. Carbohydr Polym 82:136–142

Lucas WJ, Ham LK, Kim JY (2009) Plasmodesmata—bridging the gap between neighboring plant cells. Trends Cell Biol 19:495–503

Lukaszewicz M, Feuermann M, Jerouville B, Stas A, Boutry M (2000) In vivo evaluation of the context sequence of the translation initiation codon in plants. Plant Sci 154:89–98

Malmstrom CM, Melcher U, Bosque-Pérez NA (2011) The expanding field of plant virus ecology: historical foundations, knowledge gaps, and research directions. Virus Res 159:84
Mandal B, Mandal S, Csinos AS, Martinez N, Culbreath AK, Pappu HR (2008) Biological and molecular analyses of the acibenzolar-S-methyl-induced systemic acquired resistance in flue-cured tobacco against Tomato spotted wilt virus. Phytopathology 98:196–204
Martinez-Gil L, Johnson AE, Mingarro I (2010) Membrane insertion and biogenesis of the Turnip Crinkle Virus p9 movement protein. J Virol 84:5520–5527
Martinez-Priego L, Donaire L, Barajas D, Llave C (2008) Silencing suppressor activity of the Tobacco rattle virus-encoded 16-kDa protein and interference with endogenous small RNA-guided regulatory pathways. Virology 376:346–356
Mauck KE, De Moraes CM, Mescher MC (2010) Deceptive chemical signals induced by a plant virus attract insect vectors to inferior hosts. Proc Natl Acad Sci USA 107:3600–3605
Mazarei M, Teplova I, Hajimorad MR, Stewart CN (2008) Pathogen phytosensing: plants to report plant pathogens. Sensors 8:2628–2641
Melcher U (2000) The 30 K superfamily of viral movement proteins. J Gen Virol 81:257–266
Meng BZ, Li CH (2010) The capsid protein of Grapevine rupestris stem pitting-associated virus contains a typical nuclear localization signal and targets to the nucleus. Virus Res 153:212–217
Miller WA, Kraft J, Wang Z, Fan Q (2011) Roles of cis-acting elements in translation of viral RNAs. In: Aranda MA, Tepfer M, López-Moya JJ, Caranta C (eds) Recent advances in plant virology. Caister Academic Press, Norfolk, pp 1–17
Mills-Lujan K, Deom CM (2010) Geminivirus C4 protein alters Arabidopsis development. Protoplasma 239:95–110
Mitsuya Y, Takahashi Y, Berberich T, Miyazaki A, Matsumura H, Takahashi H, Terauchi R, Kusano T (2009) Spermine signaling plays a significant role in the defense response of Arabidopsis thaliana to cucumber mosaic virus. J Plant Physiol 166:626–643
Mlotshwa S, Pruss GJ, Vance V (2008) Small RNAs in viral infection and host defense. Trends Plant Sci 13:375–382
Molnar A, Melnyk CW, Bassett A, Hardcastle TJ, Dunn R, Baulcombe DC (2010) Small silencing RNAs in plants are mobile and direct epigenetic modification in recipient cells. Science 328:872–875
Murphy AM, Otto B, Brearley CA, Carr JP, Hanke DE (2008) A role for inositol hexakisphosphate in the maintenance of basal resistance to plant pathogens. Plant J 56:638–652
Nagy PD, Pogany J (2011) Replication of plant RNA viruses. In: Caranta C, Aranda MA, Tepfer M, López-Moya JJ (eds) Recent advances in plant virology. Caister Academic Press, Norfolk, pp 19–46
Ngumbi E, Eigenbrode SD, Bosque-Perez NA, Ding H, Rodriguez A (2007) Myzus persicae is arrested more by blends than by individual compounds elevated in headspace of PLRV-Infected potato. J Chem Ecol 33:1733–1747
Niu SN, Wang BS, Guo XF, Yu JL, Wang XB, Xu K, Zhai YF, Wang JH, Liu ZX (2009) Identification of two RNA silencing suppressors from Banana bunchy top virus. Arch Virol 154:1775–1783
Olspert A, Paves H, Toomela R, Tamm T, Truve E (2010) Cocksfoot mottle sobemovirus coat protein contains two nuclear localization signals. Virus Genes 40:423–431
Padmanabhan MS, Gorepoker SP, Golem S, Shiferaw H, Culver JN (2005) Interaction of the tobacco mosaic virus replicase protein with the Aux/IAA protein PAPI/IAA26 is associated with disease development. J Virol 79:2549–2558
Pallás V, Genovés A, Sánchez-Pina MA, Navarro JA (2011) Systemic movement of viruses via the plant phloem. In: Caranta C, Aranda MA, Tepfer M, López-Moya JJ (eds) Recent advances in plant virology. Caister Academic Press, Norfolk, pp 75–102
Palukaitis P, Carr JP (2008) Plant resistance responses to viruses. J Plant Pathol 90:153–171
Park SW, Kaimoyo E, Kumar D, Mosher S, Klessig DF (2007) Methyl salicylate is a critical mobile signal for plant systemic acquired resistance. Science 318:113–116
Park J, Hwang HS, Buckley KJ, Park JB, Auh CK, Kim DG, Lee S, Davis KR (2010) C4 protein of Beet severe curly top virus is a pathomorphogenetic factor in Arabidopsis. Plant Cell Rep 29:1377–1389

Pazhouhandeh M, Dieterle M, Marrocco K, Lechner E, Berry B, Brault V, Hemmer O, Kretsch T, Richards KE, Genschik P, Ziegler-Graff V (2006) F-box-like domain in the polerovirus protein P0 is required for silencing suppressor function. Proc Natl Acad Sci USA 103:1994–1999

Piroux N, Saunders K, Page A, Stanley J (2007) Geminivirus pathogenicity protein C4 interacts with Arabidopsis thaliana shaggy-related protein kinase AtSK eta, a component of the brassinosteroid signalling pathway. Virology 362:428–440

Quecini V, Lopez ML, Pacheco FTH, Ongarelli MDG (2007) Tomato spotted wilt virus triggers specific and shared defense mechanisms in hypersensitive and susceptible Solanaceae hosts. Physiol Mol Plant Pathol 70:189–197

Quilis J, Penas G, Messeguer J, Brugidou C, Segundo BS (2008) The Arabidopsis AtNPR1 inversely modulates defense responses against fungal, bacterial, or viral pathogens while conferring hypersensitivity to abiotic stresses in transgenic rice. Mol Plant–Microbe Interact 21:1215–1231

Reichel C, Beachy RN (2000) Degradation of Tobacco mosaic virus movement protein by the 26S proteasome. J Virol 74:3330–3337

Ren T, Qu F, Morris TJ (2000) HRT gene function requires interaction between a NAC protein and viral capsid protein to confer resistance to turnip crinkle virus. Plant Cell 12:1917–1925

Ren B, Guo YY, Gao F, Zhou P, Wu F, Meng Z, Wei CH, Li Y (2010) Multiple functions of Rice dwarf phytoreovirus Pns10 in suppressing systemic RNA silencing. J Virol 84:12914–12923

Roberts K, Love AJ, Laval V, Laird J, Tomos AD, Hooks MA, Milner JJ (2007) Long-distance movement of Cauliflower mosaic virus and host defence responses in Arabidopsis follow a predictable pattern that is determined by the leaf orthostichy. New Phytologist 175:707–717

Robert-Seilaniantz A, Navarro L, Bari R, Jones JD (2007) Pathological hormone imbalances. Curr Opin Plant Biol 10:372–379

Roy P (2008) Bluetongue virus: dissection of the polymerase complex. J Gen Virol 89:1789–1804

Rubino L, Vernile P, Russo M (2008) Analysis of sequences in the Cymbidium ringspot virus replicase protein p33 that promote targeting and anchoring to endoplasmic reticulum membranes. J Plant Pathol 90:575–580

Ruiz-Medrano R, Moya JH, Xoconostle-Cazares B, Lucas WJ (2007) Influence of Cucumber mosaic virus infection on the mRNA population present in the phloem translocation stream of pumpkin plants. Funct Plant Biol 34:292–301

Salanki K, Gellert A, Naray-Szabo G, Balazs E (2007) Modeling-based characterization of the elicitor function of amino acid 461 of Cucumber mosaic virus 1a protein in the hypersensitive response. Virology 358:109–118

Santos AA, Lopes KVG, Apfata JAC, Fontes EPB (2010) NSP-interacting kinase, NIK: a transducer of plant defence signalling. J Exp Bot 61:3839–3845

Sarowar S, Kim YJ, Kim KD, Hwang BK, Ok SH, Shin JS (2009) Overexpression of lipid transfer protein (LTP) genes enhances resistance to plant pathogens and LTP functions in long-distance systemic signaling in tobacco. Plant Cell Rep 28:419–427

Sasaki N, Ogata T, Deguchi M, Nagai S, Tamai A, Meshi T, Kawakami S, Watanabe Y, Matsushita Y, Nyunoya H (2009) Over-expression of putative transcriptional coactivator KELP interferes with Tomato mosaic virus cell-to-cell movement. Mol Plant Pathol 10:161–173

Shang J, Xi DH, Xu F, Wang SD, Cao S, Xu MY, Zhao PP, Wang JH, Jia SD, Zhang ZW, Yuan S, Lin HH (2011) A broad-spectrum, efficient and nontransgenic approach to control plant viruses by application of salicylic acid and jasmonic acid. Planta 233:299–308

Shin H-I, Kim HY, Cho T-J (2010) The Pro/Hel region is indispensable for packaging non-replicating Turnip yellow mosaic virus RNA, but not replicating viral RNA. Mol Cells 29:463–469

Shung CY, Sunter G (2009) Regulation of Tomato golden mosaic virus AL2 and AL3 gene expression by a conserved upstream open reading frame. Virology 383:310–318

Siddiqui SA, Sarmiento C, Kiisma M, Koivumaki S, Lemmetty A, Truve E, Lehto K (2008) Effects of viral silencing suppressors on Tobacco ringspot virus infection in two Nicotiana species. J Gen Virol 89:1502–1508

Takabatake R, Ando Y, Seo S, Katou S, Tsuda S, Ohashi Y, Mitsuhara I (2007) MAP kinases function downstream of HSP90 and upstream of mitochondria in TMV resistance gene N-mediated hypersensitive cell death. Plant Cell Physiol 48:498–510

Tameling WIL, Joosten M (2007) The diverse roles of NB-LRR proteins in plants. Physiol Mol Plant Pathol 71:126–134

Teng KL, Chen H, Lai JB, Zhang ZH, Fang YY, Xia R, Zhou XP, Guo HS, Xie Q (2010) Involvement of C4 protein of Beet severe curly top virus (family Geminiviridae) in virus movement. PLoS One 5:e11280

Tilsner J, Cowan GH, Roberts AG, Chapman SN, Ziegler A, Savenkov E, Torrance L (2010) Plasmodesmal targeting and intercellular movement of Potato mop-top pomovirus is mediated by a membrane anchored tyrosine-based motif on the lumenal side of the endoplasmic reticulum and the C-terminal transmembrane domain in the TGB3 movement protein. Virology 402:41–51

Vannini C, Iriti M, Bracale M, Locatelli F, Faoro F, Croce P, Pirona R, Di Maro A, Coraggio I, Genga A (2006) The ectopic expression of the rice Osmyb4 gene in Arabidopsis increases tolerance to abiotic, environmental and biotic stresses. Physiol Mol Plant Pathol 69:26–42

Vasyukova NI, Ozeretskovskaya OL (2007) Induced plant resistance and salicylic acid: a review. Appl Biochem Microbiol 43:367–373

Venugopal SC, Jeong RD, Mandal MK, Zhu SF, Chandra-Shekara AC, Xia Y, Hersh M, Stromberg AJ, Navarre D, Kachroo A, Kachroo P (2009) Enhanced disease susceptibility 1 and salicylic acid act redundantly to regulate resistance gene-mediated signaling. PLoS Genet 5:e1000545

Verchot-Lubicz J, Torrance L, Solovyev AG, Morozov SY, Jackson AO, Gilmer D (2010) Varied movement strategies employed by triple gene block-encoding viruses. Mol Plant–Microbe Interact 23:1231–1247

Vlot AC, Dempsey DA, Klessig DF (2009) Salicylic acid, a multifaceted hormone to combat disease. Annu Rev Phytopathol 47:177–206

Vogler H, Kwon MO, Dang V, Sambade A, Fasler M, Ashby J, Heinlein M (2008) Tobacco mosaic virus movement protein enhances the spread of RNA silencing. PLoS Pathog 4: e1000038

Wakeman CA, Winkler WC, Dann CE 3rd (2007) Structural features of metabolite-sensing riboswitches. Trends Biochem Sci 32:415–424

Wang MB, Bian XY, Wu LM, Liu LX, Smith NA, Isenegger D, Wu RM, Masuta C, Vance VB, Watson JM, Rezaian A, Dennis ES, Waterhouse PM (2004) On the role of RNA silencing in the pathogenicity and evolution of viroids and viral satellites. Proc Natl Acad Sci USA 101: 3275–3280

Whitham SA, Quan S, Chang HS, Cooper B, Estes B, Zhu T, Wang X, Hou YM (2003) Diverse RNA viruses elicit the expression of common sets of genes in susceptible *Arabidopsis thaliana* plants. Plant J 33:271–283

Whitham SA, Yang CL, Goodin MM (2006) Global impact: elucidating plant responses to viral infection. Mol Plant–Microbe Interact 19:1207–1215

Witzany G (2008) Bio-communication of bacteria and its evolutionary interrelations to natural genome editing competences of viruses. Open Evol J 2:44–54

Wright KM, Cowan GH, Lukhovitskaya NI, Tilsner J, Roberts AG, Savenkov EI, Torrance L (2010) The N-terminal domain of PMTV TGB1 movement protein is required for nucleolar localization, microtubule association, and long-distance movement. Mol Plant–Microbe Interact 23:1486–1497

Wu JS, Wang L, Baldwin IT (2008) Methyl jasmonate-elicited herbivore resistance: Does MeJA function as a signal without being hydrolyzed to JA? Planta 227:1161–1168

Xia ZL, Wang Y, Du ZQ, Li JM, Zhao RY, Wang DW (2008) A potential nuclear envelope-targeting domain and an arginine-rich RNA binding element identified in the putative movement protein of the GAV strain of Barley yellow dwarf virus. Funct Plant Biol 35:40–50

Yaegashi H, Takahashi T, Isogai M, Kobori T, Ohki S, Yoshikawa N (2007) Apple chlorotic leaf spot virus 50 kDa movement protein acts as a suppressor of systemic silencing without interfering with local silencing in *Nicotiana benthamiana*. J Gen Virol 88:316–324

Yaegashi H, Tamura A, Isogai M, Yoshikawa N (2008) Inhibition of long-distance movement of RNA silencing signals in *Nicotiona benthamiana* by Apple chlorotic leaf spot virus 50 kDa movement protein. Virology 382:199–206

Yamakawa H, Kamada H, Satoh M, Ohashi Y (1998) Spermine is a salicylate-independent endogenous inducer for both tobacco acidic pathogenesis-related proteins and resistance against tobacco mosaic virus infection. Plant Physiol 118:1213–1222

Yang H, Wang SD, Xi DH, Yuan S, Wang JH, Xu MY, Lin HH (2010) Interaction between Cucumber mosaic virus and Turnip crinkle virus in *Arabidopsis thaliana*. J Phytopathol 158:833–836

Yang H, Huang YP, Zhi HJ, Yu DY (2011) Proteomics-based analysis of novel genes involved in response toward Soybean mosaic virus infection. Mol Biol Rep 38:511–521

Yoda H, Yamaguchi Y, Sano H (2003) Induction of hypersensitive cell death by peroxide produced through polyamine degradation in tobacco plants. Plant Physiol 132:1973–1981

Zavaliev R, Ueki S, Epel BL, Citovsky V (2011) Biology of callose (beta-1,3-glucan) turnover at plasmodesmata. Protoplasma 248:117–130

Zhang BH, Wang QL, Pan XP (2007) MicroRNAs and their regulatory roles in animals and plants. J Cell Physiol 210:279–289

Zhou P, Ren B, Zhang XM, Wang Y, Wei CH, Li Y (2010) Stable expression of Rice dwarf virus Pns10 suppresses the post-transcriptional gene silencing in transgenic *Nicotiana benthamiana* plants. Acta Virol 54:99–104

Zhu SF, Gao F, Cao XS, Chen M, Ye GY, Wei CH, Li Y (2005) The Rice dwarf virus P2 protein interacts with ent-kaurene oxidases in vivo, leading to reduced biosynthesis of gibberellins and rice dwarf symptoms. Plant Physiol 139:1935–1945

Molecular Signals and Receptors: Communication Between Nitrogen-Fixing Bacteria and Their Plant Hosts

Ann M. Hirsch and Nancy A. Fujishige

Abstract Our understanding of the extent of communication taking place between the plant and its underground microbiome (rhizosphere microbes) as well as with other soil organisms has grown exponentially in the last decade. Much of this information has been obtained from studies of nitrogen-fixing organisms, particularly members of the family *Rhizobiaceae* (Alphaproteobacteria) that establish nodules on legume roots in which atmospheric nitrogen is converted to plant-utilizable forms. Signals exchanged among organisms in the rhizosphere via quorum sensing (QS) and the responses to these signals have been identified, but it is unclear how they influence the downstream stages of nodulation and nitrogen fixation. An exchange of signal molecules ensures that a high level of specificity takes place to optimize the nitrogen-fixing interaction between host legume and symbiont. Chitin-related molecules appear to be the microbial currency for communication between the symbiotic partners in both mutualistic and pathogenic interactions. Exceptions to the paradigms based on the legume-*Rhizobium* interaction, including the discovery of Betaproteobacteria (now called beta-rhizobia) that nodulate and fix nitrogen with legumes and the lack of nodulation (*nod*) genes in certain alpha-rhizobia, particularly those that nodulate *Aeschynomene* and *Arachis*, bring into question the universality of some of the previous models. Moreover, new frontiers have opened that examine the coordination of information exchange that is needed for the induction and maintenance of nitrogen fixation and for bacteroid differentiation. Nevertheless, nitrogen-fixing organisms are just one small part of a highly interactive rhizosphere community. The challenge of the

A.M. Hirsch (✉)
Department of Molecular, Cell and Developmental Biology, University of California-Los Angeles, Los Angeles, CA, USA

Molecular Biology Institute, University of California-Los Angeles, Los Angeles, CA, USA
e-mail: ahirsch@ucla.edu

N.A. Fujishige
Department of Molecular, Cell and Developmental Biology, University of California-Los Angeles, Los Angeles, CA, USA

G. Witzany and F. Baluška (eds.), *Biocommunication of Plants*,
Signaling and Communication in Plants 14,
DOI 10.1007/978-3-642-23524-5_14, © Springer-Verlag Berlin Heidelberg 2012

next decade will be to understand in greater depth the community dynamics that occur in soil, one of our planet's most precious yet limited resources, in the hopes of maintaining the key signal webs that are critical not only for the promotion of agriculture but also for the preservation of the environment overall.

1 Introduction

When we reviewed this subject in 2003 (Hirsch et al. 2003), we presented a broad overview of the molecular interactions between several diverse rhizosphere organisms and their plant hosts. We described the molecular signals and receptors, where known, for mutualists—for example, members of the *Rhizobiaceae* and their legume hosts; for plant pathogens such as *Phytophthora sojae*; and for organisms that parasitize a number of plants such as nematodes as well as the plant parasites, *Striga* and *Orobanche*. Since that time, the amount of information concerning communication among organisms in the rhizosphere has increased greatly. Several recent reviews have dealt with this topic (Bouwmeester et al. 2007; Badri et al. 2009; Ortíz-Castro et al. 2009; Lanou et al. 2010; and others). In this chapter, we restrict our discussion to nitrogen-fixing microbes and expand upon the knowledge accrued in the last decade.

Besides the well-known *Rhizobiaceae* (Alphaproteobacteria), several bacteria that are Betaproteobacteria have been identified as capable of establishing nitrogen-fixing nodules on legumes, including bacteria in the genus *Cupriavidus* and the genus *Burkholderia* (Chen et al. 2003a, b, 2006, 2007, 2008; Moulin et al. 2001, 2002). These beta-rhizobia not only nodulate legumes but also possess genes similar to those employed by the Alphaproteobacteria to nodulate legumes via the root hair nodulation pathway (Fig. 1). This is one of three ways by which rhizobial bacteria enter the host root (Sprent 2007). We recently proposed that the plant-associated *Burkholderia* spp. be transferred to a new genus, *Cabelleronia*, due to their phylogenetic distinction based on a concatenate tree of four housekeeping genes and 16S RNA as well as the difference in G + C content from the mammalian and human-associated *Burkholderia* spp. (Estrada-de los Santos et al., ms. In prep.). In addition, the plant-associated species differ from the *Burkholderia* pathogens in lacking certain protein secretion systems and the ability to induce pathogenicity in various assays, including HeLa cells (A.A. Angus and A.M. Hirsch, unpublished results).

The interactions taking place between plant and microbe occur in soil, an environment composed of particles that are aggregates of inorganic and organic materials suspended in an aqueous medium. However, soil is not a homogenous mixture, nor are the various organisms that inhabit this environment equally distributed. Much of the soil fauna (earthworms, protozoans, nematodes, etc.) is motile and influences soil decomposition only in situ (Scheu 2001). Although numerous geochemical and biological reactions take place over time to modify bedrock into soil, few of the changes result from long-distance diffusion of biologically active molecules from the organisms that produce them. The vast majority of the biochemical processes that occur in soil are due to the neighboring bacteria and other organisms, which live close to or on

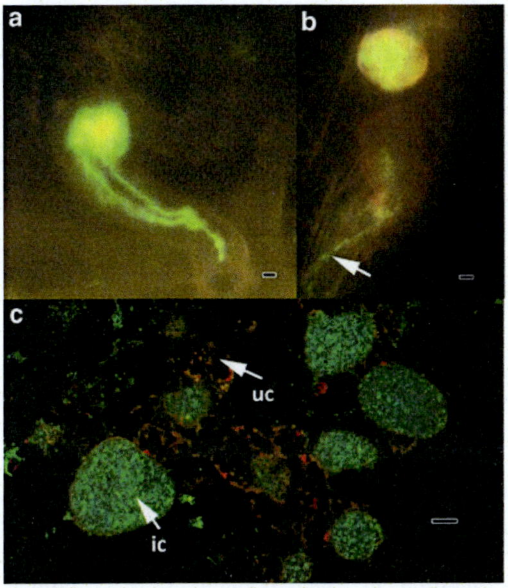

Fig. 1 Nodulating *B. tuberum* STM678 inoculated onto roots of siratro (*Macroptilium atropurpureum*). (**a** and **b**) Epifluorescent micrographs of infection threads harboring *green* fluorescent protein-labeled *Burkholderia tuberum* STM678 (*arrow*) within the root hairs of two separate siratro plants. Bar, 1.5 μm. (**c**) Confocal microscopy of siratro nodule cells containing *B. tuberum* STM678 cells labeled with the LIVE-DEAD BacLight Bacterial Viability kit (Molecular Probes). ic, infected cell; uc, uninfected cell. Bar, 10 μm. (Courtesy of M.R. Lum)

plant roots. The rhizosphere, the 1–3-mm region adjacent to the outside surface of the root, is the most significant region for exchanges of molecular signals. However, some interactions extend beyond the root, in the region called the exorhizosphere, whereas others take place in the root's intercellular or apoplastic spaces (endorhizosphere) just outside of the single-layered endodermis (Foster et al. 1983).

Earlier, we described the interchange of signals at two different stages of the interaction: (1) before the encounter between the microbe and the plant and (2) once recognition has occurred. We will utilize this same temporal/spatial division, updating what has been learned since 2003. In addition, we include data on the signaling that takes place after the symbiosis has been established. This topic is rarely discussed in reviews dealing with plant-microbe communication.

2 Chemical Signaling Before the Encounter

Microbes frequently cluster on root surfaces in biofilms where they metabolize the exudates secreted by root cells and perform a number of chemical reactions that alter the physical properties of soil. Many factors are known to be important for bacterial attachment including pili and flagella, exopolysaccharides, lipopolysaccharides, cellulose fibrils, and a number of proteins (see Hirsch et al. 2009; Downie 2010). Once attached to the root surface, the microbes signal to one another to elicit behavioral changes, some of which result in a closer association with the plant root itself. For example, plant pathogens may secrete virulence factors only when a certain cell density is achieved on the plant (von Bodman et al. 2003). This type of organism-to-organism signaling is known as quorum sensing.

2.1 *Intermicrobial Signaling: Quorum Sensing*

Although bacteria are unicellular organisms, their ability to coordinate their behavior and function as a group allows them to inhabit their ecological niche successfully. A successful outcome requires intercellular communication within the bacterial population (Lazdunski et al. 2004). Quorum sensing (QS) is a signaling mechanism that enables bacteria to assess the population density and react coordinately by synchronizing the expression of specific genes throughout the population. In QS systems, a low-molecular-weight signal molecule, known as autoinducer, is produced and secreted into the cell's environment. As the population grows, the autoinducer accumulates in the local environment. When a threshold concentration is achieved, the autoinducer activates a transcriptional regulator, which controls a set of target genes. This activation occurs in cells throughout the local environment, and ultimately, the whole population acts in a concerted manner (Waters and Bassler 2005; Wisniewski-Dye and Downie 2002; Teplitski et al. 2011).

In Gram-negative bacteria, the most widespread autoinducer is N-acyl-homoserine lactone (AHL). Different species produce specific AHLs that vary in the length of their N-linked side chains, the degree of saturation within that side chain, and the type of substitutions at the 3-carbon position (Brelles-Mariño and Bedmar 2001). The AHL QS systems consist of two major components: (1) the AHL synthase enzyme (a LuxI homologue) that catalyzes the formation of an amide bond between S-adenosyl methionine (SAM) and an acyl-acyl carrier protein (acyl ACP) and (2) the AHL receptor protein (a LuxR homologue) that regulates the transcription of target genes. The N-terminal domain of LuxR specifically binds to its cognate AHL. Binding to AHL is thought to induce a structural change that leads to protein oligomerization and unmasks the C-terminal helix-turn-helix DNA binding domain. The LuxR-AHL complex binds to palindromic sequences known as "*lux* boxes," located in the promoter regions of quorum sensing–regulated genes. LuxR then recruits RNA polymerase, thus activating transcription of the targeted genes (Egland and Greenberg 1999).

Plant-associated bacteria commonly produce AHL QS signals. Indeed, AHLs are more prevalent in plant-associated bacteria than in bacteria living in the bulk soil (Elasri et al. 2001). AHL-based QS systems are found throughout the Alpha- and Betaproteobacteria. These QS systems are remarkably diverse, in terms of both the types of AHLs produced and the processes that they control. Table 1 summarizes the quorum sensing systems found in both the alpha-rhizobia (*Rhizobiaceae* sensu stricto)

Table 1 Quorum sensing in α- and β-rhizobia

Organism	Genes	Signal	Phenotypes	References
α-rhizobia				
Agrobacterium tumefaciens	*traR*/*traI* (pTi)	3-oxo-C8-HSL	Plasmid transfer	Piper et al. (1993) and Hwang et al. (1995)
Bradyrhizobium japonicum Strain USDA110	Unknown	Bradyoxetin	*nod* gene control	Loh et al. (2002), Loh and Stacey (2003) and Jitacksorn and Sadowsky (2008)
Bradyrhizobium japonicum and *Bradyrhizobium elkanii* (multiple strains)	Unknown	Various AHLs (as detected by a biosensor)	Unknown	Pongsilp et al. (2005)
Bradyrhizobium sp. Strains ORS278 and BTAi1	*braR*/*braI* (chromosome)	Cinnamoyl-HSL	Unknown	Ahlgren et al. (2011)
Mesorhizobium huakuii Strain 93	Unknown	Uncharacterized, non-AHL molecule	Growth rate, root hair attachment, nodulation efficiency	Gao et al. (2006)
Mesorhizobium loti				
Strain R7A	*traR*/*traI1*/*traI2* (chromosome)	3-oxo-C6-HSL	Symbiotic island transfer	Ramsey et al. (2006, 2009)
Strain NZP2213	*mrlI1*/*mrlI2*	C12-HSL, 3-oxo-C6-HSL, C8-HSL, C10-HSL	Nodulation efficiency	Yang et al. (2009)
Mesorhizobium tianshanense				
Strain CCBAU 3306	*mrtR*/*mrtI*	Unknown	Nodulation efficiency, root hair attachment	Zheng et al. (2006)
Strain CCBAU 060A	*mtqR*/*mtqI*/*mtqS*	Unknown	Nodulation efficiency, growth rate	Cao et al. (2009)

(continued)

Table 1 (continued)

Organism	Genes	Signal	Phenotypes	References
Rhizobium etli				
Strain CNPAF512	*cinR/cinI* (chromosome)	3-OH-(slc)-HSL	Growth inhibition, nitrogen fixation, symbiosome development	Daniels et al. (2002)
	raiR/raiI (chromosome)	Short chain AHLs	Nitrogen fixation, growth inhibition	Rosemeyer et al. (1998) and Daniels et al. (2002)
Strain CFN42	*traR/traI* (p42a)	3-oxo-C8-HSL, 3-OH-C8-HSL	Plasmid transfer	Tun-Garrido et al. (2003)
Rhizobium leguminosarum bv. *viciae*	*cinR/cinI* (chromosome)	3-OH-C14:1-HSL	Growth inhibition	Lithgow et al. (2000)
	rhiR/rhiI (pRL1JI)	C6-HSL, C7-HSL, C8-HSL	Nodulation efficiency	Cubo et al. (1992) and Rodelas et al. (1999)
	traR/traI (pRL1JI)	2-oxo-C8-HSL, C8-HSL	Plasmid transfer	Wilkinson et al.(2002) and Danino et al. (2003)
	expR (chromosome)	Unknown	Unknown	Wisniewski-Dye and Downie (2002)
bv. *phaseoli*	*raiR/raiI* (plasmid)	3-OH-C8-HSL, C8-HSL	Unknown	Wisniewski-Dye et al. (2002)
Rhizobium sp.				
Strain NGR234	*traR/traI* (pNGR234a)	3-oxo-C8-HSL	Plasmid transfer	He et al. (2003)

Sinorhizobium meliloti				
Strain Rm1021	*sinR/sinI* (chromosome)	3-oxo-C14-HSL, C16:1-HSL	Succinoglycan synthesis, EPSII production, swarming	Marketon and González (2002), Marketon et al. (2002),Teplitski et al. (2003), Gao et al. (2005), Glenn et al. (2007), Bahlawane et al. (2008) and McIntosh et al. (2008)
	expR (chromosome)	C16:1-HSL	EPSII production, swarming, succinoglycan synthesis	Pellock et al. (2002), Marketon et al. (2003), Gao et al. (2005), Glenn et al. (2007), Bahlawane et al. (2008), Hoang et al. (2008) and McIntosh et al. (2008)
Strain Rm41	*traR/traI* (pRm41a)	3-oxo-C8-HSL	Plasmid transfer	Marketon and González (2002)
Strain RU10/406	*visN/visR* (chromosome)	Unknown	Motility (*fli, mot, fla, che* genes)	Sourjik et al. (2000)
β-rhizobia				
Burkholderia kururiensis	*braR/braI* (chromosome)	3-oxo-C12-HSL	EPS synthesis, endophytic rice colonization	Suarez-Moreno et al. (2008) and Suarez-Moreno et al. (2010)
Burkholderia unamae	*braR/braI* (chromosome)	3-oxo-C6-HSL, 3-oxo-C8-HSL, 3-oxo-C6-HSL, 3-oxo-C8-HSL	EPS synthesis, phenol degradation Negative regulation of biofilm	Suarez-Moreno et al. (2010)
Burkholderia xenovorans	*braR/braI* (chromosome)	3-oxo-C6-HSL; 3-oxo-C8-HSL	EPS synthesis	Suarez-Moreno et al. (2010)
	xenR2/xenI2	3-oxo-C10-HSL; 3-oxo-C12-HSL		
	bxeR	3-oxo-C14-HSL		

and some plant-associated, nitrogen-fixing Betaproteobacteria. For detailed descriptions of these QS systems, see reviews by Sanchez-Contreras et al. (2007) and Downie and González (2008). In the alpha-rhizobia, a single strain can contain as many as four different LuxI-type AHL synthases, their associated LuxR-type regulators, as well as some "orphan" LuxR-type regulators that lack an associated LuxI. Furthermore, each rhizobial strain has a different set of QS systems (Downie and González 2008).

Although the QS systems are not yet as thoroughly studied in the plant-associated, nitrogen-fixing *Burkholderia* species, a similar picture of diversity is emerging, which features multiple AHLs produced by a single species. However, in all plant-associated *Burkholderia* species studied so far, a highly conserved system, called BraI/R, has been found. This system produces an AHL (3-oxo-C12-HSL) that is unique to the plant-associated *Burkholderia* and distinct from the CepI/R AHLs produced by pathogenic *Burkholderia* species. The BraI/R system appears to be involved in exopolysaccharide biosynthesis, but mutating *braR* or *braI* has no effect on multiple phenotypes commonly regulated by QS, including nitrogen fixation, siderophore production, lipase activity, motility, swimming, and swarming. A second QS system as well as an "orphan" LuxR-type regulator was identified in some *Burkholderia* species. In these species, multiple AHLs have been isolated. Future work will focus on understanding the target genes controlled by these different QS systems (Suarez-Moreno et al. 2008, 2010).

In alpha-rhizobia, the diverse set of QS systems controls a wide array of physiological processes that improve survival in the rhizosphere, including growth inhibition, swarming, motility, biofilm formation, exopolysaccharide production, transfer of plasmids and symbiotic islands, and the ability to establish a symbiosis. However, the function of a given QS system often depends on the species. For instance, in *Rhizobium leguminosarum* biovar *viciae*, the *cinI/R* system is involved in plasmid transfer and growth, but mutation of *cinI* had no effect on symbiosis (Lithgow et al. 2000). However, in *R. etli*, mutation of the orthologous *cinI* or *cinR* genes resulted in a reduction in nitrogen fixation and the development of abnormal symbiosomes with reduced numbers of bacteroids in *Phaseolus* bean nodules (Daniels et al. 2002). In *Mesorhizobium tianshanensis*, mutation of the orthologous *mrtI* or *mrtR* genes completely blocked nodulation on licorice (*Glycyrrhiza uralensis*) (Zheng et al. 2006). These examples illustrate that the same QS system, most likely using the same AHL, 3-OH-C14:1-HSL, induces markedly different responses in the different species (Lithgow et al. 2000; Daniels et al. 2002).

Quorum sensing affects multiple aspects of symbiotic development: root colonization, cell division, Sym plasmid transfer, symbiosome development, nitrogen fixation, and nodule number. However, with the exception of the *Mesorhizobium* symbioses described above, QS has not yet been shown to be absolutely necessary for successful nodulation and infection. Rather, QS appears to optimize conditions for a successful symbiosis. When symbiosis is affected, mutation of QS genes typically results in delayed or reduced infection. In *R. leguminosarum* and *S. meliloti*, QS improves the rate of infection (Cubo et al. 1992; Gao et al. 2005). In *R. etli*, QS improves the accumulation of bacteroids in the symbiosome (Daniels et al. 2002).

Although no single model for quorum sensing regulation has been proposed, emerging proteomic and gene expression studies have identified a wide range of genes that are regulated by quorum sensing. In *S. meliloti*, microarray studies show that when cells are grown under batch culture, ca. 140 genes are observed to be regulated by the ExpR (Hoang et al. 2004; Gurich and González 2009). Proteome studies have identified another 50 proteins (Gao et al. 2005). These genes are important for metabolism, regulation, transport, transposition, motility, and symbiosis. Surprisingly, little overlap exists between the genes identified in the various studies, perhaps highlighting the *S. meliloti* QS system's sensitivity to cultural conditions. If QS is so strongly influenced by such conditions, then the next crucial challenge will be to identify QS-regulated genes in the rhizosphere and in the host plant (Teplitski et al. 2011).

Indeed, the host plant may actively influence bacterial QS systems. Recent publications have illustrated various ways in which the legume host reacts to a QS signal. Based upon proteomic analyses, exposure of *M. truncatula* roots to bacterial AHLs results in an accumulation of >7% of root proteins. The functions of these proteins include host defense, hormonal response, metabolism, and cytoskeletal elements, to name a few (Mathesius et al. 2003). Furthermore, host plants may also manipulate the bacterial QS system by producing compounds that mimic autoinducers. *Delisea pulchra*, a marine red alga and the first eukaryote shown to secrete such compounds, produces a set of 20–30 halogenated furanones that have structural similarities to AHLs. These furanones strongly inhibit the QS response in many Gram-negative bacteria by binding to the AHL receptor thereby promoting proteolytic degradation (Givskov et al. 1996; Manefield et al. 1999, 2002). Higher plants, including legume hosts such as pea, vetch, and *M. truncatula*, produce an array of 10–20 compounds that both stimulate and inhibit QS responses. The secretion and activity of the AHL-mimic molecules change depending on the developmental age of the plant (Daniels et al. 2002; Teplitski et al. 2000; Gao et al. 2003). This exchange of rhizobial AHL signals and legume AHL-mimic compounds may represent yet another level of regulation in establishing the *Rhizobium*-legume symbiosis.

In addition to producing autoinducers for intraspecies communication, many bacteria produce another signal, autoinducer-2 (AI-2), which is believed to be involved in cross-species signaling (Waters and Bassler 2005). In the diverse soil community, rhizobia are exposed to a mélange of signal molecules, including AI-2. So far, no evidence has been generated to suggest that rhizobia produce AI-2. However, *S. meliloti* responds to AI-2 signals from other bacteria. Although it does not produce its own AI-2, *S. meliloti* carries a functional AI-2 transporter protein that internalizes the AI-2 molecules produced by other microbes in the community (Pereira et al. 2008). The benefit of internalizing AI-2 remains to be seen. However, it has been hypothesized that this system allows *S. meliloti* to "eavesdrop" on the signaling conversations of other microbes in the rhizosphere, and perhaps interfere with AI-2-regulated behaviors such as virulence, thus benefiting potential host plants. Further analysis of this system will help elucidate the role played by *Sinorhizobium* in the soil microbial community.

2.2 *Host-Microbe Signaling Between the Symbiotic Partners*

The *Rhizobium*-legume symbiosis that results in the formation of nitrogen-fixing nodules on legume roots is one of the best-studied interactions in terms of plant-microbe communication. It is well known that the interaction between nitrogen-fixing rhizobia and legumes results from an exchange of signals from the host to the microbe and back again. For the plant, the signaling starts with the release from the host legume seed coats and roots of flavonoids and related molecules, which activate the regulatory gene *nodD*. These flavonoids are diverse and vary among legume hosts. Thus, flavonoid perception is one of the first levels of specificity between hosts and nitrogen-fixing rhizobia.

The flavonoid-induced signals from the bacterium to the plant are variations on chitin molecules, common to fungi and invertebrates but not generally found in bacteria (Fig. 2a). Upon interacting with a flavonoid, the NodD protein changes its conformation so that it can bind to *nod* boxes in the promoters of *nod* genes, leading to their expression and ultimately to the synthesis of Nod factor, a tetrameric or pentameric lipochitooligosaccharide (LCO) molecule with various substitutions on the reducing and nonreducing ends (Fig. 2b). These assorted *nod* genes include the common *nodABC* genes, which are responsible for production of the *N*-acyl glucosamine oligomer acylated by a default fatty acid, e.g., cis-vaccenic acid, as well as several host specificity *nod* genes, which are needed for the "molecular decorations" found on both the reducing and nonreducing ends of the molecule. The substitutions on the chitin-like backbone are important for a second level of specificity that is exhibited between the host legume and its particular symbiont. For example, *S. meliloti* Nod factors are sulfated on the reducing end of the LCO, enabling these symbionts to nodulate species of *Medicago*, *Melilotus*, and *Trigonella* but not *Pisum*. Proteins that are postulated to function as receptors for Nod factor (LysM-RLK/NFP in *Medicago truncatula* and NFR1/NFR5 in *Lotus japonicus*) have been identified (Jones et al. 2007; Madsen et al. 2010). Although it has not been explicitly demonstrated that Nod factor binds to these receptors, they have a LysM domain, which is likely to be the binding site for the chitin part of the LCO. Perception of the signal triggers the expression of downstream genes that encode a number of proteins, several of which are also important for establishment of the mycorrhizal symbiosis (Bonfante and Requena 2011), as well as proteins important for the development and ultimate functioning of nodules. Studies in the last decade have led to the identification at least 16 different plant receptors in legumes that are important for rhizobial infection and nodule organogenesis (Madsen et al. 2010).

How critical are the various host specificity decorations present in the Nod factor backbone? Earlier, *S. meliloti* strains with mutations in host specificity genes (*nodF*, *nodL*, *nodFL*, and *nodFE*) were shown to be altered in root hair deformation and infection thread formation in alfalfa, but the final phenotypes of the mutant-induced nodules were $Nod^{+}Fix^{+}$ (Ardourel et al. 1994). Using in vitro assays, Fujishige et al. (2008) determined that several of the same mutant rhizobia, namely *nodF*, *nodL*,

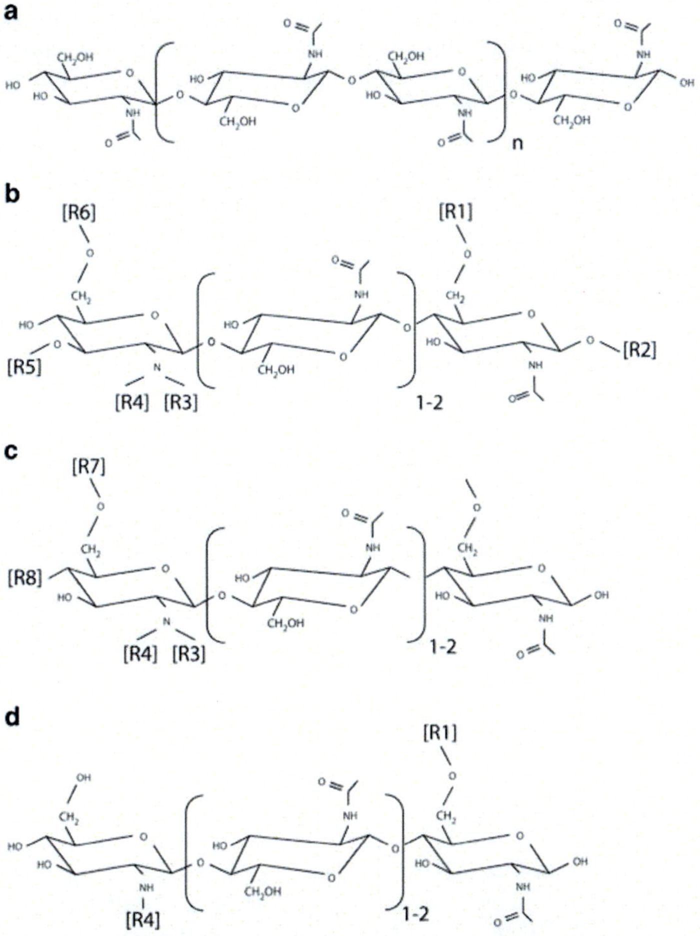

Fig. 2 Chitooligosaccharide structures. (**a**) Chitin. Each N is acetylated. Chitosan results from the deacetylation of various N residues. (**b**) Generalized structure of *Rhizobium* Nod factor, derived from the expression of the common and host-specific *nod* genes. The number of glucosamine residues in the backbone is typically four or five. The substitutions vary depending on the rhizobial strain. R_1 is H, sulfate, sulfo-methylfucose, D-arabinose, fucose, or other fucose derivatives. R_2 can be H or glycerol, whereas R_3 is either H or CH_3. R_4 is a C16, C18, or C20 fatty acid with different levels of unsaturation, whereas R_5 can be either H or a carbamoyl group. R_6 can be either H or an acetyl or carbomyl group. (**a**) and (**b**) redrawn from Hamel and Beaudoin (2010). (**c**) Structure determined for the Nod factor of *Bradyrhizobium aspalati* later identified as *Burkholderia tuberum* STM678. R_4 can be a C16, C18, C19, or C20 fatty acid. R_7 and R_8 are carbomyl groups. Redrawn from Boone et al. (1999). (**d**) Generalized LCO structure from the mycorrhizal symbiont *G. intraradices*. R_1 is H or sulfate, and R_4 is either a C16 or C18 fatty acid. Redrawn from Maillet et al. (2011).

nodFL, as well as *nodH*, established normal biofilms. In contrast, mutations in genes important for core Nod factor synthesis (*nodD1ABC*) resulted in Nod^- phenotypes. Deletions of the common *nod* genes yield rhizobia that do not trigger Ca^{2+} spiking, root hair deformation, or any of the other downstream responses important for eliciting nodule formation. Interestingly, such mutants also do not form robust biofilms under either in vivo or in vitro conditions (Fujishige et al. 2008). We hypothesized that this phenotype may be a consequence of the lack of adhesion of *nodD1ABC* mutants to each other or to other rhizobial cells in a mixed inoculum (Fujishige et al. 2008). Core Nod factor (the oligomer of N-glucosamine residues plus the default fatty acid) may thus be necessary for rhizobia to adhere to another to ensure that a certain threshold or quorum of cells is present on the root to elicit the chain of events leading to nodule formation. It is not surprising that the core glucosamine oligomer could "glue" cells together because *Caulobacter crescentus* holdfasts have an adhesive which, based on lectin-binding assays, contains glucosamine residues (Merker and Smit 1988; Ong et al. 1990). If sufficient rhizobia are tightly bound together to the host surface in soil, then following perception of the plant's flavonoid signals, Nod factor is synthesized. With higher levels of Nod factor, more profound effects occur in terms of the host's response. For example, at very low concentrations of Nod factor, Ca^{2+} spiking or root hair deformation may take place, but much higher levels are required for the initiation of cell divisions giving rise to nodule primordia. The identity of the signals produced in response to the quorum of rhizobia living in biofilms is not known at this time, but our preliminary results using RT-PCR analysis indicate that *nod* gene expression is not affected by QS (Fujishige and Hirsch unpubl.). This finding confirms earlier studies that used microarray analysis (Hoang et al. 2004; Gurich and González 2009).

The plant-associated *Burkholderia* spp. that nodulate legumes via the root hair nodulation pathway have *nod* genes similar to those found in the Alphaproteobacteria for nodulating legumes (Moulin et al. 2001, 2002). However, so far, only one Nod factor structure has been determined, which was isolated from a strain originally identified as *Bradyrhizobium aspalati* (Boone et al. 1999). This strain was later found from 16S RNA sequencing to be in the genus *Burkholderia* (Moulin et al. 2001) and was given the name *Burkholderia tuberum* (Vandamme et al. 2002). The *B. tuberum* Nod factors differ from those of alpha-rhizobia in that no substitutions occur at the reducing end of either the tetrameric or pentameric molecule (Fig. 2c). Rather, the Nod factors are highly substituted on the nonreducing end of the molecule (Boone et al. 1999). Current evidence suggests that the *nod* genes, which encode this Nod factor, are not a result of horizontal gene transfer from the alpha-rhizobia but rather the result of vertical descent (Bontemps et al. 2010). Indeed, nodulation in *Burkholderia* spp. appears as an ancient and stable ecological trait, with a possible age of 50 million years (Bontemps et al. 2010). This suggests that nodulation in the beta-rhizobia evolved at the same time that legumes exhibited sufficient changes in their genetic repertoire to become nodulated (Sprent 2007). It also suggests that the alpha- and beta-rhizobia probably gained the ability to nodulate legumes at a similar point in geological time. Because legumes are hosts for beta- as well as alpha-rhizobia, it also seems likely that the same receptors and downstream pathways are activated for nodulation.

3 Signaling Between Nitrogen-Fixing Bacteria and Plant Roots upon Contact

3.1 Other Modes of Recognition

Root hair invasion and dependence on Nod factor as just described is not the only mechanism by which rhizobia enter plant roots. Some rhizobia enter via a "crack entry" mode whereby infection threads may be formed internally after the rhizobia enter the root through intercellular spaces. However, in other symbioses such as in peanut (*Arachis hypogea*), infection threads are not formed, and rhizobia enter the interstices of the root strictly by intercellular invasion followed by entry through structurally altered cell walls (Sprent 2007; Uheda et al. 2001). For *Aeschynomene indica*, which shows a "crack entry" mode of invasion, the ingress of the rhizobial strain appears to be Nod factor independent because no *nod* genes were found in the sequenced genomes of the photosynthetic *Bradyrhizobium* strains that nodulate this plant (Giraud et al. 2007). These types of interactions are considered to be under less strict control than root hair invasion, not only because no infection thread formation occurs, but also because Nod factor also appears to be unnecessary for triggering cell divisions. As yet, it is unclear as to the identity of the signal molecules exchanged by the host legume and the symbiont, but a recent study suggests that other factors could be involved. Transposon mutagenesis of the photosynthetic *Bradyrhizobium* sp. strain ORS278 that nodulates *Aeschynomene* led to the identification of a number of mutants altered in nodule development as well as in nodule function. However, so far, no nodulation-deficient (Nod$^-$) mutants were uncovered even though an extensive genetic screen was performed (Bonaldi et al. 2010). Interestingly, mutants altered in the synthesis of purines and pyrimidines gave rise to nodules with altered development. This finding is noteworthy in part because the plant hormone cytokinin has been strongly implicated as a downstream mediator of nodule formation (Madsen et al., 2010). However, it is not known which type of signaling pathway these bradyrhizobia employ to elicit nodule formation on *Aeschynomene* roots. Nor is it known whether peanut rhizobia lacking *nod* genes would be capable of nodulating *A. hypogea*.

In any case, some proteins of the nodulation-signaling pathway are required for rhizobial crack entry into *Aeschynomene* and other legumes. RNAi knock down of CCaMK expression in *Aeschynomene* led to a major reduction (>90%) in nodule number, and the nodules had aberrant symbiosome formation (Sinharoy and DasGupta, 2009). It would be instructive to repeat these experiments with *Bradyrhizobium* sp. strain ORS278 because Sinharoy and DasGupta (2009) did their experiments with a Nod factor–producing rhizobial strain. Nevertheless, the mechanism whereby *Aeschynomene* and related legumes recognize and select their symbiotic partner from other rhizobial strains is not well understood. Are host-specific flavonoids critical for recognition and downstream gene expression in these legumes? If so, which transcriptional factors do they activate? Phenolic compounds released by the plant are known to bind to NodD and NodV proteins in certain rhizobia. This binding can subsequently activate the Tts1 protein and lead to the transcription of genes in the type III secretion (T3SS) pathway (Deakin and Broughton 2009). Unlike the T3SS of bacterial pathogens, however, the effectors secreted by the *Rhizobium* T3SS modulate legume host range either positively or negatively (Deakin and Broughton 2009). Nonetheless, T3SS are not found in all rhizobia, making it less likely that these proteins are involved in the symbiotic interactions of rhizobial strains that do not make Nod factor.

Recently, a transcriptomics analysis of *Casuarina glauca* and *Alnus glutinosa*, two different actinorhizal plants nodulated by the Gram-positive *Frankia*, demonstrated that the host plant signaling pathway for nodulation is conserved with that from legumes (Hocher et al. 2011). However, so far, no canonical *nod* genes have been found in the nitrogen-fixing symbiont *Frankia* (Normand et al. 2007). Nevertheless, the fact that similar receptors and signal-transducing proteins exist for both legume and actinorhizal nodule development strongly suggests that the downstream genes that need to be expressed for nodulation are well conserved among higher vascular plants. Remarkably, these genes are also conserved in some of the lower vascular plants (Wang et al. 2010; see later section). This finding also implies that factors important for symbiont recognition in actinorhizal plants may be conserved in *Aeschynomene*.

The similarities in the products of the downstream genes—*DMI1, doesn't make infections 1 (CASTOR/POLLUX), DMI2, doesn't make infections 2* (*SYMRK*), and *DMI3, doesn't make infections 3* (*CCaMK*)—in *Medicago truncatula* and *Lotus japonicus*, respectively, which encode Nod factor signal transduction proteins, to plant genes required for mycorrhizal formation as well as the older evolutionary age of the mycorrhizal symbiosis have prompted hypotheses that the root nodulation signaling cascade evolved from the mycorrhizal pathway (LaRue and Weeden 1994; Hirsch and Kapulnik 1998; Bucher et al. 2009; Ercolin and Reinhardt 2011). Supporting that theory is the recent identification of sulfated and nonsulfated, simple fatty acid–bearing LCOs from the exudates of mycorrhizal roots infected by *Glomus intraradices* and from germinating spore exudates (Maillet et al. 2011) (Fig. 2d). Not only do these structures resemble LCO Nod factors but also their application to *M. truncatula* roots triggered root hair deformation and the upregulation of *ENOD11*, an early nodulin gene that is expressed in developing nodules. Moreover, such treatment also boosted mycorrhizal colonization of nonlegume roots, and application of even the nonsulfated LCOs gave rise to an increase in total root length, which was dependent on the presence of the *DMI3* (*CCaMK*) gene (Maillet et al. 2011). Interestingly, Myc factor perception also required the NFP protein, which is encoded by a gene downstream of common Sym pathway genes, *DMI1*, *DMI2*, and *DMI3/CCaMK*, at least for the root branching phenotype.

Besides the fact that LCOs in mycorrhizal fungi and rhizobia are conserved, the structural relationship of Nod factor to chitin-like (short chitin oligosaccharides; COS) or chitosan molecules used in plant defense indicates that these molecules are all evolutionarily conserved (Bonfante and Requena 2011). Chitin and COS trigger what is known as PAMP (Pathogen-Associated Molecular Patterns)-triggered immunity (PTI; Jones and Dangl 2006), a response that leads to the activation of MAP kinase cascades eliciting the expression of defense mechanisms against invading pathogens (Wan et al. 2008). Like the Nod factor receptors, namely LysM-RLK/NFP (*M. truncatula*) and NFR1/NFR5 (*L. japonicus*), which are characterized by LysM domains, chitin receptor proteins also have two or three LysM domains and either active or inactive kinase domains (Hamel and Beaudoin 2010). However, unlike Nod or Myc (C16 or C18 fatty acid tail and occasional presence of a sulfate) factors, no substitutions occur on the COS backbone, and the molecule can be quite long (Fig. 2a). In *Arabidopsis*, chitin and COS of longer than eight glucosamine residues bind to a CERK1/LysM RLK1 receptor (Iizasa et al. 2010). On the other hand, Nod and Myc factors bind to LysM-RLK/NFP (NFR1/NFR5) and NFP (NFR5) receptors, respectively (Bonfante and Requena 2011), strongly suggesting that the decorations camouflage the Nod and Myc factor backbones and block binding to a chitin-binding protein, in this way repressing plant defense mechanisms by bypassing PTI. In support of the idea that Myc and Nod factors bind to a receptor distinct from defense receptor proteins, the nonlegume *Parasponia*, which is nodulated by rhizobia, has only one Nod factor receptor (NFP), which interacts with both Myc and Nod factors (Op den Camp et al. 2011). A similar situation may exist for *Gleditsia triacanthos*, a basal caesalpinioid legume (Fujishige and Hirsch, unpubl. results).

If the substitutions on the N-glucosamine backbone are indeed the reason that Nod and Myc factors are not detected as pathogenic molecules, this suggests that Myc factors from other mycorrhizal fungi will also have substitutions and that the ability to conceal a chitin-type molecule such that it is not recognized as an elicitor was an important step in the evolutionary history of the mycorrhizal symbiosis. Parallels can be observed in the human gut microbiota where a symbiosis factor, polysaccharide A (PSA), a zwitterionic galactosamine polymer, which is part of the capsule polysaccharide of the very common gut bacterium *Bacteroides fragilis*, activates the toll-like receptor pathway (TLR) to suppress the host's immunological responses (Round et al. 2011). Mutants lacking PSA activate the T-helper cell responses and are also unable to colonize the mucosal crypts. Molecules such as Nod factor and PSA would therefore qualify as SAMPs, symbiotic-associated molecular patterns (Hirsch 2004; Round et al. 2011).

Plant SAMPs and PAMPs are both likely to have evolved more than 400 million years ago concomitant with the evolution of land plants. Plant fossils containing arbuscules have been described to be of early Devonian age (Remy et al. 1994), indicating that the mycorrhizal symbiosis was very likely already established at this time. Similarly, fossils containing fungal pathogens, particularly chytrids, have also been described from the Devonian (Taylor et al. 1992). The plants inhabiting the land at this time were rootless and leafless, but both subterranean and aerial stems had vascular tissues. Many of these lower Devonian plants are equivalent to modern-day lycopods, but fossils of liverworts and hornworts were also represented (Taylor et al. 1992). Recent-day liverworts and hornworts possess genes encoding three of the proteins in the common symbiotic pathways, namely DMI1, DMI3, and IDP3 (Interacting Protein of DMI3, also known as CYCLOPs) (Wang et al. 2010).

3.2 *Signaling During the Symbiosis*

Once nodulation ensues, signaling and perception continue not only to cue the start of nitrogen fixation but also to maintain or eventually break down the coordination between host and symbiont. Although reviews on this topic are not as common as on the earlier stages of the interaction, a few recent ones have dealt with some of the factors that mediate the ongoing success of the nodulation/nitrogen fixation process (Prell and Poole 2006; Downie 2010). As mentioned earlier, rhizobial T3SS are upregulated in cis by the perception of plant flavonoids. The production of specific effectors then leads to either a successful or unsuccessful symbiotic interaction (Deakin and Broughton 2009). In general, however, the concept of signaling relative to the onset and maintenance of nitrogen fixation differs from the situation in nodulation where a highly elaborated signal-receptor complex system is utilized. Nodulation evolved later in geological time than nitrogen fixation—with the evolution of the angiosperms (Sprent 2007)—and thus may have recruited a variety of mechanisms whereby the host and symbiont establish an interaction (crack entry, root hair invasion, nodulation with or without Nod factor, etc.). In contrast, nitrogen fixation is energetically expensive and requires stricter environmental controls. Even a poorly developed nodule may house some nitrogen-fixing bacteria, thus keeping both the plant and the rhizobia alive long enough to reproduce. On the other hand, if nitrogen fixation is not functioning properly, the plant will die due to nitrogen starvation long before it flowers and sets seed. Rhizobia will not make the transition from vegetative to bacteroid state, particularly, if the latter stage of differentiation results in lethality (see later section). In any case, bacterial numbers will not increase to the extent that they would have if a well-developed nodule had formed.

4 Nitrogen Fixation

Many studies have demonstrated that low nitrogen and oxygen concentrations elicit the expression of the *nif* and ancillary genes involved in the synthesis of the enzyme nitrogenase (see Dixon and Kahn 2004; Prell and Poole 2006). It is also well established that adding nitrogen to inoculated legumes inhibits both nodulation and nitrogen fixation, and that environmental factors such as salt or phosphate stress and flooding may limit or halt nitrogen fixation. The mechanisms whereby nitrogen fixation is turned on and off have been studied for some time, yielding a great deal of information about *nif* gene regulation (Dixon and Kahn 2004). Signaling via nitrogen, oxygen, and redox sensing initiates the process of nitrogen fixation by triggering the phosphorylation and dephosphorylation of a number of two-component regulatory systems that activate the *nif* operon and other genes through the transcriptional activator NifA. In some free-living bacteria (e.g., *Klebsiella pneumoniae*), NifL binds to NifA, forming an inhibitor complex that shuts down nitrogen fixation if utilizable nitrogen levels are elevated (Dixon and Kahn 2004). The regulation of nitrogen fixation in both free-living and nodule-inhabiting bacteria overlaps to some extent, although it is decidedly more complex in nodular bacteria because they must sense not only the nutritional status of the plant cell environment but also its oxygen and redox levels. For example, rhizobial cell NifA is indirectly inactivated by oxygen in nodules by FixL/J. Variations on this theme exist in different species of bacteria as well as the type of nodules developed by the plant, resulting in many more switches that need to be triggered in nodule-inhabiting bacteria compared to free-living bacteria before nitrogen fixation occurs.

Two major types of nodule developmental patterns occur in the papilionoid legumes, either indeterminate or determinate nodules. They are differentiated from one another by (1) the site of the initiation of the first cortical cell divisions (inner cortex for indeterminate nodules versus outer cortex for determinate nodules), (2) the continued growth of the apical meristem in indeterminate nodules resulting in their cylindrical shape at maturity versus the early loss of meristematic activity in determinate nodules, which explains their spherical shape, and (3) the elongated shape and terminal differentiation of the bacteroids in indeterminate nodules, which contrasts with determinate-nodule bacteroids, which do not differentiate to the same extent and generally remain viable (Hirsch 1992). With regard to physiology, indeterminate nodules accumulate amides following the production of ammonium from dinitrogen, whereas determinate nodules convert ammonium into ureides (Prell and Poole 2006). Typically, determinate nodules are found in soybean, *Lotus* sp., and common bean, whereas indeterminate nodules develop in the IRLC (Inverted Repeat-Lacking Clade). Examples of the latter are pea, vetch, alfalfa, and clover.

4.1 Amino Acid Cycling

For nitrogen fixation to continue, an exchange of N and C between the plant and its symbionts must take place. In the nodule, plant-derived dicarboxylic acids, particularly malate, activate the nitrogen fixation process. After transport through DctA, the amides, alanine and aspartate, are thought to originate from malate through pyruvate either via malate dehydrogenase or malic acid (Prell and Poole 2006) (Fig. 3). In pea, aspartate and alanine are exported from the bacteroids into the plant cytosol through low-specificity amino acid ABC transporters, namely AapJQMP and BraDEGBC, where they are converted into asparagine and alanine, the latter accumulating in nodules (Fig. 3). Double mutants in both *aap* and *bra*, but not single mutants, result in nitrogen-starved bacteroids that accumulate a great deal of polyhydroxybutyrate (PHB) granules, indicating an imbalance in the C:N ratio. Normally, indeterminate-nodule bacteria store PHB only when they are in the infection thread. Nodules induced by *aap*/*bra* double mutants also have reduced levels of fixed nitrogen as measured by dry weight accumulation. A second component to this model is that the amino acids cycle back into the bacteroids to

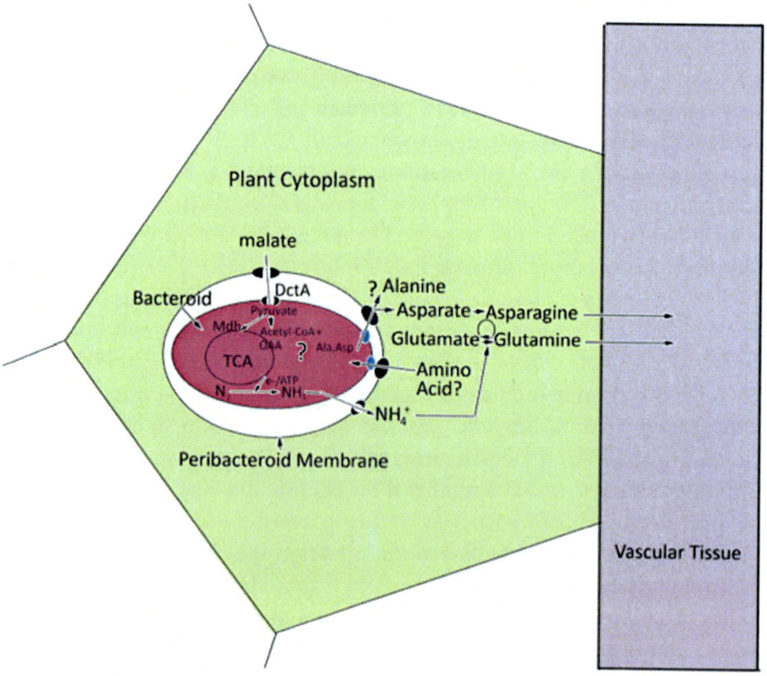

Fig. 3 Carbon and nitrogen exchange in the nodule. In the classic model, malate from the plant is transported into the bacteroids, where it is oxidized into oxaloacetate (OAA) by malate dehydrogenase (Mdh) or converted into pyruvate via malic acid and then oxidized into acetyl CoA, condensed with OAA, and incorporated into the TCA cycle. Energy is produced for converting N_2 to NH_3, which later is utilized for the synthesis of glutamine. Alanine is stored within the plant cytoplasm, whereas aspartate will be converted to asparagine by plant enzymes. In the new model, Aap/Bra (blue circles) are proposed to transport alanine and aspartate into the plant cytosol via a still not completely understood mechanism. Amino acids are postulated to be transported into the bacteroid cytosol, where they transaminate OAA or pyruvate to produce aspartate or alanine. Disruption of the cycling by mutating both *aap* and *bra* results in the reduction of fixed nitrogen in pea nodules and the accumulation of PHB in *R. leguminosarum* bv. *viciae* bacteroids. Redrawn and simplified from a model proposed by Prell and Poole (2006).

replenish the secreted aspartate and alanine. Taken together, these data strongly suggest that amino acid cycling is important for nodule function.

Meanwhile, enough energy is utilized to generate sufficient electrons and ATP (16–18) to power the nitrogen fixation process (Fig. 3). Ammonium, the direct product of nitrogenase, is transported from the bacteroids into the cytoplasm of the nodule, where in indeterminate nodules such as *Pisum sativum* (pea), it is used for the synthesis of asparagine and glutamine, via plant enzymes. These amino acids, which are localized in the plant cytosol, are subsequently transported into the vascular tissues of the host plant to be mobilized to other parts of the plant or to be cycled back into the bacteroids (Prell and Poole 2006) (Fig. 3).

Rhizobium leguminosarum bv. *viciae* and other indeterminate nodule bacteroids appear to depend on their host for branched-chain amino acid biosynthesis because they are terminally differentiated (see next section). To test whether or not this state of symbiotic auxotrophy is applicable to rhizobia that establish symbioses with determinate-nodule forming legumes, Prell et al. (2010) constructed *aap*/*bra* double mutants of *R. leguminosarum* bv. *phaseoli*, which nodulates *Phaseolus vulgaris*. As in pea, the French bean nodules, after inoculation with the double mutant, exhibited a reduction in dry weight accumulation, which correlated with reduced nitrogen fixation, compared to controls. Moreover, the bacteroids within the bean nodules accumulated considerably more PHB than the wild-type bacteroids. Thus, neither nodule morphology nor bacteroid differentiation state influenced the nitrogen fixation status of the double *R. leguminosarum* mutants.

In contrast, *aap*/*bra* double mutants of *S. meliloti* did not significantly affect nitrogen fixation in alfalfa, another example of an indeterminate nodule host, even though amino acid transport levels were reduced to background levels (Prell et al. 2010). The exact reason for this difference between the two genera of indeterminate-nodulating rhizobia is not known, but it may reflect the highly coordinated interaction between host plants and their respective symbionts. This remarkable complexity may also explain in part, why unlike the common nodulation genes that can complement mutations across species, mutations in *nif* genes do not appear to be complemented by *nif* genes from other species (Innes 1988). Coevolution of a host and its symbiont for the nitrogen fixation process thus appears to be under much more stringent control than nodulation is, most likely because it is critical for the survival of both partners. Clearly, more studies are required to determine the extent of symbiotic auxotrophy in different nitrogen-fixing associations and also the controls that modulate nitrogen fixation in other rhizobia.

4.2 Bacteroid Senescence

It has been known for some time that the type of nodule developed by a legume host is under plant and not rhizobial control (Dart 1977). As described above, legumes have either indeterminate or determinate nodules, except for *Lupinus* species, which develop nodules that are intermediate in origin and structure (González-Sama et al. 2004). Another difference between the two major types of nodule morphologies is that bacteroids maintain their viability in determinate nodules and may also fix nitrogen *ex planta*, whereas in indeterminate nodules, the bacteria elongate and then differentiate into bacteroids. Although many factors influence bacteroid differentiation, nodule-specific cysteine-rich repeat proteins that function as antimicrobial peptides (AMPs) and are delivered by plant's secretory pathway are responsible for indeterminate nodule bacteroid mortality (Van de Velde et al. 2010). AMPs are not present in determinate-nodule-forming legumes, but engineering *Lotus japonicus* to express the NCR035 gene resulted in the terminal differentiation of these bacteroids. Interestingly, even when the NCRs were added to *S. meliloti* cultures, the free-living cells underwent changes in their DNA content and eventually exhibited symptoms of cell death (Van de Velde et al. 2010).

5 Concluding Remarks

The last decade of research has yielded considerable information about the mechanisms of signaling between nitrogen-fixing bacteria and their hosts, especially with regard to the nodulating rhizobia. However, the discovery of the beta-rhizobia that fix nitrogen and nodulate legumes also demonstrates how limited our knowledge is of the dynamics and scope of biological nitrogen fixation (BNF) and nodulation, and also illustrates how important the process and the microbes that perform it are for agriculture and the environment. Much emphasis has been recently placed on producing more food and fuel for our ever-expanding and, in many countries, still hungry population, but in all these discussions, little mention is made of where the nitrogen to fertilize all of the crops will come from, especially if these plants are grown in soils that are becoming increasingly unfertile (Banwart 2011). Moreover, we are still woefully ignorant of how other residents of the soil interact with each other to promote plant growth by functions other than nitrogen fixation. It is well known that plant-associated microbes contribute to plant nutrition, health, and development by acquiring phosphate and iron, by secreting plant hormones, and by protecting their hosts against pathogens. We have only just begun to get a better idea of the plant's microbiome through metagenomic analyses that have told us much about "who's present" in the rhizosphere. We now need to learn the mechanisms whereby these organisms help plants survive and thrive, especially in challenged environments. After water, soil is our planet's most precious resource, and contamination, erosion, desertification, and concomitant loss of fertility threaten not only our food supply but also, more importantly, the health of our planet, which affects every living being. Soils store carbon, purify water, and sustain biodiversity, and the soil microbes transform solid rock into nutrients (Banwart 2011). By focusing more on the soil and the important contributions made by its living components, such as nutrient procurement, preservation of biodiversity, and sustainability, and also by recognizing that the plant microbiome is both highly diverse and integrated, we will be better prepared to preserve our planet's thin and quickly eroding surface for the generations that follow us.

Acknowledgments. Research in the Hirsch laboratory is funded by grants from the National Science Foundation and from the support of the Shanbrom Family Foundation. We are grateful to Philip Poole and Roger Innes for answering our e-mails about nitrogen fixation. We thank Stefan J. Kirchanski thanked for his helpful comments on a draft of the manuscript. We also thank Michelle Lum for Fig. 1c and Benny Gee for Fig. 2.

We dedicate this chapter to the late W. Dietz Bauer, one of the pioneers in the field of plant-microbe communication.

References

Ahlgren NA, Harwood CS, Schaefer AL, Giraud E, Greenberg EP (2011) Aryl-homoserine lactone quorum sensing in stem-nodulating photosynthetic bradyrhizobia. Proc Natl Acad Sci USA 17:7183–7188

Ardourel M, Demont N, Debellé F, Maillet F, de Billy F, Promé J-C, Dénarié J, Truchet G (1994) *Rhizobium meliloti* lipochitooligosaccharide nodulation factors: Different structural requirements for bacterial entry to target root hair cells and induction of plant symbiotic developmental responses. Plant Cell 6:1357–1374

Badri DV, Weir TL, van der Lelie D, Vivanco JM (2009) Rhizosphere chemical dialogues: plant–microbe interactions. Curr Opin Biotechnol 20:642–650

Bahlawane C, McIntosh M, Krol E, Becker A (2008) *Sinorhizobium meliloti* regulator MucR couples exopolysaccharide synthesis and motility. Mol Plant–Microbe Interact 21:1498–1509

Banwart S (2011) Save our soils. Nature 474:151–152

Bonaldi K, Gourion B, Fardoux J, Hannibal L, Cartieaux F, Boursot M, Vallenet D, Chaintreuil C, Prin Y, Nouwem N, Giraud E (2010) Large-scale transposon mutagenesis of photosynthetic *Bradyrhizobium* sp. strain ORS278 reveals new genetic loci putatively important for *nod*-independent symbiosis with *Aeschynomene indica*. Mol Plant–Microbe Interact 23:760–770

Bonfante P, Requena N (2011) Dating in the dark: how roots respond to fungal signals to establish arbuscular mycorrhizal symbiosis. Curr Opin Plant Biol 14:1–7

Bontemps C, Elliott GN, Simon MF, dos Reis Junior FB, Gross E, Lawton RC, Neto NE, Loureiro MF, de Faria SM, Sprent JI, James EK, Young JPW (2010) *Burkholderia* species are ancient symbionts of legumes. Mol Ecol 19:44–52

Boone CM, Olsthoorn MMM, Dakora FD, Spaink HP, Thomas-Oates JE (1999) Structural characterization of lipo-oligosaccharides isolated from *Bradyrhizobium aspalati*, microsymbionts of commercially important South African legumes. Carbohydr Res 317:155–163

Bouwmeester HJ, Roux C, Lopez-Raez JA, Bécard G (2007) Rhizosphere communication of plants, parasitic plants and AM fungi. Trends Plant Sci 12:224–230

Brelles-Mariño G, Bedmar EJ (2001) Detection, purification and characterisation of quorum-sensing signal molecules in plant-associated bacteria. J Biotechnol 91:197–209

Bucher M, Wegmüller S, Drissner D (2009) Chasing the structures of small molecules in arbuscular mycorrhizal signaling. Curr Opin Plant Biol 12:500–507

Cao H, Yang M, Zheng H, Zhang J, Zhong Z, Zhu J (2009) Complex quorum-sensing regulatory systems regulate bacterial growth and symbiotic nodulation in *Mesorhizobium tianshanense*. Arch Microbiol 191:283–289

Chen W-M, James EK, Prescott AR, Kierans M, Sprent JI (2003a) Nodulation of *Mimosa* spp. by the β-proteobacterium *Ralstonia taiwanensis*. Mol Plant–Microbe Interact 16:1051–1061

Chen W-M, Moulin L, Bontemps C, Vandamme P, Béna G, Boivin-Masson C (2003b) Legume symbiotic nitrogen fixation by β-proteobacteria is widespread in nature. J Bacteriol 185:7266–7272

Chen W-M, James EK, Coenye T, Chou J-H, Barrios E, de Faria SM, Elliott GN, Sheu S-Y, Sprent JI, Vandamme P (2006) *Burkholderia mimosarum* sp. nov., isolated from root nodules of *Mimosa* spp. from Taiwan and South America. Int J Syst Evol Microbiol 56:1847–1851

Chen W-M, de Faria SM, James EK, Elliott GN, Lin K-Y, Chou J-H, Sheu S-Y, Cnockaert M, Sprent JI, Vandamme P (2007) *Burkholderia nodosa* sp. nov., isolated from root nodules of the woody Brazilian legumes *Mimosa bimucronata* and *Mimosa scabrella*. Int J Syst Evol Microbiol 57:1055–1059

Chen W-M, de Faria SM, Chou J-H, James EK, Elliott GN, Sprent JI, Bontemps C, Young JPW, Vandamme P (2008) *Burkholderia sabiae* sp. nov., isolated from root nodules of *Mimosa caesalpiniifolia*. Int J Syst Evol Microbiol 58:2174–2179

Cubo MT, Economou A, Murphy G, Johnston AW, Downie JA (1992) Molecular characterization and regulation of the rhizosphere-expressed genes *rhiABCR* that can influence nodulation by *Rhizobium leguminosarum* biovar *viciae*. J Bacteriol 174:4026–4035

Daniels R, De Vos DE, Desair J, Raedschelders G, Luyten E, Rosemeyer V, Verreth C, Schoeters E, Vanderleyden J, Michiels J (2002) The *cin* quorum sensing locus of *Rhizobium etli* CNPAF512 affects growth and symbiotic nitrogen fixation. J Biol Chem 277:462–468

Danino VE, Wilkinson A, Edwards A, Downie JA (2003) Recipient-induced transfer of the symbiotic plasmid pRL1JI in *Rhizobium leguminosarum* bv. *viciae* is regulated by a quorum-sensing relay. Mol Microbiol 50:511–525

Dart P (1977) Infection and development of leguminous nodules. In: Hardy RWF (ed) A treatise on biological nitrogen fixation. Wiley, New York, pp 367–472

Deakin WJ, Broughton WJ (2009) Symbiotic use of pathogenic strategies: rhizobia protein secretion systems. Nat Rev Microbiol 7:3312–3320

Dixon R, Kahn D (2004) Genetic regulation of biological nitrogen fixation. Nat Rev Microbiol 2:621–631

Downie JA (2010) The roles of extracellular proteins, polysaccharides and signals in the interactions of rhizobia with legume roots. FEMS Microbiol Rev 34:150–170

Downie JA, González JE (2008) Cell-to-cell communication in rhizobia: Quorum sensing and plant signaling. In: Winans SC, Bassler BL (eds) Chemical communication among bacteria. ASM Press, Washington, DC, pp 213–232

Egland KA, Greenberg EP (1999) Quorum sensing in *Vibrio fischeri*: elements of the *luxI* promoter. Mol Microbiol 31:1197–1204

Elasri M, Delorme S, Lemanceau P, Stewart G, Laue B, Glickmann E, Oger PM, Dessaux Y (2001) Acyl-homoserine lactone production is more common among plant-associated *Pseudomonas* spp. than among soilborne *Pseudomonas* spp. Appl Environ Microbiol 67:1198–1209

Ercolin F, Reinhardt D (2011) Successful joint ventures of plants: Arbuscular mycorrhiza and beyond. Trends Plant Sci 16:356

Foster RC, Rovira AD, Cock TW (1983) Ultrastructure of the root–soil interface. The American Phytopathological Society, St. Paul, pp 5–11

Fujishige NA, Lum MR, De Hoff PL, Whitelegge JP, Faull KF, Hirsch AM (2008) *Rhizobium* common *nod* genes are required for biofilm formation. Mol Microbiol 67:504–515

Gao M, Teplitski M, Robinson JB, Bauer WD (2003) Production of substances by *Medicago truncatula* that affect bacterial quorum sensing. Mol Plant–Microbe Interact 16:827–834

Gao M, Chen H, Eberhard A, Gronquist MR, Robinson JB, Rolfe BG, Bauer WD (2005) *sinI*- and *expR*-dependent quorum sensing in *Sinorhizobium meliloti*. J Bacteriol 187:7931–7944

Gao Y, Zhong Z, Sun K, Wang H, Zhu J (2006) The quorum-sensing system in a plant-bacterium *Mesorhizobium huakuii* affects growth rate and symbiotic nodulation. Plant Soil 286:53–60

Giraud E, Moulin L, Vallenet D, Barbe V, Cytryn E et al (2007) Legume symbioses: absence of *nod* genes in photosynthetic bradyrhizobia. Science 316:1307–1312

Givskov M, de Nys R, Manefield M, Gram L, Maximilien R, Eberl L, Molin S, Steinberg PD, Kjelleberg S (1996) Eukaryotic interference with homoserine lactone mediated prokaryotic signaling. J Bacteriol 178:6618–6622

Glenn SA, Gurich N, Feeney MA, González JE (2007) The ExpR/Sin quorum sensing system controls succinoglycan production in *Sinorhizobium meliloti*. J Bacteriol 189:7077–7088

González-Sama A, Lucas MM, de Felipe MR, Pueyo JJ (2004) An unusual infection mechanism and nodule morphogenesis in white lupin (*Lupinus albus*). New Phytologist 163:371–380

Gurich N, González JE (2009) Role of quorum sensing in the *Sinorhizobium meliloti*-alfalfa symbiosis. J Bacteriol 191:4372–4382

Hamel L-P, Beaudoin N (2010) Chitooligosaccharide sensing and downstream signaling: contrasted outcomes in pathogenic and beneficial plant–microbe interactions. Planta 232:787–806

He X, Chang W, Pierce DL, Seib LO, Wagner J, Fuqua C (2003) Quorum sensing in *Rhizobium* sp. strain NGR234 regulates conjugal transfer (*tra*) gene expression and influences growth rate. J Bacteriol 185:809–822

Hirsch AM (1992) Tansley Review No. 40. Developmental biology of legume nodulation. New Phytologist 122:211–237

Hirsch AM (2004) Plant-microbe symbioses: a continuum from commensalism to parasitism. Symbiosis 37:345–363

Hirsch AM, Kapulnik Y (1998) Signal transduction pathways in mycorrhizal associations: comparisons with the *Rhizobium*-legume symbiosis. Fungal Genet Biol 23:205–212

Hirsch AM, Bauer WD, Bird DM, Cullimore J, Tyler B, Yoder JI (2003) Molecular signals and receptors: controlling rhizosphere interactions between plants and other organisms. Ecology 84:858–868

Hirsch AM, Lum MR, Fujishige NA (2009) Microbial encounters of a symbiotic kind—attaching to roots and other surfaces. In: Emons AMC, Ketelaar T (eds) Root hairs. Plant cell monographs, vol 12. Springer, Berlin/Heidelberg, pp 295–314

Hoang HH, Becker A, González JE (2004) The LuxR homolog ExpR, in combination with the Sin quorum sensing system, plays a central role in *Sinorhizobium meliloti* gene expression. J Bacteriol 186:5460–5472

Hoang HH, Gurich N, González JE (2008) Regulation of motility by the ExpR/Sin quorum sensing system in *Sinorhizobium meliloti*. J Bacteriol 190:861–871

Hocher V, Alloisio N, Auguy F, Fournier P, Doumas P, Pujic P et al (2011) Transcriptomics of actinorhizal symbioses reveals homologs of the whole common symbiotic signal cascade. Plant Physiol 156:700–711

Hwang I, Cook DM, Farrand SK (1995) A new regulatory element modulates homoserine lactone-mediated autoinduction of Ti plasmid conjugal transfer. J Bacteriol 177:449–458

Iizasa E, Mitsutomi M, Nagano Y (2010) Direct binding of a plant LysM receptor kinase, LysM RlK1/CERK1, to chitin in vitro. J Biol Chem 285:2996–3004

Innes Roger William (1988). A molecular genetic analysis of host-range control in *Rhizobium trifolii*. Ph.D. dissertation, University of Colorado at Boulder, United States, Colorado. Retrieved June 3, 2011, from Dissertations & Theses: A&I. (Publication No. AAT 8819666)

Jitacksorn S, Sadowsky MJ (2008) Nodulation gene regulation and quorum sensing control density dependent suppression and restriction of nodulation in the *Bradyrhizobium japonicum*-soybean symbiosis. Appl Environ Microbiol 74:3749–3756

Jones JD, Dangl J (2006) The plant immune system. Nature 444:323–329

Jones KM, Kobayashi H, Davies BW, Taga ME, Walker GC (2007) How rhizobial symbionts invade plants: the *Sinorhizobium-Medicago* model. Nat Rev Microbiol 5:619–633

Lanou A, Burlat V, Schurr U, Röse USR (2010) Induced root-secreted phenolic compounds as a belowground plant defense. Plant Signal Behav 5:1037–1038

LaRue TA, Weeden NF (1994). The symbiosis genes of the host. In: Kiss GB, Endre G (eds) Proceedings of the 1st European nitrogen fixation conference, Officia Press, Szeged, 147–151

Lazdunski AM, Ventre I, Sturgis JN (2004) Regulatory circuits and communication in Gram-negative bacteria. Nat Rev Microbiol 2:581–592

Lithgow JK, Wilkinson A, Hardman A, Rodelas B, Wisniewski-Dye F, Williams P, Downie JA (2000) The regulatory locus *cinRI* in *Rhizobium leguminosarum* controls a network of quorum sensing loci. Mol Microbiol 37:81–97

Loh J, Stacey G (2003) Nodulation gene regulation in *Bradyrhizobium japonicum*: a unique integration of global regulatory circuits. Appl Environ Microbiol 69:10–17

Loh J, Carlson RW, York WS, Stacey G (2002) Bradyoxetin, a unique chemical signal involved in symbiotic gene regulation. Proc Natl Acad Sci USA 99:14446–14451

Madsen LH, Tirichine L, Jurkiewicz A, Sullivan JT, Heckman AB, Bek AS, Ronson CW, James EK, Stougaard J (2010) The molecular network governing nodule organogenesis and infection in the model legume *Lotus japonicus*. Nat Commun 1:10

Maillet F, Poinsot V, André L, Puech-Pagè V, Haouy A, Gueunier M, Cromer L, Giraudet D, Formey D, Niebel A, Martinez EA, Driguez H, Bécard G, Dénarié J (2011) Fungal lipochitooligosaccharide symbiotic signals in arbuscular mycorrhizae. Nature 469:58–64

Manefield M, de Nys R, Kumar N, Read R, Givskov M, Steinberg P, Kjelleberg S (1999) Evidence that halogenated furanones from *Delisea pulchra* inhibit acylated homoserine lactone (AHL)-mediated gene expression by displacing the AHL signal from its receptor protein. Microbiology 145:283–291

Manefield M, Rasmussen TB, Hentzer M, Andersen JB, Steinberg P, Kjelleberg S, Givskov M (2002) Halogenated furanones inhibit quorum sensing through accelerated LuxR turnover. Microbiology 148:119–1127

Marketon MM, González JE (2002) Identification of two quorum-sensing systems in *Sinorhizobium meliloti*. J Bacteriol 185:3466–3475

Marketon MM, Gronquist MR, Eberhard A, González JE (2002) Characterization of the *Sinorhizobium meliloti sinR/sinI* locus and the production of novel N-acyl homoserine lactones. J Bacteriol 184:5686–5695

Marketon MM, Glenn SA, Eberhard A, González JE (2003) Quorum sensing controls exopolysaccharide production in *Sinorhizobium meliloti*. J Bacteriol 185:325–331

Mathesius U, Mulders S, Gao M, Teplitski M, Caetano-Anolles G, Rolfe RB, Bauer WD (2003) Extensive and specific responses of a eukaryote to bacterial quorum-sensing signals. Proc Natl Acad Sci USA 100:1444–1449

McIntosh M, Krol E, Becker A (2008) Competitive and cooperative effects in quorum sensing regulated galactoglucan biosynthesis in *Sinorhizobium meliloti*. J Bacteriol 190:5308–5317

Merker R, Smit J (1988) Characterization of the adhesive holdfast of marine and freshwater caulobacters. Appl Environ Microbiol 54:2078–2085

Moulin L, Mundive A, Dreyfus B, Boivin-Masson C (2001) Nodulation of legumes by the beta-subclass of Proteobacteria. Nature 411:948–950

Moulin L, Chen WM, Béna G, Dreyfus B, Boivin-Masson C (2002). Rhizobia: the family is expanding. In: Finan T, O'Brian M, Layzell D, Vessey K, Newton, W (eds) Nitrogen fixation: global perspectives, CAB International, Wallingford, UK, New York, NY, pp 61–65

Normand P, Lapierre P, Tisa LS, Gogarten JP, Alloisio N et al (2007) Evaluation of *Frankia* strains isolated from provenances of two *Alnus* species. Genome Res 17:7–15

Ong CJ, Wong MLY, Smit J (1990) Attachment of the adhesive holdfast organelle to the cellular stalk of *Caulobacter crescentus*. J Bacteriol 172:1448–1456

Op den Camp R, Streng A, De Mita S, Cao Q, Polone E, Liu W, Ammiraju JS, Kudrna D, Wing R, Untergasser A, Bisseling T, Geurts R (2011) LysM-type mycorrhizal receptor recruited for rhizobium symbiosis in nonlegume *Parasponia*. Science 331:909–912

Ortíz-Castro R, Contreras-Cornejo HA, Macías-Rodríquez L, López-Bucio L (2009) The role of microbial signals in plant growth and development. Plant Signal Behav 4:701–712

Pellock BJ, Teplitski M, Boinay RP, Bauer WD, Walker GC (2002) A LuxR homolog controls production of symbiotically active extracellular polysaccharide II by *Sinorhizobium meliloti*. J Bacteriol 184:5067–5076

Pereira CS, McAuley JR, Taga ME, Xavier KB, Miller ST (2008) *Sinorhizobium meliloti*, a bacterium lacking autoinducer-2 (AI-2) synthase responds to AI-2 supplied by other bacteria. Mol Microbiol 70:1223–1235

Piper KR, von Bodman SB, Farrand SK (2004) Conjugation factor of *Agrobacterium tumefaciens* regulates Ti plasmid transfer by autoinduction. Nature 362:448–450

Pongsilp N, Triplett EW, Sadowsky MJ (2005) Detection of homoserine lactone-like quorum sensing molecules in *Bradyrhizobium* strains. Curr Microbiol 51:250–254

Prell J, Poole P (2006) Metabolic changes of rhizobia in legume nodules. Trends Microbiol 14:161–168

Prell J, Bourdès A, Kumar S, Lodwig E, Hosie A, Kinghorn S, White J, Poole P (2010) Role of symbiotic auxotrophy in the *Rhizobium*-legume symbiosis. PLoS One 5:e13933

Ramsay JP, Sullivan JT, Stuart GS, Lamont IL, Ronson CW (2006) Excision and transfer of the *Mesorhizobium loti* R7A symbiosis island requires an integrase IntS, a novel recombination directionality factor RdfS, and a putative relaxase RlxS. Mol Microbiol 62:723–734

Ramsay JP, Sullivan JT, Jambari N, Ortori CA, Heeb S, Williams P, Barrett DA, Lamont IL, Ronson CW (2009) A LuxRI-family regulatory system controls excisions and transfer of the *Mesorhizobium loti* R7A symbiosis island by activating expression of two conserved hypothetical genes. Mol Microbiol 73:1141–1155

Remy W, Taylor TN, Hass H, Kerp H (1994) Four hundred-million year old vesicular arbuscular mycorrhizae. Proc Natl Acad Sci USA 91:11841–11843

Rodelas B, Lithgow JK, Wisniewski-Dye F, Hardman A, Wilkinson A, Economou A, Williams P, Downie JA (1999) Analysis of quorum-sensing-dependent control of rhizosphere expressed (*rhi*) genes in *Rhizobium leguminosarum* bv. *viciae*. J Bacteriol 181:3816–3823

Rosemeyer V, Michiels J, Verreth C, Vanderleyden J (1998) *luxI*- and *luxR*-homologous genes of *Rhizobium etli* CNPAF512 contribute to the synthesis of autoinducer molecules and nodulation of *Phaseolus vulgaris*. J Bacteriol 180:815–821

Round JL, Lee SM, Li J, Tran G, Jabri B, Chatila TA, Mazmanian SK (2011) The Toll-like receptor 2 pathway establishes colonization by a commensal of the human microbiota. Science 332:974–977

Sanchez-Contreras M, Bauer WD, Gao M, Robinson JB, Downie JA (2007) Quorum-sensing regulation in rhizobia and its role in symbiotic interactions with legumes. Phil Trans R Soc B 362:1149–1163

Scheu S (2001) Plants and generalist predators as links between the below-ground and above-ground system. Basic Appl Ecol 2:3–13

Sinharoy S, DasGupta M (2009) RNA interference highlights the role of CCaMK in dissemination of endosymbionts in the Aeschynomeneae legume *Arachis*. Mol Plant–Microbe Interact 22:1466–1475

Sourjik V, Muschler P, Scharf B, Schmitt R (2000) VisN and VisR are global regulators of chemotaxis, flagellar, and motility genes in *Sinorhizobium (Rhizobium) meliloti*. J Bacteriol 182:782–788

Sprent JI (2007) Tansley review. Evolving ideas of legume evolution and diversity: a taxonomic perspective on the occurrence of nodulation. New Phytologist 174:11–25

Suarez-Moreno ZR, Caballero-Mellado J, Venturi V (2008) The new group of non-pathogenic plant-associated *Burkholderia* spp. shares a conserved quorum sensing system which is tightly regulated by the RsaL repressor. Microbiology 154:2048–2059

Suarez-Moreno ZR, Devescovi G, Myers M, Hallack L, Mendonca-Previato L, Caballero-Mellado J, Venturi V (2010) Commonalities and differences in regulation of N-acyl homoserine lactone quorum sensing in the beneficial plant associated *Burkholderia* species cluster. Appl Environ Microbiol 76:4302–4317

Taylor TN, Remy W, Hass H (1992) Fungi from the lower Devonian Rhynie chert: Chytridiomycetes. Am J Bot 79:1233–1241

Teplitski M, Robinson JB, Bauer WD (2000) Plants secrete substances that mimic bacterial N-acyl homoserine lactone signal activities and affect population density-dependent behaviors in associated bacteria. Mol Plant–Microbe Interact 13:637–648

Teplitski M, Eberhard A, Gronquist MR, Gao M, Robinson JB, Bauer WD (2003) Chemical identification of N-acyl homoserine lactone quorum sensing signals produced by *Sinorhizobium meliloti* strains in defined medium. Arch Microbiol 180:494–497

Teplitski M, Merighi M, Gao M, Robinson J (2011) Interaction of cell-to-cell signals in soil bacterial communities. In: Witzany G (ed) Biocommunication in soil microorganisms, soil biology, vol 23. Springer, Berlin/Heidelberg, pp 369–401

Tun-Garrido C, Bustos P, Gonzalez V, Brom S (2003) Conjugative transfer of p42a from *Rhizobium etli* CFN42, which is required for mobilization of the symbiotic plasmid, is regulated by quorum sensing. J Bacteriol 185:1681–1692

Uheda E, Daimon H, Yoshizako F (2001) Colonization and invasion of peanut (*Arachis hypogea* L.) roots by *gusA*-marked *Bradyrhizobium* sp. Can J Bot 79:733–739

Van de Velde W, Zehirov G, Szatmari A, Debreczeny M, Ishihara H et al (2010) Plant peptides govern terminal differentiation of bacteria in symbiosis. Science 327:1122–1126

Vandamme P, Goris J, Chen WM, de Vos P, Willems A (2002) *Burkholderia tuberum* sp. nov. and *Burkholderia phymatum* sp. nov., nodulate the roots of tropical legumes. Syst Appl Microbiol 5:507–512

von Bodman SB, Bauer WD, Coplin DL (2003) Quorum sensing in plant-pathogenic bacteria. Annu Rev Phytopathol 41:455–482

Wan J, Zhang XC, Neece D, Ramonell KM, Clough S, Kim SY, Stacey MG, Stacey G (2008) A LysM receptor-like kinase plays a critical role in chitin signalling and fungal resistance in Arabidopsis. Plant Cell 20:471–481

Wang B, Yeun LH, Xue J-Y, Liu Y, Ané JM, Qiu Y-L (2010) Presence of three mycorrhizal genes in the common ancestor of land plants suggests a key role of mycorrhizas in the colonization of land by plants. New Phytologist 186:514–525

Waters CM, Bassler BL (2005) Quorum sensing: Cell-to-cell communication in bacteria. Annu Rev Cell Dev Biol 21:319–346

Wilkinson A, Danino V, Wisniewski-Dye F, Lithgow JK, Downie JA (2002) N-acyl-homoserine lactone inhibition of rhizobial growth is mediated by two quorum sensing genes that regulate plasmid transfer. J Bacteriol 184:4510–4519

Wisniewski-Dye F, Downie JA (2002) Quorum sensing in *Rhizobium*. Antonie Van Leeuwenhoek 81:397–407

Wisniewski-Dye F, Jones J, Chhabra SR, Downie JA (2002) *raiR* genes are part of the quorum sensing network controlled by *cinI* and *cinR* in *Rhizobium leguminosarum*. J Bacteriol 184:1597–1606

Yang M, Sun K, Zhou L, Yang R, Zhong Z, Zhu J (2009) Functional analysis of three AHL autoinducer synthase genes in *Mesorhizobium loti* reveals the important role of quorum sensing in symbiotic nodulation. Can J Microbiol 55:210–214

Zheng H, Zhong Z, Lai X, Chen W-X, Li S, Zhu J (2006) A LuxR/LuxI-type quorum sensing system in a plant bacterium *Mesorhizobium tianshanense* controls symbiotic nodulation. J Bacteriol 188:1943–1949

Recent Trends in the Olfactory Responses of Insect Natural Enemies to Plant Volatiles

Gadi V.P. Reddy

Abstract The area of plant volatile signaling in multitrophic interactions has developed one of the fascinating and fastest growing fields of research. It has been reported that plant leaves generally release minor quantities of volatile compounds, but when a plant is damaged by insects, several more volatiles are released. Numerous studies have demonstrated the dynamic role of herbivore-damaged plants in the attraction of natural enemies (predators and parasitoids). Volatile plant compounds released in response to insect feeding serve as a chemical signal for herbivore natural enemies. Volatiles released by insect-damaged plants function as attractants and affect the behavior of the natural enemies of herbivorous insects. They also display diverse effects on insect behaviors and are also used as foraging cues by parasitoids and predators. After damaged by phytophagous insects, some host plants could attract parasitoids and predators as an indirect defense. They can also induce defense responses in adjacent plants. Trees of some species are reported to produce volatile signals that affect the behavior of natural enemies. A summary on the recent trends published since 2000 to date on the plant volatiles in relation to insect natural enemies was specified. The use of plant volatiles in integrated pest management programs was also discussed.

1 Introduction

The constant struggle for survival between plants, herbivores, and natural enemies has produced many highly specialized defense and attack strategies (Reddy and Guerrero 2010). Plants have evolved themselves against herbivores that pose a

G.V.P. Reddy (✉)
Western Pacific Tropical Research Center, College of Natural and Applied Sciences
University of Guam, Mangilao, GU, USA
e-mail: reddy@uguam.uog.edu

G. Witzany and F. Baluška (eds.), *Biocommunication of Plants*,
Signaling and Communication in Plants 14,
DOI 10.1007/978-3-642-23524-5_15,

potential threat (Connor et al. 2007). The complexity in the interactions between plant and multiple attacks was reviewed (Poelman et al. 2008; Dicke et al. 2009). Management may be by means of manipulation of the environment of the pest for population suppression or for enhancement of natural enemies. Interactions between herbivores, their hosts plants, and natural enemies are increasingly understood in a chemical tritrophic context (Tapia et al. 2010). Factors influencing the search behavior of natural enemies include habitat characteristics such as crop, associated plants and plant assemblages, host plant characteristics, influences of associated organisms, and characteristics of the searching entomophage (Inbar and Gerling 2008). Recent studies have shown potential for simultaneous management of a pest species and enhancement of natural enemies using pest pheromones.

The concept that the host selection process involves responses to a composite of stimuli has been addressed by a number of authors, and the important role of chemical cues is well documented (Reddy and Guerrero 2004, 2010). Thus, the searching natural enemy will encounter a variety of cues, most of which are indirect, that vary in nature and reliability with the distance from the hosts (Cory and Hoover 2006). At great distance, the chemical cues may convey only the information that a habitat is available and is likely to contain suitable hosts. As the natural enemy gets closer to the host, different semiochemicals from damaged plants, feces, or other host by-products give a much more direct and reliable indication of the availability and location of the host (Dixon 2000; Oppenheim and Gould 2002; De Boer and Dicke 2006). In fact, the searching natural enemy utilizes semiochemicals, as well as visual cues, to locate and exploit her hosts. Volatiles from plants represent cues for phytophagous insects that can mediate the relationship between predators and prey (Reddy et al. 2002; Hatano et al. 2008). On the other hand, the natural enemy can also learn combinations of chemical and visual cues to further enhance their foraging success (Costa and Reeve 2011; Reddy and Raman 2011).

The natural enemy preference for certain plant community can be a response to wide botanical diversity or to plant status as affected by allelobiosis (Pettersson et al. 2008). Although the current gained knowledge is limited to certain crops and invasive weeds, research is required on these areas. Considerable progress was made by several workers (Shimoda et al. 2002; Mithöfer et al. 2005) in exploring differences in volatile emission from lima bean plants damaged mechanically compared with herbivore-damaged plants, which revealed a systemic response in the absence of natural enemy elicitors. Similarly, in the case of tritrophic system of *Brassica oleracea* and their herbivores (Shiojiri et al. 2000; Reddy et al. 2002), the parasitoid response to plant volatiles were studied in detail. However, this system is well established in the study of parasitoid response to plants and particularly chemically mediated interactions. It is also known from the literature that the insect natural enemies were attracted to herbivore-damaged plants over mechanically damaged and live healthy plants. Most of the bioassays conducted so far provided evidence that the plant plays an important role in the chemical interactions. The aim of this chapter is to review the recent trends on some chemical interactions between plant volatiles and insect natural enemies.

2 Interaction Between Insect Predators and Plant Volatiles

The olfactory responses of the predatory insects to plant volatiles were mentioned in Table 1. Influence on behavior of predaceous insects and mites are an operative method for improving the efficacy of natural enemies of pest herbivores (Symondson et al. 2002). It is reported that many predacious insects exploit a variety of chemicals either from their prey or the host plants of prey (De Boer and Dicke 2006; Dicke et al. 2009). It is also known that predacious insects use different semiochemicals or infochemicals emitted by plants and insects to mediate in a series of key processes during foraging behavior (Tapia et al. 2010). On the other hand, the authors further reported that presence of predators on the foliage could favor emission of aphid alarm pheromones, which could attract *Eriopis connexa* and *Hippodamia variegata* (Coleoptera: Coccinellidae).

The active movement of predacious insects characterized by a high locomotor activity plays an important role in the searching behavior (Bell 1990). However, little research on the effect of volatile semiochemicals was done on this aspect. However, the locomotor activity configuration of *Cycloneda sanguinea* (Coleoptera: Coccinellidae) to *Capsicum annuum* (Solanaceae) substrates or infested with *Myzus persicae* (Hemiptera: Aphididae) was reported by Heit et al. (2007). The authors also described that the volatile chemicals and tactile cues from preys or host plants or their interaction must be obligatory. These authors' results indicated that the individual occurrence of an olfactory stimulus could not be adequate to modulate a different locomotor pattern of *C. sanguinea*. They assume that this could be due to short acclimation time to the odor sources. The change in the environment also can affect the volatile composition and its interaction with the natural enemies. For example, the generalist predator *Podisus maculiventris* discriminated only between the odors of intact and *P. xylostella*–damaged plants grown at ambient CO_2 concentration, preferring the odor of the damaged plants (Vuorinen et al. 2004).

One of the most established chemically mediated interactions was Coccinellid predators and aphids and various host plants. These predators are important predators of aphids and various sucking insect pests, actively moving in the environment in search of food by using visual and olfactory cues (Raymond et al. 2000). Various authors reported that the Coccinellid predators use mostly olfactory cues to find the food source and induce response to volatiles released by host plants (Schaller and Nentwig 2000; Zhu and Park 2005). Olfactory cues such as plant-based semiochemicals are chemical messages crucial for survival of the predatory species. Plant stress responses to herbivores may cause variations in the volatile profile that makes the plant more attractive to predators (Pettersson et al. 2008). Previous olfactory studies by Ninkovic et al. (2001) and Ninkovic and Pettersson (2003) showed that *Coccinella septempunctata* (Coleoptera: Coccinellidae) prefer plants previously attacked by the aphid *Rhopalosiphus padi* (Homoptera: Aphididae) compared to live healthy plants. This indicates that insect attack induces variations in the plant volatile profile that can be a prime for the searching behavior of the predator.

Another such established chemically mediated interactions is predatory mites on lima beans where it has been shown that beans attacked by phytophagous mites are

Table 1 Insect predators reported to be mediating with plant species

Predator	Order: family	Target organisms	Host plant species	Reference
Adalia bipunctata (L.)	Coleoptera: Coccinellidae	*Myzus persicae* (Sulzer), *Acyrthosiphon pisum* (Harris), and *Brevicoryne brassicae* (L.)	*Vicia faba* (L.), *Brassica napus* L., and *Sinapis alba* L. Engl.	Francis et al. (2004) and Raymond et al. (2000)
Aiolocaria hexaspilota (Hope)	Coleoptera: Coccinellidae	*Plagiodera versicolora* (Laicharting)	*Salix eriocarpa* Fr. and Sav.	Yoneya et al. (2009)
Amblyseius womersleyi Schicha	Acari: Phytoseiidae	*Tetranychus urticae* Koch	*Phaseolus vulgaris* L.	Maeda et al. (2000)
Anthocoris nemoralis (F.)	Hemiptera: Anthocoridae	*Cacopsylla pyricola* (Förster)	*Pyrus communis* L.	Drukker et al. (2000)
Chrysoperla carnea Stephens	Neuroptera: Chrysopidae	Sucking pests	*Solanum melongena*, L. *Abelmoschus esculents*, L. *Capsicum annum*, L. *Lycopersicum esculentum* Mill.	Reddy (2002)
		Plutella xylostella (L.)	*B. oleracea* L. subsp. *capitata*, *B. oleracea* L. subsp. *botrytis*, *B. oleracea* L. subsp. *gongylodes*, *B. oleracea* L. subsp. *italica*	Reddy et al. (2002, 2004)
Coccinella septempunctata L.	Coleoptera: Coccinellidae	*Diuraphis noxia* (Mordvilko)	*Triticum* spp.	Liu et al. (2005)
		M. persicae	*B. juncea* L., *B. napus* L., and *Arabidopsis thaliana* (L.) Heynh	Girling and Hassall (2008)
		Aphis glycines Matsumura	*Glycine max* (L.) Merr.	Zhu and Park (2005)
		Rhopalosiphum padi (L.)	*Hordeum vulgare* L. and cultivars	Glinwood et al. (2009), Ninkovic et al. (2001), and Ninkovic and Pettersson 2003)
Dactylosternum abdominale (F.)	Coleoptera: Hydrophilidae	*Cosmopolites sordidus* (Germar)	*Musa* spp.	Tinzaara et al. (2005)

Dicyphus hesperus Knight	Heteroptera: Miridae	*Trialeurodes vaporariorum* Westwood, *M. persicae*, and *T. urticae*	*Lycopersicon esculentum L.*	McGregor and Gillespie (2004)
		Ephestia kuehniella Zeller	*Verbascum thapsus* L, *Nicotiana tabacum* L., and *Stachys albotomentosa* L.	Sanchez et al. (2004)
Episyrphus balteatus De Geer	Diptera: Syrphidae	*Aphis fabae* Scopoli, *M. persicae*	*Solanum nigrum* L. and *S. tuberosum* L. (Solanaceae)	Almohamad et al. (2007)
		M. persicae	*L. esculentum*	Verheggen et al. (2005)
Eriopis connexa (Germar) and	Coleoptera: Coccinellidae	*Acyrthosiphon pisum* Harris	*Vicia faba* (L.)	Tapia et al. (2010)
Exochomus flaviventris Mader	Coleoptera: Coccinellidae	*Phenacoccus manihoti* Matile-Ferrero	*Manihot esculenta Crantz* (var. Zanaga)	Le Rü and Makosso (2001)
Geocoris pallens Stål	Heteroptera: Lygaeidae	*Manduca sexta (L.)*	*Nicotiana attenuata* Torr. ex S. Wats	Halitschke et al. (2008)
Hippodamia variegata (Goeze)	Coleoptera: Coccinellidae	*Acyrthosiphon pisum* Harris	*Vicia faba* (L.)	Tapia et al. (2010)
Macrolophus caliginosus Wagner	Heteroptera: Miridae	*T. urticae* and *M. persicae*	*Capsicum* spp.	Moayeri et al. (2007)
Neoseiulus californicus (McGregor)	Acari: Phytoseiidae	*T. urticae*	*Phaseolus lunatus* L. and *Phaseolus vulgaris* L.	Shimoda (2010)
Neoseiulus cucumeris (Oudemans)	Acari: Phytoseiidae	*Thrips tabaci*	*Cucumis sativus L.*	Tatemoto and Shimoda (2008)
Neoseiulus womersleyi (Schicha)	Acari: Phytoseiidae	*T. urticae*, *Tetranychus kanzawai* Kishida	*P. vulgaris*	Ishiwari et al. (2007) and Maeda and Liu (2006)
		T. kanzawai	*Camellia sinensis* L.	Maeda et al. (2006)
Oligota kashmirica benefica Naomi	Coleoptera: Staphylinidae	*T. urticae* and *Mythimna separata* (Walker)	*P. lunatus*	Shimoda et al. (2002)
		T. urticae		Shimoda and Takabayashi (2001)
		T. kanzawai Kishida		Takahashi et al. (2001)
Orius albidipennis Reut	Heteroptera: Anthocoridae	*T. urticae* Koch	*Fragaria virginiana* Mill. and *Cucumis sativus* L.	Karimy et al. (2006)

(continued)

Table 1 (continued)

Predator	Order: family	Target organisms	Host plant species	Reference
Orius strigicollis (Poppius)	Heteroptera: Anthocoridae	*Thrips tabaci* Lindeman	*Cucumis sativus* L	Tatemoto and Shimoda (2008)
Orius sauteri (Poppius)	Heteroptera: Anthocoridae	*Thrips palmi* Karny	*Solanum melongena* L.	
Perillus bioculatus (Fabricius)	Hemiptera: Pentatomidae	*Leptinotarsa decemlineata* Say	*Solanum tuberosum* L.	van Loon et al. (2000)
Pheidole megacephala (Fabricius)	Hymenoptera: Formicidae	*Cosmopolites sordidus* (Germar)	*Musa* spp.	Tinzaara et al. (2005)
Platysoma cylindrica (Paykull)	Coleoptera: Histeridae	*Ips pini* (Say)	*Pinus strobus* L. and *Pinus banksiana* Lamb	Erbilgin and Raffa (2000)
Podisus maculiventris (Say)	Heteroptera: Pentatomidae	*P. xylostella*	*B. oleracea*	Vuorinen et al. (2004)
Propylaea japonica Thunberg	Coleoptera: Coccinellidae	*D. noxia*	*Triticum* spp.	Liu et al. (2005)
Scolothrips takahashii Priesner	Thysanoptera: Thripidae	*T. urticae* and *Mythimna separata* (Walker)	*P. lunatus* L.	Shimoda et al. (2002)
		T. kanzawai	*P. vulgaris*	Takahashi et al. (2001)
Stethorus gilvifrons (Muls.)	Coleoptera: Coccinellidae	*T. urticae* and *Panonychus ulmi* (Koch)	*P. vulgaris* and *Malus domestica* Borkh.	Gencer et al. (2009)
Stethorus japonicus Kamiya	Coleoptera: Coccinellidae	*T. kanzawai*	*P. vulgaris*	Takahashi et al. (2001)
Thanasimus dubius (Fabricius)	Coleoptera: Cleridae	*Ips* and *Dendroctonus* spp.	*Pinus* spp.	Costa and Reeve (2011)
Trirammatus striatula (Fabricius)	Coleoptera: Carabidae	*Acyrthosiphon pisum* Harris	*V. faba*	Tapia et al. (2010)
Wollastoniella rotunda Yasunaga and Miyamoto	Hemiptera: Anthocoridae	*Thrips palmi* Karny and *T. kanzawai*	*S. melongena*	Uefune et al. (2010)

attractive to predatory mites (Bruin and Dicke 2001). Shimoda et al. (2005) have demonstrated that odors from *T. urticae*–infested lima bean leaves, including herbivore-induced plant volatiles (HIPVs) and green leaf volatiles (GLVs), strongly attract *Neoseiulus californicus* while odors from physically damaged lima bean leaves are slightly attractive to the predators. However, Shimoda (2010) reported that methyl salicylate is a strong predator attractant, and its potential attractiveness almost equaled that of the blend of HIPVs from *T. urticae*–infested leaves. Further, this author's results suggest that a single compound of methyl salicylate or mixtures of this compound and methyl salicylate + linalool are good candidates for the use in manipulating foraging behavior of *N. californicus* in a field. On the other hand, methyl salicylate at very high concentrations was not attractive (even repellent) to *Phytoseiulus persimilis* (van Wijk et al. 2008). Shimoda and Dicke (2000) demonstrated that this predator is attracted to volatiles from bean plants infested with *Spodoptera exigua* caterpillars, but that this attraction is affected by predator starvation and host plant experience. Also, several studies have suggested that application of only this synthetic compound may be less efficient or ineffective in attracting specialist predators of *Tetranychus* spider mites (e.g., *N. womersleyi* (Maeda et al. 2006) and *Oligota kashmirica benefica* (Shimoda et al. 2002). Predatory mites also can learn to respond to volatile blends from certain prey–plant combinations (van Wijk et al. 2008). For example, Ishiwari et al. (2007) reported that the predatory mites *N. womersleyi* reared on *T. kanzawai*–infested tea leaves exhibited a strong preference for a mixture of three synthetic HIPVs included in the infested tea leaves [(3E)-4,8-dimethyl-1,3,7-nonatriene, (E)-b-Ocimene, and (E, E)-a-farnesene], while they were strongly attracted to a mixture of four synthetic HIPVs included in *T. urticae*-infested kidney bean leaves [methyl salicylate, (3E)-4,8-dimethyl-1,3,7-nonatriene, (E,E)-4,8,12-trimethyl-1,3,7,11-tridecatetraene, and b-caryophyllene] after the rearing on the infested bean leaves. Interestingly, they also reported that, in the absence of any one of the four or three HIPVs, each blend was ineffective in attracting the predators with different odor experience. Remarkably, novel attractants have been identified in *P. persimilis* using transgenic plants of *Arabidopsis thaliana*, such as 2-butanone (De Boer et al. 2004), (3S)-(E)-nerolidol (or a mixture of this compound and (3E)-4,8-dimethyl-1,3,7-nonatriene; Kappers et al. 2005), and octan-1-ol (van Wijk et al. 2008). Shimoda (2010) findings indicate that *T. urticae*–infested *Satsuma mandarin* leaves have diverse volatile compound(s) that elicit a strong response. Predator response to linalyl acetate or (E, E)-a-farnesene is fascinating because it has not been studied on predatory mites (e.g., De Boer et al. 2005).

Another such predatory response to plant volatiles was demonstrated studies of lacewing *Chrysoperla carnea* (Neuroptera: Chrysopidae). The volatiles from eggplant, okra, and peppers highly attracted *C. carnea* adults in an olfactometer (Reddy 2002). Interestingly, both sexes of the predator highly preferred the odors emanating from eggplant followed by okra and peppers. Reddy et al. (2002) observed that the generalist predator *C. carnea* is attracted to the odor of (Z)-3-hexenyl acetate, which is released in large amounts from herbivore-damaged cabbages. However, Raina et al. (2004) reported that the response of males to several of the compounds, particularly to the terpenoids, was higher than that of females. Similarly, the adults of both male and female *C. carnea* showed typical electrophysiological response to kairomonal substance of cotton leaf and boll extract. Among the sexes of *C. carnea*, higher EAG response was recorded in mated females than mated males (Hanumantharaya et al. 2010).

Volatiles have been demonstrated to protect plants by attracting herbivore enemies, such as parasitic wasps, predatory arthropods, and possibly even insectivorous birds (Unsicker et al. 2009). Even belowground, herbivory results in the release of volatiles that attract herbivore enemies. The process of infested plants attracting natural enemies can reduce and even eliminate herbivore pressure, in that the predators acquire information on the location of its prey (Kessler and Baldwin 2001; Sznajder et al. 2010). In this regard, the predators evolved behavioral responses to plant-produced volatiles induced by herbivore feeding (Sznajder et al. 2010). Furthermore, these authors reported that predators evolved genetically determined preferences for plant volatiles induced by herbivorous prey, in the past generations, predators innately reacting to such volatiles had higher fitness than those that did not show such behavior.

3 Interaction Between Insect Parasitoids and Plant Volatiles

Various olfactory responses of the insect parasitoids to plant volatiles were quoted in Table 2.

Generally, plant reactions can either directly affect the herbivore by way of higher production of toxins (Roda and Baldwin 2003) or changes in plant volatile emission (Hern and Dorn 2002) or indirect encouragement of the efficiency of parasitoids (Dorn et al. 2002). Numerous examples are available on the response of *Cotesia* spp. to volatiles from different plant species particularly *Brassica* spp. For example, Connor et al. (2007) reported that plant's response to progressive mechanical damage was more similar to herbivore damage, regardless of damage duration, and *C. glomerata* did not significantly discriminate between progressive damage and herbivore damage. This result is stimulating, as many previous publications have shown that mechanical damage alone does not elicit a strong response in parasitoids (e.g., Van Poecke et al. 2001). However, recent studies have revealed that progressive mechanical damage over a period of time can provoke a physiological response in the plant, resulting in a variation in volatile production (Mithöfer et al. 2005; Röse and Tumlinson 2005).

It is known that insect parasitoids use volatiles from the plants infested by plant feeding insects to find host herbivores, but their behavioral response to such semiochemicals is exceedingly variable (Wang et al. 2003). According to these authors, prior exposure to the semiochemicals significantly enhanced the subsequent response of female *C. glomerata*, independent of genetic differences, and their results suggest that both genetic component and environmental conditioning have played an important role in the evolution of host selection and utilization by the parasitoid in a tritrophic system. Vet (2001) reported that a high positive

Table 2 Insect parasitoids reported to be mediating with plant species

Parasitoid	Order: family	Target organisms	Host plant species	Reference
Anaphes iole (Girault)	Hymenoptera: Mymaridae	*Lygus hesperus Knight*	*Gossypium hirsutum* L.	Manrique et al. (2005)
Aphidius colemani (Viereck)	Hymenoptera: Braconidae	*Aphis gossypii* Glover	*Cucumis sativa* L.	Pinto et al. (2004)
		M. persicae and *Brevicoryne brassicae* (Linnaeus)	*B. oleracea* var *capitata*	Kalule and Wright (2004)
		Rhopalosiphum padi (L.)	*Hordeum vulgare* L. cultivars	Glinwood et al. (2009)
Aphidius ervi (Haliday)	Hymenoptera: Braconidae	*Acyrthosiphon pisum* (Harris)	*Vicia faba* L. (cv. "the Sutton")	Takemoto et al. (2009)
		A. pisum (Harris) and *Sitobion avenae* (Fabricius)	*Medicago sativa* L. and *Triticum aestivum* L.	Ojeda-Camacho et al. (2001)
Aphidius gifuensis (Ashmead)	Hymenoptera: Braconidae	Aphid pests	*N. tabacum*	Dong et al. (2008)
Asobara anastrephae (Muesebeck)	Hymenoptera: Braconidae	*Ceratitis capitata* (Wied.) and *Anastrepha fraterculus* (Wied.)	*Psidium guajava* (L.)	Silva et al. (2007)
Binodoxys communis (Gahan)	Hymenoptera: Braconidae	*Aphis glycines* Matsumura	*Glycine max* (L.) Merr.	Wyckhuys and Heimpel (2007)
Campoletis sonorensis (Cameron)	Hymenoptera: Ichneumonidae	*Spodoptera littoralis* Boisduval	*Gossypium herbaceum* (Goss) and *Vigna unguiculata* (L.)	Tamò et al. (2006)
Cardiochiles nigriceps Vierick	Hymenoptera : Braconidae	*Heliothis virescens* (Fabricius) and *Helicoverpa zea* (Boddie)	*Nicotiana tabacum* L., *Gossypium* spp., and *Zea mays* L.	De Moraes et al. (1998)
		H. zea, *Heliothis subflexa* Gn., *H. virescens*	*N. tabacum*, *Physalis angulata* L., and *Gossypium* spp.	Oppenheim and Gould (2002)
Chrysonotomyia ruforum (Krausse)	Hymenoptera: Eulophidae	*Diprion pini* (L.)	*Pinus sylvestris* L.	Hilker et al. (2002) and Mumm and Hilker (2005)
Cotesia flavipes Cameron	Hymenoptera: Braconidae	*Chilo partellus* (Swinhoe)	*Hemizygia petiolata* Ashby	Ngumbi et al. (2005)
Cotesia glomerata (L.)	Hymenoptera: Braconidae	*Delia radicum* (L.)	*Brassica nigra* (L.)	Soler et al. (2007)

(continued)

Table 2 (continued)

Parasitoid	Order: family	Target organisms	Host plant species	Reference
		Pieris brassicae (L.) and *P. rapae* (L.)	*B. oleracea*	Gu and Dorn (2000), Shiojiri et al. (2000, 2001)
		Pieris brassicae (L.)	*B. oleracea* var. *gemmifera*	Connor et al. (2007)
Cotesia kariyai (Watanabe)	Hymenoptera: Braconidae	*Mythimna separata* (Walker)	*Vicia faba* (L.)	Fukushima et al. (2002)
Cotesia marginiventris (Cresson)	Hymenoptera: Braconidae	*Spodoptera* spp	*Vigna* spp. and *Z. mays*	Hoballah et al. (2002), Hoballah and Turlings (2005)
		Spodoptera littoralis	*Gossypium herbaceum* (Goss) and *Vigna unguiculata* (L.)	Tamò et al. (2006)
Cotesia plutellae Kurdjumov	Hymenoptera: Braconidae	*Plutella xylostella* (L.) and *Pieris rapae* (L.)	*Brassica oleracea* L.	Shiojiri et al. (2000, 2001)
		P. xylostella	*Brassica nigra* L. and *B. oleracea*	Seenivasagan and Paul (2011)
			Brassica oleracea (L.)	Vuorinen et al. (2004) and Ibrahim et al. (2005)
Cotesia rubecula (Marshall)	Hymenoptera: Braconidae	*P. brassicae*	*B. oleracea*	Smid et al. (2002) and Wang et al. (2003)
		Pieris rapae (L.)	*Arabidopsis thaliana* (L.) Heynh	Van Poecke et al. (2001)
Cotesia vestalis (Haliday)	Hymenoptera: Braconidae	*P. xylostella*	*B. oleracea*	Girling et al. (2011)
Diachasmimorpha juglandis (Muesebeck)	Hymenoptera: Braconidae	*Rhagoletis* spp.	*Juglans nigra* L	Henneman et al. (2002)
Diachasmimorpha kraussii	Hymenoptera: Braconidae	*Bactrocera jarvisi* (Tryon) and *B. tryoni* (Froggatt)	*Psidium guajava* L., *Prunis persica* L., *Malus domestica* Borkh., *Pyrus communis* L., and *Citrus sinensis* L.	Ero et al. (2011)
	Hymenoptera: Braconidae	*C. capitata* and *A. fraterculus*	*P. guajava*	Silva et al. (2007)

Diachasmimorpha longicaudata (Ashmead)				
Diadegma semiclausum (Hellen)	Hymenoptera: Ichneumonidae	*P. xylostella*	*Hordeum vulgare* L., *Sinapis alba* L., and *B. oleracea*	Bukovinszky et al. (2005)
Diaeretiella rapae (McIntosh)	Hymenoptera: Aphidiidae	*M. persicae*	*Arabidopsis thaliana* (L.) Heynh	Girling et al. (2006)
Doryctobracon areolatus (Szépligeti)	Hymenoptera: Braconidae	*C. capitata* and *A. fraterculus*	*P. guajava*	Silva et al. (2007)
Encarsia formosa Gahan	Hymenoptera:Aphelinidae	*Trialeurodes vaporariorum* (West.)	*P. vulgaris*	Birkett et al. (2003)
Exorista japonica Townsend	Diptera: Tachinidae	*Mythimna separata* (Walker)	*Zea mays* L.	Ichiki et al. (2011)
Gonatocerus ashmeadi Girault	Hymenoptera: Mymaridae	*Homalodisca vitripennis* (Germar)	*Citrus* spp., Vitis vinifera L., *Photinia fraseri* Dress, and *Lagerstroemia* spp	Krugner et al. (2008)
Lysiphlebus testaceipes Cresson	Hymenoptera: Braconidae	*Aphis gossypii* Glover	*Cucumis sativa* L	Pinto et al. (2004)
			Solanum melongena L	Pérez et al. (2007)
Microplitis mediator Haliday	Hymenoptera: Braconidae	*Gossypium* spp.	*Helicoverpa armigera* (Hübner)	Yu et al. (2010) and Wen-xia et al. (2000)
Microplitis rufiventris Kokujev	Hymenoptera: Braconidae	*Spodoptera littoralis* Boisduval	*Z. mays*	Hoballah and Turlings (2005)
.			*Gossypium herbaceum* (Goss) and *Vigna unguiculata* (L.)	Tamò et al. (2006)
Opius dissitus Muesebeck	Hymenoptera: Braconidae	*Liriomyza huidobrensis* (Blanchard) and *L. sativae* Blanchard	*P. vulgaris*	Wei and Kang (2006)
Phradis interstitialis (Thomson) *P. morionellus* (Holmgren)	Hymenoptera: Ichneumonidae	*Meligethes aeneus* F. and *Spodoptera littoralis* (Boisduval)	*Brassica napus* L. and *Gossypium hirsutum* L	Jönsson (2005)
Psyllaephagus pistaciae Ferričre	Hymenoptera: Encyrtidae	*Agonoscena pistaciae* Burckhardt and Lauterer	*Pistacia vera* L. Pistah	Mehrnejad and Copland (2006)
Roptrocerus mirus (Walker)		*Ips typographus* L.	*Picea abics* L. Karst	Pettersson (2001)

(continued)

Table 2 (continued)

Parasitoid	Order: family	Target organisms	Host plant species	Reference
Rhopalicus tutela (Walker)	Hymenoptera: Pteromalidae			
Roptrocerus xylophagorum (Ratzeburg)				
Tersilochus heterocerus Thomson	Hymenoptera: Ichneumonidae	*Meligethes aeneus* F. and *Spodoptera littoralis* (Boisduval)	*Brassica napus* L. and *Gossypium hirsutum* L.	Jönsson (2005)
Trichogramma chilonis Ishii.	Hymenoptera: Trichogrammatidae	*P. xylostella*	*B. oleracea* subsp. *capitata*, *B. oleracea* subsp. *botrytis*, *B. oleracea* subsp. *gongylodes*, and *B. oleracea* subsp. *italica*	Reddy et al. (2002)
Trybliographa rapae Westwood	Hymenoptera: Figitidae	*Pieris brassicae* Linnaeus	*Brassica* spp and *Delia radicum* (L.)	Pierre et al. (2011)

response of foraging parasitoids to volatiles from herbivore-infested plants would contribute to high reproductive success, and natural selection should favor the genotypes with a strong response to these semiochemicals. As *C. glomerata* is a generalist parasitoid and attacks on various *Pieris* species, these insect species also feed on various *Brassica* species. When infested by different herbivore species, a single *Brassica* species can emit diverse volatile chemicals, since the plant volatiles are often herbivore specific (Shiojiri et al. 2001).

This specialist parasitoid species also responds differently to changes in the environment. Vuorinen et al. (2004) reported that the *C. plutellae* preferred the odor of damaged *Brassica oleracea* ssp. *capitata* plants of both cultivators (Lennox and Rinda) grown at ambient CO_2 but did not detect damaged cv Lennox plants grown at elevated CO_2. Their results suggest that elevated atmospheric CO_2 concentration could weaken the plant response induced by insect herbivore feeding and thereby lead to a disturbance of signaling to the third trophic level. Vuorinen et al. (2004) reported that *C. plutellae* shows a specific response toward the host plant complex, unlike *C. glomerata*, and the presence of the nonhost affects the specificity of the response of the wasps (Shiojiri et al. 2000). Liu and Jiang (2003) showed that volatile compounds from Chinese cabbage were more attractive to female *C. plutellae* than those from white cabbage when both plant species were either intact or infested with *P. xylostella*. Homoterpene (E)-4,8-dimethyl-1,3,7-nonatriene and other terpenes seemed to be important cues for orientation of *C. plutellae* to *P. xylostella*-damaged plants (Shiojiri et al. 2001). Therefore, *C. plutellae* and other Braconids at the top of the food chains maintain important position in terms of global biodiversity (Dolphin and Quicke 2001). Soler et al. (2005) showed that *C. glomerata*, parasitoid of caterpillars of the cabbage butterfly *Pieris brassicae*, developed significantly slower and adults were smaller when roots of *Brassica nigra* (Brassicaceae) plants were damaged by larvae of the cabbage root fly, *Delia radicum* (Diptera: Anthomyiidae).

Maximum studies have been primarily based on aboveground interactions, but similar plant-induced indirect defense responses have also been perceived belowground (Van Tol et al. 2001; Rasmann et al. 2005). The root-associated organisms can distress the development of leaf-associated herbivores sharing the host plant (Bezemer et al. 2003), and higher trophic levels including parasitoids, and even hyperparasitoids of the fourth trophic level (Soler et al. 2005). Therefore, any changes in the plant volatile blend induced by root-feeding insects may enable the aboveground parasitoids to be vigilant about the occurrence of the root herbivores on the host plant, which has possibly negative consequences for offspring fitness of the parasitoid. Rasmann et al. (2005) demonstrated that root-feeding insects induce a volatile signal in the soil that attracts entomopathogenic nematodes, while simultaneously inducing the release of the same volatile compound aboveground from the leaves of the plant. Likewise, the damage triggered in the roots as a consequence of feeding offers a point of entry for subsequent infection by endemic root rot pathogens (Soroka et al. 2004). Soler et al. (2007) provided evidence that the foraging behavior of a *C. glomerata* of an aboveground herbivore can be influenced by belowground herbivores through changes in the plant volatile blend. This kind of indirect interactions may have reflective significances for the evolution of host selection behavior in parasitoids and may play a vital role in the constituting and functioning of communities.

At low herbivore densities, only parasitoids with a larger foraging radius could take advantage of plant cues (Puente et al. 2008). While preference for herbivore-induced volatiles was not always beneficial for a parasitoid, under the most likely natural conditions, it is believed that parasitoids such as *C. rubecula* gain fitness from plant cues. Similarly, several braconid aphid parasitoid species have been reported as responding to a variety of olfactory cues linked with the host or with the host's habitat (Jang et al. 2000; Carver and Franzmann 2001). Pinto et al. (2004) showed that the two parasitoid species *Lysiphlebus testaceipes* and *Aphidius colemani* (Hymenoptera: Braconidae) responded to stimuli from the host plants of *A. gossypii* in a similar way to parasitoids of aphid pests in other crops. Also, Silva et al. (2007) showed that *Doryctobracon areolatus* and *D. longicaudata* females responded to the odors of uninfested rotting guavas, although *D. areolatus* was also attracted to fruits at the initial maturation stage. However, females were not attracted toward fruits on the ground in the shade house, regardless of host, suggesting that this parasitoid does not forage on fallen fruits.

Herbivore-induced cues are also vital for the foraging success of egg parasitoids and for plant defense (Hilker et al. 2000). For instance, the egg parasitoid *Trissolcus basalis* (Hymenoptera: Scelionidae) depends largely on olfactory cues released from its adult host, *Nezara viridula* (Heteroptera: Pentatomidae), such as the male sex pheromone (Conti et al. 2003). However, for long-range attraction, the parasitoid uses plant volatile chemicals induced by host feeding and oviposition (Colazza et al. 2004). The work of Silva et al. (2006) shows sensory stimuli originating from *Euschistus heros* (Heteroptera: Pentatomidae) females were weakly active to the parasitoid *Telenomus podisi* (Hymenoptera: Scelionidae), but in combination with males, the behavior of the parasitoid changed significantly. *E. heros* males differ chemically from the females because of sex pheromones, and the parasitoid may have learned to associate sex pheromone with the presence of host eggs. These authors concluded that in its foraging behavior, *T. podisi* uses sensory stimuli from male *Euschistus heros*; at long distances, females and egg masses alone are inadequately attractive to the parasitoid when searching for the host.

4 Conclusions

To conclude, olfactory responses of insect natural enemies using volatiles from plants as stimuli suggest that the complex mixture constituting odor compounds is processed in a unique way due to the high behavior relevance. The effect of plant-induced responses on plant herbivores and their natural enemies may have a potential use in pest management. However, this causes the importance of performing both chemical analyses and behavioral bioassays in order to fully understand the ecological processes and to relate minor differences in plant physiological responses to the plant's natural enemy interactions (Connor et al. 2007). The volatile-based lures can be used alone or in combination with other sources of attractants in control strategies such as mass trapping, attract and kill, push-pull, and disruption of host finding (Reddy and Guerrero 2010). Plant-based volatiles in most cases synergize with sex pheromones, and biological control therefore will have an important role in integrated pest management of programs (Reddy and Guerrero 2004).

The use of plant-based volatile technology is one of the important tools in integrated pest management programs, which would offer a novel and ecologically sound tactic to control insect pests. This practice includes the optimization of lures that attract herbivore natural enemies against economically important pests. Future research that will address plant herbivores in the framework of multitrophic-level interactions is greatly encouraging. Such interactions will possibly comprise the plant (and its properties), other herbivores, natural enemies, and microorganisms. The examples given above exemplify the capacity of predators and parasitoids to use information on essential food sources. Present information is far from wide ranging and presents a challenge to seek additional knowledge of the mechanisms of adaptive capacity and foraging strategies of insect natural enemies. Integrating the function of natural enemy attraction with other volatile functions that can prove to diminish herbivore density on crops is probable to be a productive area of upcoming research. Nevertheless, additional studies are needed to explore novel volatile semiochemicals that play important roles in attracting insect natural enemies.

Acknowledgments This work was supported by Professional Development Program (PDP) of the USDA-Western SARE project #2009-EW09-012/Utah State University sub award # 090757010; FY 2009 Pacific Islands Area Conservation Innovation Grants (PIA-CIG) Program, Grant Agreement No. 69-9251-9-822, The Natural Resources Conservation Service (NRCS)-USDA; Western Integrated Pest Management Center (WIPMC) Award # 2007-51120-03885/ University of California, Davis sub award # 07 -001492-GUAM3; and USDA Hatch funds (Project# GUA0561) and W-2185 (Project # GUA0612). In accordance with federal law and USDA policy, this institution is prohibited from discrimination on the basis of race, color, national origin, sex, age, or disability.

References

Almohamad R, Verheggen FJ, Francis F, Haubruge E (2007) Predatory hoverflies select their oviposition site according to aphid host plant and aphid species. Entomol Exp Appl 125:13–21

Bell WJ (1990) Searching behavior pattern in insects. Annu Rev Entomol 35:447–467

Bezemer TM, Wagenaar R, Van Dam NM, Wackers FL (2003) Interactions between above- and belowground insect herbivores as mediated by the plant defense system. Oikos 101:555–562

Birkett MA, Chamberlain K, Guerrieri E, Pickett JA, Wadhams LJ, Yasuda T (2003) Volatiles from whitefly-infested plants elicit a host-location response in the parasitoid, *Encarsia formosa*. J Chem Ecol 29:1589–1600

Bruin J, Dicke M (2001) Chemical information transfer between wounded and unwounded plants: backing up future. Biochem Syst Ecol 29:1103–1113

Bukovinszky T, Gols R, Posthumus MA, Vet LEM, Van Lenteren JC (2005) Variation in plant volatiles and attraction of the parasitoid *Diadegma semiclausum* (Hellén). J Chem Ecol 31:461–480

Carver M, Franzmann B (2001) *Lysiphlebus* Forster (Hymenoptera: Braconidae: Aphidiinae) in Australia. Aust J Entomol 40:198–201

Colazza S, McElfresh JS, Millar JG (2004) Identification of volatile synomones, induced by *Nezara viridula* feeding and oviposition on bean spp., that attract the egg parasitoid *Trissolcus basalis*. J Chem Ecol 30:945–964

Connor EC, Rott AS, Samietz J, Dorn S (2007) The role of the plant in attracting parasitoids: response to progressive mechanical wounding. Entomol Exp Appl 125:145–155

Conti E, Salerno G, Bin F, Williams HJ, Vinson SB (2003) Chemical cues from *Murgantia histrionica* eliciting host location and recognition in the egg parasitoid *Trissolcus brochymenae*. J Chem Ecol 29:115–130

Cory JS, Hoover K (2006) Plant-mediated effects in insect–pathogen interactions. Trends Ecol Evol 21:278–286

Costa A, Reeve JD (2011) Upwind flight response of the bark beetle predator *Thanasimus dubius* towards olfactory and visual cues in a wind tunnel. Agric Forest Entomol 13. doi: 10.1111/j.1461-9563.2011.00519.x

De Boer JG, Dicke M (2006) Olfactory learning by predatory arthropods. Anim Biol 56:143–155

De Boer JG, Posthumus MA, Dicke M (2004) Identification of volatiles that are used in discrimination between plants infested with prey or nonprey herbivores by a predatory mite. J Chem Ecol 30:2215–2230

De Boer JG, Snoeren TAL, Dicke M (2005) Predatory mites learn to discriminate between plant volatiles induced by prey and nonprey herbivores. Anim Behav 69:869–879

De Moraes CM, Lewis WJ, Paré PW, Alborn HT, Tumlinson JH (1998) Herbivore-infested plants selectively attract parasitoids. Nature 393:570–573

Dicke M, van Loon JJA, Soler R (2009) Chemical complexity of volatiles from plants induced by multiple attack. Nat Chem Biol 5:317–324

Dixon FG (2000) Insect predator–prey dynamics: ladybird beetles and biological control. Cambridge University Press, UK

Dolphin K, Quicke DLJ (2001) Estimating the global species richness of an incompletely described taxon: an example using parasitoid wasps (Hymenoptera: Braconidae). Biol J Linn Soc Lond 73:279–286

Dong WX, Zhang F, Fang YL, Zhang ZN (2008) Electroantennogram responses of aphid parasitoid *Aphidius gifuensis* to aphid pheromones and host–plant volatiles. Chinese J Ecol 27:591–595

Dorn S, Hern A, Mattiacci L (2002) Time course of induced volatile emission of mature fruits upon herbivory, and response of conspecific adult herbivores and of a natural antagonist. IOBC/WPRS Bull 25:99–102

Drukker B, Bruin J, Sabelis MW (2000) Anthocorid predators learn to associate herbivore-induced plant volatiles with presence or absence of prey. Physiol Entomol 25:260–265

Erbilgin N, Raffa KF (2000) Effects of host tree species on attractiveness of tunnelling pine engravers, *Ips pini* (Coleoptera: Scolytidae), to conspecifics and insect predators. J Chem Ecol 26:823–840

Ero MM, Neale CJ, Hamacek E, Peek T, Clarke AR (2011) Preference and performance of *Diachasmimorpha kraussii* (Fullaway) (Hymenoptera: Braconidae) on five commercial fruit species. J Appl Entomol 135:214–224

Francis F, Lognay G, Haubruge E (2004) Olfactory responses to aphid and host plant volatile releases: (E)-b-Farnesene an effective kairomone for the predator *Adalia bipunctata*. J Chem Ecol 30:741–755

Fukushima J, Kainoh Y, Honda H, Takabayashi J (2002) Learning of herbivore-induced and nonspecific plant volatiles by a parasitoid, *Cotesia kariyai*. J Chem Ecol 28:579–586

Gencer NS, Kumral NA, Sivritepe HO, Seidi M, Susurluk H, Senturk B (2009) Olfactory response of the ladybird beetle *Stethorus gilvifrons* to two preys and herbivore-induced plant volatiles. Phytoparasitica 37:217–224

Girling RD, Hassall M (2008) Behavioural responses of the seven-spot ladybird *Coccinella septempunctata* to plant headspace chemicals collected from four crop *Brassicas* and *Arabidopsis thaliana*, infested with *Myzus persicae*. Agric Forest Entomol 10:297–306

Girling RD, Hassall M, Turner JG, Poppy GM (2006) Behavioural responses of the aphid parasitoid *Diaeretiella rapae* to volatiles from Arabidopsis thaliana induced by *Myzus persicae*. Entomol Exp Appl 120:1–9

Girling RD, Stewart-Jones A, Dherbecourt J, Staley JT, Wright DJ, Poppy GM (2011) Parasitoids select plants more heavily infested with their caterpillar hosts: a new approach to aid interpretation of plant headspace volatiles. Proc R Soc B 278(1718):2646–2653

Glinwood R, Ahmed E, Qvarfordt E, Ninkovic V, Pettersson J (2009) Airborne interactions between undamaged plants of different cultivars affect insect herbivores and natural enemies. Arthropod–Plant Interact 3:215–224

Gu H, Dorn S (2000) Genetic variation in behavioral response to herbivore-infested plants in the parasitic wasp, *Cotesia glomerata* (L.) (Hymenoptera: Braconidae). J Insect Behav 13:141–156

Halitschke R, Stenberg JA, Kessler D, Kessler A, Baldwin IT (2008) Shared signals—"alarm calls" from plants increase apparency to herbivores and their enemies in nature. Ecol Lett 11:24–34

Hanumantharaya L, Basavana goud K, Krishna Naik L, Kulkarni KA (2010) Electroantennogram responses of *Chrysoperla carnea* (Stephens) and *Helicoverpa armigera* (Hubner) to volatiles of different cultivars of cotton. Karnataka J Agric Sci 23:123–126

Hatano E, Kunert G, Michaud JP, Weisser WW (2008) Chemical cues mediating aphid location by natural enemies. Eur J Entomol 105:797–806

Heit GE, Sardoy P, Cohen GR, Mareggiani G (2007) Locomotor activity of *Cycloneda sanguinea* (Coleoptera: Coccinellidae) exposed to volatile semiochemicals and to direct contact with the odour source. Rev Soc Entomol Argent 66:197–203

Henneman ML, Dyreson EG, Takabayashi J, Raguso RA (2002) Response to walnut olfactory and visual cues by the parasitic wasp *Diachasmimorpha juglandis*. J Chem Ecol 28:2221–2244

Hern A, Dorn S (2002) Induction of volatile emissions from ripening apple fruits infested with *Cydia pomonella* and the attraction of adult females. Entomol Exp Appl 102:145–151

Hilker M, Bläske V, Kobs C, Dippel C (2000) Kairomonal effects of sawfly sex pheromones on egg parasitoids. J Chem Ecol 26:2591–2601

Hilker M, Kobs C, Varma M, Schrank K (2002) Insect egg deposition induces *Pinus sylvestris* to attract egg parasitoids. J Exp Biol 205:455–461

Hoballah ME, Turlings TCJ (2005) The role of fresh versus old leaf damage in the attraction of parasitic wasps to herbivore-induced maize volatiles. J Chem Ecol 31:2003–2018

Hoballah MEF, Tamò C, Turlings TCJ (2002) Differential attractiveness of induced odors emitted by eight maize varieties for the parasitoid *Cotesia marginiventris*: is quality or quantity important? J Chem Ecol 28:951–968

Ibrahim MA, Nissinen A, Holopainen JK (2005) Response of *Plutella xylostella* and its parasitoid *Cotesia plutellae* to volatile compounds. J Chem Ecol 31:1969–1984

Ichiki RT, Kainoh Y, Yamawaki Y, Nakamura S (2011) The parasitoid fly *Exorista japonica* uses visual and olfactory cues to locate herbivore-infested plants. Entomol Exp Appl 138:175–183

Inbar M, Gerling D (2008) Plant-mediated interactions between whiteflies, herbivores, and natural enemies. Annu Rev Entomol 53:431–448

Ishiwari H, Suzuki T, Maeda T (2007) Essential compounds in herbivore-induced plant volatiles that attract the predatory mite *Neoseiulus womersleyi*. J Chem Ecol 33:1670–1681

Jang EB, Messing RH, Klungness LM, Carvalho LA (2000) Flight tunnel responses of *Diachasmimorpha longicaudata* (Ashmead) (Hymenoptera: Braconidae) to olfactory and visual stimuli. J Insect Behav 13:525–538

Jönsson M (2005) Responses to oilseed rape and cotton volatiles in insect herbivores and parasitoids. PhD thesis, Swedish University of Agricultural Sciences, Alnarp, Sweden.

Kalule T, Wright DJ (2004) The influence of cultivar and cultivar-aphid odours on the olfactory response of the parasitoid *Aphidius colemani*. J Appl Entomol 128:120–125

Kappers IF, Aharoni A, van Herpen TWJM, Luckerhoff LLP, Dicke M, Bouwmeester HJ (2005) Genetic engineering of terpenoid metabolism attracts bodyguards to Arabidopsis. Science 309:2070–2072

Karimy A, Ashouri A, Kharazi Pakdel A, Goldansaz SH, Kosari AA, Moayeri HR (2006) Olfactory response of the predatory bug, *Orius albidipennis* to volatile blends of strawberry and cucumber infested by spider mite. Commun Agric Appl Biol Sci 71:399–402

Kessler A, Baldwin IT (2001) Defensive function of herbivore-induced plant volatile emissions in nature. Science 291:2141–2144

Krugner R, Johnson MW, Daane KM, Morse JG (2008) Olfactory responses of the egg parasitoid, *Gonatocerus ashmeadi* Girault (Hymenoptera: Mymaridae), to host plants infested by *Homalodisca vitripennis* (Germar) (Hemiptera: Cicadellidae). Biol Control 47:8–15

Le Rü B, Makosso MJP (2001) Prey habitat location by the cassava mealybug predator *Exochomus flaviventris*: Olfactory responses to odor of plant, mealybug, plant-mealybug complex, and plant-mealy-natural enemy complex. J Insect Behav 14:557–572

Liu SS, Jiang LH (2003) Differential parasitism of *Plutella xylostella* (Lepidoptera: Plutellidae) larvae by the parasitoid *Cotesia plutellae* (Hymenoptera: Braconidae) on two host plant species. Bull Entomol Res 93:65–72

Liu Y, Guo GX, Chen JL, Ni HX (2005) Behavioral and electrophysiological responses of four predatory insect species to semiochemicals of wheat. Acta Entomol Sin 48:161–165

Maeda T, Liu Y (2006) Intraspecific variation in the olfactory response of the predatory mite *Neoseiulus womersleyi* Schicha (Acari: Phytoseiidae) to different amount of spider mite-infested plant volatiles. Appl Entomol Zool 41:209–215

Maeda T, Takabayashi J, Yano S, Takafuji A (2000) The effects of rearing conditions on the olfactory response of predatory mites, *Phytoseiulus persimilis* and *Amblyseius womersleyi* (Acari: Phytoseiidae). Appl Entomol Zool 35:345–351

Maeda T, Liu Y, Ishiwari H, Shimoda T (2006) Conditioned olfactory responses of a predatory mite *Neoseiulus womersleyi*, to volatiles from prey-infested plants. Entomol Exp Appl 121:167–175

Manrique V, Jones WA, Williams LH III, Bernal JS (2005) Olfactory responses of *Anaphes iole* (Hymenoptera: Mymaridae) to volatile signals derived from host habitats. J Insect Behav 18:89–104

McGregor RR, Gillespie DR (2004) Olfactory responses of the omnivorous generalist predator *Dicyphus hesperus* to plant and prey odours. Entomol Exp Appl 112:201–205

Mehrnejad RM, Copland MJW (2006) Behavioral responses of the parasitoid *Psyllaephagus pistaciae* (Hymenoptera: Encyrtidae) to host plant volatiles and honeydew. Entomol Sci 9:31–37

Mithöfer A, Wanner G, Boland W (2005) Effects of feeding *Spodoptera littoralis* on lima bean leaves. II. Continuous mechanical wounding resembling insect feeding is sufficient to elicit herbivory-related volatile emission. Plant Physiol 137:1160–1168

Moayeri HRS, Ashouri A, Poll L, Enkegaard A (2007) Olfactory response of a predatory mirid to herbivore induced plant volatiles: multiple herbivory vs. single herbivory. J Appl Entomol 131:326–332

Mumm R, Hilker M (2005) The significance of background odour for an egg parasitoid to detect plants with host eggs. Chem Senses 30:337–343

Ngumbi EN, Ngi-Song AJ, Njagi ENM, Torto R, Wadhams LJ, Birkett MA, Pickett JA, Overholt WA, Torto B (2005) Responses of the stem borer larval endoparasitoid *Cotesia flavipes* (Hymenoptera: Braconidae) to plant derived synomones: laboratory and field cage experiments. Biocontrol Sci Technol 15:271–279

Ninkovic V, Pettersson J (2003) Searching behaviour of the seven spotted ladybird, *Coccinella septempunctata*—effects of plant–plant odour interaction. Oikos 100:65–70

Ninkovic V, Al Abassi S, Pettersson J (2001) The influence of aphid-induced plant volatiles on ladybird beetle searching behavior. Biol Control 21:191–195

Ojeda-Camacho M, Rodríguez LC, Niemeyer MH (2001) Evaluación olfactométrica del parasitoide *Aphidius ervi* (Hymenoptera: Braconidae), de diferentes proveniencias y niveles de experiencia de oviposición, frente a volátiles de plantas de complejos planta-hospedero. Rev Chil Entomol 28:63–69

Oppenheim SJ, Gould F (2002) Is attraction fatal? The effects of herbivore-induced plant volatiles on herbivore parasitism. Ecology 83:3416–3425

Pérez ML, Argdín MF, Powell W (2007) Foraging behaviour of the parasitoid *Lysiphlebus testaceipes* (Hymenoptera: Braconidae) in response to plant volatiles, with reference to biocontrol of aphids in peri-urban vegetable production systems. Biocontrol Sci Technol 7:677–686

Pettersson EM (2001) Volatile attractants for three Pteromalid parasitoids attacking concealed spruce bark beetles. Chemoecology 11:89–95

Pettersson J, Ninkovic V, Glinwood R, Al Abassi S, Birkett MA, Pickett JA, Wadhams L (2008) Chemical stimuli supporting foraging behavior of *Coccinella septempunctata* L. (Coleoptera: Coccinellidae): volatiles and allelobiosis. Appl Entomol Zool 43:315–321

Pierre PS, Dugravot S, Ferry A, Soler R, van Dam NM, Cortesero A-M (2011) Aboveground herbivory affects indirect defences of brassicaceous plants against the root feeder *Delia radicum* Linnaeus: laboratory and field evidence. Ecol Entomol 36:326–334

Pinto ML, Wajnberg E, Colazza S, Curty C, Fauvergue X (2004) Olfactory response of two aphid parasitoids, *Lysiphlebus testaceipes* and *Aphidius colemani*, to aphid-infested plants from a distance. Entomol Exp Appl 110:159–164

Poelman EH, van Loon JJ, Dicke M (2008) Consequences of variation in plant defense for biodiversity at higher tropic levels. Trends Plant Sci 13:534–54D

Puente M, Magori K, Kennedy GG, Gould F (2008) Impact of herbivore-induced plant volatiles on parasitoid foraging success: a spatial simulation of the *Cotesia rubecula*, *Pieris rapae*, and *Brassica oleracea* system. J Chem Ecol 34:959–70

Raina R, Joseph M, Avalokiteswar S (2004) Electroantennogram responses of *Chrysoperla carnea* (Stephens) to volatiles. Indian J Exp Biol 42:1230–1234

Rasmann S, Kollner TG, Degenhardt J, Hiltpold I, Toepfer S, Kuhlmann U, Gershenzon J, Turlings TCJ (2005) Recruitment of entomopathogenic nematodes by insect-damaged maize roots. Nature 434:732–737

Raymond B, Darby AC, Douglas AE (2000) The olfactory responses of coccinellids to aphids on plants. Entomol Exp Appl 95:113–117

Reddy GVP (2002) Plant volatiles mediate orientation and plant preference by the predator *Chrysoperla carnea* Stephens (Neuroptera: Chrysopidae). Biol Control 25:49–55

Reddy GVP, Guerrero A (2004) Interactions of insect pheromones and plant semiochemicals. Trends Plant Sci 9:253–261

Reddy GVP, Guerrero A (2010) New pheromones and insect control strategies. Vitam Horm 83:493–519

Reddy GVP, Raman A (2011) Visual cues are relevant in behavioral control measures for *Cosmopolites sordidus* (Coleoptera: Curculionidae). J Econ Entomol 104:436–442

Reddy GVP, Holopainen JK, Guerrero A (2002) Olfactory responses of *Plutella xylostella* natural enemies to host pheromone, larval frass, and green leaf cabbage volatiles. J Chem Ecol 28:131–143

Reddy GVP, Tabone E, Smith MT (2004) Mediation of host selection and oviposition behavior in the diamondback moth *Plutella xylostella* and its predator *Chrysoperla carnea* by chemical cues from cole crops. Biol Control 29:270–277

Roda AL, Baldwin IT (2003) Molecular technology reveals how the induced direct defenses of plants work. Basic Appl Ecol 4:15–26

Röse USR, Tumlinson JH (2005) Systemic induction of volatile release in cotton: how specific is the signal to herbivory? Planta 222:327–335

Sanchez JA, Gillespie DR, McGregor RR (2004) Plant preference in relation to life history traits in the zoophytophagous predator *Dicyphus hesperus*. Entomol Exp Appl 112:7–19

Schaller M, Nentwig W (2000) Olfactory orientation of the seven-spot ladybird beetle, *Coccinella septempunctata* (Coleoptera: Coccinelidae): attraction of adults to plants and conspecific females. Eur J Entomol 97:155–159

Seenivasagan T, Paul AVN (2011) Electroantennogram and flight orientation response of *Cotesia plutellae* to hexane extract of cruciferous host plants and larvae of *Plutella xylostella*. Entomol Res 41:7–17

Shimoda T (2010) A key volatile infochemical that elicits a strong olfactory response of the predatory mite *Neoseiulus californicus*, an important natural enemy of the two-spotted spider mite *Tetranychus urticae*. Exp Appl Acarol 50:9–22

Shimoda T, Dicke M (2000) Attraction of a predator when can it be adaptive? To chemical information related to nonprey. Behav Ecol 11:606–613

Shimoda T, Takabayashi J (2001) Response of *Oligota kashmirica benefica*, a specialist insect predator of spider mites, to volatiles from prey-infested leaves under both laboratory and field conditions. Entomol Exp Appl 101:41–47

Shimoda T, Ozawa R, Arimura G, Takabayashi J, Nishioka T (2002) Olfactory responses of two specialist insect predators of spider mites toward plant volatiles from lima bean leaves induced by jasmonic acid and/or methyl salicylate. Appl Entomol Zool 37:535–541

Shimoda T, Ozawa R, Sano K, Yano E, Takabayashi J (2005) The involvement of volatile infochemicals from spider mites and from food–plants in prey location of the generalist predatory mite *Neoseiulus californicus*. J Chem Ecol 31:2019–2032

Shiojiri K, Takabayashi J, Yano S, Takafuji A (2000) Flight response of parasitoids toward plant-herbivore complexes: a comparative study of two parasitoid-herbivore systems on cabbage plants. Appl Entomol Zool (Jpn) 35:87–92

Shiojiri K, Takabayashi J, Yano S, Takafuji A (2001) Infochemically mediated tritrophic interaction webs on cabbage plants. Popul Ecol 43:23–29

Silva CC, Moraes MCB, Laumann RA, Borges M (2006) Sensory response of the egg parasitoid *Telenomus podisi* to stimuli from the bug *Euschistus heros*. Pesq Agropec Bras 41:1093–1098

Silva JWP, Bento JMS, Zucchi RA (2007) Olfactory response of three parasitoid species (Hymenoptera: Braconidae) to volatiles of guavas infested or not with fruit fly larvae (Diptera: Tephritidae). Biol Control 41:304–311

Smid HM, van Loon JJA, Posthumus MA, Vet LEM (2002) GC-EAG-analysis of volatiles from Brussels sprouts plants damaged by two species of Pieris caterpillars: olfactory receptive range of a specialist and a generalist parasitoid wasp species. Chemoecology 12:169–176

Soler R, Bezemer TM, Van der Putten WH, Vet LEM, Harvey JA (2005) Root herbivore effects on aboveground herbivore, parasitoid and hyperparasitoid performance via changes in plant quality. J Anim Ecol 74:1121–1130

Soler R, Harvey JA, Kamp AFD, Vet LEM, Van der Putten WH, Van Dam NM, Stuefer JF, Gols R, Hordijk CA, Bezemer MT (2007) Root herbivores influence the behaviour of an aboveground parasitoid through changes in plant-volatile signals. Oikos 116:367–376

Soroka JJ, Dosdall LM, Olfert OO, Seidle E (2004) Root maggots (*Delia* spp., Diptera: Anthomyiidae) in prairie canola (*Brassica napus* L. and *B. rapa* L.): spatial and temporal surveys of root damage and prediction of damage levels. Can J Plant Sci 84:1171–1182

Symondson WOC, Sunderland KD, Greenstone MH (2002) Can generalist predators be effective biocontrol agents? Annu Rev Entomol 47:561–594

Sznajder B, Sabelis MW, Egas M (2010) Response of predatory mites to a herbivore-induced plant volatile: genetic variation for context-dependent behaviour. J Chem Ecol 36:680–688

Takahashi H, Takafuji A, Takabayashi J, Yano S, Shimoda T (2001) Seasonal occurrence of specialist and generalist insect predators of spider mites and their response to volatiles from spider-mite-infested plants in Japanese pear orchards. Exp Appl Acarol 25:393–402

Takemoto H, Powell W, Pickett J, Kainoh Y, Takabayashi J (2009) Learning is involved in the response of parasitic wasps *Aphidius ervi* (Haliday) (Hymenoptera: Braconidae) to volatiles from a broad bean plant, *Vicia faba* (Fabaceae), infested by aphids *Acyrthosiphon pisum* (Harris) (Homoptera: Aphididae). Appl Entomol Zool 44:23–28

Tamò C, Ricard I, Held M, Davison AC, Turlings TCJ (2006) A comparison of naïve and conditioned responses of three generalist endoparasitoids of lepidopteran larvae to host-induced plant odours. Anim Biol 56:205–220

Tapia DH, Morales F, Grez AA (2010) Olfactory cues mediating prey-searching behavior in interacting aphidophagous predators: are semiochemicals key factors in predator-facilitation. Entomol Exp Appl 137:28–35

Tatemoto S, Shimoda T (2008) Olfactory responses of the predatory mites (*Neoseiulus cucumeris*) and insects (*Orius strigicollis*): to two different plant species infested with onion thrips (*Thrips tabaci*). J Chem Ecol 34:605–613

Tinzaara W, Gold CS, Dicke M, van Huis A (2005) Olfactory responses of banana weevil predators to volatiles from banana pseudostem tissue and synthetic pheromone. J Chem Ecol 31:1537–1553

Uefune M, Nakashima Y, Tagashira E, Takabayashi J, Takagi M (2010) Response of *Wollastoniella rotunda* (Hemiptera: Anthocoridae) to volatiles from eggplants infested with its prey *Thrips palmi* and *Tetranychus kanzawai*: prey species and density effects. Biol Control 54:19–22

Unsicker SB, Kunert G, Gershenzon J (2009) Protective perfumes: the role of vegetative volatiles in plant defense against herbivores. Curr Opin Plant Biol 12:479–85

Van Loon JJA, De Vos EW, Dicke M (2000) Orientation behaviour of the predatory hemipteran *Perillus bioculatus* to plant and prey odors. Entomol Exp Appl 96:51–58

Van Poecke RMP, Posthumus MA, Dicke M (2001) Herbivore-induced volatile production by *Arabidopsis thaliana* leads to attraction of the parasitoid *Cotesia rubecula*: Chemical, behavioural, and gene-expression analysis. J Chem Ecol 27:1911–1928

Van Tol RWHM, van der Sommen ATC, Boff MIC, van Bezooijen J, Sabelis MW, Smits PH (2001) Plants protect their roots by alerting the enemies of grubs. Ecol Lett 4:292–294

Van Wijk M, De Bruijn PJA, Sabelis MW (2008) Predatory mite attraction to herbivore-induced plant odor is not a consequence of attraction to individual herbivore-induced plant volatiles. J Chem Ecol 34:791–803

Verheggen F, Arnaud L, Capella Q, Francis F, Haubruge E (2005) Perception of aphid infested tomato plant volatiles by the predator Episyrphus balteatus. Comp Biochem Physiol 141A:225–236

Vet LEM (2001) Parasitoid searching efficiency links behavior to population processes. Appl Entomol Zool 36:399–408

Vuorinen T, Nerg A-M, Ibrahim MA, Reddy GVP, Holopainen JK (2004) Emission of *Plutella xylostella*–induced compounds from cabbages grown at elevated CO_2 and orientation behavior of the natural enemies. Plant Physiol 135:1984–1992

Wang Q, Gu H, Dorn S (2003) Selection on olfactory response to semiochemicals from a plant–host complex in a parasitic wasp. Heredity 91:430–435

Wei JN, Kang L (2006) Electrophysiological and behavioral responses of a parasitic wasp to plant volatiles induced by two leaf miner species. Chem Senses 31:467–477

Wen-xia D, Zhong-ning Z, Yu-ling F, Feng Z, Wei K (2000) Response of parasitoid *Microplitis mediator* to plant volatiles in an olfactometer. Insect Sci 7:344–350

Wyckhuys KA, Heimpel GE (2007) Response of the soybean aphid parasitoid *Binodoxys communis* to olfactory cues from target and non-target host-plant complexes. Entomol Exp Appl 123:149–158

Yoneya K, Kugimiya S, Takabayashi J (2009) Can herbivore-induced plant volatiles inform predatory insect about the most suitable stage of its prey? Physiol Entomol 34:379–386

Yu H, Zhang Y, Wyckhuys KAG, Wu K, Gao X, Guo Y (2010) Electrophysiological and behavioral responses of *Microplitis mediator* (Hymenoptera: Braconidae) to caterpillar-induced volatiles from cotton. Environ Entomol 39:600–609

Zhu J, Park KC (2005) Methyl salicylate, a soybean aphid induced plant volatile attractive to the predator *Coccinella septempunctata*. J Chem Ecol 31:1733–1746

Plant Resistance to Insect Herbivory

Jurgen Engelberth

Abstract Plants are the major food source for most insects. While insects have developed various feeding strategies, plants respond by activating distinct signaling pathways resulting in the production of defensive compounds. Important regulators in this signaling system are compounds in the insect saliva, which are often modified plant molecules. The perception of these elicitor initiates signaling events like calcium release, oxidative burst, and several protein kinases, resulting in the activation of the octadecanoid signaling pathway with jasmonic acid (JA) as the major regulator of herbivore-specific defense response. JA is essential in inducing the production of toxic secondary metabolites, volatile organic compounds, and antidigestive proteins like proteinase inhibitors and polyphenol oxidases. Additionally, natural enemies of the attacking insect herbivore are attracted by volatiles release or the production of extrafloral nectar. Taken together, these measures provide a broad protection against insect herbivores. A detailed understanding of the underlying mechanisms will give us new insights into the coevolutionary processes that govern plant-insect interactions and may also lead to new approaches for the development of more ecological pest management strategies in an increasing agricultural environment.

1 Introduction

Plants in their natural or agricultural habitats are constantly exposed to a plethora of pest and pathogens. Among the pests, insect herbivores in particular have learned in over 350 million years of coevolution to identify appropriate host plants for feeding and oviposition. Since about half of the one million insects on this planet are herbivores, the survival of plants strongly depends on their ability to respond

J. Engelberth (✉)
Department of Biology, University of Texas at San Antonio, San Antonio, TX, USA
e-mail: jurgen.engelberth@utsa.edu

G. Witzany and F. Baluška (eds.), *Biocommunication of Plants*,
Signaling and Communication in Plants 14,
DOI 10.1007/978-3-642-23524-5_16,

distinctly but flexibly to this threat (Gatehouse 2002, Howe and Jander 2008, Wu and Baldwin 2010).

Insects feed on all parts of a plant, from flowers to leaves and from stems to roots, and in doing so have developed into specialized groups. Generalists among the herbivorous insects feed on many different plant species but generally avoid those with higher toxicity. Specialists, on the other side, can only exist on one or few species within one plant family but are often resistant to the toxic metabolites of the plant. Insects have also developed a great variety of feeding mechanisms. Chewing insects like caterpillars and certain beetles use their mandibles to remove relatively large chunks of plant material. Leaf miners prefer to eat on the mesophyll of leaves but leave the epidermis intact. Mites and thrips are piercing/sucking herbivores and use needle-like mouthparts to suck the liquid cell content from damaged cells. While these types of herbivory cause significant cell damage, other insects have developed a feeding strategy that avoids cell damage. Aphids and whiteflies, for example, use their stylets to access the phloem, thereby avoiding any actual cell damage.

Under this pressure, plants were forced to develop strategies on their own to prevent or reduce damage by this diverse array of insect herbivores. Physical barriers like thorns, trichomes, and a thick cuticle often provide a first line of defense and help to reduce damage significantly. Additionally, plants have developed strategies to recognize and respond to movement, mechanical damage, and factors in the oral secretions of insect herbivores. Upon receiving one or more of these stimuli, plants activate a complex regulatory network resulting in the production of metabolites and proteins that help to protect them. Plants produce an astonishing number of more than 500,000 secondary metabolites (Mendelsohn and Balick 1995), which are of crucial importance in plant-insect interactions. Some of these compounds are part of a constitutive defense, which is based on the permanent presence of toxic compounds. However, while this provides some basic protection, insects may adapt quickly to tolerate these compounds, as it is often the case with specialist herbivores. Likely to be more effective are therefore inducible defenses, and it is evident that most plants start producing toxic, repelling, and antidigestive compounds only upon actual insect-herbivore damage. And it is in this context that many secondary metabolites exhibit their biological and ecological function. Another form of direct defense is provided by proteins, which inhibit the digestion of nutrients in the insect gut like proteinase inhibitors.

Besides these direct defenses, plants have developed an additional system to reduce herbivore damage, in which natural enemies of the attacking insect are attracted by the release of volatile organic compounds or the provision of food in form of extrafloral nectar. Parasitic wasps and ants, for example, are thus attracted to plants under attack by these cues and help to reduce damage.

But how do plants recognize insect herbivory and activate these defenses? Since investing in defense is costly, plants have to make sure that the effort is worth the investment of resources in the form of defensive chemicals. In the following, some of the strategies used by plants to fend off herbivores are reviewed. Also, the active or involuntary roles that herbivores play in this interaction are described.

The resulting picture is that of a multilayered interaction, which allows them to coexist even though both may take some damage in the process.

2 Recognition of Herbivory/Hostility

2.1 Discerning Between Mechanical Wounding and Herbivory

While the infliction of mechanical damage is the obvious way in which plants recognize insect herbivory, structures on the surface of the plant like glandular trichomes may already sense the presence of potential herbivores. Glandular trichomes, which can be found, for example, on the leaf surface of tomato plants, are very sensitive structures and break at the slightest touch. Besides instantly releasing stored volatiles, this specific kind of mechanical damage initiates signaling processes and the upregulation of distinct defenses (Peiffer et al. 2009). More often, however, plants recognize insect herbivory only when those start feeding. Insects often feed in a very distinct manner, which is characterized by the way they use their mouthparts as well as their typical feeding pattern. This form of repeated wounding in a spatial and temporal context may allow plants already to distinguish between insect herbivory and simple mechanical damage as it may be caused by wind or hail. A striking example for the effects of a feeding pattern as the inducer of antiherbivore defenses was provided by Mithoefer et al. (2005). By designing a technical device named MecWorm, they were able to remove tissue portions from a leaf of a lima bean in a way that was comparable to actual insect feeding. As a result, these plants showed almost identical defense responses as those actually damaged by a caterpillar. Thus far, several plant species have been identified that may detect insect herbivory simply by the distinct pattern of mechanical damage. However, in most examples described to date, recognition of insect herbivory is detected by a combination of mechanical damage and the simultaneous application of elicitor compounds abundant in the insect saliva.

2.2 Recognition of Compounds in the Insect Saliva

Although the vast majority of insects feed on plants, only few insect-herbivore-derived elicitors are known. However, those that have been identified show some intriguing features with regard to their specific activity, but also with regard to their biochemical origin (Fig. 1).

Best characterized among the known elicitors from insect herbivores are the fatty acid-amino acid conjugates (FAC). FAC were first identified by Alborn et al. (1997) in the oral secretions of *Spodoptera exigua*. When plants were mechanically damaged and extracts or fractions of the oral secretions (OS) applied to the damage

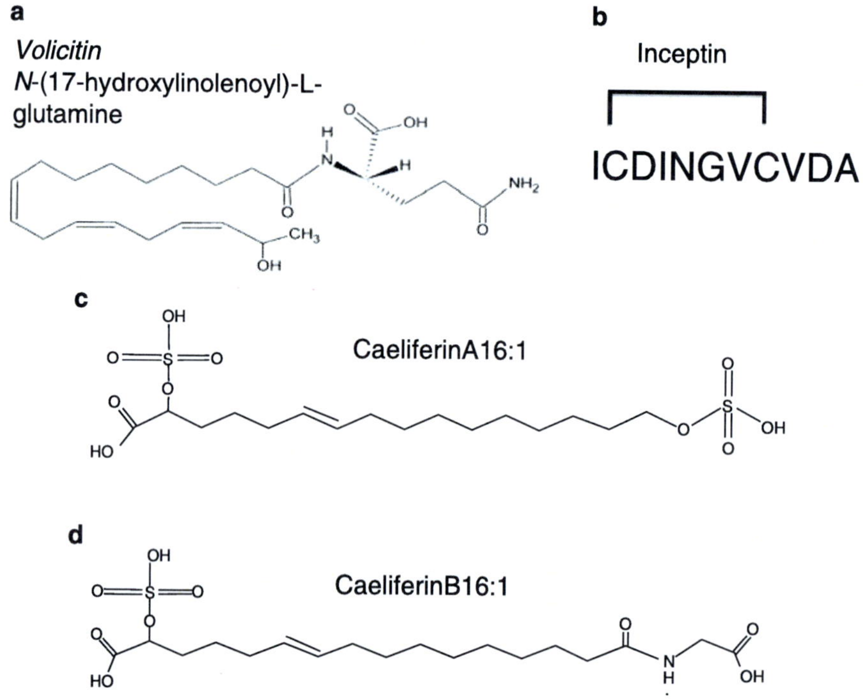

Fig. 1 Structures of major insect-derived elicitors. Linolenic acid-amino acid conjugates (**a**: volicitin) that have been found to be active as elicitors of VOC release in corn seedlings. These compounds and their linoleic acid analogs have been found in the regurgitant of the larvae of numerous lepidopteran species and more recently in crickets and *Drosophila* larvae. Inceptin (**b**) is a proteolytic fragment of the γ-subunit of the chloroplastic ATP synthase and was isolated from oral secretions of *Spodoptera frugiperda*. Caeliferins were isolated and identified from regurgitant of *Schistocerca Americana*. Caeliferins in the A group with hydroxyls in the α and ϖ position are sulfated (**c**). Caeliferins of type B are diacids with a sulfate in the α position and a glycine conjugated to the ϖ carboxyl (**d**). Little is known to date about the biological activity of B-type caeliferins

site, the release of volatiles was almost identical when compared to actual insect damage (Fig. 2). In contrast, mechanical wounding alone was not sufficient to induce comparable qualities and quantities (Alborn et al. 1997; Baldwin 1990; Halitschke et al. 2001; Reymond et al. 2004; Wu et al. 2007; Engelberth et al. 2007). Further investigations into the composition of OS lead to the identification of volicitin, named after its capacity to induce volatile release from corn (Fig. 1a) (Alborn et al. 1997). Volicitin is composed of linolenic acid, which conjugated to glutamine. Furthermore, the linolenic acid portion is hydroxylated in position 17. Since its initial discovery, a great variety of different FAC have been identified not only in different lepidopteran species but also in crickets and fruit flies, and most of them were found to exhibit elicitor-like activities when applied to plants (Halitschke et al. 2001; Pohnert et al. 1999; Spiteller and Boland 2003; Spiteller et al. 2004). Common to most FAC is that the fatty acid part is either linoleic or

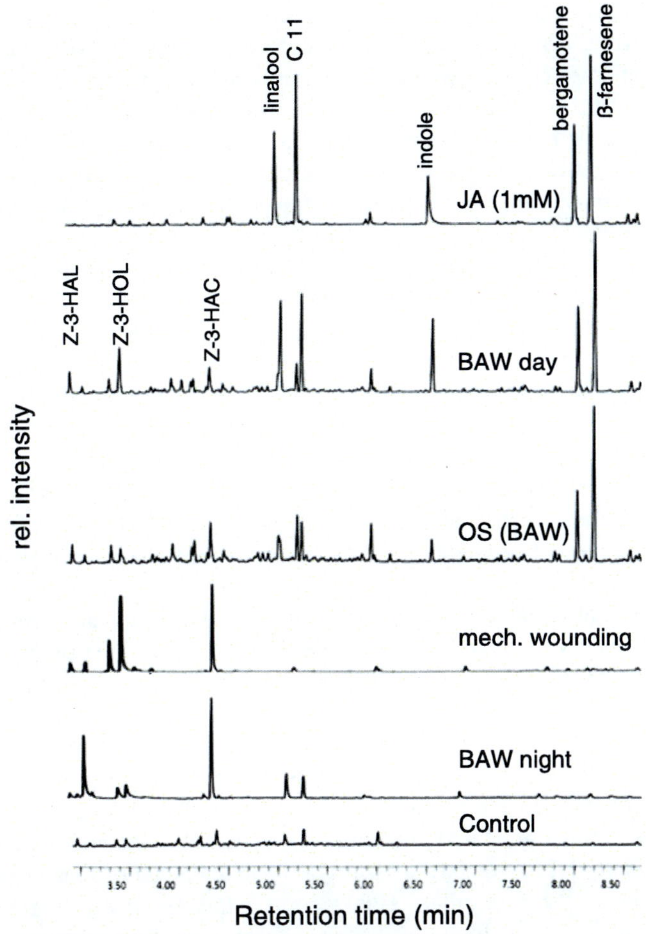

Fig. 2 Volatile profiles of corn (*Zea mays* var. Delprim) treated with insect herbivore's (*Spodoptera exigua*) larvae during daytime and nighttime, oral secretions (*OS*) from the same caterpillar, mechanical damage, jasmonic acid (*JA*), and untreated. Daytime herbivory, OS, and JA induce almost similar profiles. Nighttime herbivory and mechanical damage (within first hour) mainly results in green leaf volatile (*GLV*) release. Z-2-HAL Z-3-hexenal, Z-2-HOL Z-3-hexenol, Z-2-HAC Z-3-hexenyl acetate, C11 3*E*-4,8-dimethyl-1,3,7-nonatriene (DMNT)

linolenic acid, whereas most variations concern the associated amino acid. For example, both glutamine and glutamate conjugates have been identified in addition to several other amino acids.

While FAC exhibit a broad range of activity among various plant species, little is known about the immediate signaling events elicited by these compounds. Truitt et al. (2004) found that volicitin binds rapidly to plasma membranes isolated from corn leaves in a typical receptor-ligand fashion. While this implies the existence of a specific FAC receptor on the cell surface, no such protein has been identified to date.

An alternative mode of activity was proposed by Maischak et al. (2007), who described the channel-forming properties of OS resulting in distinct ion fluxes and depolarization of cells. However, FAC themselves are physically not able to form stable ion channels, and therefore, the described channel-forming activity has to be attributed to other components in the OS. Also, if this would be the general principle for FAC activity responses to this class of elicitors, they should be more widely distributed among different plant species. However, FAC were found to have no effect, for example, in *Arabidopsis*, lima beans, cotton, and tomato. Therefore, the existence of a specific receptor appears to be the most likely option to explain FAC-induced biological activities.

The biosynthesis of FAC provides an intriguing example for the complexity of plant-insect interactions. FAC are synthesized enzymatically in the midgut of the caterpillar (Pare et al. 1998). However, in order to perform the synthesis, the caterpillar first has to ingest the fatty acid from the plant tissue. The free fatty acid is then rapidly conjugated to an amino acid, preferably glutamine or glutamate, which is provided by the insect. The newly synthesized FAC is then regurgitated and thus applied to the current damage site of the plant, where it exhibits its signaling activity. The interdependence of plant- and insect-derived substrates in the biosynthesis of FAC raised the question as to why insects produce these compounds that are evidently harmful to them. Yoshinaga et al. (2008) found that some FAC may be involved in the nitrogen metabolism of the caterpillar. Through feeding experiments with radiolabeled nitrogen (ammonia) and fatty acids, it was found that the presence of fatty acids in the diet increased the efficiency of nitrogen assimilation in the insect gut by more than 20%. According to their hypothesis, glutamate is first conjugated to fatty acids in the lumen of midgut cells. In the presence of ammonia and the enzyme glutamine synthetase, the conjugated glutamate is then transformed into glutamine and exported to the gut lumen. There, the glutamine-fatty acid conjugate is hydrolyzed and the glutamine reabsorbed into the hemolymph of the caterpillar. This process explains why in previous studies glutamine in FAC was found to originate from the caterpillar and not from the plant. While nitrogen gets fixated in the process, some of the produced FAC are regurgitated and thus become activators of the plant defense responses.

While FAC seem to be the major class of elicitors, other types of insect-derived activators of defense responses have been identified in recent years. A bioassay-based approach to identify ethylene-inducing factors in the oral secretions of *Spodoptera frugiperda* led to the discovery of a proteolytic fragment of the chloroplastic ATP synthase γ-subunit that was named inceptin (Fig. 1b) (Schmelz et al. 2006; Schmelz et al. 2007). Inceptins are small peptides consisting of 11 amino acids and are characterized by a disulfide bond between two cysteines. The sequence of this peptide may differ slightly depending on the plant species on which the caterpillar is feeding. When applied at minute amounts to cowpea (*Vigna unguiculata*) or common bean (*Phaseolus vulgaris*) leaves, inceptin induced a significant release of ethylene but also caused the accumulation of JA and salicylic acid (SA). As for FAC, inceptins are produced from a plant-derived substrate through the proteolytic activity in the gut and are then regurgitated back to the damage site. However, in contrast to FAC, which exhibit their activity in many different plant species (Schmelz et al. 2009), inceptins are quite limited in activating defense signaling in plants other than those described above. For example, soybean, lima bean, tobacco, *Arabidopsis*, maize, and tomato did not respond to this elicitor.

As described for FAC, the activity of inceptin to induce plant defense signaling is consistent with the "guard hypothesis" of plant immunity (Jones and Dangle 2006). In this hypothesis, a modified or damaged "self" is recognized by the host organism rather than compounds from the attacking organism. As such, both FAC and inceptin fulfill this requirement. However, since certain insect herbivores produce these elicitors independent of whether they may be active in a certain plant species, the chemical interaction between plants and their attackers are more complex and seem to be regulated on multiple levels.

A novel class of insect-derived elicitors was isolated and characterized from the oral secretions of a grasshopper (*Schistocerca americana*) by Alborn et al. (2007). Since these elicitors were thus far only found in the suborder Caelifera, they were named caeliferins (Fig. 1c and d). Caeliferins are also fatty acid–based compounds with a chain length between 15 and 19 carbons and are usually saturated or monounsaturated. For caeliferins in the A group, hydroxyls in the α and ϖ position are sulfated (Fig. 1C). Caeliferins of type B are diacids with a sulfate in the α position and a glycine conjugated to the ϖ carboxyl (Fig. 1d). By using a volatile-based bioassay with corn seedlings, caeliferin A 16:1 was found to be the most active compound among this group of elicitors. In a comparatistic study by Schmelz et al. (2009), it was also found that caeliferin A 16:0 was active in *Arabidopsis*. Application of this compound to a wounding site induced a transient ethylene emission and significantly higher JA accumulation when compared to mechanical wounding alone. Thus far, caeliferin A 16:0 is the only insect-derived elicitor with biological activity in this model plant. Otherwise, it has to be mentioned that as for inceptin, the biological activity of caeliferins appears to be very limited. Neither legumes nor solanaceous plants responded to an application of this elicitor with increased defense signaling. In contrast to FAC and inceptin, caeliferins do not seem to be plant-derived compounds. Irregular chain lengths as well as a *trans*-configurated double bond make this rather unlikely.

Although phloem feeders like aphids and whiteflies cause little actual damage, plants have developed mechanisms to recognize this type of herbivory. In contrast to chewing and piercing/sucking insects, which inflict severe tissue damage resulting in the activation of the JA signaling pathway, phloem feeders avoid this kind of defense response by activating salicylic acid–related defense pathways. This kind of signaling is usually associated with pathogen infections, and interestingly, many phloem feeders seem to be recognized by the same plant detection system. Evidence from several plant species like rice, melon, and tomato suggests that R-genes recognize secreted compounds from the herbivore and activate defenses accordingly. For example, *Mi-1* in tomato confers resistance to aphids and whiteflies (Nombela et al. 2003; Rossi et al. 1998), *Bph14* in rice confers resistance to the brown plant hopper (Du et al. 2009), and the *Vat* gene in melon confers resistance to the cotton aphid (Dogimont et al. 2007). All of these R-genes belong to the group of nucleotide binding site-leucine-rich repeat (NBS-LRR) proteins. While this mechanism suggests a gene for gene resistance, as it has been described for plant defenses against pathogens, none of the corresponding virulence factors from the herbivore has been identified.

Other insect-derived effectors of plant defense responses have been described, but very little is known about the signaling pathways they invoke. Bruchins, for example, are long-chain α, ϖ diols esterified on one or both ends with 3-hydroxy propanoic acid. These compounds are part of the oviposition fluid of the pea weevil (*Bruchus pisorum* L.). Upon contact with the plant, bruchins induce the formation of neoplasms, a small tumorlike structure beneath the eggs, which elevates them and inhibits the entry of the larvae into the plant (Doss et al. 2000). Often, these structures simply fall off the plant thereby removing the eggs entirely.

Several proteinaceous effectors have also been characterized. A β-glucosidase from the OS of *Pieris brassicae* hydrolyzes glycosides of terpenes and causes *Arabidopsis* plants to release volatiles (Mattiacci et al. 1995). While this enzyme does not actually elicit defense responses, it contributes to the plant's reservoir of effective defense strategies.

A different type of elicitor was characterized from the OS of *Helicoverpa zea* and *Spodoptera exigua* (Musser et al. 2005). The protein, a glucose oxidase (GOX), suppresses plant defense responses. For the GOX from *Spodoptera exigua*, it was shown that the protein causes SA accumulation, which is thought to suppress JA-related defenses.

It is obvious from those examples that the interactions of insect-derived elicitors and effectors with their host plants are quite complex. Considering the fact that only few of these defense-affecting compounds have been identified to date, one can expect many more of these interactions to be discovered in the future.

2.3 Early Signaling Events Associated with Insect Herbivory

In most instances, insect herbivory is characterized by two distinct events. First, and probably most prominent, is the mechanical damage inflicted to the plant tissue under attack. Second, the application of elicitors abundant in the insect saliva, as they have been described previously, are known to affect plants in a way that is, at least on the physiological level, comparable to actual herbivory. Despite the fact that several classes of insect-derived elicitors have been identified and characterized over the last 15 years, surprisingly little is known about the immediate signaling events triggered by these compounds. Research on insect elicitors' activity during this period was mostly focused on describing effects by these compounds in comparison to mechanical wounding alone. And although differences are quite obvious, the signaling events leading to these differences are only poorly understood and are in dire need of further studies. Nonetheless, a picture is beginning to emerge from multiple studies providing evidence for the involvement of certain signaling pathways in the immediate response to insect herbivory and in particular to the activity of insect elicitors.

But what are those rapid signaling events that are elicited when caterpillar saliva gets into contact with the plant's damage site?

The calcium ion (Ca^{2+}) has been implicated to act as a major signal in the mediation of insect elicitor-induced responses. Ca^{2+} is a ubiquitous second messenger in multiple cellular responses of all eukaryotic cell systems. Under normal conditions, Ca^{2+} levels are usually very low in the cytosol of cells (~ 100 nM). When stimulated, Ca^{2+} is rapidly released into the cytosol from storage compartments like mitochondria, endoplasmic reticulum, vacuole, and the extracellular space, where concentrations of Ca^{2+} can be up to 1 mM. Higher Ca^{2+} levels in the cytosol then activate an array of target proteins like calmodulin, Ca^{2+}-dependent protein kinases, and many other Ca^{2+}-binding proteins, which in turn activate downstream targets of the respective signaling pathway. This may include protein phosphorylation and transcriptional activation of stimulus-specific responses. Although many of the cellular responses involve Ca^{2+}, cells can very well integrate different stimuli by recognizing different frequencies of Ca^{2+} spikes in the cytosol.

In lima beans (*Phaseolus lunatus*), the most significant increases in cytosolic Ca^{2+} levels were found in cell layers adjacent to the herbivore damage site but were also detectable in more distant tissues, albeit somewhat less prominent (Maffei et al. 2004). Compared to mechanical wounding, herbivore-induced levels in cytosolic Ca^{2+} were much higher and suggest that factors in the insect saliva play an important role in the regulation of Ca^{2+} influxes into the cytosol. While this strongly suggests an active role for Ca^{2+} in herbivory-induced signaling, Ca^{2+} receptors mediating this interaction have yet to be identified.

For other plant-insect interactions, downstream signaling units have been characterized. In *Arabidopsis*, IQD1 binds calmodulin in a Ca^{2+}-dependent manner and activates genes involved in glucosinolate biosynthesis (Levy et al. 2005). Also, overexpression of IQD1 negatively affected herbivore performance. In addition to the results describing increased cytosolic Ca^{2+} concentrations in leaf areas adjacent to active herbivory, these results support the importance of Ca^{2+} signaling in the regulation of antiherbivore defenses. However, more research is necessary to further characterize this important signaling pathway, also with regard to the different types of elicitors and their species-specific effects.

Reactive oxygen species (ROS) like superoxide anion ($O2^{-}$), singlet oxygen ($^1O_2$), hydroxyl radical ($^{\cdot}OH$), and hydrogen peroxide (H_2O_2) are often produced by plants in response to various stresses. Probably best characterized in plant-pathogen interaction (Lamb and Dixon 1997), they also seem to play a significant role in herbivore-induced defense responses (Bi and Felton 1995; Leitner et al. 2005; Maffei et al. 2006; Orozco-Cardenas and Ryan 1999; Orozco-Cardenas et al. 2001). ROS can be produced in mitochondria, plastids, peroxisomes, and on the external surface of the plant cell. Defense-related ROS are produced by a multienzyme complex generally referred to as NADPH oxidase, which is located in the plasma membrane of cells. The NADPH oxidase transfers electrons from NADPH to molecular oxygen thereby generating a highly reactive product. There is evidence that the NADPH oxidase is activated by phosphorylation through a calcium-dependent protein kinase resulting in an enhanced activity of the enzyme (Sagi and Fluhr 2001; Keller et al. 1998). On the other hand,

ROS are also known to activate Ca^{2+} channels thereby increasing cytosolic Ca^{2+} concentrations. However, how this process is actually involved in anti-herbivore defense signaling remains unclear. Also, it cannot be excluded that ROS may have a direct effect on the attacking herbivore or play a critical role in the avoidance of secondary pathogen infections as a consequence of herbivore-inflicted damage. In addition, cross-linking of cell wall components like extensins as well as the production of lignin strongly depend on the production of ROS in the cell wall. Interestingly, while insect herbivory induces a strong oxidative burst (often measured as ROS production) at the actual damage site, insect-derived elicitors do not. This suggests that components other than the FAC, inceptins, or caeliferins in the insect saliva are responsible for the activation of this process.

While mitogen-activated protein kinases or MAPK are established as important signaling system in plant-pathogen interactions, little is known about these pathways as regulators of antiherbivore defense responses. However, from the little data available, it seems to be clear that insect-herbivore-induced defense signaling involves several types of MAPK. For example, in tobacco virus-induced gene silencing (VIGS) of the wound-induced protein kinase (WIPK) and the SA-induced protein kinase (SIPK), both members of the MAPK family demonstrated a central role of these enzymes in the signaling process induced after insect herbivory and also after treatment with FAC (Kandoth et al. 2007; Keller et al. 1998). Additionally, it was shown that herbivory and FAC significantly induced the gene expression for these two kinases. Also, in tomato, VIGS studies showed that at least three different MAPK are necessary to fully activate systemin-induced defenses against *Manduca sexta* caterpillars (Kandoth et al. 2007). However, to date studies demonstrating the direct effect of MAPK on JA accumulation and signaling as well as general regulation of defense gene activation are still missing.

3 Regulation of Defense by Jasmonate Signaling

3.1 The Jasmonate Pathway

Most of the countermeasures plants initiate when under insect-herbivore attack are signaled through the octadecanoid signaling pathway with JA and JA-isoleucine (JA-Ile) as the main regulators. But although the main interest for JA arose from its predominant role in the regulation of plant defense responses, it is also an important developmental signal and, for example, regulates pollen development and maturation.

The biosynthesis of JA begins in the chloroplast by incorporating molecular oxygen into α-linolenic acid by a 13-lipoxygenase (LOX), resulting in 13-hydroperoxy linolenic acid (13-HPLA). 13-HPLA is then converted to an unstable allene oxide by the allene oxide synthase (AOS), which represents the bottleneck enzyme for this pathway. The allene oxide undergoes a rapid cyclization by the allene oxide cyclases (AOC). This step also establishes the correct stereochemistry of the resulting 9*S*, 13*S*-12-oxo phytodienoic acid (or *cis*-OPDA). *Cis*-OPDA is an important intermediate of the pathway for it has been demonstrated to exhibit its own JA-independent biological activity. This is partially attributed to a distinct structural feature of the molecule, which contains an α,β-unsaturated carbonyl moiety. This makes it a potential target for nucleophilic attack by $-NH_2$ or –SH groups, thereby forming a stable Michael adduct. This form of protein modification has been shown in the animal system to significantly alter the biochemical properties of enzymes. However, it is unclear if this form of protein modification occurs in plants as a means of biosynthetic regulation.

OPDA is produced in the chloroplast and has to be transferred to the peroxisome for further processing. While the export system from the chloroplast has not yet been identified, import into the peroxisome is facilitated by an ABC transport system. There, the olefinic bond in the pentacyclic system is reduced by the enzyme 12-oxo phytodienoate reductase (OPR). Interestingly, plant genomes often contain several different homologues of this gene (*Zea mays* 8, *Arabidopsis* 6), but usually only one of these OPR genes is involved in the JA biosynthetic pathway. A potential function for the other OPRs may be the more general reduction of the olefinic bond in α,β-unsaturated carbonyls as they occur in many other oxylipin-derived compounds like traumatin, *E*-2-hexenal, and certain phytoprostanes.

After being reduced, the resulting 12-oxo phytodienoic acid undergoes 3 cycles of β-oxidation eventually yielding (+)- *iso* JA (or *cis* (*epi*) JA). While JA was long thought to be the most active jasmonate, it is now clear that for most responses, JA first needs to be conjugated to an amino acid, for example, isoleucine (Kang et al. 2006; Staswick and Tiryaki 2004). This conjugate is then recognized by its receptor and activates JA-related gene expression. The signaling mechanism of JA appears to be quite conserved and bears close resemblance to those activated by other plant hormones like auxin and gibberellins. Best studied so far is the mechanism for JA-Ile signaling in *Arabidopsis*. JA-Ile binds to its receptor COI1 (Thines et al. 2007), which is an essential part of a SCF-protein complex (SCF^{COI1}). The target for this complex is a JAZ protein, which acts as a suppressor of JA-activated transcription factors (Chini et al. 2007; Thines et al. 2007). The binding of the $SCF^{COI1\text{-}JA\text{-}ILE}$-protein complex to JAZ leads to the polyubiquitination and subsequent degradation of the JAZ repressor in a 26S proteasome. Transcription factors like MYC2 then initiate the transcription of typical JA-inducible genes (Chini et al. 2007). Interestingly, among the genes activated by this mechanism are also those for the JAZ proteins, which provides a negative feedback loop in this signaling system.

Wang et al. (2008) reported that the conjugation of JA to amino acids like isoleucine is not the only active jasmonate in tobacco (*Nicotiana attenuata*). By silencing the genes that are primarily responsible for the conjugation (JAR4 and/or JAR6), they found that major defenses against insect herbivory are strongly suppressed. However, adding JA-Ile to a *lox*-silenced plant, which is JA deficient, did not restore full resistance, indicating that JA itself or some other oxylipin is also significantly involved in defense gene regulation. This is further supported by the fact that *jar1* mutants are not male sterile as it has been described for other jasmonate-biosynthesis mutants.

3.2 *Jasmonates-Inducible Coregulation of Metabolic Pathway Genes*

The mechanisms by which JA signals massive reprogramming of gene expression are beginning to resolve, as described above. In the context of plant defense responses, JA provides a main switch that shuts down growth and activates those genes that provide attack-specific protection. In this, JA is often aided by other signaling compounds like ethylene. In response to insect-herbivore damage, JA induces the synthesis of a diverse array of proteinase inhibitors but also genes leading to the production of toxic or deterring secondary metabolites like terpenes, alkaloids, phenylpropanes, and glucosinolates. It is also characteristic for JA to coregulate the complete set of genes required for the respective pathway instead of upregulating just one bottleneck enzyme (Pauwels et al. 2009 and references therein). Consequently, complex mixtures of diverse secondary metabolites can be produced. Often, several pathways for secondary metabolites exist within one plant species and can differentially be activated by JA. In *Arabidopsis*, JA can activate several classes of secondary metabolite including phenylpropanoids, glucosinolates, anthocyanins, and isoprenoids. The induction of either pathway or combinations of several depends on the context in which JA accumulates or is exogenously applied. For example, a cell culture of *Arabidopsis* responds differently to JA treatment than young seedlings growing on an artificial substrate. Interestingly, this coregulatory activity of JA also includes the genes for its own biosynthetic pathway. These pathways are regulated through transcriptional cascades, which suggest the existence of common regulatory elements. The best-characterized activator in this context is the above-described transcription factor MYC2, which appears to initiate many of these regulatory units. This transcriptional regulation also requires common *cis*-elements among the JA-regulated genes, meaning that similar regulatory sequences within the respective promoter regions exist. These sequential functional similarities are not limited to one species but must have evolved in almost all plant species with regard to JA-activated metabolic pathways. In fact, functional orthologs of the activator MYC2, its suppressor JAZ1, and corresponding *cis*-regulatory elements have been identified in *Arabidopsis*, tomato, tobacco, and periwinkle, and appear to be quite conserved. This capacity of JA to assemble complete metabolic pathways is the likely reason why it has become one of the most efficient defense signals in plants.

4 Systemic Signaling

The induction of defenses is not limited to the area of actual damage. Within hours, many inducible defenses are also activated in other undamaged parts of the plant and aid to the protective measures plants undertake to fend off insect herbivores. But although the phenomenon has been known for more than 35 years, little is known about the signaling involved in this process (Green and Ryan 1972; Karban and

Baldwin 1997). In tomato, a series of experiments have shown that JA production at the damage site and JA perception in distal leaves are necessary requirements for long-distant signaling. This suggests that JA or one of its derivatives is actively transported through the phloem (Schilmiller and Howe 2005). While systemin has been thought for long to be the mobile signal, it is clear now that it is only required to potentiate the wound signal but is not actively involved in the long-distance signaling. FAC also seem to play an important part in long-distance signaling in those plants that can recognize these elicitors. In *Nicotiana attenuata*, FAC elicited a rapid activation of MAPK activity in undamaged areas of the same leaf (Wu et al. 2007). In corn (*Zea mays*), treatment with FAC induced JA in distal tissues of the damaged leaf, but no increases in JA were found in basal or systemic tissue (Engelberth et al. 2007). Evidence for the existence of systemic signaling was further provided by gene expression analyses in undamaged leaves. For example, Shen et al. (2000) described an increased accumulation of a sesquiterpene cyclase in systemic leaves after treatment with elicitor, and Park et al. (2010) found a lipoxygenase (LOX 5) induced in systemic parts of the plant. However, in all these cases, little to nothing is known about the actual signaling pathways that enable the plant to alert distant tissues. Electric and hydraulic signaling represents options for long-distance signaling (Malone et al. 1994: Stankovic and Davies 1997). In fact, recent studies on broad bean and barley provided evidence for electric signaling as the mechanism by which these plants facilitate systemic signaling (Zimmermann et al. 2009).

5 Direct Defenses

The defensive measures that can be activated by JA are quite complex. Among the more direct strategies are the biosynthesis of toxic or deterring secondary metabolites and the production of proteins that reduce the nutritional value of the consumed plant material. Secondary metabolites are common to all plants, and it can be assumed that a large portion of these compounds is either constitutively or inducibly involved in some kind of defense response. For example, all plants can produce terpenoids since they are also essential parts of the primary metabolism like carotenoids in photosynthesis or gibberellins and abscisic acid as major plant hormones. Terpenes are by far the most metabolic diverse group among the secondary metabolites and have been shown to play essential roles in many antiherbivore defense strategies, but also as active components in the defense against pathogens. Alkaloids like caffeine, morphine, nicotine, and cocaine are probably best known for the effects they have on humans. However, they are also essential as harmful substances in the plant's toxic repertoire. Phenylpropanes as the third major group of secondary metabolites also play multiple roles in plants under attack by pests and pathogens, for example, by serving as substrates for lignin biosynthesis in damaged tissues, or for tannins, which may inhibit the digestibility of plant proteins, and by forming toxic compounds like flavonoids or furanocoumarins. Other major classes of secondary metabolites involved in antiherbivore defenses comprise glucosinolates and cyanogenic glycosides, which are both hydrolyzed upon tissue damage and release toxic products. Since many of these compounds would also be toxic for the plant, they are often stored in an inactive form, for example, as a glycoside, in the vacuole. Only when tissue damage occurs do these conjugates get in contact with an enzyme that releases the aglycone.

For some of these pathways, all genes involved in the biosynthesis have already been identified. For 2,4-dihydroxy-1,4-benzoxazin-3-one (DIBOA) biosynthesis, Frey et al. (1997) characterized all five genes responsible for the production of this compound, which is a toxin found in maize and other Gramineae. Likewise, all genes required for glucosinolate biosynthesis in *Arabidopsis* have been identified (Halkier and Gershenzon 2006). For many secondary metabolites, it has also been demonstrated that lacking the ability to produce these compounds makes the plants much more vulnerable for herbivore and pathogen attack. On the other hand, herbivores have found ways to tolerate these compounds by means of detoxification. Often, harmful secondary metabolites are oxidized through the activity of distinct P450 enzymes, which introduce oxygen into the molecule. This leads then to an inactivation of the toxic properties of the secondary metabolite.

Besides secondary metabolites, plants produce defensive proteins that interfere with the digestion of the ingested plant material in the insect gut (Green and Ryan 1972; Ryan 1990). Protease inhibitors (PI) are rapidly produced by plants in response to herbivory and are mostly regulated by JA. PI block the degradation of the proteinaceous part of the food and thereby significantly reduce caterpillar growth (Lison et al. 2006; Zavala et al. 2004). Other plant defensive proteins that interfere with the digestion of proteins and amino acids in the caterpillar gut are arginases and threonine deaminases (Chen et al. 2005). These proteins also reduce the nutritional value of the plant for the caterpillar by removing the nitrogen portion from essential amino acids like arginine and threonine. Interestingly, threonine deaminase has to be proteolytically activated by removing a C-terminal regulatory domain (Chen et al. 2007). Polyphenol oxidases and lipoxygenases other than those involved in JA and GLV biosynthesis further contribute to the massive attack launched by the plant to reduce its nutritional value for the herbivore (Constabel et al. 1995; Felton et al. 1994; Wang and Constabel 2004). By producing reactive *o*-quinones and lipid hydroperoxides, dietary proteins become covalently modified, which reduces their availability for the caterpillar digestive system.

A more direct attack on the herbivore's digestive system is performed by some plants through the production of a specific cysteine protease, which disrupts the peritrophic membrane that protects the gut epithelium (Konno et al. 2004; Mohan et al. 2006). While none of these genes are essential for the vegetative growth of the plant, they have likely evolved from normal housekeeping genes during the coevolution of plants and their insect herbivores.

Most of the genes coding for enzymes involved in the biosynthesis of induced secondary metabolites as well as those involved in the inhibition of digestion are regulated by JA. As described above, JA does not just induce one bottleneck enzyme but rather a whole set of genes necessary to activate the whole pathway, and it may very well be this capacity that makes JA such an effective defense-signaling compound.

6 Herbivore-Induced Volatiles and Indirect Defenses

In response to insect herbivory, many plants emit a complex bouquet of volatile organic compounds (VOC), also often referred to as herbivore-induced plant volatiles (HIPV). These VOC may be derived from all pathways for secondary metabolites. Most prominent among those VOC are terpenes, in particular mono- and sesquiterpenes. But also irregular terpenes like 3*E*-4,8-dimethyl-1,3,7-nonatriene (DMNT) and 3*E*,7*E*-4,8,12-trimethyl −1,3,7,11-trideca-tetraene (TMTT), both of which are also referred to as homoterpenes, are often found among the emitted VOC. Additionally, alkaloids like indole or phenylpropanes like methyl salicylate or methyl anthranilate may also contribute to the composition of the bouquet. Another class of important components of most HIPV are the green leaf volatiles (GLV). Pare and Tumlinson (1997) showed by stable isotope labeling experiments that in corn seedling most HIPV are synthesized *de novo* with the exception of GLV, which are cleavage products of readily available fatty acids. Also, while it usually takes several hours for most herbivore-induced volatiles to be emitted, GLV are released immediately after wounding.

Upon herbivory, damaged leaves are usually the first to emit volatiles, but systemic leaves may, after some delay, also contribute to the overall emission of these compounds. In those plants that are covered with glandular trichomes, all herbivore-induced volatiles may be emitted immediately after upon damaging from this stored source, although often, de novo synthesis sets in shortly thereafter. The quality and quantities of HIPV can vary tremendously even within one species. In corn, for example, HIPV profiles differ significantly between different hybrid and inbred lines (Schmelz et al. 2009).

The release of these volatile secondary metabolites in response to insect herbivory provides multiple advantages for the plant. Probably the first function found to be associated with the release of HIPV was the attraction of natural enemies of the attacking herbivore (Kessler and Baldwin 2001). For example, parasitic wasps home in on their prey by following these volatile cues. By parasitizing the herbivore, they play an important role in the plant's defensive repertoire. Interestingly, these wasps can be trained on specific volatiles if prey is associated with it.

HIPV are mostly emitted during the photoperiod (Arimura et al. 2008) since they require a significant amount of energy and substrate input, which seems to be provided by photosynthesis. During nighttime, HIPV emissions are strongly reduced, with the exception of GLV (Fig. 2). While during daytime natural enemies are attracted to the damaged plants, nighttime volatiles seem to have a different function. De Moraes and coworkers (2001) found that the specific nighttime bouquet mostly consisting of GLV-derived compounds repelled conspecific moths from further egg deposition. While this results in the avoidance of further infestation, it may also benefit the moth. By avoiding already infested plants, they provide a better environment for their offspring by selecting defensively inactive plants as a starting point. Different effects of day- and nighttime volatiles were also found with regard to the feeding behavior of caterpillars. Shiojiri et al. (2006) studied why certain caterpillars preferred to feed on plants during nighttime. While this was first thought to provide better protection against natural enemies, which are mostly active during daytime, it was the lack of HIPV during nighttime that stimulated feeding. Caterpillars simply did not like the taste of plants that emitted HIPV.

Another intriguing example for the ecological function of HIPV release by plants was described by Rasmann et al. (2005). In a study focused on belowground herbivory, they found that roots of corn plants infested by larvae of the western corn rootworm (*Diabrotica virgifera virgifera*) emitted HIPV, in particular *E*-β-caryophyllene, which attracted entomopathogenic nematodes. This added another group of organisms to those that can recognize plant volatiles and use them as cues to find their prey.

However, these are not the only ways plants utilize indirect defenses to protect themselves. Lima beans attract ants by producing extrafloral nectar when under herbivore attack (Heil and Bueno 2007). These ants then remove the herbivore and as a reward are provided with nectar.

7 Inter- and Intraplant Communication by Volatiles

The bouquet of plant volatiles emitted in response to insect herbivory plays important roles: as a mediator of tritrophic interactions, as a repellant for other herbivorous insects, or as a feeding deterrent. Another role for volatiles has emerged in recent years when it was discovered that undamaged neighboring plants can "smell" some of these compounds. Several components of the often-complex blends have been described to date to induce genes related to defense responses against insect herbivores in various plant species. MeJA, several terpenes like DMNT and ocimene, as well as GLV were shown to have a significant effect on plants exposed to these compounds (Tumlinson and Engelberth 2008). However, since many plants including maize do not emit MeJA in response to herbivore damage and the composition of herbivore-induced VOC varies enormously, GLV have emerged as novel volatile signals that are common to all plant species. GLV are released immediately in significant amounts by plants when under insect-herbivore attack, but can also be produced and released systemically (Turlings and Tumlinson 1992; Roese et al. 1996; Pare and Tumlinson 1997). This makes them ideal rapid and universal candidates for volatile signaling. The biosynthetic pathway starts with 13-hydroperoxy linolenic acid and is catalyzed by the enzyme hydroperoxide lyase (HPL). Major products of this pathway are Z-3-hexenal, Z-3-hexenol, and Z-3-hexenyl acetate and their respective E-2-enantiomers. Additionally, this pathway also produces 12-oxo-Z-9-decenoic acid, the natural precursor of traumatin, the first wound hormone described for plants. HPL, like AOS, belongs to the family of P450 enzymes and show a high degree of sequence similarities among each other. In fact, the exchange of just one amino acid in AOS converted the protein into a HPL (Lee et al. 2008). Although the HPL pathway was already characterized 100 years ago, it has only recently gained significance when it was shown that the volatile products of this pathway serve as potent signals in inter- and intraplant signaling.

Communication between plants through the release of volatiles was first described by Rhoades (1983) and Baldwin and Schultz (1983). They found that plants exposed to volatiles from damaged neighboring plants were less attractive to insect herbivores. More than 15 years later, it was found that plants exposed to volatiles from herbivore-infested plants accumulate transcription of defense genes that were previously described to be important in the insect-herbivore defense (Arimura et al. 2000). While in all those cases complex blends of volatiles were emitted by the source plants, Bate and Rothstein (1998) demonstrated that GLV, when applied as pure chemicals, also induced defense-related genes in *Arabidopsis*.

However, while GLV exposure induced defense gene expression and volatile release, these responses were always incomplete or less prominent when compared to actual herbivory. In a study by Engelberth et al. (2004), it was shown that GLV may have a function apart from providing direct protection. Corn seedlings that were previously exposed to GLV from neighboring plants produced significantly more JA and volatile sesquiterpenes when mechanically damaged and induced with elicitors (Fig. 2) when compared to controls. Also, caterpillar-induced nocturnal volatiles, which are enriched in GLV, also exhibited a strong priming effect, inducing production of larger amounts of JA and release of greater quantities of sesquiterpenes after subsequent elicitor application (Engelberth et al. 2004). This was the first report on priming against insect herbivory signaled by GLV and it was demonstrated that this effect is specifically linked to defense response.

Since its initial discovery, the priming effect of GLV has been confirmed in a more natural environment (Kessler et al. 2006). By using a microarray enriched in tobacco genes related to insect herbivory, this study showed increased transcriptional responses in the plants growing adjacent to clipped sagebrush. Although no detectable increases in direct defenses like nicotine or proteinase inhibitors were found, however, when *Manduca sexta* caterpillars started feeding on these primed plants, an accelerated production of trypsin proteinase inhibitor occurred. This primed state of tobacco plants exposed to clipped sagebrush also resulted in lower herbivore damage and higher mortality rate of young *Manduca* caterpillars. Among the volatiles responsible for this priming effect were *E*-2-hexenal, methacrolein, and methyl jasmonate. In a more recent study (Ton et al. 2007), the effect of priming by herbivore-induced volatiles on direct and indirect resistance in corn was shown on a molecular, chemical, and behavioral level. By a differential hybridization screen ten defense-related genes were identified, which were inducible by caterpillar feeding, mechanical wounding, application of elicitors, and JA. Exposure to volatiles from herbivore-infested plants did not activate these genes directly, but primed a subset of them for stronger and/or earlier induction upon subsequent defense elicitation, resulting in reduced caterpillar damage and increased attraction to the natural enemies of the caterpillar, the parasitic wasp *Cotesia marginiventris*.

Although GLV received most attention for their potential role in interplant signaling, other studies revealed that HIPV may also serve as signals in intraplant communication. Karban et al. (2006) investigated the role of volatiles as inducers of resistance between different branches of sagebrush (*Artemisia tridentata*). It was found that airflow was essential for the induction of induced resistance. Sagebrush, like other desert plants, is highly sectorial, and does not allow for a free transport of signaling molecules between different parts of the plant through vascular connections. Instead, volatiles are used to overcome these constraints and provide systemic signaling. A similar effect was observed for lima beans and the induction of extrafloral nectar. Besides providing a signal for neighboring plants, infested plants may very well send a volatile signal to other parts of themselves (Heil and Bueno 2007). From an evolutionary point of view, this may in fact be the original function of those volatiles that can be recognized by plants and used to enhance their own defenses.

Priming plant defense responses to diseases resulting in an accelerated and/or enhanced reaction is well established (Conrath et al. 2002). Although precisely how priming agents regulate subsequent responses is unknown, they appear to work through one or more of the commonly studied defense-signaling pathways (SA-, JA-, ethylene-mediated) or subsets of genes these signals normally regulate, without influencing the concentrations of the signals themselves. Often, low concentrations of signaling compounds can cause a priming effect thereby potentiating the response to subsequent elicitation, while higher concentrations are responsible for the direct induction of defense-related measures. The result of priming can be a unique response, a more rapid response, and/or a stronger response upon subsequent challenge.

The HPL pathway also appears to be important in the context of direct plant defense response. Tobacco plants depleted in HPL were more susceptible to insect herbivory than control plants (Halitschke et al. 2004). Also, potato plants depleted in HPL were more susceptible to aphid attack (Vancanneyt et al. 2001). There is evidence that GLV trigger the production of phytoalexins (Zeringue 1992), reduce insect feeding rates (Hildebrand et al. 1993), reduce germination frequency in soybean (Gardener et al. 1990), and have antimicrobial activity (Juttner and Slusarenko 1993).

How GLV signal is still unknown. It seems clear, however, that for GLVs to fully exhibit their activity, a functioning JA signaling pathway is required. But while in corn and other monocots JA accumulates during the initial exposure to these compounds, no such effect has been reported for dicot plants albeit the fact that they also recognize these signaling compounds and in most cases this recognition primes JA-regulated defense responses. Considering the conserved nature of the GLV signal emission among various plant species, it can be hypothesized that common signaling mechanisms exist for the perception of GLVs, but these have yet to be discovered. Nonetheless, GLV signaling appears to be closely associated with the JA signaling pathway. Further exploration of the molecular mechanisms of priming might eventually lead to the development of environmentally sound pest management strategies.

8 Conclusions

The defense responses plants activate when under insect-herbivore attack are a result of 350 million years of coevolution between the two life forms. During this time, plants have learned to recognize distinct temporal and spatial feeding patterns that are clear indications of species-specific herbivory. Also, plants have developed mechanisms to recognize components from the insect saliva to boost their own defenses. Several of these elicitors represent modified plant compounds like FAC and inceptin and thus, fit into the guard hypothesis of plant immunity. However, to date, no corresponding receptor has been identified.

Signaling mechanisms that are initiated by these elicitors include Ca^{2+} signaling, production of ROS, and MAP kinase cascades. While some of these mechanisms were characterized in some plant species, it is yet unclear as to what degree these represent a general response of plants to insect herbivory. More detailed studies on the cellular responses to herbivore are necessary to gain a better picture of these rapid signaling events. Additionally, the localization of these signaling events need to be further investigated and may help in the characterization of long-distant defense signaling.

A major mediator of herbivore-activated defense responses in the plant is JA. While signaling mechanisms have now been revealed, certain aspects of JA signaling are still elusive. For example, systemic signaling appears to depend on JA, but detailed knowledge about how JA is either transported or creates long-distant signaling is missing. Grafting experiments may provide an important tool to access this problem.

Plant defenses are very sophisticated in their complexity and target-directed effectivity. Direct defenses address the problems at the immediate damage site but are also rapidly upregulated in other parts of the plant. Indirect defenses mediated by volatiles or extrafloral nectar demonstrate the complexity of herbivore-activated defenses in the plant by recruiting natural enemies of the attacking herbivore. Additional roles for volatiles in inter- and intraplant signaling further emphasize the ecological function of secondary metabolites as communicative means that help govern community responses to insect-herbivore attacks.

Acknowledgments This work was supported by grants from the National Science Foundation (IOS-0925615) and USDA, NIFA (Grant No.03836).

References

Alborn HT, Turlings TCJ, Jones TH, Stenhagen G, Loughrin JH, Tumlinson JH (1997) An elicitor of plant volatiles from beet armyworm oral secretion. Science 276:945–949

Alborn HT, Hansen TV, Jones TH, Bennett DC, Tumlinson JH, Schmelz EA, Teal PE (2007) Disulfooxy fatty acids from the American bird grasshopper schistocerca Americana, elicitors of plant volatiles. Proc Natl Acad Sci USA 104:12976–12981

Arimura G, Ozawa R, Shimoda T, Nishioka T, Boland W, Takabayashi J (2000) Herbivory-induced volatiles elicit defense genes in lima bean leaves. Nature 406:512–515

Arimura G, Köpke S, Kunert M, Volpe V, David A, Brand P, Dabrowska P, Maffei ME, Boland W (2008) Effects of feeding spodoptera littoralis on lima bean leaves: IV. Diurnal and nocturnal damage differentially initiate plant volatile emission. Plant Physiol 146:965–973

Baldwin IT (1990) Herbivory simulations in ecological research. Trends Ecol Evol 5:91–93

Baldwin IT, Schultz JC (1983) Rapid changes in tree leaf chemistry induced by damage-evidence for communication between plants. Science 221:277–279
Bate NJ, Rothstein SJ (1998) C-6-volatiles derived from the lipoxygenase pathway induce a subset of defense-related genes. Plant J 16:561–569
Bi JL, Felton GW (1995) Foliar oxidative stress and insect herbivory—primary compounds, secondary metabolites, and reactive oxygen species as components of induced resistance. J Chem Ecol 21:1511–1530
Chen H, Wilkerson CG, Kuchar JA, Phinney BS, Howe GA (2005) Jasmonate-inducible plant enzymes degrade essential amino acids in the herbivore midgut. Proc Natl Acad Sci USA 102:19237–19242
Chen H, Gonzales-Vigil E, Wilkerson CG, Howe GA (2007) Stability of plant defense proteins in the gut of insect herbivores. Plant Physiol 143:1954–1967
Chini A, Fonseca S, Fernández G, Adie B, Chico JM, Lorenzo O, García-Casado G, López-Vidriero I, Lozano FM, Ponce MR, Micol JL, Solano R (2007) The JAZ family of repressors is the missing link in jasmonate signalling. Nature 448:666–671
Conrath U, Pieterse CMJ, Mauch-Mani B (2002) Priming in plant–pathogen interactions. TIPS 7:210–216
Constabel CP, Bergey DR, Ryan CA (1995) Systemin activates synthesis of wound-inducible tomato leaf polyphenol oxidase via the octadecanoid defense signaling pathway. Proc Natl Acad Sci USA 92:407–411
Croft KPC, Juttner F, Slusarenko AJ (1993) Volatile products of the lipoxygenase pathway evolved from *phaseolus vulgaris* (L.) leaves inoculated with pseudomonas syringae pv phaseolicola. Plant Physiol 101:13–24
De Moraes CM, Mescher MC, Tumlinson JH (2001) Caterpillar-induced nocturnal plant volatiles repel conspecific females. Nature 410:577–580
Dogimont C, Bendahmane A, Pitrat M, Burget-Bigeard E, Hagen L, et al. (2007) U.S. Patent No. 20,070,016,977
Doss RP, Oliver JE, Proebsting WM, Potter SW, Kuy S, Clement SL, Williamson RT, Carney JR, DeVilbiss ED (2000) Bruchins: insect-derived plant regulators that stimulate neoplasm formation. Proc Natl Acad Sci USA 97:6218–6223
Du B, Zhang W, Liu B, Hu J, Wei Z, Shi Z, He R, Zhu L, Chen R, Han B, He G (2009) Identification and characterization of Bph14, a gene conferring resistance to brown plant hopper in rice. Proc Natl Acad Sci USA 106:22163–22168
Engelberth J, Alborn HT, Schmelz EA, Tumlinson JH (2004) Airborne signals prime plants against insect herbivore attack. Proc Natl Acad Sci USA 101:1781–1785
Engelberth J, Seidl-Adams I, Schultz JC, Tumlinson JH (2007) Insect elicitors and exposure to green leafy volatiles differentially upregulate major octadecanoids and transcripts of 12-oxo phytodienoic acid reductases in *Zea Mays*. MPMI 20:707–716
Felton GW, Bi JL, Summers CB, Mueller AJ, Duffey SS (1994) Potential role of lipoxygenases in defense against insect herbivory. J Chem Ecol 20:651–666
Frey M, Chomet P, Glawischnig E, Stettner C, Grun S, Winklmair A, Eisenreich W, Bacher A, Meeley RB, Briggs SP, Simcox K, Gierl A (1997) Analysis of a chemical plant defense mechanism in grasses. Science 277:696–699
Gardener HW, Dornbos DL, Desjardins AE (1990) Hexanal, trans-2-hexenal, and trans-2-nonenal inhibit soybean, glycine max, seed germination. J Agri Food Chem 38:1316–1320
Gatehouse JA (2002) Plant resistance towards insect herbivores: a dynamic interaction. New Phytol 156:145–169
Green TR, Ryan CA (1972) Wound-induced proteinase inhibitor in plant leaves—possible defense mechanism against insects. Science 175:776–777
Halitschke R, Schittko U, Pohnert G, Boland W, Baldwin IT (2001) Molecular interactions between the specialist herbivore *manduca sexta* (Lepidoptera, sphingidae) and its natural host nicotiana attenuata. III. Fatty acid-amino acid conjugates in herbivore oral secretions

are necessary and sufficient for herbivore-specific plant responses. Plant Physiol 125:711–717

Halitschke R, Ziegler J, Keinänen M, Baldwin IT (2004) Silencing of hydroperoxide lyase and allene oxide synthase reveals substrate and defense signaling crosstalk in *nicotiana attenuata*. Plant J 40:35–46

Halkier BA, Gershenzon J (2006) Biology and biochemistry of glucosinolates. Annu Rev Plant Biol 57:303–333

Heil M, Bueno CS (2007) Within-plant signaling by volatiles leads to induction and priming of an indirect plant defense in nature. Proc Natl Acad Sci USA 104:5467–5472

Hildebrand DF, Brown GC, Jackson DM, Hamilton-Kemp TR (1993) Effects of some leaf-emitted volatile compounds on aphid population increase. J Chem Ecol 19:1875–1887

Howe GA, Jander G (2008) Plant immunity to insect herbivores. Annu Rev Plant Biol 59:41–66

Jones JDG, Dangle JL (2006) The plant immune system. Nature 444:323–329

Kandoth PK, Ranf S, Pancholi SS, Jayanty S, Walla MD, Miller W, Howe GA, Lincoln DE, Stratmann JW (2007) Tomato MAPKs LeMPK1, LeMPK2, and LeMPK3 function in the systemin-mediated defense response against herbivorous insects. Proc Natl Acad Sci USA 104:12205–12210

Kang JH, Wang L, Giri A, Baldwin IT (2006) Silencing threonine deaminase and JAR4 in *nicotiana attenuata* impairs jasmonic acid-isoleucine-mediated defenses against *manduca sexta*. Plant Cell 18:3303–3320

Karban R, Baldwin IT (1997) Induced responses to herbivory. University Chicago Press, Chicago

Karban R, Shiojiri K, Huntzinger M, McCall AC (2006) Damage-induced resistance in sagebrush: volatiles are key to intra- and interplant communication. Ecology 87:922–930

Keller T, Damude HG, Werner D, Doerner P, Dixon RA, Lamb C (1998) A plant homolog of the neutrophil NADPH oxidase gp91phox subunit gene encodes a plasma membrane protein with Ca^{2+} binding motifs. Plant Cell 10:255–266

Kessler A, Baldwin IT (2001) Defensive function of herbivore-induced plant volatile emissions in nature. Science 291:2141–2144

Kessler A, Halitschke R, Diezel C, Baldwin IT (2006) Priming of plant defense responses in nature by airborne signaling between *Artemisia tridentata* and *nicotiana attenuata*. Oecologia 148:280–292

Konno K, Hirayama C, Nakamura M, Tateishi K, Tamura Y, Hattori M, Kohno K (2004) Papain protects papaya trees from herbivorous insects: role of cysteine proteases in latex. Plant J 37:370–378

Lamb C, Dixon RA (1997) The oxidative burst in plant disease resistance. Annu Rev Plant Physiol Plant Mol Biol 48:251–275

Lee DS, Nioche P, Hamberg M, Raman CS (2008) Structural insights into the evolutionary paths of oxylipin biosynthetic enzymes. Nature 455:363–368

Leitner M, Boland W, Mithoefer A (2005) Direct and indirect defences induced by piercing-sucking and chewing herbivores in *medicago truncatula*. New Phytol 167:597–606

Levy M, Wang Q, Kaspi R, Parrella MP, Abel S (2005) Arabidopsis IQD1, a novel calmodulin-binding nuclear protein, stimulates glucosinolate accumulation and plant defense. Plant J 43:79–96

Lison P, Rodrigo I, Conejero V (2006) A novel function for the cathepsin D inhibitor in tomato. Plant Physiol 142:1329–1339

Maffei M, Bossi S, Spiteller D, Mithoefer A, Boland W (2004) Effects of feeding *spodoptera littoralis* on lima bean leaves. I. Membrane potentials, intracellular calcium variations, oral secretions, and regurgitate components. Plant Physiol 134:1752–1762

Maffei ME, Mithoefer A, Arimura G, Uchtenhagen H, Bossi S, Bertea CM, Cucuzza LS, Novero M, Volpe V, Quadro S, Boland W (2006) Effects of feeding *spodoptera littoralis* on lima bean leaves. III. Membrane depolarization and involvement of hydrogen peroxide. Plant Physiol 140:1022–1035

Maischak H, Grigoriev PA, Vogel H, Boland W, Mithoefer A (2007) Oral secretions from herbivorous lepidopteran larvae exhibit ion channel-forming activities. FEBS Lett 581:898–904

Malone M, Alarcon JJ, Palumbo L (1994) An hydraulic interpretation of rapid, long-distance wound signaling in the tomato. Planta 193:181–185

Mattiacci L, Dicke M, Posthumous MA (1995) Beta-glucosidase: an elicitor of herbivore-induced plant odor that attracts host-searching parasitic wasps. Proc Natl Acad Sci USA 92:2036–2040

Mendelsohn R, Balick MJ (1995) The value of undiscovered pharmaceuticals in tropical forests. Econ Bot 49:223–228

Mithoefer A, Wanner G, Boland W (2005) Effects of feeding *Spodoptera littoralis* on lima bean leaves. II. Continuous mechanical wounding resembling insect feeding is sufficient to elicit herbivory-related volatile emission. Plant Physiol 137:1160–1168

Mohan S, Ma PWK, Pechan T, Bassford ER, Williams WP, Luthe DS (2006) Degradation of the S. Frugiperda peritrophic matrix by an inducible maize cysteine protease. J Insect Physiol 52:21–28

Musser RO, Cipollini DF, Hum-Musser SM, Williams SA, Brown JK, Felton GW (2005) Evidence that the caterpillar salivary enzyme glucose oxidase provides herbivore offense in solanaceous plants. Archives Insect Biochem Physiol 58:128–137

Nombela G, Williamson VM, Muniz M (2003) The root-knot nematode resistance gene Mi-1.2 of tomato is responsible for resistance against the whitefly *bemisia tabaci*. MPMI 16:645–649

Orozco-Cardenas M, Ryan CA (1999) Hydrogen peroxide is generated systemically in plant leaves by wounding and systemin via the octadecanoid pathway. Proc Natl Acad Sci USA 96:6553–6557

Orozco-Cardenas ML, Narvaez-Vasquez J, Ryan CA (2001) Hydrogen peroxide acts as a second messenger for the induction of defense genes in tomato plants in response to wounding, systemin, and methyl jasmonate. Plant Cell 13:179–191

Pare PW, Tumlinson JH (1997) De novo biosynthesis of volatiles induced by insect herbivory in cotton plants. Plant Physiol 114:1161–1167

Pare PW, Alborn HT, Tumlinson JH (1998) Concerted biosynthesis of an insect elicitor of plant volatiles. Proc Natl Acad Sci USA 95:13971–13975

Park YS, Kunze S, Ni X, Feussner I, Kolomiets MW (2010) Comparative molecular and biochemical characterization of segmentally duplicated 9-lipoxygenase genes ZmLOX4 and ZmLOX5 of maize. Planta 231:1425–1437

Pauwels L, Inze D, Goossens A (2009) Jasmonate-inducible gene: what does it mean? Trends Plant Sci 14:87–91

Peiffer M, Tooker JF, Luthe DS, Felton GW (2009) Plants on early alert: glandular trichomes as sensors for insect herbivores. New Phytol 184:644–656

Pohnert G, Jung V, Haukioja E, Lempa K, Boland W (1999) New fatty acid amides from regurgitant of lepidopteran (noctuidae, geometridae) caterpillars. Tetrahedron 55:11275–11280

Rasmann S, Köllner TG, Degenhardt J, Hiltpold I, Toepfer S, Kuhlmann U, Gershenzon J, Turlings TC (2005) Recruitment of entomopathogenic nematodes by insect-damaged maize roots. Nature 434:732–737

Reymond P, Bodenhausen N, Van Poecke RM, Krishnamurthy V, Dicke M, Farmer EE (2004) A conserved transcript pattern in response to a specialist and a generalist herbivore. Plant Cell 16:3132–147

Rhoades DF (1983) Responses of alder and willow to attack by tent caterpillars and webworms-evidence for pheromonal sensitivity of willows. ACS Symp Ser 208:55–68

Roese USR, Manukian A, Heath RR, Tumlinson JH (1996) Volatile semiochemicals released from undamaged cotton leaves: a systemic response of living plants to caterpillar damage. Plant Physiol 111:487–495

Rossi M, Goggin FL, Milligan SB, Kaloshian I, Ullman DE, Williamson VM (1998) The nematode resistance gene Mi of tomato confers resistance against the potato aphid. Proc Natl Acad Sci USA 95:9750–9754

Ryan CA (1990) Protease inhibitors in plants: genes for improving defenses against insects and pathogens. Annu Rev Phytopathol 28:425–449

Sagi M, Fluhr R (2001) Superoxide production by plant homologues of the gp91(phox) NADPH oxidase. Modulation of activity by calcium and by tobacco mosaic virus infection. Plant Physiol 126:1281–1290

Schilmiller AL, Howe GA (2005) Systemic signaling in the wound response. Curr Opin Plant Biol 8:369–377

Schmelz EA, Carroll MJ, LeClere S, Phipps SM, Meredith J, Chourey PS, Alborn HT, Teal PE (2006) Fragments of ATP synthase mediate plant perception of insect attack. Proc Natl Acad Sci USA 103:8894–8899

Schmelz EA, LeClere S, Carroll MJ, Alborn HT, Teal PE (2007) Cowpea chloroplastic ATP synthase is the source of multiple plant defense elicitors during insect herbivory. Plant Physiol 144:793–805

Schmelz EA, Engelberth J, Alborn HT, Tumlinson JH, Teal PE (2009) Phytohormone-based activity mapping of insect herbivore-produced elicitors. Proc Natl Acad Sci USA 106:653–657

Shen B, Zheng Z, Dooner HK (2000) A maize sesquiterpene cyclase gene induced by insect herbivory and volicitin: characterization of wild-type and mutant alleles. Proc Natl Acad Sci USA 97:14807–14812

Shiojiri K, Ozawa R, Takabayashi J (2006) Plant volatiles, rather than light, determine the nocturnal behavior of a caterpillar. PLoS Biol 4:e164

Spiteller D, Boland W (2003) N-(15,16-dihydroxylinoleoyl)-glutamine and N-(15,16-epoxylinoleoyl)-glutamine isolated from oral secretions of lepidopteran larvae. Tetrahedron 59:135–139

Spiteller D, Oldham NJ, Boland W (2004) N-(17-phosphonooxylinolenoyl)glutamine and N-(17-phosphonooxylinoleoyl)glutamine from insect gut: the first backbone-phosphorylated fatty acid derivatives in nature. J Org Chem 69:1104–1109

Stankovic B, Davies E (1997) Intercellular communication in plants: electrical stimulation of proteinase inhibitor gene expression in tomato. Planta 202:402

Staswick PE, Tiryaki I (2004) The oxylipin signal jasmonic acid is activated by an enzyme that conjugates it to isoleucine in Arabidopsis. Plant Cell 16:2117–2127

Thines B, Katsir L, Melotto M, Niu Y, Mandaokar A, Liu GH, Nomura He SY, Howe GA, Browse J (2007) JAZ repressor proteins are targets of the SCFCOI1 complex during jasmonate signalling. Nature 448:661–665

Ton J, D'Alessandro M, Jourdie V, Jakab G, Karlen D, Held M, Mauch-Mani B, Turlings TCJ (2007) Priming by airborne signals boosts direct and indirect resistance in maize. Plant J 49:16–26

Truitt CL, Wei HX, Pare PW (2004) A plasma membrane protein from *Zea Mays* binds with the herbivore elicitor volicitin. Plant Cell 16:523–532

Tumlinson JH, Engelberth J (2008) Fatty acid derived signals that induce or regulate plant defenses against herbivory. In: Schaller A (ed) Induced plant resistance to herbivory. Springer, Amsterdam, The Netherlands

Turlings TC, Tumlinson JH (1992) Systemic release of chemical signals by herbivore-injured corn. Proc Natl Acad Sci USA 89:8399–8402

Vancanneyt G, Sanz C, Farmaki T, Paneque M, Ortego F, Castanera P, Sanchez-Serrano JJ (2001) Hydroperoxide lyase depletion in transgenic potato plants leads to an increase in aphid performance. Proc Natl Acad Sci USA 98:8139–8144

Wang JH, Constabel CP (2004) Polyphenol oxidase overexpression in transgenic populus enhances resistance to herbivory by forest tent caterpillar (*malacosoma disstria*). Planta 220:87–96

Wang L, Allmann S, Wu J, Baldwin IT (2008) Comparisons of LIPOXYGENASE3- and JASMONATE-RESISTANT4/6-silenced plants reveal that jasmonic acid and jasmonic acid-amino acid conjugates play different roles in herbivore resistance of *nicotiana attenuata*. Plant Physiol 146:904–915

Wu J, Baldwin IT (2010) New insights into plant responses to the attack from insect herbivores. Annu Rev Genet 44:1–24

Wu J, Hettenhausen C, Meldau S, Baldwin IT (2007) Herbivory rapidly activates MAPK signaling in attacked and unattacked leaf regions but not between leaves of *nicotiana attenuata*. Plant Cell 19:1096–1122

Yoshinaga N, Aboshi T, Abe H, Nishida R, Alborn HT, Tumlinson JH, Mori N (2008) Active role of fatty acid amino acid conjugates in nitrogen metabolism in *spodoptera lituralis* larvae. Proc Natl Acad Sci USA 105:18058–18063

Zavala JA, Patankar AG, Gase K, Hui DQ, Baldwin IT (2004) Manipulation of endogenous trypsin proteinase inhibitor production in *nicotiana attenuata* demonstrates their function as antiherbivore defenses. Plant Physiol 134:1181–1190

Zeringue HJ (1992) Effects of C6-C10 alkenals and alkanals on eliciting a defence response in the developing cotton boll. Phytochemistry 31:2305–2308

Zimmermann MR, Maischak H, Mithoefer A, Boland W, Felle HH (2009) System potentials, a novel electrical long-distance apoplastic signal in plants, induced by wounding. Plant Physiol 149:1593–1600

Bacterial Volatiles Mediating Information Between Bacteria and Plants

Katrin Wenke, Teresa Weise, Rene Warnke, Claudio Valverde, Dierk Wanke, Marco Kai, and Birgit Piechulla

Abstract At present, more than 400 volatiles are known to appear in bacterial headspace samples, but more are expected as more bacteria will be investigated and several identification technologies will be applied. A comprehensive list of bacteria and their respective effects on plants were presented. The volatiles emitted from *Serratia plymuthica* HRO-C48 and *Stenotrophomonas maltophilia* R3089 retarded leaf and root development of *Arabidopsis thaliana* starting at day 2 of cocultivation, while first signs of activation of stress promoters appeared already after 18 h. Most *A. thaliana* ecotypes reacted similar to the volatiles of *S. plymuthica*, but a stronger root growth inhibition was observed for the accession C24. β-Phenyl-ethanol was identified as one compound of the *S. plymuthica* volatile mixture inhibiting the growth of *Arabidopsis thaliana*.

1 Introduction

Most of the compounds of fragrances known today originate from plants and animals. It is not commonly realized that also prokaryotes produce and emit an enormous diversity of volatiles, although the aromas of cheese and wine are well known (e.g., Urbach 1997; Schreier 1980). Furthermore, it is not very evident that the earthy smell in forests is primarily due to the emission of volatiles synthesized

---

K. Wenke • T. Weise • R. Warnke • M. Kai • B. Piechulla (✉)
University of Rostock, Institute of Biological Sciences (IfBi), Rostock, Germany
e-mail: birgit.piechulla@uni-rostock.de

C. Valverde
Department of Science and Technology, National University of Quilmes, Buenos Aires, Argentina

D. Wanke
Center for Plant Molecular Biology, University Tübingen, Tübingen, Germany

G. Witzany and F. Baluška (eds.), *Biocommunication of Plants*,
Signaling and Communication in Plants 14,
DOI 10.1007/978-3-642-23524-5_17,

by bacteria, e.g., actinomycetes emit the typical earth odor geosmin (Gerber and Lechevalier 1965). Volatiles are chemicals with low molecular masses (<300 Da), low polarity, and high vapor pressure (0.01 kPa or higher at 20°C). Together, these features facilitate evaporation. Typical volatiles are monoterpenes, aromatic compounds, and fatty acid derivatives. They appear in the atmosphere and act over long distances. Besides aboveground volatile-based exchanges, also below-ground volatile interactions have to be considered. The biological and ecological roles of bacterial volatiles were so far underestimated, and it is a future task to unravel their action potentials. In this chapter, we focus on the interactions between bacteria and plants that are solely based on volatile compounds; bacterial interactions based on nonvolatile metabolites were not considered. The latter activate plant defense mechanisms and stimulate signal transduction pathways, such as SAR (systemic acquired resistance) or ISR (induced systemic resistance) with salicylic acid and jasmonic acid as key components. It is a goal of upcoming research to unravel whether and which responses or signaling chains are activated in plants after bacterial volatile perception. The processes of volatile perception and the conversion of information remain so far elusive.

This chapter describes first the state of the art regarding the wealth and distribution of bacterial volatiles including information about collection and detection. Thereafter, the cellular and molecular alterations in plants due to bacterial volatile administration are addressed. Finally, an ecological aspect was taken into consideration.

2 The Wealth of Bacterial Volatiles

Microorganisms, including bacteria, are everywhere on the earth, in the air, in the water, in the soil, in extreme localizations (in hot springs, in arctic regions, several 1,000 m deep in the ocean), as well as in and on organisms. They produce a large spectrum of volatiles, inorganic as well as organic compounds. Often, these volatiles contribute to the characteristic aroma of foodstuffs, such as vine and beer, cheese and other milk products, sour cabbage, or other fermented eatables. The qualitative and quantitative volatile compound compositions of aromas are primarily determined by the bacterial species and their growth conditions. The availability of substrates and the metabolic capabilities and capacities of the bacteria are decisive for product formation, including volatile emission (Stotzky and Schenck 1976; Fiddaman and Rossall 1994).

The first publication that indicated the emission of volatile fatty acids from *Dysenteria* bacteria appeared in 1921 (Zoller and Mansfield Clark 1921). Our recent literature search included 336 bacterial species that produce volatile organic compounds (VOCs). In total, ca. 770 different VOCs are released by bacteria. These compounds were grouped into ca. 50 classes, such as acids, alcohols and aldehydes (Fig. 1). The dominant compound groups were alcohols, alkenes, ketones, and terpenoids (comprising 120–190 different substances) followed by acids, benzenoids, esters, or pyrazines (comprising 60–80 different compounds),

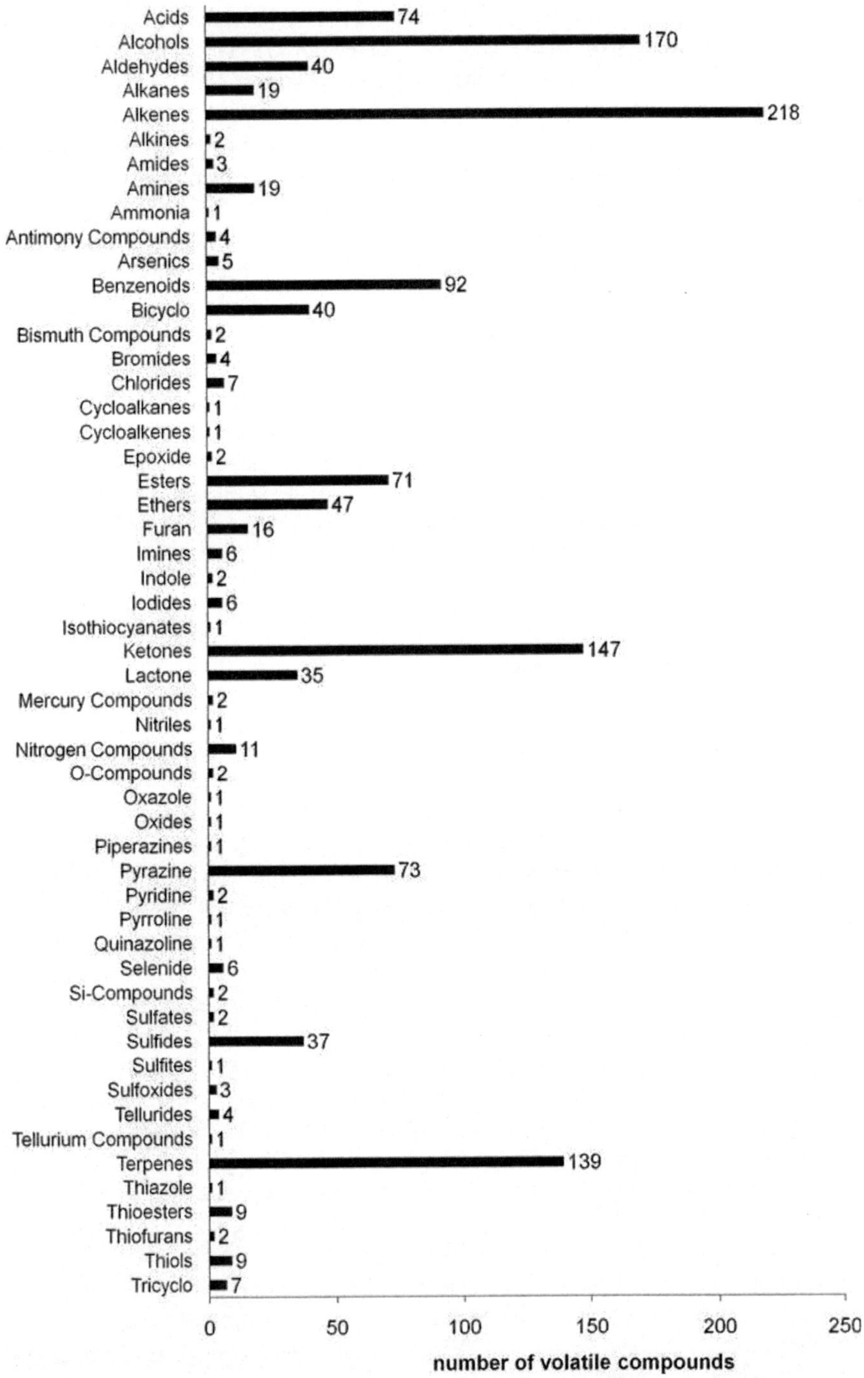

Fig. 1 Distribution of bacterial volatiles in chemical classes. Presently known bacterial volatiles are assigned to different chemical classes

and aldehydes, ethers, and lactones (comprising 30–40 compounds). *Chondromyces crocatus*, *Carnobacterium divergens* 9P, *Streptomyces* sp. GWS-BW-H5, and *Serratia odorifera* 4Rx13 are the bacteria with the largest VOC emission spectra, ca. 75–100 compounds were emanated by each species (Schulz et al. 2004; Ercolini et al. 2009; Dickschat et al. 2005; Kai et al. 2010). Seven hundred seventy bacterial VOCs were incorporated into the SuperScent database, which is open for public access (http://bioinf-applied.charite.de/superscent/ index.php?site = home). Besides the VOCs with identified structure, numerous bacterial volatiles and their isomers remain to be structurally elucidated. Recently, we successfully isolated and characterized a new compound from *Serratia odorifera* 4Rx13 (Kai et al. 2010). Its extraordinary chemical structure is new to science, and it was named "sodorifen" (von Reuß et al. 2010).

The VOC profiles of ca. 340 bacterial strains were analyzed so far, which represent a rather small number compared to species and isolates existing on earth. Therefore, more VOC spectra from prokaryotes need to be investigated in the future to identify and estimate the potential of these natural compounds. To define the VOC spectra of bacteria as complete as possible, several methods have to be applied.

3 Methods to Collect and Detect Volatiles

The techniques described below are suitable to collect and investigate volatiles, which are emitted into the headspace of bacterial cultures. Bacterial volatiles can be captured in open or closed airflow systems. The volatiles of this dynamic headspace are trapped on polymeric adsorption matrices (SuperQ, Tenax, Lewatit, and activated charcoal). In open volatile collection systems (Ryu et al. 2003; Kai et al. 2007; Kai et al. 2010), purified, sterile air enters the test vessel. Half of the influx air is sucked out and is delivered to an adsorption trap; consequently, a defined volume of excess air escapes. Therefore, external gaseous compounds and bacterial contaminations can be avoided. In closed systems, the total headspace air is analyzed since the airflow circulates continuously through the bacterial culture and through the trap (e.g., Dickschat et al. 2004; Schulz et al. 2004). This "closed-loop-stripping apparatus" (CLSA) was established by Boland et al. (1984). An alternative without continuous airflow is the analysis of the waste air of a bioreactor containing *Streptomyces citreus* by direct adsorption on a Lewatit-filled glass tube (Pollak et al. 1996). Compounds trapped in open or closed systems are either eluted with a solvent (methanol, dichloromethane, pentane) and analyzed using gas chromatography/mass spectrometry (GC/MS) or directly thermally desorbed.

Another possibility to extract bacterial volatiles encounters the static headspace of bacterial cultures using solid-phase microextraction (SPME). SPME was introduced in 1990 (Arthur and Pawliszyn 1990). A thin film of an extracting phase immobilized over the surface of a fused silica fiber facilitates the adsorption of compounds present in the headspace. According to the properties of expected volatiles, different coatings are available for extraction, e.g., polydimethylsiloxan, carboxen, and divinylbenzene or combinations of these adsorbents. The SPME technique provides advantages, e.g., the method is solventless, simple in situ sampling, and a short analytical time. Till now, several bacterial headspace-SPME investigations have been performed (e.g., Vergnais et al. 1998; Kataoka et al. 2000; Chuankun et al. 2004; Schulz et al. 2004; Farag et al. 2006; Zou et al. 2007; Ercolini et al. 2009; Preti et al. 2009). Other static approaches (diffusive sampling) were established (Larsen and Frisvad 1994) using polymeric substances (Carbon black, Tenax). They were filled into stainless steel tubes and directly placed into the Petri dishes to capture volatiles from the headspace of different bacterial cultures (Schöller et al. 1997), or activated charcoal was placed in the lid of the Petri dishes (Gust et al. 2003).

All volatile collection methods mentioned above were combined with GC/MS techniques. Instead of GC/MS, the collection system can also be attached to proton transfer reaction/mass spectrometer (PTR/MS) (Mayr et al. 2003; Bunge et al. 2008; Kai et al. 2010) or selected ion flow tube/mass spectrometer (SIFT/MS) (Carrol et al. 2005; Allardyce et al. 2006; Thorn et al. 2010). While GC/MS depicts volatile profiles that are based on the analyses of defined retention times, PTR/MS and SIFT/MS allow continuous monitoring of volatile emission. Another substantial benefit of PTR/MS and SIFT/MS is that prior to analysis no preconcentration step or chromatography is needed. PTR/MS determines the m/z ratio of a molecule and no fragmentation pattern; therefore, the use of natural isotopic ratios and literature search are necessary to make an educated guess to identify the compounds. To overcome this limitation, an alternative method can be used to detect and characterize volatiles: secondary electron spray ionization/mass spectrometry (SESI/MS) (Zhu et al. 2010). It has to be realized that all specific techniques mentioned here only allows the detection and determination of a certain spectrum of volatiles emitted from the bacteria. To get a comprehensive compilation of volatiles, it is inevitable to combine the different volatile collection methods.

4 Bacterial Volatiles Mediating Interactions with *Arabidopsis thaliana*

4.1 Observations at the Level of Phenotype

In contrast to the large number of bacterial volatiles that have been described so far, not many details are known about their ecological and biological functions. This issue is difficult to approach because bacterial volatiles can act as individual compounds or in mixtures of different compositions. Another drawback is that often the complete volatile spectra of bacteria are not known, or the contributions of individual compounds in mixtures have yet not been determined. Furthermore, the biologically active compound(s) and relevant concentration(s) are not known. Dual cultures where only volatiles can act as a functional agents are simple test systems.

In one compartment of bipartite or tripartite Petri dishes, bacteria were plated, and in the other compartment(s), young plant seedlings, (*Arabidopsis thaliana* or *Physcomitrella patens*) were planted. Only volatiles can diffuse through the atmosphere from one side to the other side of the Petri dish. The growth of the plants during cocultivation was followed by photographic documentation or determination of, e.g., fresh weight, leaf length, or root length. Figure 2a summarizes the experiment performed with the volatiles of 11 bacterial strains and isolates acting on *A. thaliana* (Vespermann et al. 2007; Kai et al. 2008). While *A. thaliana* develops normally in coculture with *Bacillus subtilis*, *Burkholderia cepacia*, *Staphylococcus epidermidis*, and *Escherichia coli*, weak growth or no growth was obtained with *Pseudomonas fluorescens*, *Pseudomonas trivialis*, *Serratia odorifera*, *Serratia plymuthica*, *Stenotrophomonas maltophilia*, and *Stenotrophomonas rhizophila*. Phenotypical changes that appeared during cocultivation with S. *plymuthica* HRO-C48 and *S. maltophilia* R3089 were visible after 5 days (Fig. 2c). Dual culture assays with application of increasing cell numbers of *S. plymuthica* HRO-C48 (Fig. 2b) resulted in significant effects on green plant parts and roots. The more bacterial cells were applied at the beginning of the experiment, the more dramatic phenotypic effects were observed at *A. thaliana*. A stronger effect on the relative root growth could be observed compared to the inhibition of cotyledons. This difference between the effects on belowground and aboveground plant parts is presumable due to faster elongation growth of root cells. It also should be considered that the diffusion of volatiles is different in the agar versus in the air of the Petri dish; it is a consequence of different polarity and volatility of individual compounds. Also, the mode of perception as well as the mode of action *in planta* (direct or indirect) is until now an open question. The presented experiments, however, clearly demonstrate that the highest tested number of 10^7 CFU of *S. plymuthica* HRO-C48 caused significant retardation of root and leaf growth within 2 days of cocultivation. These cell numbers are ecologically relevant because at strawberry roots under field conditions, *S. plymuthica* HRO-C48 reached up to 10^7 CFU per g (Kurze et al. 2001), and in potato and oilseed rape rhizospheres, 10^8 CFU per g root fresh weight was determined (Berg et al. 2002). Furthermore, formation of microbial biofilms on root surfaces was also reported with locally high densities of rhizobacteria (Bloemberg et al. 2000; Bais et al. 2004; Walker et al. 2004).

4.2 Alterations at the Physiological and Molecular Level

Exposure to bacterial volatiles resulted in phenomenological alterations, which are the most likely consequences of changes at the cellular and physiological levels. Cotyledons of seedlings of *A. thaliana* were incubated with Evans blue dye, which is an indicator for cell vitality. The blue color accumulates only in dead cells without intact cellular membranes (Kim et al. 2003). Leaf growth arrested between the third and fourth day in dual culture of *S. maltophilia* R3089 and *S. plymuthica* HRO-C48 (Fig. 2b). In the same time frame, Evans blue staining leads to weak local

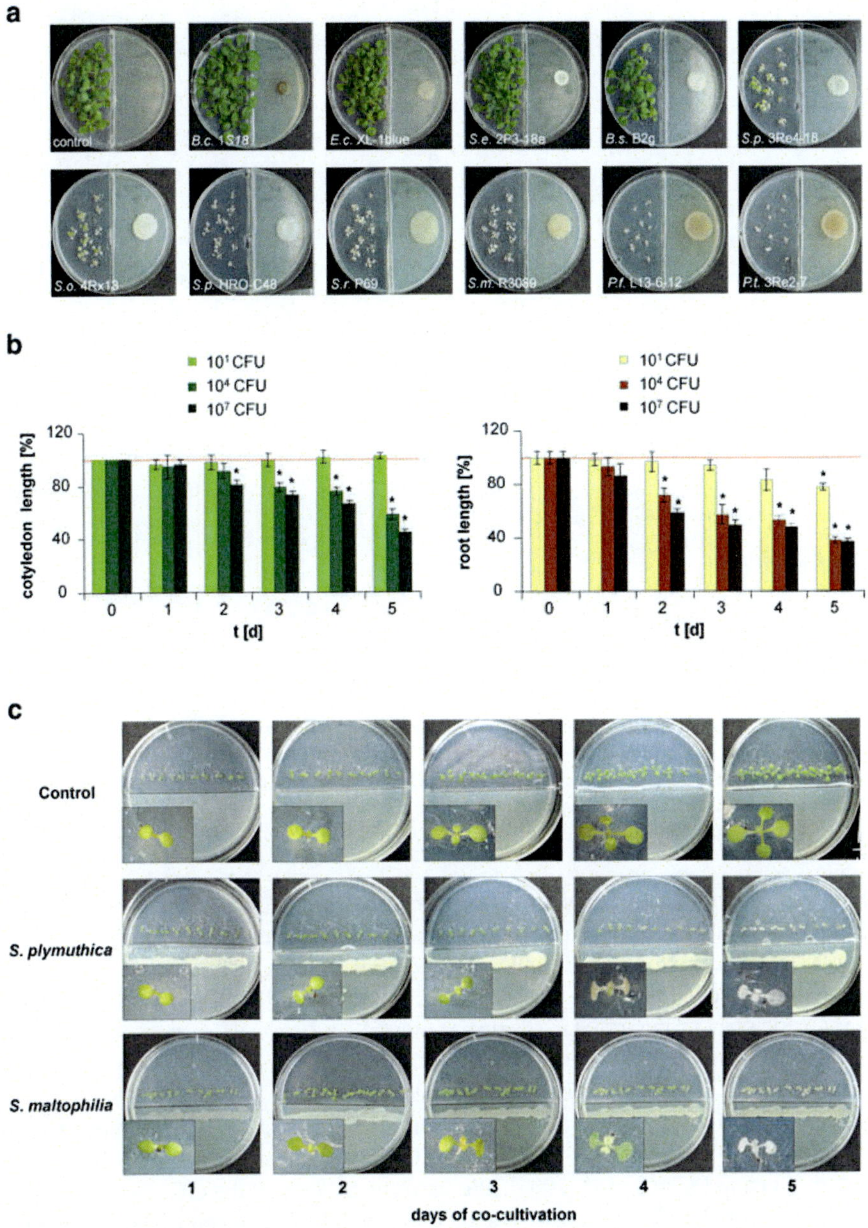

Fig. 2 Bacterial volatiles affect the growth of plants. (**a**) *Arabidopsis thaliana* in coculture with several rhizobacteria (*Bacillus subtilis* B2g; *Burkholderia cepacia* 1S18; *Escherichia coli* XL-1blue; *Pseudomonas fluorescens* L13-6-12; *P. trivialis* 3Re2-7; *Serratia odorifera* 4Rx13; *S. plymuthica* HRO-C48; *S. plymuthica* 3Re4-18; *Stenotrophomonas rhizophila* P69; *S. maltophilia* R3089; *S. epidermidis* 2P3-18a). (**b**) Bacterial cell number–dependent growth inhibitions of *Arabidopsis thaliana* cocultivated with *S. plymuthica* HRO-C48 ($n = 3$; $p \leq 0.01$), cotyledon length (*left*) and primary root length (*right*). (**c**) Photographic documentation of *A. thaliana* growth in dual cultures *S. plymuthica* HRO-C48 and *S. maltophilia* R3089 compared to control.

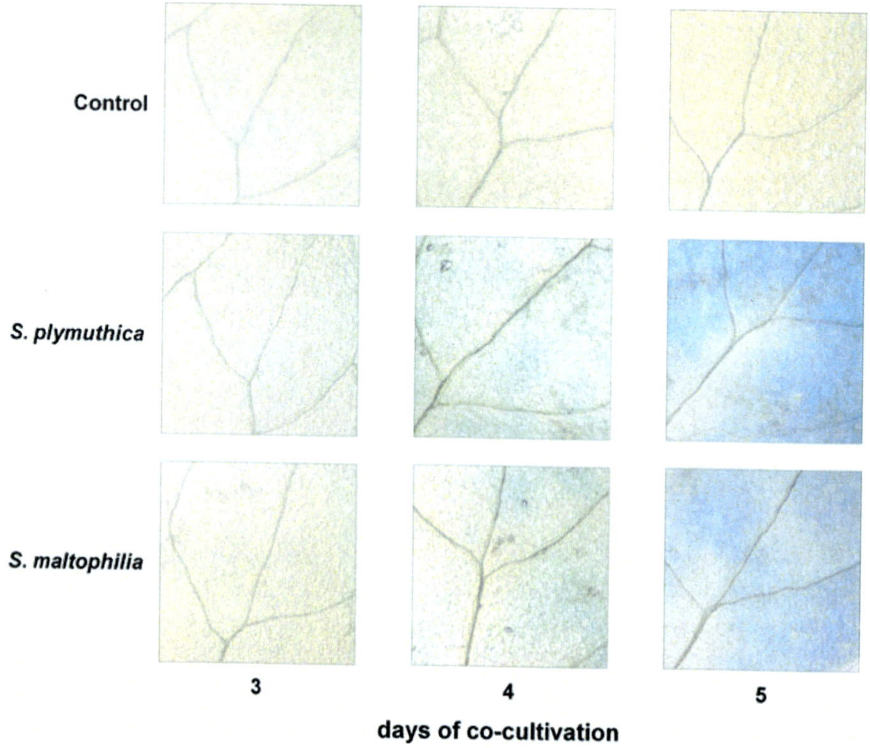

Fig. 3 Bacterial volatiles induce cell death in plants. *Serratia plymuthica* HRO-C48 and *Stenotrophomonas maltophilia* R3089 volatiles induce cell death of *Arabidopsis thaliana* cotyledons. Evans blue stains dead cells

blue signals in the cotyledons with both rhizobacteria (Fig. 3). These results show that the vitality of the leaf cells was significantly reduced after the application of volatiles of *S. maltophilia* R3089 and *S. plymuthica* HRO-C48.

These observations were further substantiated by studies at the molecular level. Synthetic plant promoter/GUS (β-glucuronidase) constructs containing defined regulatory elements (e.g., S-box, GCC-box) (Rushton et al. 2002) allow a simple and easy detection of altered gene expression due to pathogen response. The GCC-box (AGCCGCC) is often found in promoter regions of defense genes (Ohme-Takagi and Shinsi 1995), and the S-box (AGCCACC) directs gene expression upon fungal elicitor action (Kirsch et al. 2001). We used the S-box and the GCC-box promoter/GUS constructs to detect gene activation after bacterial volatile emission. Qualitative determination of the GUS activity by using 5-bromo-4-chloro-3-indolyl glucuronide (X-gluc) as substrate for the β-glucuronidase revealed an unspecific activation of the ethylene-inducible GCC-box in control experiments and *S. plymuthica* HRO-C48 cocultivated seedlings (Fig. 4a). The unregular activation/nonactivation of the GCC-box in response to the bacterial volatiles underlines the absence of ethylene in the volatile blend of *S. plymuthica* HRO-C48, which was verified by laser-based analysis

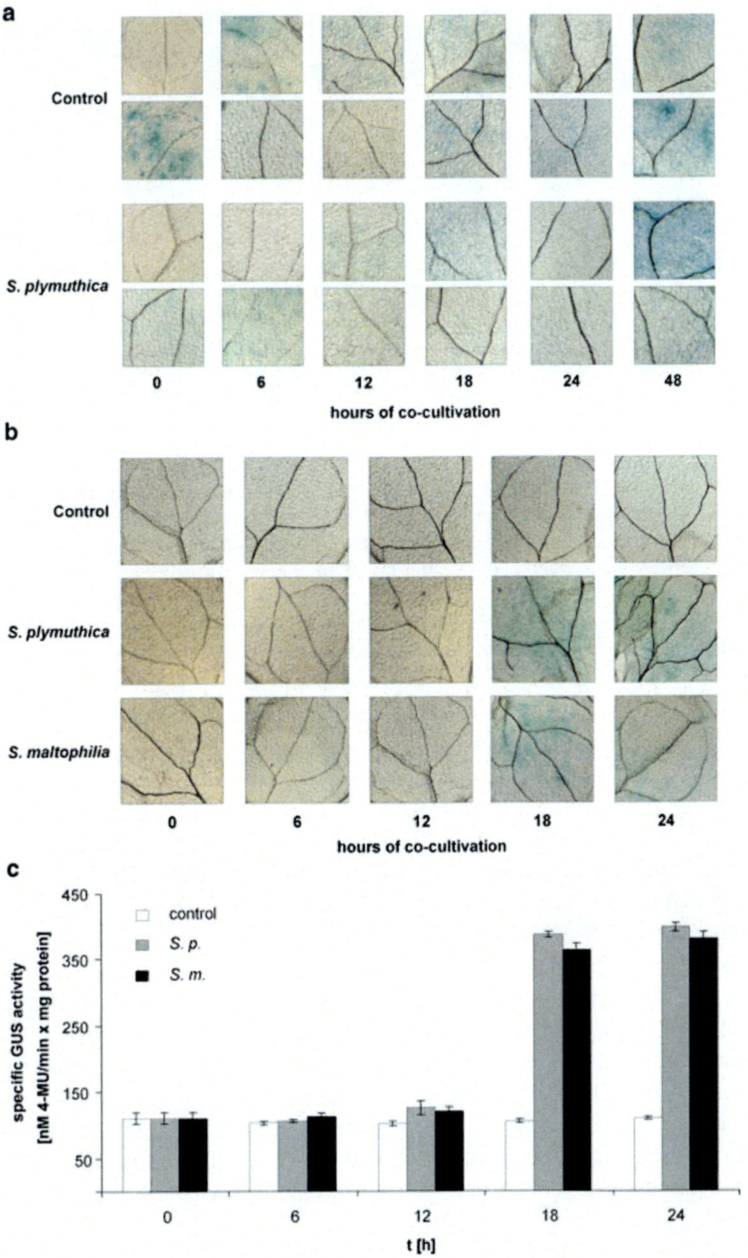

Fig. 4 Bacterial volatiles activate plant promoters. *Serratia plymuthica* HRO-C48 and *Stenotrophomonas maltophilia* R3089 activate stress-inducible promoter elements fused to the β-glucuronidase (GUS) marker gene. Induction of GUS gene expression is visualized by formation of a blue product of degraded 5-bromo-4-chloro-3-indolyl glucuronide (X-Gluc). (**a**) Activation of the GCC-box in *Arabidopsis thaliana* cocultivated with the rhizobacterial strain *S. plymuthica* HRO-C48 compared to control. (**b**) Activation of the S-box in *A. thaliana* cocultivated with the rhizobacterial strain *S. plymuthica* HRO-C48 and *S. maltophilia* R3089 compared to control. (**c**) Quantification of S-box-dependent GUS activity with 4-methylumbelliferyl-β-D-glucuronide (MUG)

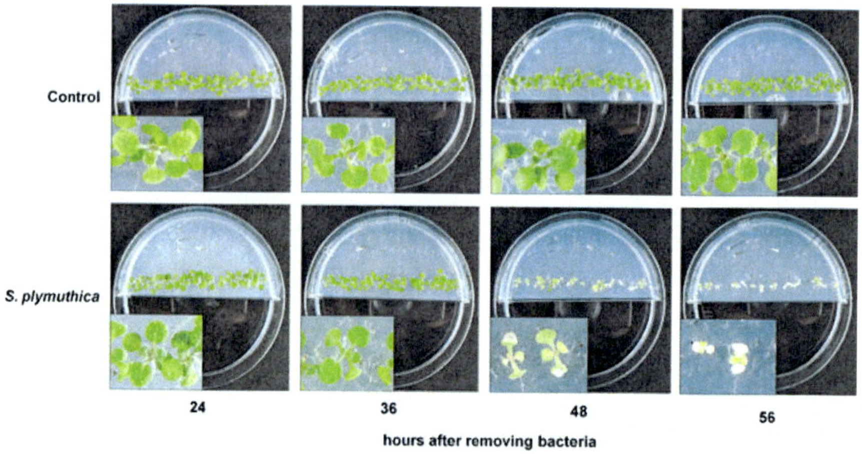

Fig. 5 Plant recovery after elimination of bacterial volatile exposure. Growth of *Arabidopsis thaliana* recovered after removal of *Serratia plymuthica* HRO-C48 within 36 h of cocultivation. Longer periods of cocultivations (>8 h) lead to growth inhibition and plant death. The growth of the seedlings was documented at day 6 after initiation of cocultivation

for the closely related *Serratia odorifera* 4Rx13 (Kai et al. 2010). In contrast, other rhizobacteria such as *Pseudomonas syringae* pv. *glycinea*, pv. *phaseolicola* (Weingart and Völksch 1997) are indeed able to produce ethylene. The promoter GUS assays with the S-box indicated volatile-dependent regulation of gene expression in dual culture with *S. plymuthica* HRO-C48 and *S. maltophilia* R3089 (Fig. 4b). The activity of the S-box/GUS element was quantitatively determined 6, 12, 18, and 24 h after initiating the cocultivation by application of bacteria (Fig. 4c). A twofold increase of GUS activity was detected 18 h after starting cocultivation. These data show that the volatiles of both bacteria have the capability to activate genes in plants via stress-responsive promoters, and furthermore, primary gene activations were detectable within one day after *A. thaliana* was exposed to bacterial volatiles.

To attest that bacterial volatiles are the causing agents, two different approaches were used. When the third compartment in a tripartite Petri dish was filled with charcoal, the plant growth could be restored because volatiles bind to charcoal (Vespermann et al. 2007). In another set of experiments, bacteria were removed after 1, 2, 3, and 4 days of cocultivation to allow recovery of *A. thaliana* (Fig. 5). The plants have the capacity to regrow when the bacteria are removed within 36 h of cocultivation. Longer exposures (48 and 56 h) to the bacterial blends dramatically reduced the recovery capacity; apparently, cell damage was too severe, and/or cell death processes had been initiated.

4.3 *Bacterial Volatiles Cause Plant Growth Inhibitions*

Volatiles emitted by bacteria are usually very complex mixtures (Kai et al. 2007). The observed growth promotions and inhibitions of *A. thaliana* in the dual culture assays

Fig. 6 Cocultivation of *A. thaliana* with *P. fluorescens* HCN-emitting CHA0 wild type (*left*) and HCN-negative CHA207 mutant (*centre*), and global regulatory CHA1144 mutant (right) strains (14 days of cocultivation)

are therefore due to the overall action of different compounds of which the causing agents and their relevant concentrations are often not known. As a first step to determine which bacterial volatiles have the potential to affect the growth, individual compounds like ammonia, HCN, and dimethyl disulfide (DMDS), were tested with *A. thaliana* (Fig. 6). Different concentrations of commercially available substances were applied on one side of the bipartite Petri dish, while *A. thaliana* was growing in the other compartment. DMDS exerts insecticidal activity via cytochrome oxidase in the mitochondrial electron transport system and potassium channel blockage (Dugravot et al. 2003; Gautier et al. 2008). The amount of DMDS with an inhibiting effect of 50% on *A. thaliana* seedlings was recently determined to be 20 μmol (Kai et al. 2010). Furthermore, Blom et al. (2011) described a negative effect of HCN on *A. thaliana* growth; 1 μmol HCN reduced plant growth ca. fourfold. Hydrogen cyanide is a volatile produced by *Pseudomonas*, *Chromobacterium*, and *Rhizobium* (Blumer and Haas 2000; Kai et al. 2010; Blom et al. 2011). The wild type of *Pseudomonas fluorescens* (CHA0) exhibited a strong volatile-dependent retardation of *A. thaliana* fresh weight, which was partially reestablished in cocultures with the HCN negative mutant *P. fluorescens* CHA207 (Fig. 6) Other volatiles than HCN also contribute to seedling growth retardation because co-cultivation with a global regulatory *P. fluorescens* mutant (CHA1144), affected in the synthesis of several secondary metabolites (Valverde and Haas 2008), fully reestablished seedling growth (Fig. 6). In addition to HCN, the CHA1144 mutant emits much less DMDS (data not shown). Additionally, reduced root length was observed in response to CHA0 and the cyanogenic *P. aeruginosa* PAO1, but no inhibition in response to respective noncyanogenic mutants (Rudrappa et al. 2008). *Serratia odorifera* 4Rx13 does not produce HCN (Kai et al. 2010), and therefore, growth inhibitions of *A. thaliana* by volatiles of *S. plymuthica* HRO-C48 also may not relate to HCN. *S. odorifera* 4Rx13, however, is able to emit ammonia at concentrations <1 μmol. At least 2.5 μmol of ammonia is necessary to inhibit plant growth in the Petri dish test system (Kai et al. 2010). A toxic effect of ammonia results in decoupling of the electron transport (Losada und Arnon 1963), which causes chlorosis and ultimately growth inhibitions (Britto und Kronzucker 2002). Ammonia and DMDS, may act additively or synergistically on plants coculturing with *S. plymuthica*. Experiments with volatile compounds applied individually or in mixtures with different ratios need be performed in the future to understand the action potential of complex bacterial blends.

4.4 Bacterial Volatiles Cause Plant Growth Promotions

Beside bacterial volatiles exerting growth inhibitions on *A. thaliana*, also growth promotions were observed, e.g., cocultivation with *Bacillus amyloliquefaciens* IN937a and *B. subtilis* GB03 (Table 1). These bacteria are known as plant growth promoting rhizobacteria (PGPR), which support plant growth by mechanisms and agents such as (1) synthesis and release of plant hormones by bacteria (e.g., indole-3-acetic acid, cytokinin, gibberellin), (2) increasing the availability of soil minerals (e.g., Fe), (3) fixation of airborne nitrogen (N_2), and (4) release of antibiotics (e.g., antifungal metabolites AFMs), toxins, or biosurfactants (Raaijmakers et al. 2002). Bacterial volatiles apparently add another facet to the multitude of plant growth promoting mechanisms. Several publications summarized in Table 1 appeared that describe the positive growth effects in *A. thaliana* due to bacterial volatile emissions.

Bacillus subtilis GB03 is the prominent bacterium which was often used for plant growth promoting experiments. In dual culture systems, the volatile mixture of *B. subtilis* effected the auxin homeostasis; augmented photosynthetic capacity, chloroplast number, chlorophyll content, starch accumulation, and iron uptake; increased tolerance to osmotic, salt, and drought stress; reduced severity of disease symptoms; and increased resistance against pathogens of the model plant *A. thaliana* (Table 1). These induced alterations improved and stimulated the plant growth and established an additional function for volatiles as signaling molecules mediating plant-microbe interactions. The volatiles emitted by *B. subtilis* GB03 seem to influence numerous and various physiological processes. It has to be considered that GB03 emits 38 different VOCs (Farag et al. 2006). Each compound could have the potential to influence cellular or molecular processes individually. So far, only the two characteristic volatiles of bacilli, 2,3-butanediol or acetoin or the racemic mixture of 2,3-butanediol were applied individually. In these test systems, 2,3-butanediol could verify some results obtained with the bacterial volatile mixtures (leaf growth stimulation and decrease of disease symptoms); however, 2,3-butanediol was excluded to improve photosynthetic efficiency. Therefore, other compounds of the volatile blend of *B. subtilis* may be the causing agents for the latter (Farag et al. 2006). Besides the organic volatile compounds, also CO_2 emission due to metabolic reactions (e.g., tricarbonic acid cycle) has to be considered. In sealed Petri dishes, the CO_2 concentrations reached levels that were eightfold compared to ambient concentrations (3,000 ppm) (Kai and Piechulla 2009) and therefore may very well play a role in plant growth stimulations under respective test conditions. Surprisingly, out of 15 publications regarding plant growth promotions due to bacterial volatile fumigation, only one, Ezquer et al. 2010, discussed the possibility that CO_2 may affect the plants positively in the used experimental setup. Ezquer et al. (2010), however, theoretically excluded that the increased starch accumulation might be a consequence of bacterial

Table 1 Bacterial volatiles mediating plant growth promotions

Bacterial volatiles	Plants	Test system	Effects	References
B.s. GB03, *B.a.* IN937a, *B.p.* T4, *B.p.* C9, *P.f.* 89B/61, *S.m.* 90–166, *E.c.* DH5α	*A. thaliana* Col-0, C24, WS, Ler	TSA, bipartite Petri dishes sealed with parafilm	Leaf area, promotion: *B.s.*, GB03 *B.a.* IN937a	Ryu et al. (2003)
2,3-Butanediol	*A. thaliana* Col-0	TSA, bipartite Petri dish, sealed	Leaf area, promotion	Ryu et al. (2003)
B.s. GB03, *B.a.* IN937a, *B.p.* T4, *B.p.* C9, *P.f.* 89B/61, *S.m.* 90–166, *E.c.* DH5α, *E.c.* JM22, *B.p.* SE34, *B.s.* 168, *B.s.* BSIP1171	*A. thaliana* Col-0	TSA, bipartite Petri dish, sealed	*E.c.* leaf symptoms decreased, disease resistance increased: *B.s.* GB03, *B.a.*, IN937a *B.p.* T4 + C9, *P.f.*89B61	Ryu et al. (2004)
2,3-Butanediol	*A. thaliana* Col-0	TSA, bipartite Petri dish, sealed	Leaf symptoms decrease	Ryu et al. (2004)
B.s. GB03, *B.a.* IN937a, *P.f.*89B61 *E. c.* DH5α		TSA, *S. t.* (potato) slices, sealed	Volatile profiles with SPME	Faraq et al. (2006)
B.s. GB03	*A. thaliana* Col-0	TSA, bipartite Petri dish, o/c? RNA extraction 48 and 72 hours for microarray analysis	Around 350 upregulated genes, auxin synthesis upregulated, auxin accumulation down in leaves and up in roots, cell wall–loosening enzymes upregulated correlates with cell expansion	Zhang et al. (2007)
B.s. GB03	*A. thaliana* Col-0	TSA, bipartite Petri dish, o/c?	More chloroplasts, photosynthetic efficiency increased, sugar accumulation elevated, sugar sensing repressed, inhibition of hypocotyl elongation and seed germination, ABA down regulation	Zhang et al. (2008a)
2,3-Butanediol	*A. thaliana* Col-0		No effect on photosynthetic efficiency	Zhang et al. (2008a)
	A. thaliana Col-0	TSA, bipartite Petri dish, o/c? Leaf area and root mass	Salt tolerance, high-affinity K^+ transporter, HKT expression high in shoots and low in roots, lower Na^+ accumulation in whole plant	Zhang et al. (2008b)

(continued)

Table 1 (continued)

Bacterial volatiles	Plants	Test system	Effects	References
2R,3R-butanediol *P.c.* O6 and gacS mutant	*A. thaliana* Col-0, Ler, root colonization	King's medium, sealed	Induced drought resistance, decrease of stomatal apertures	Cho et al. (2008)
S.o. 4Rx13	*A. thaliana* Col-0	NB, bipartite Petri dish, sealed	Fresh weight increases due to CO_2 accumulation	Kai and Piechulla (2009)
B.s. GB03	*A. thaliana* Col-0	TSA, o/c? Magenta boxes	Growth promotion, increased inflorescences and silique number, chlorophyll, quantum yield increased, iron uptake and iron transporter upregulated	Xie et al. (2009)
B.s. GB03	*A. thaliana* Col-0	TSA, bipartite Petri dish, o/c?	Increase of iron accumulation, upregulation of Fe-deficient-induced transcription factor, root ferric reductase activity, increased acidification of rhizosphere by proton release, increased photosynthetic capacity	Zhang et al. (2009)
B.s. GB03	Ocimum basilicum	TSA, bipartite Petri dish, and Magenta boxes, o/c?	α-Terpineol and eugenol emission and essential oil increased, root and shoot biomass increase	Banchio et al. (2009)
B.s. FB17, acetoin	*A. thaliana* Col-0	LB, Magenta boxes, root inoculation, o/c?	Reduced disease severity against *P.s.* systemic resistance, ethylene-responsive gene expression increased	Rudrappa et al. (2010)
B.s. GB03	*A. thaliana* Col-0	TSA, bipartite Petri dish, o/c?	Choline and betaine synthesis enhanced, increased tolerance to osmotic stress, improved	Zhang et al. (2010)

			drought tolerance in soil-grown plants	
B.s. GB03	*A. thaliana* Col-0	TSA, bipartite Petri dish, o/c?	Proteome analysis, upregulation of ROS scavenging, ethylene biosynthesis, TCA cycle, gluconeogenesis enzymes	Kwon et al. (2010)
B.s. 168, *E. c.* BW25113, *P.s.* several isolates, *P. c.*, *P. a.*, *S. e.* LT2, *A. t.* EHA105 + GV2260	*A. thaliana* Col-0,	M9 minimal medium + 50 mM glucose, Petri dish in plastic box, o/c?	Starch accumulation on M9 medium but not on LB, consider ammonia and CO_2!	Ezquer et al. (2010)

A. thaliana Arabidopsis thaliana, *A.t.* EHA105 *Agrobacterium tumefaciens* EHA105, *A.t.* GV2260 *Agrobacterium tumefaciens* GV2260, *B.a.* IN937a *Bacillus amyloliquefaciens* IN937a, *B.p.* C9 *Bacillus pasteurii* C9, *B.p.* T4 *Bacillus pumilus* T4, *B.p.* SE168 *Bacillus pumilus* SE168, *B.s.* GB03 *Bacillus subtilis* GB03, *B.s.* 168 *Bacillus subtilis* 168, *B.s.* BSIP1171 *Bacillus subtilis* BSIP1171, *B.s.* FB17 *Bacillus subtilis* FB17, *E.c. Erwinia carotovora*, *E.c.* DH5α *Escherichia coli* DH5α, *E.c.* BW25113 *Escherichia coli* BW25113, *E.c.* JM22 *Enterobacter cloacae* JM22, *P.a. Pseudomonas aurantiogriseum*, *P.c. Pseudomonas charlesii*, *P.c.* O6 *Pseudomonas chlororaphis* O6, *P.f.*89B61 *Pseudomonas fluorescens* 89B61, *P.s. Pseudomonas syringae*, *S.e.* LT2 *Salmonella enterica* LT2, *S.m.* 90–116 *Serratia marcescens* 90–116, *S.o.* 4R×13 *Serratia odorifera* 4R×13, *S.t. Solanum tuberosum*

ROS reactive oxygen species, LB Luria-Bertani broth, TSA tryptic soy agar, ABA absisic acid, SPME solid phase micro extraction

o/c? not indicated whether open or closed (=sealed) test system was used to perform dual culture assays

CO_2 production; however, our experimental experiences with sealed plastic containers (Kai and Piechulla 2009) would argue against their theoretical considerations. Only experimental proofs could eliminate doubts.

5 Plant Volatiles Affecting *Arabidopsis thaliana* Ecotypes

Naturally occurring plant variations result from genetic diversity and epigenetic processes and occur even within one species. Besides genetic, also phenotype association studies are important to understand underlying ecological and evolutionary forces. *Arabidopsis thaliana* is an ideal candidate because of the known whole genome sequence and the availability of up to 750 accessions. It can be envisioned that *A. thaliana* had to cope with different bacterial volatile exposures under certain natural circumstances and that after long times of adaptations, different ecotypes evolved. Here we used 21 accessions of *A. thaliana* and performed cocultivations with *S. plymuthica* HRO-C48. The effects of bacterial volatiles were registered by fresh weight and root length determinations (Fig. 7a, b and c, d, respectively). The effects of *S. plymuthica* HRO-C48 volatiles on the accessions C24, Col-0, and Ler were exemplified in Fig. 7a and c. The results of all 21 accessions were summarized in Fig. 7b, d. No significant variations of fresh weight reductions (90%) were measured after exposure of the different *A. thaliana* accessions to *S. plymuthica* HRO-C48 (Fig. 7a, b). The inhibition of root growth of most accessions varied between 50% and 60%, except accession C24 (inhibition of 82%) and Ler (inhibition of 42%) (Fig. 7c, d). These results verify the higher sensitivity of primary root growth compared to the growth of green plant parts already described in Fig. 2. The hints for accession-dependent variation of root growth inhibitions correlate with experiments made by Walch-Liu et al. (2006). Concentrations of 50 μM L-glutamate lead to a similar range of inhibitions of primary roots (ca. 80% and 40% of C24 and Ler, respectively), and alterations of root branching. The latter effect was not observed in our experiments, indicating that the mode of action of L-glutamate is different to the effect of the volatiles of *S. plymuthica* HRO-C48. *A. thaliana* C24 presumably developed under laboratory conditions and Ler were isolated from the natural habitat in Landsberg (Germany). Apparently, Ler and also many other ecotypes adapted to growth inhibitions in their original locations, including to volatiles emitted by rhizobacteria, while C24 obviously did not experience such inhibitory pressures in the laboratory and therefore expresses higher sensitivity to volatiles of *S. plymuthica*.

6 Outlook

Volatile emissions of bacteria are more widespread and complex than previously thought. Comprehensive emission patterns of bacteria can only be determined when several different methods are applied and bacteria are tested under different growth

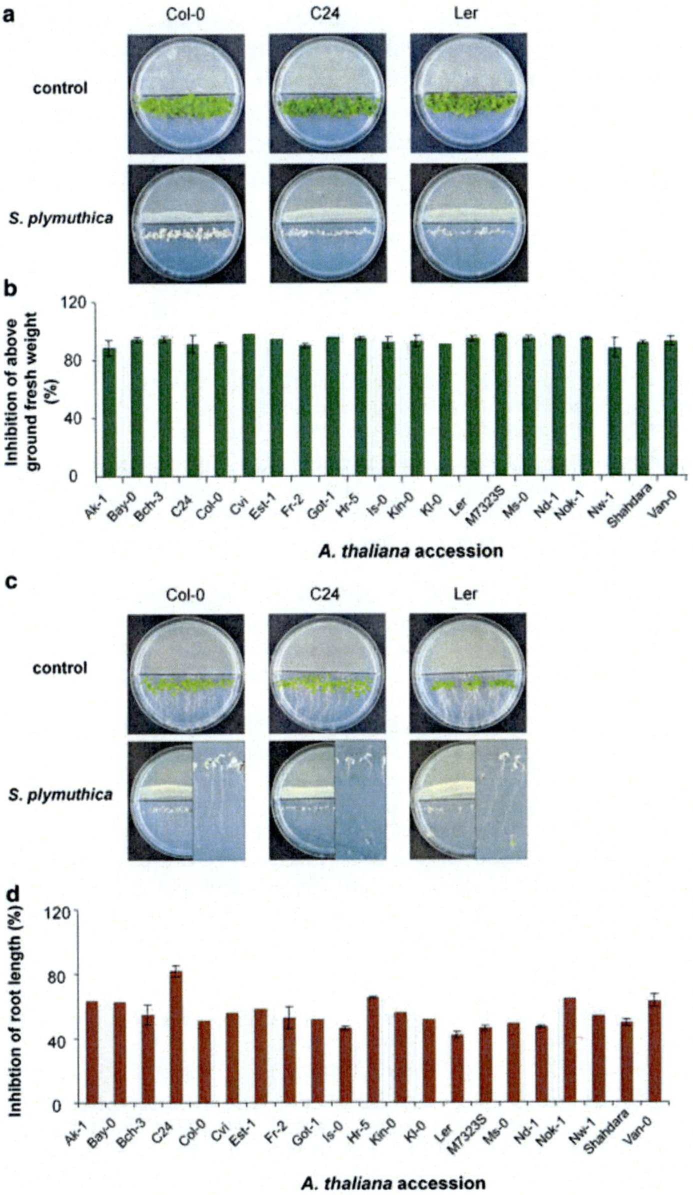

Fig. 7 Bacterial volatiles affect growth of *Arabidopsis thaliana* accessions. *Serratia plymuthica* HRO-C48 was cocultivated with various *A. thaliana* ecotypes. Fresh weight of aboveground plant parts (**a**, **b**) and roots (**c**, **d**) were determined after 10 days of cocultivation. Achkarren/DE Ak-1; Bayreuth/DE Bay-0; Buchen/DE Bch-3; unknown location C24; Columbia/MO Col-0; Cape Verde Islands Cvi; Estonia/EE Est-1; Frankfurt/DE Fr-2; Goettingen/DE Got-1; Isenburg/DE Is-0; United Kingdom/location unknown Hr-5; Kendallville/MI Kin-0; Kaiserslautern/DE Kl-0; Landsberg/DE Ler; unknown location M7323S; Moscow/RU Ms-0; Niederlenz/DE Nd-1; Noordwijk/NL Nok-1; Neuweilnau/DE Nw-1; Pamiro-Alay/TJ Shahdara; Vancouver/BC Van-0

conditions. It is a future research task to unravel biological and ecological effects of individual compounds as well as volatile mixtures at their relevant concentrations to elucidate the communication highway between bacteria and plants. Furthermore, it will be important to investigate the biosynthetic pathways and regulations of volatile syntheses in bacteria (emitter) and the perceptions and signal transductions in plants (receiver).

References

Allardyce RA, Langford VS, Hill AL, Murdoch DR (2006) Detection of volatile metabolites produced by bacterial growth in blood culture media by selected ion flow tube mass spectrometry (SIFT-MS). J Microbiol Methods 65:361–365

Arthur CL, Pawliszyn J (1990) Solid phase microextraction with thermal desorption using fused silica optical fibers. Anal Chem 62:2145–2148

Bais HP, Fall R, Vivanco JM (2004) Biocontrol of *Bacillus subtilis* against infection of Arabidopsis roots by *Pseudomonas syringae* is facilitated by biofilm formation and surfactin production. Plant Physiol 134:307–319

Banchio E, Xie X, Zhang H, Pare PW (2009) Soil bacteria elevate essential oil accumulation and emissions in sweet basil. J Agric Food Chem 57:653–657

Berg G, Roskot N, Steidle A, Eberl L, Zock A, Smalla K (2002) Plant-dependent genotypic and phenotypic diversity of antagonistic rhizobacteria isolated from different *Verticillium* host plants. Appl Environ Microbiol 68:3328–3338

Bloemberg GV, Wijfjes AHM, Lamers GEM, Stuurman N, Lugtenberg BJJ (2000) Simultaneous imaging of *Pseudomonas fluorescens* WCS365 populations expressing three different autofluorescent proteins in the rhizosphere: new perspectives for studying microbial communities. Mol Plant Microbe Interact 13:1170–1176

Blom D, Fabbri C, Eberl L, Weisskopf L (2011) Volatile-mediated killing of *Arabidopsis thaliana* by bacteria is mainly mediated due to hydrogen cyanide. Appl Environ Microbiol 77:1000–1008

Blumer C, Haas D (2000) Mechanism, regulation, and ecological role of bacterial cyanide biosynthesis. Arch Microbiol 173:170–177

Boland W, Ney P, Jaenicke L, Gassmann G (1984) A "closed-loop-stripping" technique as a versatile tool for metabolic studies of volatiles. In: Schreier P (ed) Analysis of volatiles. Walter De Gruyter & Co, D-Berlin, New York, pp 371–380

Britto DT, Kronzucker HJ (2002) NH_4—toxicity in higher plants: a critical review. J Plant Physiol 159:567–584

Bunge M, Araghipour N, Mikoviny T, Dunkl J, Schnitzhofer R, Hansel A, Schinner F, Wisthaler A, Margesin R, Märk TD (2008) On-line monitoring of microbial volatile metabolites by proton transfer reaction-mass spectrometry. Appl Environ Microbiol 74:2179–2186

Carroll W, Lenney W, Wang TS, Spanel P, Alcock A, Smith D (2005) Detection of volatile compounds emitted by *Pseudomonas aeruginosa* using selected ion flow tube mass spectrometry. Pediatr Pulmonol 39:452–456

Cho SM, Kang BR, Han SH, Anderson AJ, Park JY, Lee YH, Cho BH, Yang KY, Ryu CM, Kim YC (2008) 2R,3R-Butanediol, a bacterial volatile produced by *Pseudomonas chlororaphis* O6, is involved in induction of systemic tolerance to drought in *Arabidopsis thaliana*. Mol Plant–Microbe Interact 21:1067–1075

Chuankun X, Minghe M, Leming Z, Kegin Z (2004) Soil volatile fungistasis and volatile fungistatic compounds. Soil Biol Biochem 36:1997–2004

Dickschat JS, Wenzel SC, Bode HB, Müller R, Schulz S (2004) Biosynthesis of volatiles by the myxobacterium *Myxococcus Xanthus*. Chem Biol Chem 5:778–787

Dickschat JS, Martens R, Brinkhoff T, Simon M, Schulz S (2005) Volatiles released by a *Streptomyces* species isolated from the North Sea. Chem Biodivers 2:837–865

Dugravot S, Grolleau F, Macherel D, Rochetaing A, Hue B, Stankiewicz M, Huignard J, Lapied B (2003) Dimethyl disulfide exerts insecticidal neurotoxicity through mitochondrial dysfunction and activation of insect K_{ATP} channels. J Neurophysiol 90:259–270

Ercolini D, Russo F, Nasi A, Ferranti P, Villani F (2009) Mesophilic and psychrotrophic bacteria from meat and their spoilage potential in vitro and in beef. Appl Environ Microbiol 75:1990–2001

Ezquer I, Li J, Ovecka M, Baroja-Fernandez E, Munoz FJ, Montero M, Diaz de Cerio J, Hildago M, Sesma MT, Bahaji A, Etxeberria E, Pozueta-Romero J (2010) Microbial volatile emissions promote accumulation of exceptionally high levels of starch in leaves of mono- and dicotyledonous plants. Plant Cell Physiol 51:1674–1693

Farag MA, Ryu CM, Summer LW, Pare PW (2006) GC-MS SPME profiling of rhizobacterial volatiles reveals prospective inducers of growth promotion and induced systemic resistance in plants. Phytochemistry 67:2262–2268

Fiddaman PJ, Rossall S (1994) Effect of substrate on the production of antifungal volatiles from *Bacillus Subtilis*. J Appl Bacteriol 76:395–405

Gautier H, Auger J, Legros C, Lapied B (2008) Calcium-activated potassium channels in insect pacemaker neurons as unexpected target site for the novel fumigant dimethyl disulfide. J Pharmacol Exp Ther 324:149–159

Gerber NN, Lechevalier HA (1965) Geosmin, an earthy-smelling substance isolated from actinomycetes. Appl Microbiol 13:935–938

Gust B, Challis GL, Fowler K, Kieser T, Chater KF (2003) PCR-targeted Streptomyces gene replacement identifies a protein domain needed for biosynthesis of the sesquiterpene soil odor geosmin. Proc Natl Acad Sci USA 100:1541–1546

Kai M, Piechulla B (2009) Plant growth promotions due to rhizobacterial volatiles—an effect of CO_2? FEBS Lett 583:3473–3477

Kai M, Effmert U, Berg G, Piechulla B (2007) Volatiles of bacterial antagonists inhibit mycelial growth of the plant pathogen *Rhizoctonia solani*. Arch Microbiol 187:351–360

Kai M, Vespermann A, Piechulla B (2008) The growth of fungi and *Arabidopsis thaliana* is influenced by bacterial volatiles. Plant Signal Behav 3:1–3

Kai M, Crespo E, Cristescu SM, Harren FJM, Piechulla B (2010) *Serratia odorifera*: analysis of volatile emission and biological impact of volatile compounds on *Arabidopsis thaliana*. Appl Microbiol Biotechnol 88:965–976

Kataoka H, Lord HL, Pawliszyn J (2000) Applications of solid-phase microextraction in food analysis. J Chromatogr A 880:35–62

Kim M, Ahn JW, Jin UH, Choi D, Paek KH, Pai HS (2003) Activation of the programmed cell death pathway by inhibition of proteasome function in plants. J Biol Chem 278:19406–19415

Kirsch C, Logemann E, Lippok B, Schmelzer E, Hahlbrock K (2001) A highly specific pathogen-responsive promoter element from the immediate-early activated CMPG1 gene in *Petroselinum crispum*. Plant J 26:1–12

Kurze S, Dahl R, Bahl H, Berg G (2001) Biological control of soil-borne pathogens in strawberry by *Serratia plymuthica* HRO-C48. Plant Dis 85:529–534

Kwon YS, Ryu CM, Lee S, Park HB, Han KS, Lee JH, Lee K, Chung WS, Jeong MJ, Kim HK, Bae DW (2010) Proteome analysis of Arabidopsis seedlings exposed to bacterial volatiles. Planta 232:1355–1370

Larsen TO, Frisvad JC (1994) A simple method for collection of volatile metabolites from fungi based on diffusive sampling from Petri dishes. J Microbiol Methods 19:297–305

Losada M, Arnon DJ (1963) Selective inhibitors of photosynthesis. In: Hochster RM and Quasted JH (eds), Metabolic Inhibitors, vol 2. Academic, New York, pp 503–611

Mayr D, Margesin R, Klingsbichel E, Hartungen E, Jenewein D, Märk TD Schinner F und (2003) Rapid detection of meat spoilage by measuring volatile organic compounds by using proton transfer reaction mass spectrometry. Appl Environ Microbiol 69:4697–4705

Ohme-Takagi M, Shinsi H (1995) Ethylene-inducible DNA binding proteins that interact with an ethylene-responsive element. Plant Cell 7:173–182

Pollak FC, Berger RG (1996) Geosmin and related volatiles in bioreactor-cultured *Streptomyces citreus* CBS 109.60. Appl Environ Microbiol 62:1295–1299

Preti G, Thaler E, Hanson CW, Troy M, Eades J, Gelperin A (2009) Volatile compounds characteristic of sinus-related bacteria and infected sinus mucus: analysis by solid-phase microextraction and gas chromatography–mass spectrometry. J Chromatogr B 877:2011

Raaijmakers JM, Vlami M, de Sou JT (2002) Antibiotic production by bacterial biocontrol agents. Antonie Van Leeuwenhoek 81:537–547

Rudrappa T, Splaine RE, Biedrzycki ML, Bais HP (2008) Cyanogenic *Pseudomonads* influence multitrophic interactions in the rhizosphere. PLoS One 30:3(4), e2073

Rudrappa T, Biedrzycki ML, Kunjeti SG, Donofrio NM, Czymmek KJ, Paré PW, Bais HP (2010) The rhizobacterial elicitor acetoin induces systemic resistance in *Arabidopsis thaliana*. Commun Integr Biol 3:130–138

Rushton PJ, Reinstädler A, Lipka V, Lippok B, Somssich IE (2002) Synthetic plant promoters containing defined regulatory elements provide novel insights into pathogen- and wound-induced signaling. Plant Cell 14:749–762

Ryu CM, Farag MA, Hu CH, Reddy MS, Wie HX, Pare PW, Kloepper JW (2003) Bacterial volatiles promote growth in Arabidopsis. Proc Natl Acad Sci USA 100:4927–4932

Ryu CM, Farag MA, Hu CH, Reddy MS, Kloepper JW, Pare PW (2004) Bacterial volatiles induce systemic resistance in Arabidopsis. Plant Physiol 134:1017–1026

Schöller CEG, Molin S, Wilkins K (1997) Volatile metabolites from some gram-negative bacteria. Chemosphere 35:1487–1495

Schreier P (1980) Wine aroma composition: identification of additional volatile constituents in red wine. J Agric Food Chem 28:926–928

Schulz S, Fuhlendorff J, Reichenbach H (2004) Identification and synthesis of volatiles released by the myxobacterium *Chondromyces crocatus*. Tetrahedron 60:3863–3872

Stotzky G, Schenck S (1976) Volatile organic compounds and microorganisms. CRC Critical Rev Microbiol 4:333–382

Thorn RMS, Reynolds DM, Greenman J (2010) Multivariate analysis of bacterial volatile compound profiles for discrimination between selected species and strains in vitro. J Microbiol Methods 84:258–264

Urbach G (1997) The flavour of milk and dairy products: II. Cheese: contribution of volatile compounds. Intern J Dairy Technol 50:79–89

Valverde C, Haas D (2008) Small RNAs controlled by two-component systems. Adv Exp Med Biol 631:54–79

Vergnais L, Masson F, Montel MC, Berdagué JL, Talon R (1998) Evaluation of solid-phase microextraction for analysis of volatile metabolites produced by Staphylococci. J Agric Food Chem 46:228–234

Vespermann A, Kai M, Piechulla B (2007) Rhizobacterial volatiles affect the growth of fungi and *Arabidopsis thaliana*. Appl Environ Microbiol 73:5639–5641

von Reuß S, Kai M, Piechulla B, Francke W (2010) Octamethylbicyclo(3.2.1)octadienes from *Serratia odorifera*. Angew Chem Int Ed 49:2009–2010

Walch-Liu P, Liu LH, Remans T, Tester M, Forde BG (2006) Evidence that L-glutamate can act as endogenous signal to modulate root growth and branching in *Arabidopsis thaliana*. Plant Cell Physiol 47:1045–1057

Walker TS, Bais HP, Deziel E, Schweitzer HP, Rahme LG, Fall R, Vivanco JM (2004) *Pseudomonas aeruginosa*-plant root interactions. Pathogenicity, biofilm formations, and root exudation. Plant Physiol 134:3210–3331

Weingart H, Völksch B (1997) Ethylene production by *Pseudomonas syringae* pathovars in vitro and in planta. Appl Environ Microbiol 63:156–161

Xie X, Zhang H, Pare PW (2009) Sustained growth promotion in Arabidopsis with long-term exposure to the beneficial soil bacterium *Bacillus subtilis* (GB03). Plant Signal Behav 4:948–953

Zhang H, Kim MS, Krishnamachari V, Payton P, Sun Y, Grimson M, Farag MA, Ryu CM, Allen R, Melo IS, Pare PW (2007) Rhizobacterial volatile emissions regulate auxin homeostasis and cell expansion in Arabidopsis. Planta 226:839–851

Zhang H, Kim MS, Sun Y, Dowd SE, Shi H, Pare PW (2008a) Soil bacteria confer plant salt tolerance by tissue-specific regulation of the sodium transporter HKT1. Mol Plant–Microbe Interact 21:737–744

Zhang H, Xie X, Kim MS, Kornyeyev DA, Holaday S, Pare RW (2008b) Soil bacterium augment Arabidopsis photosynthesis by decreasing glucose sensing and abscisic acid levels in planta. Plant J 56:264–273

Zhang H, Sun Y, Xie X, Kim MS, Dowd SE, Pare RW (2009) A soil bacterium regulates plant acquisition of iron via deficiency-inducible mechanisms. Plant J 58:568–577

Zhang H, Murzello C, Sun Y, Kim MS, Xie X, Jeter RM, Zak JC, Dowd SE, Pare PW (2010) Choline and osmotic-stress tolerance induced by the soil microbe *Bacillus subtilis* (GB03). Mol Plant–Microbe Interact 23:1097–1104

Zhu J, Bean HD, Kuo YM, Hill JE (2010) Fast detection of volatile organic compounds from bacterial cultures by secondary electrospray ionization–mass spectrometry. J Clin Microbiol 48:4426–4431

Zoller HF, Mansfield Clark W (1921) The production of volatile fatty acids by bacteria of the dysentery group. J Gen Physiol 3:325–330

Zou CS, Mo MH, Gu YQ, Zhou JP, Zhang KQ (2007) Possible contribution of volatile-producing bacteria in soil fungistasis. Soil Biol Biochem 39:2371–2379

Infection of Plants by the Human Pathogen *Salmonella* Typhimurium: Challenges and New Insights

Adam Schikora, Ana Victoria Garcia, Amélie Charrier, and Heribert Hirt

Abstract *Salmonella* are the causative agents of the majority of food-borne bacterial poisonings and are responsible for more than 100 million infections of humans annually. In contrast to typhoid and paratyphoid fever, salmonellosis is frequent in the developed world. This is largely contributed by changes in the nutritional behavior resulting in eating more fruits and raw vegetables. Recently, it was discovered that the colonization of plants by *Salmonella* is a highly organized process. These results indicate that plants form part of the natural life cycle of *Salmonella* and open up new strategies to understand and combat bacterial diseases.

1 Introduction

The enteric pathogens *Salmonella* are the causative agents of the majority of food-borne bacterial poisonings. They are responsible for an estimated 1 million casualties and about 100 million human infections annually. Not only in developing countries in Africa or Southeast Asia, where typhoid and paratyphoid fever are unfortunately still common, also in developed communities, salmonellosis is still not vanquished. Recently, the change in our nutritional behavior exhibited the potential of *Salmonella* to use plants as vectors for animal infections. Research on the interaction between vegetable hosts and these bacteria discovered that the colonization of plants by *Salmonella* is an active infection process. *Salmonella* change their metabolism and

---

Adam Schikora and Ana Victoria Garcia contributed equally to the work.

A. Schikora (✉)
Institute for Plant Pathology and Applied Zoology, IFZ, JL University Giessen, Giessen, Germany

A.V. Garcia • A. Charrier • H. Hirt (✉)
URGV Plant Genomics, INRA/CNRS/University of Evry, Evry, France
e-mail: heribert.hirt@univie.ac.at

G. Witzany and F. Baluška (eds.), *Biocommunication of Plants*,
Signaling and Communication in Plants 14,
DOI 10.1007/978-3-642-23524-5_18,

adjust to the plant host. On the other hand, the host plant responds to bacterial attack with defense mechanisms. The newest findings are reviewed in this chapter.

2 *Salmonella* Infect Animal and Plant Hosts

Numerous bacteria, pathogenic to humans and other mammals, are found to thrive also on plants. Among these, *Salmonella enterica*, *Pseudomonas aeruginosa*, *Burkholderia cepacia*, *Erwinia spp.*, *Staphylococcus aureus*, *Escherichia coli* O157:H7, and *Listeria monocytogenes* are able to infect both animal and plant organisms (Plotnikova et al. 2000; Prithiviraj et al. 2005; Heaton and Jones 2008; Milillo et al. 2008; Holden et al. 2009).

Salmonella are a genus of Gram-negative enteropathogenic bacteria that successfully colonize a wide range of animal hosts including humans. *Salmonella* are the causal agents of both gastroenteritis and typhoid fever. The most common mode of infection in humans is by ingestion of contaminated food or water. Many reports have now linked food poisoning with the consumption of *Salmonella*-contaminated raw vegetables and fruits (for review see Brandl 2006; Holden et al. 2009). A large study conducted in the European Union revealed that in 2007, 0.3% of products were infected with *Salmonella* bacteria (Westrell et al. 2009), during the same time in the UK, the Netherlands, Germany, and Ireland, 0.1–2.3% of precut products were contaminated (Westrell et al. 2009). In the USA, the proportion of raw food-associated salmonellosis outbreaks increased from 0.7% in the 1960s to 6% in the 1990s (Sivapalasingam et al. 2004), and crossed 25% in recent years (Rangel et al. 2005). In order to monitor the molecular subtype pattern of the outbreak strains, a national program (PulseNet) was created in the USA (Gerner-Smidt et al. 2006). This program significantly improved the identification of outbreaks and their sources.

Most studies on *Salmonella*-plant interactions suggested an epiphytic lifestyle of *Salmonella* on plants. However, a growing body of evidence points to an active process in which bacteria infect various plants and use them as viable hosts (Barak et al. 2005; Iniguez et al. 2005; Klerks et al. 2007; Saggers et al. 2008; Schikora et al. 2008; Barak et al. 2009; Kroupitski et al. 2009; Noel et al. 2010; Barak et al. 2011; Golberg et al. 2011).

3 Modification of Host Physiology Is Often Achieved Through Effectors

Salmonellosis develops after the bacteria enter epithelial cells of the intestine (Patel et al. 2005). Although a typical infection leads to self-limiting gastroenteritis, *Salmonella* cause systemic infections by invading spleen, liver, and other organs in susceptible hosts. Studies of the infection mechanisms in animals have shown that *Salmonella* actively remodel the host cell's physiology and architecture and suppress the host's immune system by injecting a cocktail of effectors delivered by type III secretion systems (T3SSs). *Salmonella enterica* subsp. *enterica* has two distinct T3SSs, T3SS-1 and T3SS-2, encoded by the *Salmonella* pathogenicity islands (SPI) SPI-1 and SPI-2, respectively (Collazo and Galan 1997; Hensel 2000). T3SS-1 secretes at least 16 proteins of which six were shown to interact with the host signaling cascades and the cytoskeleton. T3SS-2 secretes at least 19 *Salmonella enterica*–specific effector proteins that are involved in survival and multiplication within the host cell (Waterman and Holden 2003; Kuhle and Hensel 2004). The expression and the secretion of SPI-1- and SPI-2-encoded effectors are tightly regulated. Recently, a sorting platform for T3SS effectors was reported that determines the appropriate hierarchy for protein secretion (Lara-Tejero et al. 2011). In this study, the authors identified the cytoplasmic SpaO-OrgA-OrgB complex, which enables the sequential delivery of translocases before the secretion of the actual effectors. Furthermore, the authors described the role of specific chaperones in the recognition and loading of effectors into the sorting SpaO-OrgA-OrgB complex. SicA and InvE escort *Salmonella* translocases, while SicP is required for proper loading of the SptP effector. The removal of the chaperone-binding site on SptP was shown to prevent its recruitment to the SpaO complex (Lara-Tejero et al. 2011). In conclusion, it was postulated that similar sorting platforms may exist in other T3SSs as their components are widely conserved. However, such a complex has not been reported in plant pathogenic bacteria. Even though many recent reports suggest that the mechanisms used by *Salmonella* to infect animal and plant hosts might be similar, the role of *Salmonella* T3SSs and effectors in plant infections remains unclear.

4 Effector Proteins Suppress the First Layer of Immune Defenses

In the battle between pathogen and its host, the pathogen needs to suppress the host's immune system in order to establish a successful infection. The early line of immunity relies on the recognition of conserved pathogen-associated molecular patterns (PAMPs) by host-encoded pattern recognition receptors (PRRs) and thereby the activation of an array of defense responses called PAMP-triggered immunity (PTI). The best-studied PAMP in plants is flg22, a conserved 22-amino acid peptide from the bacterial flagellar protein flagellin, recognized by the PRR flagellin insensitive 2 (FLS2) (Gomez-Gomez and Boller 2000a, b)). During infection, pathogens secrete effectors with the aim to suppress PTI and cause effector-triggered susceptibility (ETS). In a second layer of defense, intracellular resistance proteins (R-proteins) recognize pathogen effectors and activate effector-triggered immunity (ETI). The plant pathogen *Pseudomonas syringae* injects about 40 effectors into plant cells. Among these, AvrPto and AvrPtoB interact with FLS2 and its coreceptor BR1-associated kinase 1 (BAK1) in *Arabidopsis thaliana* plants (Chinchilla et al. 2007; Gohre et al. 2008; Shan et al. 2008). AvrPtoB catalyzes the polyubiquitination and subsequent proteasome-dependent degradation of FLS2, which is enhanced when FLS2 binds to flg22. AvrPto interacts with BAK1 and thereby prevents its binding to FLS2 (Shan et al. 2008). In these ways, both AvrPto and AvrPtoB interrupt signaling to the downstream mitogen-activated protein kinase (MAPK) module. *P. syringae* has another effector that directly interacts with the MAPK cascade components: HopAI1 is a phosphothreonine lyase that dephosphorylates the threonine residue at which MAPKs are activated by their upstream MAPKKs (Zhang et al. 2007). When expressed *in planta*, HopAI1 directly interacts with the *Arabidopsis* MAPKs AtMPK3 and AtMPK6 attenuating flg22-induced MAPK activation and downstream defense responses. Strikingly, HopAI1 is also present in animal/human pathogens such as *Shigella spp.* (OspF) (Li et al. 2007; Zhu et al. 2007) and *Salmonella spp.* (SpvC) (Mazurkiewicz et al. 2008), where it interacts with the MAPKs ERK1/2 and p38. The role of multiple *Salmonella* effectors in animal infection has been described (reviewed in McGhie et al. 2009), but a functional proof of *Salmonella* effector action in plants is still missing. Nonetheless, several evidences point to an active interaction between these bacteria and plant hosts, and the newest development in this field shall be presented in this chapter.

5 Virulence for Plants Depends on the Ability to Attach to Plant Surfaces

Pathogen adhesion to the host's cell surface is an initial step of infection. *Salmonella enterica* serovars have been shown to bind to alfalfa sprouts efficiently and significantly better than for instance the pathogenic *E. coli* strain O157:H7 (Barak et al. 2002). Saggers et al. suggest that *Salmonella* actively attach to plant tissues and need to be viable for successful colonization (Saggers et al. 2008). In a large screen, 20 out of 6,000 *S.* Newport mutants with lower attachment ability to alfalfa sprouts were identified (Barak et al. 2005). Interestingly, some of the genes identified in this study code for the surface-exposed aggregative fimbria nucleator curli (*agfB*) and for the global stress regulator *rpoS* which regulates the production of curli, cellulose and, other adhesins that are important for animal pathogenicity. *AgfD*, which was also identified in this study, regulates the production of the lipopolysaccharide *O*-polysaccharide (also known as *O*-antigen) capsule. By regulating the *yih* operon in coordination with other extracellular matrix genes, *agfD* not only plays a central role in the ability to attach to plant surfaces (Barak et al. 2007), but also in environmental fitness and the pathogenicity toward animals (Gibson et al. 2006). In addition, Barak et al. showed that *yihO* (involved in *O*-antigen capsule formation) and *bcsA* (coding for a cellulose synthase) are important for adhesion to alfalfa sprouts (Barak et al. 2007), whereas cellulose and curli are involved in transmission of *S.* Typhimurium from water onto parsley leaves (Lapidot and Yaron 2009). In another study, two previously uncharacterized genes (STM0278 and STM0650) were characterized as important factors for the infection of alfalfa sprouts, due to their essential role in biofilm formation and swarming (Barak et al. 2009). In summary, it is becoming clear that the genetic equipment of *Salmonella* plays an important role in the infection of animals and plants alike.

6 Genetic Dependence of Plant Infection by *Salmonella*

The genus *Salmonella* is divided into two species: *Salmonella bongori* and *Salmonella enterica*, and several hundred related isolates. *S. enterica* acquired the second SPI (SPI-2) most probably through horizontal gene transfer and with it, the ability to spread systemically in infected hosts. One of its seven subspecies *Salmonella enterica* subsp. *enterica* is the major cause of food-related poisonings. Many of the hundreds of isolated serovars of *Salmonella enterica* subsp. *enterica* were identified as strains causing salmonellosis in vegetable or fruit-originated outbreaks. Subsequent studies regarding the ability to attach and infect plants on those serovars revealed divergences in the ability to infect plants. A comparative study on the internal colonization in lettuce leaves by five *S. enterica* serovars (Dublin, Enteritidis, Montevideo, Newport, and Typhimurium) indicated a significant effect of the serovar type; however, no effect was observed when different lettuce cultivars were compared (Klerks et al., 2007). This study indicates that different genetic backgrounds have an impact on the pathogenicity toward plants. A similar study conducted on the serovars Braenderup, Negev, Newport, Tennessee, and Thompson also revealed differences between serovars (Patel and Sharma 2010). Interestingly, the authors pointed out a correlation between the capacity to produce biofilms and the attachment to leaves, with *S.* Thompson producing the strongest biofilms and showing the most efficient adhesion to lettuce leaves (Patel and Sharma 2010).

Similar to other plant pathogens, not only the pathogenicity on plants but also the response of the plant host depends on the *Salmonella* genetics. Recently, Berger et al. studied the wilting and chlorosis symptoms in *Arabidopsis* plants after infiltration with different serovars of *S. enterica* subsp. *enterica*, as well as *S. enterica* subsp. *arizonae* and *diarizonae* (Berger et al. 2011). Infiltration with *S.* Senftenberg and also with *S.* Cannstatt, Krefeld, and Liverpool, all of which belong to the serogroup E_4 (O: 1, 3, 19) possessing the *O*-antigen, resulted in rapid wilting and chlorosis. In contrast, infiltration with serovars lacking the *O*-antigen did not provoke any symptoms (Berger et al. 2011). In addition, the authors stated that the response to *Salmonella* infiltration is independent of the most prominent and studied PRRs, suggesting that specific receptors for *Salmonella O*-antigen could exist in *Arabidopsis*.

7 Endophytic Lifestyle of *Salmonella* in Plants

In animals, *Salmonella* actively enter epithelial and other cell types in order to replicate and spread through the organism. The question whether *Salmonella* use a similar strategy to infect plants is therefore of great interest. Previously, we showed that the GFP-marked *S.* Typhimurium strain 14028s added to *Arabidopsis* roots was observed inside root hairs 3 h after inoculation and inside rhizodermal cells 20 h later (Fig. 1a) (Schikora et al. 2008). At that time point, large numbers of motile bacteria were observed inside host cells and *in planta* bacterial titers increased, confirming that *Salmonella* can proliferate in plants. *Salmonella* were also found to form biofilm-like structures on the surface of roots, preferentially colonizing regions around emerging lateral roots and wounded tissues (Schikora et al. 2008). The formation of biofilms of *Salmonella* on leaves was also reported. Recently, three reports presented the possible entry points of bacteria to the inner layers of mesophyll cells (Kroupitski et al. 2009; Barak et al. 2011; Golberg et al. 2011). Barak et al. postulate trichomes as preferential colonization site (Barak et al. 2011). Stomata are natural openings that are shielded by two guard cells and that are responsible for the gas exchange in leaves. Kroupitski et al. (2009) showed that *Salmonella* make use of these natural openings in order to penetrate into lettuce leaves. Moreover, bacterial aggregation near stomata occurs only under light conditions when the stomata are open. Artificial opening of stomata in the dark had no impact on the bacterial behavior, suggesting that bacteria are attracted to photosynthesis-dependent products. Additional tests revealed that motility and the ability of chemotaxis are essential for *Salmonella* to colonize the interior of lettuce

Fig. 1 *Salmonella* infection on *Arabidopsis thaliana*. (**a**) Two-week-old *Arabidopsis* plants were transferred to liquid medium and inoculated with *Salmonella enterica* serovar Typhimurium strain 14028s expressing GFP. Three hours after infection, first GFP-marked bacteria were observed in root hairs; 20 h later, numerous bacteria were present in rhizodermal cells. (**b–d**) Soil grown *Arabidopsis* plants were syringe infiltrated with mock (**b**) *S. Typhimurium* 14028s wild type (**c**) or a T3SS *Salmonella* mutant (**d**). When compared to wild-type bacteria, stronger symptoms are observed after infiltration with the T3SS mutant, suggesting that a functional T3SS is necessary for the suppression of active defense mechanisms such as hypersensitive response–associated cell death

leaves. In a follow-up report, the same group demonstrated that not all plants are equally susceptible (or resistant) to *Salmonella* internal infection. Using GFP-marked bacteria, the authors analyzed the internalization of the *S.* Typhimurium strain 1344 in many leafy vegetables and herbs (Golberg et al. 2011). Interestingly, while some plant species (e.g., arugula) allow 1344 to internalize, some others (e.g., parsley) seem to have effective means to prevent infection (Golberg et al. 2011). Studies on lettuce, cabbage, and tomatoes demonstrated significant differences in the susceptibility to *Salmonella* infection (Klerks et al. 2007; Barak et al. 2011), pointing to an important role of plant innate immunity in modulating the response to infection by these bacteria.

8 Mitogen-Activated Protein Kinases and JA/ET Signaling Pathway Are Important for *Salmonella* Infection

The first event toward activation of plant immune responses is the recognition of the pathogen. Although a variety of PAMPs are known, only few PAMP receptors have been identified so far. FLS2 (Gomez-Gomez and Boller 2000a, b) and EFR (Zipfel et al. 2006) are closely related LRR receptor kinases that recognize the bacterial PAMPs flagellin and EF-Tu, respectively. Both receptors trigger the activation of similar downstream kinases and defense responses. Activation of MAPK cascades is an essential step to induce defense reactions in response to pathogen attack, and several MAPKs are activated by bacterial pathogens and PAMPs (Nuhse et al. 2000; Desikan et al. 2001; Asai et al. 2002; Zipfel et al. 2006). *Salmonella* infection of *Arabidopsis* plants results in the activation of AtMPK3 and AtMPK6 (Schikora et al. 2008). Since AtMPK3 and AtMPK6 are implicated in various stress-induced signaling pathways, the respective complexes rather than the MAPKs themselves provide the necessary signaling specificity. A role for AtMPK6 in defense against *Salmonella* is also emphasized by the fact that *mpk6* mutant plants are less resistant to *Salmonella* attack (Schikora et al. 2008).

Arabidopsis responds to *Salmonella* infection with a rapid transcriptional induction of a number of defense genes, including the antifungal defensin gene *PDF1.2* and the pathogenesis-related genes *PR2* and *PR4* (Schikora et al. 2008). The transcription of these genes is generally activated in response to necrotrophic pathogens and depends on the plant hormones jasmonic acid (JA) and ethylene (ET) (Jung et al. 2007). The marker gene for salicylic acid (SA)-induced defenses *PR1*, normally induced during infection with biotrophs, was also upregulated after contact with *Salmonella* (Schikora et al. 2008). In a simplified view, SA and JA/ET hormones trigger mutually antagonistic pathways, where SA-dependent responses (further subdivided into NPR1-dependent and NPR1-independent reactions) are important in defense against biotrophic pathogens (Durrant and Dong 2004), whereas JA and ET are mainly involved in responses to wounding, herbivores, and necrotrophic pathogens (Zimmerli et al. 2004). The JA-insensitive *coi1-16* mutant is defective in an F-box protein required for degradation of repressors of JA-responsive genes (Chini et al. 2007; Thines et al. 2007), and it is highly susceptible to *Salmonella* attack (Schikora et al. 2008) indicating that the JA signaling pathway is required to induce downstream defense reactions against *Salmonella*. In addition, the ET-signaling impaired *ein2-1* mutant showed delayed expression of defense genes, which correlated with enhanced proliferation rates of *Salmonella* (Schikora et al. 2008). However, Iniquez et al. (2005) reported that the SA-deprived *NahG* transgenic plants (expressing a bacterial salicylate hydroxilase) and *npr1* mutants impaired in SA signaling are more susceptible to *Salmonella* (Iniquez et al. 2005), indicating that SA may also play a role in defense against *Salmonella*. Together, these data indicate that *Salmonella* attack induces in *Arabidopsis* a complex defense response similar to that observed upon attack by other plant pathogens (Jones and Dangl 2006).

9 Are *Salmonella* Effectors Functional in Plant Cells?

Many animal and plant pathogenic bacteria use T3SS effectors to suppress host's immune responses (Fig. 1b–d). *Salmonella enterica* has two different T3SSs with different functions during infection. To date, about 44 *Salmonella* effectors, many of them with known function, have been described to be injected into host cells through one or both T3SSs (reviewed in (Heffron et al. 2011). Many of these effectors target the MAPK cascades, which are important regulators of the immune response in animals and plants. As previously mentioned, SpvC from *Salmonella spp.* and OspF from *Shigella spp.* encode a phosphothreonine lyase that dephosphorylates the pTXpY double phosphorylated activation loop in the ERK1/2 kinases (Arbibe et al. 2007; Li et al. 2007; Mazurkiewicz et al. 2008). Interestingly, also *P. syringae* possesses a homologue of SpvC/OspF, HopAI1, which has phosphothreonine lyase activity and can dephosphorylate activated AtMPK3 and AtMPK6 (Arbibe et al. 2007; Li et al. 2007; Zhang et al. 2007). Besides OspF/SpvC/HopAI1, HopPtoD2 from *Pseudomonas* also has homologues in different human pathogenic bacteria. HopPtoD2 is a tyrosine phosphatase which inhibits pathogen-triggered programmed cell death (Espinosa et al. 2003), while its homologue from *Salmonella spp.* SptP inhibits phosphorylation and membrane localization of Raf kinase and therefore the activation of the downstream ERK kinase (Lin et al. 2003). Although several *Salmonella* effectors have homologues in other plant pathogenic bacteria, the function of *Salmonella* proteins in the inactivation of the plant's immune system remains undetermined. It is tempting to speculate that biochemical features of those effectors are conserved between animal and plant hosts, providing *Salmonella* (and other pathogenic bacteria) with efficient tools for suppression of the host's immune systems. Such suppression was reported during infection of tobacco plants with *S.* Typhimurium (Shirron and Yaron 2011). Authors showed that in contrast to wild-type living bacteria, dead or chloramphenicol-treated bacteria elicited oxidative burst and pH changes in tobacco cells. Similar response was provoked by the *invA* mutant, which has no functional SPI-1 T3SS (Shirron and Yaron 2011). Those results suggest that *Salmonella* depend on the secretion of effectors during plant infection and actively suppress the immune responses. Recently, the function of SseF in plant-*Salmonella* interaction was characterized. SseF and SseG are SPI-2-encoded effector proteins involved in the formation of *Salmonella*-induced aggregation of host endosomes and of the replication niche (Kuhle and Hensel 2002; Deiwick et al. 2006). When expressed in tobacco, SseF triggers a type of programmed cell death termed hypersensitive response (HR) which is normally indicative of recognition and resistance triggered by R-proteins in plant cells (Üstün et al. in preparation). This report shows that *Salmonella* effectors might be recognized not only by animal but also by plant cells.

10 Conclusions

Today, along with *Escherichia coli*, *Salmonella* belong to the best-studied bacteria. The growing knowledge about the infection process in plants points to the so far underestimated possibilities of other human pathogenic bacteria to infect and proliferate in plants. Many of these, including dangerous bacteria such as *Listeria monocytogenes* or *Pseudomonas aeruginosa*, infect and survive in plant hosts. To understand the mechanisms by which these bacteria infect plants and how plants protect themselves may offer new insight into infection mechanisms and should contribute to diminish the number of vegetable- and fruit-related infections.

Acknowledgments The work was supported from grants of the ERANET Systems Biology project SHIPREC (Salmonella Host Interaction Project European Consortium)

References

Arbibe L, Kim DW, Batsche E, Pedron T, Mateescu B, Muchardt C, Parsot C, Sansonetti PJ (2007) An injected bacterial effector targets chromatin access for transcription factor NF-kappaB to alter transcription of host genes involved in immune responses. Nat Immunol 8:47–56

Asai T, Tena G, Plotnikova J, Willmann MR, Chiu WL, Gomez-Gomez L, Boller T, Ausubel FM, Sheen J (2002) MAP kinase signalling cascade in Arabidopsis innate immunity. Nature 415:977–983

Barak JD, Whitehand LC, Charkowski AO (2002) Differences in attachment of *Salmonella enterica* serovars and *Escherichia coli* O157:H7 to alfalfa sprouts. Appl Environ Microbiol 68:4758–4763

Barak JD, Gorski L, Naraghi-Arani P, Charkowski AO (2005) Salmonella enterica virulence genes are required for bacterial attachment to plant tissue. Appl Environ Microbiol 71:5685–5691

Barak JD, Jahn CE, Gibson DL, Charkowski AO (2007) The role of cellulose and O-antigen capsule in the colonization of plants by *Salmonella enterica*. Mol Plant Microbe Interact 20:1083–1091

Barak JD, Gorski L, Liang AS, Narm KE (2009) Previously uncharacterized Salmonella enterica genes required for swarming play a role in seedling colonization. Microbiology 155:3701–3709

Barak JD, Kramer LC, Hao LY (2011) Colonization of tomato plants by Salmonella enterica is cultivar dependent, and type 1 trichomes are preferred colonization sites. Appl Environ Microbiol 77:498–504

Berger CN, Brown DJ, Shaw RK, Minuzzi F, Feys B, Frankel G (2011) Salmonella enterica strains belonging to O serogroup 1,3,19 induce chlorosis and wilting of *Arabidopsis thaliana* leaves. Environ Microbiol 13:1299–1308

Brandl MT (2006) Fitness of human enteric pathogens on plants and implications for food safety. Annu Rev Phytopathol 44:367–392

Chinchilla D, Zipfel C, Robatzek S, Kemmerling B, Nurnberger T, Jones JD, Felix G, Boller T (2007) A flagellin-induced complex of the receptor FLS2 and BAK1 initiates plant defence. Nature 448:497–500

Chini A, Fonseca S, Fernandez G, Adie B, Chico JM, Lorenzo O, Garcia-Casado G, Lopez-Vidriero I, Lozano FM, Ponce MR, Micol JL, Solano R (2007) The JAZ family of repressors is the missing link in jasmonate signalling. Nature 448:666–671

Collazo CM, Galan JE (1997) The invasion-associated type-III protein secretion system in Salmonella. Gene 192:51–59

Deiwick J, Salcedo SP, Boucrot E, Gilliland SM, Henry T, Petermann N, Waterman SR, Gorvel JP, Holden DW, Meresse S (2006) The translocated Salmonella effector proteins SseF and SseG interact and are required to establish an intracellular replication niche. Infect Immun 74:6965–6972

Desikan R, Hancock JT, Ichimura K, Shinozaki K, Neill SJ (2001) Harpin induces activation of the Arabidopsis mitogen-activated protein kinases AtMPK4 and AtMPK6. Plant Physiol 126:1579–1587

Durrant WE, Dong X (2004) Systemic acquired resistance. Ann Rev Phytopathol 42:185–209

Espinosa A, Guo M, Tam VC, Fu ZQ, Alfano JR (2003) The *Pseudomonas syringae* type III-secreted protein HopPtoD2 possesses protein tyrosine phosphatase activity and suppresses programmed cell death in plants. Mol Microbiol 49:377–387

Gerner-Smidt P, Hise K, Kincaid J, Hunter S, Rolando S, Hyytia-Trees E, Ribot EM, Swaminathan B (2006) PulseNet USA: a five-year update. Foodborne Pathog Dis 3:9–19

Gibson DL, White AP, Snyder SD, Martin S, Heiss C, Azadi P, Surette M, Kay WW (2006) Salmonella produces an O-antigen capsule regulated by AgfD and important for environmental persistence. J Bacteriol 188:7722–7730

Gohre V, Spallek T, Haweker H, Mersmann S, Mentzel T, Boller T, de Torres M, Mansfield JW, Robatzek S (2008) Plant pattern-recognition receptor FLS2 is directed for degradation by the bacterial ubiquitin ligase AvrPtoB. Curr Biol 18:1824–1832

Golberg D, Kroupitski Y, Belausov E, Pinto R, Sela S (2011) *Salmonella typhimurium* internalization is variable in leafy vegetables and fresh herbs. Int J Food Microbiol 145:250–257

Gomez-Gomez L, Boller T (2000a) FLS2: An LRR receptor-like kinase involved in the perception of the bacterial elicitor flagellin in Arabidopsis. Mol Cell 5:1003–1011

Gomez-Gomez L, Boller T (2000b) FLS2: an LRR receptor-like kinase involved in the perception of the bacterial elicitor flagellin in Arabidopsis. Mol Cell 5:1003–1011

Heaton JC, Jones K (2008) Microbial contamination of fruit and vegetables and the behaviour of enteropathogens in the phyllosphere: a review. J Appl Microbiol 104:613–626

Heffron F, Niemann G, Yoon H, Kidwai A, Brown RNE, McDermott JD, Smith R, Adkins JN (2011) Salmonella-secreted virulence factors. In: Porwollik S (ed) Salmonella: from genome to function. Caister Academic Press, San Diego, pp 187–223

Hensel M (2000) Salmonella pathogenicity island 2. Mol Microbiol 36:1015–1023

Holden N, Pritchard L, Toth I (2009) Colonization out with the colon: plants as an alternative environmental reservoir for human pathogenic enterobacteria. FEMS Microbiol Rev 33:689–703

Iniguez AL, Dong Y, Carter HD, Ahmer BM, Stone JM, Triplett EW (2005) Regulation of enteric endophytic bacterial colonization by plant defenses. Mol Plant Microbe Interact 18:169–178

Iniquez LA, Dong Y, Carter HD, Ahmer BMM, Stone JM, Triplett EW (2005) Regulation of enteric endophytic bacterial colonization by plant defenses. Mol Plant Microbe Interact 18:169–178

Jones JDG, Dangl JL (2006) The plant immune system. Nature 444:323–329

Jung C, Lyou S, Yeu S, Kim M, Rhee S, Kim M, Lee J, Choi Y, Cheong J (2007) Microarray-based screening of jasmonate-responsive genes in *Arabidopsis thaliana*. Plant Cell Rep 26:1053–1063

Klerks MM, Franz E, van Gent-Pelzer M, Zijlstra C, van Bruggen AH (2007) Differential interaction of Salmonella enterica serovars with lettuce cultivars and plant-microbe factors influencing the colonization efficiency. ISME J 1:620–631

Kroupitski Y, Golberg D, Belausov E, Pinto R, Swartzberg D, Granot D, Sela S (2009) Internalization of *Salmonella enterica* in leaves is induced by light and involves chemotaxis and penetration through open stomata. Appl Environ Microbiol 75:6076–6086

Kuhle V, Hensel M (2002) SseF and SseG are translocated effectors of the type III secretion system of Salmonella pathogenicity island 2 that modulate aggregation of endosomal compartments. Cell Microbiol 4:813–824

Kuhle V, Hensel M (2004) Cellular microbiology of intracellular *Salmonella enterica*: functions of the type III secretion system encoded by Salmonella pathogenicity island 2. Cell Mol Life Sci 61:2812–2826

Lapidot A, Yaron S (2009) Transfer of *Salmonella enterica* serovar Typhimurium from contaminated irrigation water to parsley is dependent on curli and cellulose, the biofilm matrix components. J Food Prot 72:618–623

Lara-Tejero M, Kato J, Wagner S, Liu X, Galan JE (2011) A sorting platform determines the order of protein secretion in bacterial type III systems. Science 331:1188–1191

Li H, Xu H, Zhou Y, Zhang J, Long C, Li S, Chen S, Zhou JM, Shao F (2007) The phosphothreonine lyase activity of a bacterial type III effector family. Science 315:1000–1003

Lin SL, Le TX, Cowen DS (2003) SptP, a Salmonella typhimurium type III-secreted protein, inhibits the mitogen-activated protein kinase pathway by inhibiting Raf activation. Cell Microbiol 5:267–275

Mazurkiewicz P, Thomas J, Thompson JA, Liu M, Arbibe L, Sansonetti P, Holden DW (2008) SpvC is a Salmonella effector with phosphothreonine lyase activity on host mitogen-activated protein kinases. Mol Microbiol 67:1371–1383

McGhie EJ, Brawn LC, Hume PJ, Humphreys D, Koronakis V (2009) Salmonella takes control: effector-driven manipulation of the host. Curr Opin Microbiol 12:117–124

Milillo SR, Badamo JM, Boor KJ, Wiedmann M (2008) Growth and persistence of *Listeria monocytogenes* isolates on the plant model *Arabidopsis thaliana*. Food Microbiol 25:698–704

Noel JT, Arrach N, Alagely A, McClelland M, Teplitski M (2010) Specific responses of *Salmonella enterica* to tomato varieties and fruit ripeness identified by in vivo expression technology. PLoS One 5:e12406

Nuhse TS, Peck SC, Hirt H, Boller T (2000) Microbial elicitors induce activation and dual phosphorylation of the *Arabidopsis thaliana* MAPK 6. J Biol Chem 275:7521–7526

Patel J, Sharma M (2010) Differences in attachment of *Salmonella enterica* serovars to cabbage and lettuce leaves. Int J Food Microbiol 139:41–47

Patel JC, Rossanese OW, Galan JE (2005) The functional interface between Salmonella and its host cell: opportunities for therapeutic intervention. Trends Pharmacol Sci 26:564–570

Plotnikova JM, Rahme LG, Ausubel FM (2000) Pathogenesis of the human opportunistic pathogen *Pseudomonas aeruginosa* PA14 in Arabidopsis. Plant Physiol 124:1766–1774

Prithiviraj B, Bais HP, Jha AK, Vivanco JM (2005) Staphylococcus aureus pathogenicity on *Arabidopsis thaliana* is mediated either by a direct effect of salicylic acid on the pathogen or by SA-dependent, NPR1-independent host responses. Plant J 42:417–432

Rangel JM, Sparling PH, Crowe C, Griffin PM, Swerdlow DL (2005) Epidemiology of *Escherichia coli* O157:H7 outbreaks, United States, 1982–2002. Emerg Infect Dis 11:603–609

Saggers EJ, Waspe CR, Parker ML, Waldron KW, Brocklehurst TF (2008) Salmonella must be viable in order to attach to the surface of prepared vegetable tissues. J Appl Microbiol 105:1239–1245

Schikora A, Carreri A, Charpentier E, Hirt H (2008) The dark side of the salad: *Salmonella typhimurium* overcomes the innate immune response of *Arabidopsis thaliana* and shows an endopathogenic lifestyle. PLoS One 3:e2279

Shan L, He P, Li J, Heese A, Peck SC, Nurnberger T, Martin GB, Sheen J (2008) Bacterial effectors target the common signaling partner BAK1 to disrupt multiple MAMP receptor-signaling complexes and impede plant immunity. Cell Host Microbe 4:17–27

Shirron N, Yaron S (2011) Active suppression of early immune response in Tobacco by the human pathogen *Salmonella* typhimurium. PLoS One 6:e18855

Sivapalasingam S, Friedman CR, Cohen L, Tauxe RV (2004) Fresh produce: a growing cause of outbreaks of foodborne illness in the United States, 1973 through 1997. J Food Prot 67:2342–2353

Thines B, Katsir L, Melotto M, Niu Y, Mandaokar A, Liu G, Nomura K, He SY, Howe GA, Browse J (2007) JAZ repressor proteins are targets of the SCFCOI1 complex during jasmonate signalling. Nature 448:661–665

Üstün S, Müller P, Palmisano R, Hensel M, Börnke F. SseF, a type III effector protein from the mammalian pathogen *Salmonella enterica*, requires resistance-gene mediated signalling to activate cell death in the model plant *Nicotiana benthamiana*, submitted

Waterman SR, Holden DW (2003) Functions and effectors of the Salmonella pathogenicity island 2 type III secretion system. Cell Microbiol 5:501–511

Westrell T, Ciampa N, Boelaert F, Helwigh B, Korsgaard H, Chriel M, Ammon A, Makela P (2009) Zoonotic infections in Europe in 2007: a summary of the EFSA-ECDC Annual Report. Euro Surveill 14

Zhang J, Shao F, Li Y, Cui H, Chen L, Li H, Zou Y, Long C, Lan L, Chai J, Chen S, Tang X, Zhou JM (2007) A *Pseudomonas syringae* effector inactivates MAPKs to suppress PAMP-induced immunity in plants. Cell Host Microbe 1:175–185

Zhu Y, Li H, Long C, Hu L, Xu H, Liu L, Chen S, Wang DC, Shao F (2007) Structural insights into the enzymatic mechanism of the pathogenic MAPK phosphothreonine lyase. Mol Cell 28:899–913

Zimmerli L, Stein M, Lipka V, Schulze-Lefert P, Somerville S (2004) Host and non-host pathogens elicit different jasmonate/ethylene responses in Arabidopsis. Plant J 40:633–646

Zipfel C, Kunze G, Chinchilla D, Canierd A, Jones JDG, Boller T, Felix G (2006) Perception of the bacterial PAMP EF-Tu by the receptor EFR restricts Agrobacterium-mediated transformation. Cell 125:746–760

Coadaptationary Aspects of the Underground Communication Between Plants and Other Organisms

Akifumi Sugiyama, Daniel K. Manter, and Jorge M. Vivanco

Abstract Soil microbial communities are comprised of a vast array of bacteria, fungi, nematodes, and other organisms. It is becoming increasingly clear that these communities are not passively determined but actively regulated by plants. This chapter discusses the role plant root exudates play in the active regulation of soil microbial communities. In addition, we discuss the potential role coadapted plant-soil microbial communities may play in agricultural sustainability and production. We suggest that minimal disruption in the plant microbial community should be maintained in order to achieve maximum long-term agricultural production by minimizing disease outbreaks and by reducing costly agricultural inputs such as pesticides and fertilizers.

1 Root Exudates of Plants and Their Involvement in the Underground Communication

Plant roots release a wide range of compounds that are involved in the underground communication between plants and other organisms. These compounds include proteins, sugars, polysaccharides, amino acids, fatty acids, phenolics, and more.

A. Sugiyama
Center for Rhizosphere Biology, Department of Horticulture and Landscape Architecture, Colorado State University, Fort Collins, CO, USA

Research Institute for Sustainable Humanosphere, Kyoto University, Gokasho, Uji, Japan

D.K. Manter
USDA-ARS, Soil, Plant Nutrient Research Unit, Fort Collins, CO, USA

J.M. Vivanco (✉)
Center for Rhizosphere Biology, Department of Horticulture and Landscape Architecture, Colorado State University, Fort Collins, CO, USA
e-mail: J.Vivanco@colostate.edu

G. Witzany and F. Baluška (eds.), *Biocommunication of Plants*, Signaling and Communication in Plants 14,
DOI 10.1007/978-3-642-23524-5_19,

Quantity and quality of root exudates vary depending on the plant species, plant cultivars, and also the developmental stage of the plants (Priha et al. 1999; Innes et al. 2004; Batten et al. 2006; Mazzola et al. 2004; Kowalchuka et al. 2006; Narasimhan et al. 2003; Mougel et al. 2006; Yang and Crowley 2000). Some root exudates such as mugineic acid in rice are regulated by diurnal rhythms, but most root exudates are not predominantly regulated by diurnal rhythms in *Arabidopsis* (Takagi 1976; Badri et al. 2010). The function of root exudates is diverse (Badri et al. 2009b; Badri and Vivanco 2009), but for the most part, these compounds function as chemical signals between plants and soil microbes, such as (1) rhizobia, (2) arbuscular mycorrhizal fungi, and (3) pathogens.

1.1 Root Exudates Involved in the Symbiosis with Rhizobia

Legume plants (Fabaceae), composed of approximately 700 genera and 20,000 species, are the third largest plant family next to Orchidaceae and Asteraceae, and the second most important family for crop production (Doyle and Luckow 2003). The hallmark feature of legume plants, and one of the most important plant-soil microbe interactions known in agriculture, is the fact that legume plants establish a symbiotic relationship with rhizobia, which fix atmospheric nitrogen. Legume plants secrete signaling compounds from roots which help to attract rhizobia and in the establishment of root nodules. Several flavonoids were identified as signaling compounds, for example, luteolin from alfalfa (*Medicago sativa*), 7,4′-dihydroxyflavone and geraldone from white clover (*Trifolium repens*), and daidzein and genistein from soybean (*Glycine max*), respectively (Peters et al. 1986; Redmond et al. 1986; Djordjevic et al. 1987; Kosslak et al. 1987). Beside these flavones and isoflavonoids, a chalcone (4,4′-dihydroxy-2′-methoxychalcone) from alfalfa (Maxwell et al. 1989), anthocyanidins (petunidin and malvidin) from common bean (*Phaseolus vulgaris*) (Hungria et al. 1991), betains (trigonelline and stachydrine) from alfalfa (Phillips et al. 1992), and aldonic acids (erythronic acid and tetronic acid) from white lupine (*Lupinus albus*) (Gagnon and Ibrahim 1998) have also been reported as signaling compounds from roots, suggesting that a structurally diverse variety of phytochemicals can function as signal molecules. These signaling compounds diffuse around plant roots and bind to the Nod receptor in the rhizobial cell surface, which induces the expression of *nod* genes and the synthesis of signaling compounds, or Nod factors, from rhizobia.

Nod factors are lipochitooligosaccharides consisting of β-1, 4-linked N-acetyl-D-glucosamine backbones, and an acyl chain at C2 in the nonreducing end with acetyl, sulfonyl, carbamoyl, fucosyl, or arabinosyl moieties at defined positions depending on the rhizobial species (D'Haeze and Holsters 2004). Nod factors secreted into the rhizosphere are perceived by the Nod receptors located at the plasma membrane of the host legume root cells, which induce drastic physiological changes in plant roots that result in the formation of nodules. It is also noteworthy that canavanine, a root exudate of various legume plants, is toxic to many soil bacteria but not to rhizobial strains that possess a specific transporter for detoxifying this compound (Cai et al. 2009). It has been postulated that canavanine might select a rhizosphere microbiome in favor of rhizobial species.

1.2 *Root Exudates Involved in the Symbiosis with Arbuscular Mycorrhizal Fungi*

Mycorrhizae are divided into two groups: endomycorrhiza (such as arbuscular, ericoid, and orchid mycorrhiza) and ectomycorrhiza. These heterogeneous fungi colonize the roots of more than 200,000 plant species in a wide range of terrestrial ecosystems. Arbuscular mycorrhiza (AM) symbiotically interact with more than 80% of species in the plant kingdom, aiding in plant uptake of nutrients and water from the soil (Parniske 2008). The AM hyphal network is extensive and can reach upward of 100 meters per cubic centimeter of soil (Miller et al. 1995). The establishment of arbuscular symbioses begins with the colonization of the root by hyphae originating in the surrounding soil, followed by appressorium formation and entrance into the cortex. AM symbiosis results in the formation of tree-shaped subcellular structures, called arbuscules, where nutrient exchange between fungi and plants occurs. The life cycle of arbuscules was estimated to be 2–3 days in rice using fluorescent proteins as tags (Kobae and Hata 2010). Fossil records show that the origin of AM symbiosis occurred more than 400 million years ago (Remy et al. 1994; Parniske 2008). The appearance of the first terrestrial plants occurred approximately at the same period of time, suggesting that arbuscular colonization may have been essential for the first plants to successfully adapt to the terrestrial ecosystem (Simon et al. 1993; Remy et al. 1994; Redecker et al. 2000).

Hyphal branching is a critical step in the development and success of AM symbioses. Similar to the rhizobia-flavonoids interaction, AM branching is controlled by a plant-derived compound or branching factor (Buee et al. 2000). The chemical structure of the branching factor was identified as strigolactone, using root exudates of *L. japonicus* (Akiyama et al. 2005). Strigolactones are short lived and fragile in the rhizosphere because a labile ether bond spontaneously hydrolyzes in the soil. This fragility of strigolactones results in a steep concentration gradient from plant roots to the surrounding soil. AM fungi are obligate biotrophs that depend on a living photoautotrophic host to complete their life cycle. Mycorrhizae can find living plant roots using the concentration gradient of strigolactone in the soil. Strigolactones were previously identified from the root exudates of a variety of plants as seed germination factors for parasitic weeds such as *Striga* and *Orobanche* (Bouwmeester et al. 2003). Parasitic weeds are also biotrophs, and it appears that they evolved to utilize these ancient signal molecules of living plants to find the roots of a suitable living host.

Strigolactones are found in the root exudates of tomato, sorghum, pea in addition to *L. japonicus*, but not in the AM-forming plants carrot, tobacco, and alfalfa (Garcia-Garrido et al. 2009), which suggest that other branching factors have yet to be discovered. Although *Arabidopsis* and lupine do not form symbiosis with AM fungi, root exudates of these plants contain strigolactones (Goldwasser et al. 2008; Yoneyama et al. 2008). It has been shown that lupine secretes pyranoisoflavones that inhibit hyphal development in arbuscular mycorrhizal fungi (Akiyama et al. 2010). Recently, the signaling molecules from arbuscular mycorrhizal fungi, called Myc factors, were identified to be a mixture of sulfated and nonsulfated lipochitooligosaccharides (Maillet et al. 2011). The structure of Myc factors, although chemically simpler, resembles that of Nod factor produced by rhizobia, suggesting a possible evolutionary linkage between Myc factors and Nod factors. It is then possible to hypothesize about the presence of receptors, or other components similar to those of rhizobia-legume symbiosis, that may help regulate AM symbioses. It is worth noting that exudates from AM fungi influence soil bacterial community composition (Sprent 2007), and that some bacteria associated with AM fungi can improve colonization, root branching, and antibiosis (Bonfante and Anca 2009; Hartmann et al. 2009).

1.3 Root Exudates Involved in the Interaction with Pathogens

Roots encounter a large number of pathogens in the soil, including fungi, oomycetes, bacteria, and nematodes. The role of root exudates in host recognition, infection, and colonization is perhaps best document in the case of oomycetes. For example, the isoflavones, daidzein, and genistein from soybean roots are known to attract *Phytophthora sojae* zoospores (Hirsch et al. 2003), while preformed antifungal secondary metabolites such as phytoanticipins act as chemical barriers against soil pathogens. Plants also biosynthesize and secrete antimicrobial secondary metabolites, called phytoalexins, in response to pathogen infection. For example, glyceollin of soybean, medicarpin of alfalfa, vestitol of *L. japonicus*, momilactone A of rice are among these phytoalexins. In *Arabidopsis* roots and leaves, indolic phytoalexins such as camalexin accumulate at the site of bacterial and fungal infection, and a putative transporter involved in the secretion of phytoalexins has been reported (Stein et al. 2006). Abcg36 (pdr8/pen3) was identified from the screening of *Arabidopsis* mutants deficient in nonhost resistance. Microscopic observation with promoter ABCG36-GFP transgenic plants revealed that ABCG36 localized to the plasma membrane of the penetration site upon fungal attack, possibly exporting toxic metabolites to the apoplast at the site of invasion (Stein et al. 2006). AtABCG36 was also reported to be induced in leaf blades upon infection by both virulent and avirulent bacterial pathogens (Kobae et al. 2006). Because soilborne pathogens account for a net loss of 10–20% of potential crop production all over the world (Raaijmakers et al. 2009), there is a need for better understanding how these secondary metabolites act to protect plants from infection.

Nematodes are complex, slender, and wormlike eukaryotic invertebrates, typically less than 2.5 millimeters long, and the most numerous multicellular animals on earth (Perry and Moens 2006). Most nematodes in soil are free living, and consume bacteria, fungi, and other nematodes, but some nematodes parasitize plant roots. Root-knot and cyst nematodes tend to cause the most damage in crops (Bird 2004). Root-knot nematodes are thought to perceive root exudates from host plants to locate and penetrate the roots and to establish a permanent feeding site. Within the feeding site, nematodes secrete cytokinins to induce root cell growth and proliferation. *Medicago sativa* roots emit a volatile dimethyl sulfide, which can attract nematodes. One interesting side benefit of nematode attraction to roots may be in its ability to facilitate root-bacterial interactions. For example, nematodes often have various rhizobia attached to their cuticle (Horiuchi et al. 2005), and/or nematodes (*C. elegans*) secrete compounds such as organic acids, amino acids, and sugars that have been shown to function in the attraction of bacterial species and the inhibition of quorum sensing that regulate bacterial virulence (Kaplan et al. 2009).

2 Coadaptation Between Plants and Soil Microbes

Plants provide large and diverse habitats for a wide variety of soil microbes. It has been reported that 1 g of soil contains as many as 1,000,000 bacteria and fungi from thousands of different species (Trevors 2010), and these soil microbial communities have reciprocal interactions with resident plant species (Klironomos 2003; Morgan et al. 2005; Reinhart and Callaway 2006; Badri and Vivanco 2009). Rhizosphere microbial communities have been shown to differ between plant species (Priha et al. 1999; Innes et al. 2004; Batten et al. 2006), cultivars (Mazzola et al. 2004), chemotypes (Kowalchuka et al. 2006), developmental stages (Narasimhan et al. 2003; Mougel et al. 2006), and nutrient conditions of a given plant (Yang and Crowley 2000). It has also been reported that different root types within the same plant cultivate specific soil microbes (Liljeroth et al. 1991; Yang and Crowley 2000; Baudoin et al. 2002), which has been attributed to the difference in microenvironments around various root types such as root tips and lateral roots. All these examples highlight the tight association between plants and soil microbes.

2.1 *Involvement of Root Exudates in the Coadaptation Between Plants and Other Organisms*

Broeckling et al. (Broeckling et al. 2008) performed a detailed analysis of the effects of plant species and their root exudates on the soil fungal community. In this study, natural soils under *Arabidopsis thaliana* and *Medicago truncatula* were collected and used for fungal community analysis. When *Arabidopsis* and

Medicago were grown on their respective resident soils, the soil fungal communities remained relatively unchanged for several plant generations; however, when *Arabidopsis* was grown on soils from *Medicago* or *Medicago* was grown on soils from *Arabidopsis*, soil fungal communities changed dramatically. The changes in the fungal communities included both a decrease in microbial richness and diversity. The same trend was observed when root exudates from *Arabidopsis* or *Medicago* were applied to the soil, suggesting that the effect on fungal communities is mediated, at least in part, by root exudates.

Badri et al. (2009a) analyzed the effect of gene mutations in *Arabidopsis* on root exudate composition and ultimately on soil microbial communities. Among the 25 ABC transporter genes highly expressed in the roots, eight genes representing various subfamilies were chosen, and knockout mutants for each gene were grown on natural *Arabidopsis* soil under greenhouse conditions. Both bacterial and fungal communities did not differ after one generation of plant growth, but after the second generation, both bacterial and fungal communities under *abcg30* (previously called *pdr2*), plants showed significant differences from those under the wild-type plants. Interestingly, when ribosomal DNA sequences were analyzed by 454 pyrosequencing, it appeared that this mutant cultivated a relatively greater abundance of beneficial bacteria, such as plant-growth-promoting rhizobacteria, nitrogen-fixing bacteria, and bacteria involved in heavy metal remediation. Root exudates from *abcg30* mutant contained more phenolic compounds and fewer sugars than that of the wild type. The drastic changes in root exudate composition were apparently not due to the direct effect of the mutation but by the pleiotropic consequence caused by the gene mutation. Microarray analysis showed that hundreds of genes were up- or downregulated in the roots of *abcg30* mutants (Badri et al. 2009a), thus giving further support to the pleiotropic hypothesis. In a different study, it was reported that different ecotypes of *Arabidopsis* that show different profiles of root exudates can culture distinct bacterial communities after only one generation of plant growth (Micallef et al. 2009). There are many experimental differences between these two studies, but the biggest difference seems to be the soils, that is, Badri et al. used soils with a history of *Arabidopsis* growth (coadapted soil) while Micallef used artificial soils with no history of *Arabidopsis* coadaptation. Considering the coadaptationary aspect of root-soil microbiome interactions, it seems that the artificial soils are more prone to be influenced by the difference of plant ecotype or root exudates.

2.2 Coadaptation Between Plants and Other Organisms in the Field

Greenhouse experiments described in the previous section suggested a tight coadaptationary association between a given plant and soil microbes; however, in natural systems, soil microbes encounter a variety of plant species and changing weather conditions. A study was performed under natural conditions to determine the effect of the exotic invasive plant species spotted knapweed (*Centaurea stoebe*) on the native soil microbiome (Broz et al. 2007). Soils collected from high-density stands of spotted knapweed in Montana had significantly less diversity of fungal communities as compared to soils collected from low-density stands (Broz et al. 2007). Spotted knapweed is a native plant in Eurasia; therefore, it is reasonable to assume that it has not yet evolved coevolutionary links with the native soil microbiota in Montana and that the lack of coevolved signals in the root exudates of spotted knapweed may have accounted for the negative effects on the native soil fungal communities. It is also possible that fungal diversity in these fields was supported via the natural communities of various plant species, and that the near monocultures of spotted knapweed may not be adequate to support this microbial diversity.

The above is an example of how an invasive plant can disrupt the coadaptationary interactions between plants and native soil microbes; we suggest that conventional farming may also cause similar effects. Conventional farming often combines the use of high-yielding crop varieties, chemical fertilizers, and pesticides in order to maximize yields and feed the needs of an ever-growing world population. However, this altruistic strategy has posed severe environmental problems such as soil degradation, increased use of fossil fuels, water pollution, and the development of species that show resistance to the pesticides (Reganold et al. 1987; Kaufman and Franz 1993). Organic farming has the potential of becoming an alternative to conventional farming and has the possibility of reducing the negative effects of conventional agriculture. More than 32.3 million hectares of agricultural lands were maintained organically worldwide in 2007, and this trend seems to be increasing with higher support from customers in developed countries that are aware of the health and environmental risks of intensive agriculture and are willing to pay a premium prices for organic products (Helga and Lukas 2009). Organic farming is frequently touted as being environmentally friendly and beneficial to soil health, although there have been contradictory results showing either higher microbial diversity in the organic farms (Mader et al. 2002; Rangarajan et al. 2002; Oehl et al. 2004) or similar levels of diversity in both types of farms (Esperschutz et al. 2007; Wu et al. 2008; Micallef et al. 2009). Both culture-dependent and culture-independent methods have been used to analyze the soil microbial communities in these experiments; however, it is still difficult to obtain a comprehensive picture of soil microbial communities by such methods as phospholipid fatty acid (PLFA) profile, fatty acid methyl ester (FAME) profile, denaturing gradient gel electrophoresis (DGGE), terminal-restriction fragment length polymorphism (T-RFLP), and length heterogeneity PCR (LH-PCR).

Pyrosequencing or metagenomic approaches can provide a more comprehensive picture of the microbial community in environment samples from soils, deep ocean sediments, and the human intestine (Kirk et al. 2004; Liu et al. 2007). A pilot study was performed to analyze the soil fungal communities from conventional and organic potato farms in southern Colorado using pyrosequencing (Sugiyama et al. 2010). In this study, it was revealed that both conventional and organic farms contained similar richness of fungi; organic farms showed slightly higher fungal diversity (shown by Simpson's Reciprocal Index) but significantly higher fungal community evenness

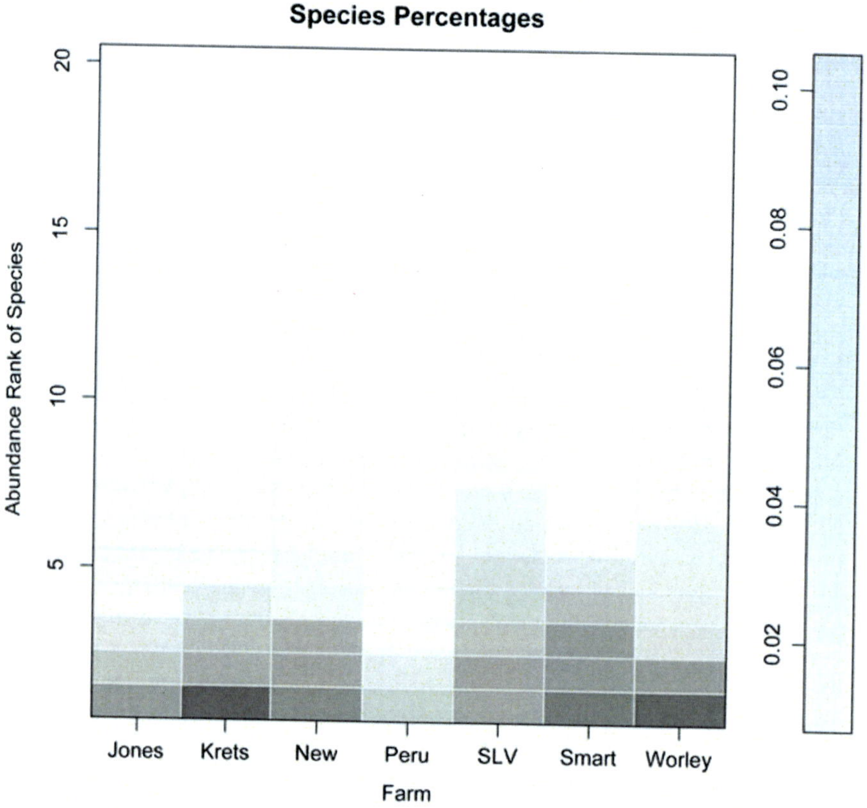

Fig. 1 Visual representation of soil microbial evenness. Each block represents one microbial species that is not necessarily the same in the different farms. Relative abundance of a given microbe is depicted by the color scale where *darker color* represents the highest abundance. Therefore, a farm that has darker colored blocks signifies a farm where the microbiome is not even. Jones, Krets, and New are organic potato farms; Peru depicts a native subsistence agricultural site in the Andes of Huancayo, Peru; and SLV, Smart, and Worley are intensive agricultural potato farms

(shown by Simpson's Evenness) compared with conventional farms. This study provides the possibility of using pyrosequencing approaches to analyze the soil microbial communities and the effect of chemicals on coadaptationary link between plants and soil microbes in other systems. The same approach was employed to analyze the soil fungal communities from potato farms in the Andes of Peru, the center of origin of the potato, where potatoes have been grown organically for more than 5,000 years due to the lack of intensive products in those communities. Interestingly, the soil fungal communities belonging to those potato plants grown in the Andes showed even higher evenness than farms under organic agriculture regimes, suggesting that stable interaction between plants and soil fungal species have a coevolutionary link (Fig. 1).

3 Coadaptation with Soil Microbes and Plant Disease

Plants have evolved strategies to cope with various pathogens, including cuticle, cell wall, secondary metabolite, hypersensitive response, and R-gene-mediated disease resistance. In field conditions, plants encounter not only pathogenic microbes but beneficial microbes and other types in the soils. Therefore, resistance to pathogens in the field must be analyzed on the basis of plant-soil microbial community interactions. It has been hypothesized that longer coadaptationary interactions among plants and soil microbes maintain an overall balance of microbes, or evenness, which prevents any particular species from becoming dominant and causing a pathogenic outbreak (Badri and Vivanco 2009).

3.1 *Community Evenness and Pathogenic Outbreak*

A large field experiment was performed in Washington potato fields to analyze the difference of agricultural management regimes (organic vs. conventional) on the natural pest control. Under those conditions, potato beetles are attacked by predatory bugs in the foliage, and pathogenic nematodes and fungi in the soil attack pupating adults of the beetle. In these potato fields, interactions among potato, potato beetles (*Leptinotarsa decemlineata*), predatory bugs (*Nabis alternatus* and *Geocoris bullatus*), beetles (*Hippodamia convergens* and *Pterostichus melanarius*), and pathogenic nematodes (*Heterorhabditis megidis* and *Steinernema carpocapsae*) and fungi (*Beauveria bassiana*) of the pest were investigated to determine whether organic farming improves natural biocontrol of the potato beetle (Crowder et al. 2010). It was found that organic farming increased the evenness of both predator and pathogen communities (from foliage and soil, respectively) resulting in better pest control and increased potato production (Crowder et al. 2010). These results indicate the importance of organisms' evenness in ecosystems.

As described in the previous section, it was found that soil from organic farms showed a significant increase in evenness of soil fugal communities compared to conventional agriculture in Colorado potato farms (Sugiyama et al. 2010). Taxonomical information on fungal species was also analyzed in the soil of these potato farms. *Alternaria* spp., including *A. solani*, the casual agent of early blight disease, and *A. alternata*, the casual agent of brown spot (Rotem 1994), were the most common pathogen present in this region and were detected in all farms; however, their relative abundance was significantly lower in organic farms as compared to conventional farms. In addition, *A. solani* was not detected in the fungal communities of the Andes of Peru, which had the highest level of fungal species' evenness. *Ulocladium* spp. is the causal agent of Ulocladium blight, and its relative abundance was also lower in organic farms. In contrast, *Pythium ultimum*, which causes leak, was more abundant in organic farms, although its relative abundance was much lower than *Alternaria* spp. Other potato pathogens such as *Phoma foveata*, *Rhizoctonia solani*, *Spongospora subterranea*, and *Pythium* spp. were also detected in some or all farms, but did not differ significantly between organic and conventional farms. In these potato farms, *Alternaria* spp. was by far the most predominant pathogen. Therefore, these results similar to those of Crowder (above) suggest the potential strong contribution of community evenness in pest/pathogen suppression. This hypothesis should be tested in the future with a large number of conventional and organic farms from various climate zones and various crop species.

3.2 Disruption of the Coadapted Soil Microbial Community by Invaded Species and Pesticides

It was reported that coadaptationary interactions between plants and soil microbial communities are maintained in organic farms, where symbiotic rhizobacteria and AM fungi contribute to nutrient acquisition. Higher evenness observed in the organic farms and in subsistent agricultural farms in Peru (above) could prevent pests and/or pathogens from becoming epidemic. Drastic changes in the soil microbial communities can be observed in areas where invaded species occur and in conventional farms. Antimicrobial compounds in the root exudates of invasive species (or new species not coadapted with the microbial community of a given soil), or toxic compounds in the pesticides that are introduced into soil, could drastically change the microbial composition leading to a loss of biocontrol organisms in the soil.

Evenness of the coadapted soil microbial communities is postulated to prevent the outbreak of any particular species and thus prevent the deleterious results from pathogenic species (Badri and Vivanco 2009). In the subsistence agriculture of Andes of Peru, where higher microbial evenness was observed, there has been no outbreak of *Phytopthora infestans* to date; however, this pathogen became dominant in potato fields with no history of coadaptation between the plants and the soil microbes. This lack of coadaptation may have caused the potato famine of Ireland. In nature, there are many factors other than soil microbial evenness that impact plant disease, such as soil fertility, soil texture, climate, and most importantly, the genetic background of plant species; however, unlike these factors, coadaptationary interactions and microbial evenness had not gain particular attention, and it should be considered for future research and breeding.

3.3 Coadaptation for Future Breeding

To date, abiotic traits of the rhizosphere such as pH and minerals and particular beneficial microbes such as rhizobia, arbuscular mycorrhizal fungi, PGPRs have been targeted for rhizosphere engineering approaches (Ryan et al. 2009).

Disease-suppressive soils, defined as soils where "pathogen does not establish or persist, establishes but causes little or no damage, or establishes and causes disease for awhile but thereafter, the disease is less important, although the pathogen may persist in the soil" (Baker and Cook 1974; Cook and Baker 1983), are probably maintained by some sort of coadaptation between soil microbes and plants. It is urgent to obtain the scientific basis for the coadaptation to provide disease suppressiveness in the soil. A highly prominent approach in this direction is the finding that *Arabidopsis* mutants for *abcg30* cultivate a microbial community with a relatively higher abundance of beneficial bacteria, such as PGPRs, nitrogen-fixing bacteria, and bacteria involved in heavy metal remediation when grown in coadapted soils. Root exudates of this mutant contained higher phenolics and lower sugars. Identification of key regulatory genes for the alternation of root exudates as well as the identification of the key element(s) or mixture of root exudates to cultivate agriculturally beneficial microbes is under way in the laboratories of the authors. Thus, it is highly feasible to use the regulatory genes or root exudates profiles for selecting varieties of crop species and breeding strategies that could culture beneficial soil microbiomes. These aspects had been ignored in breeding, except for a few examples such as nitrogen fixation and responsiveness to AM fungi (Wissuwa et al. 2009; Rengel 2002).

4 Conclusions

There is no doubt that conventional farming is necessary to provide enough yields to meet the nutritional demand for humans globally. However, conventional farming usually disrupts the coadaptationary plant-microbe interactions and community diversity, and the monoculture of crops under conventional agricultural regimes contributes to further negative impacts on the soil biota, thus, affecting the functions of the agroecosystem (Postma-Blaauw et al. 2010). It is not our intent to promote organic farming to create soil microbial evenness in order to prevent epidemics, but to bring an awareness of the importance of coadaptationary interactions between plant and soil microbes. Ultimately, a better understanding of these interactions will allow us to design good agricultural practices to promote healthy soil microbiomes that in turn will develop a more sustainable and healthy agriculture.

Acknowledgments We thank Mr. Derek Sonderegger for statistical analysis for visual representation of soil microbial evenness (Fig. 1). A.S. was supported by a JSPS Postdoctoral Fellowships for Research Abroad. These studies were supported by the National Science Foundation (MCB-0950857 to JMV).

References

Akiyama K, Matsuzaki K, Hayashi H (2005) Plant sesquiterpenes induce hyphal branching in arbuscular mycorrhizal fungi. Nature 435:824–827

Akiyama K, Tanigawa F, Kashihara T, Hayashi H (2010) Lupin pyranoisoflavones inhibiting hyphal development in arbuscular mycorrhizal fungi. Phytochemistry 71:1865–1871

Badri DV, Vivanco JM (2009) Regulation and function of root exudates. Plant Cell Environ 32:666–681

Badri DV, Quintana N, El Kassis EG, Kim HK, Choi YH, Sugiyama A, Verpoorte R, Martinoia E, Manter DK, Vivanco JM (2009a) An ABC transporter mutation alters root exudation of phytochemicals that provoke an overhaul of natural soil microbiota. Plant Physiol 151:2006–2017

Badri DV, Weir TL, van der Lelie D, Vivanco JM (2009b) Rhizosphere chemical dialogues: plant–microbe interactions. Curr Opin Biotechnol 20:642–650

Badri DV, Loyola-Vargas VM, Broeckling CD, Vivanco JM (2010) Root secretion of phytochemicals in Arabidopsis is predominantly not influenced by diurnal rhythms. Mol Plant 3:491–498

Baker KF, Cook RJ (1974) Biological control of plant pathogens. Freeman and Company, San Franscisco

Batten K, Scow K, Davies K, Harrison S (2006) Two invasive plants alter soil microbial community composition in serpentine grasslands. Biol Invasions 8:217–230

Baudoin E, Benizri E, Guckert A (2002) Impact of growth stage on the bacterial community structure along maize roots, as determined by metabolic and genetic fingerprinting. Appl Soil Ecol 19:135–145

Bird DM (2004) Signaling between nematodes and plants. Curr Opin Plant Biol 7:372–376

Bonfante P, Anca IA (2009) Plants, mycorrhizal fungi, and bacteria: a network of interactions. Annu Rev Microbiol 63:363–383

Bouwmeester HJ, Matusova R, Zhongkui S, Beale MH (2003) Secondary metabolite signalling in host-parasitic plant interactions. Curr Opin Plant Biol 6:358–364

Broeckling CD, Broz AK, Bergelson J, Manter DK, Vivanco JM (2008) Root exudates regulate soil fungal community composition and diversity. Appl Environ Microbiol 74:738–744

Broz AK, Manter DK, Vivanco JM (2007) Soil fungal abundance and diversity: another victim of the invasive plant *Centaurea maculosa*. ISME J 1:763–765

Buee M, Rossignol M, Jauneau A, Ranjeva R, Becard G (2000) The pre-symbiotic growth of arbuscular mycorrhizal fungi is induced by a branching factor partially purified from plant root exudates. Mol Plant–Microbe Interact 13:693–698

Cai T, Cai W, Zhang J, Zheng H, Tsou AM, Xiao L, Zhong Z, Zhu J (2009) Host legume-exuded antimetabolites optimize the symbiotic rhizosphere. Mol Microbiol 73:507–517

Cook RJ, Baker KF (1983) The nature and practice of biological control of plant pathogens. APS Press, St. Paul

Crowder DW, Northfield TD, Strand MR, Snyder WE (2010) Organic agriculture promotes evenness and natural pest control. Nature 466:109–112

D'Haeze W, Holsters M (2004) Surface polysaccharides enable bacteria to evade plant immunity. Trends Microbiol 12:555–561

Djordjevic MA, Redmond JW, Batley M, Rolfe BG (1987) Clovers secrete specific phenolic compounds which either stimulate or repress nod gene expression in *Rhizobium trifolii*. EMBO J 6:1173–1179

Doyle JJ, Luckow MA (2003) The rest of the iceberg. Legume diversity and evolution in a phylogenetic context. Plant Physiol 131:900–910

Esperschutz J, Gattinger A, Mader P, Schloter M, Fliessbach A (2007) Response of soil microbial biomass and community structures to conventional and organic farming systems under identical crop rotations. FEMS Microbiol Ecol 61:26–37

Gagnon H, Ibrahim RK (1998) Aldonic acids: a novel family of nod gene inducers of *Mesorhizobium loti*, *Rhizobium lupini*, and *Sinorhizobium meliloti*. Mol Plant–Microbe Interact 11:988–998

Garcia-Garrido JM, Lendzemo V, Castellanos-Morales V, Steinkellner S, Vierheilig H (2009) Strigolactones, signals for parasitic plants and arbuscular mycorrhizal fungi. Mycorrhiza 19:449–459

Goldwasser Y, Yoneyama K, Xie X, Yoneyama K (2008) Production of Strigolactones by *Arabidopsis thaliana* responsible for *Orobanche aegyptiaca* seed germination. Plant Growth Regul 55:21–28

Hartmann A, Schmid M, van Tuinen D, Berg G (2009) Plant-driven selection of microbes. Plant Soil 321:235–257

Helga W, Lukas KE (2009) The world of organic agriculture—statistics and emerging trends 2009. Bonn, Germany

Hirsch AM, Dietz Bauer W, Bird DM, Cullimore J, Tyler B, Yoder JI (2003) Molecular signals and receptors: controlling rhizosphere interactions between plants and other organisms. Ecology 84:858–868

Horiuchi J, Prithiviraj B, Bais HP, Kimball BA, Vivanco JM (2005) Soil nematodes mediate positive interactions between legume plants and Rhizobium bacteria. Planta 222:848–857

Hungria M, Joseph CM, Phillips DA (1991) Anthocyanidins and flavonols, major nod gene Inducers from seeds of a black-seeded common bean (*Phaseolus vulgaris* L.). Plant Physiol 97:751–758

Innes L, Hobbs PJ, Bardgett RD (2004) The impacts of individual plant species on rhizosphere microbial communities in soils of different fertility. Biol Fertil Soils 40:7–13

Kaplan F, Badri DV, Zachariah C, Ajredini R, Sandoval FJ, Roje S, Levine LH, Zhang F, Robinette SL, Alborn HT, Zhao W, Stadler M, Nimalendran R, Dossey AT, Bruschweiler R, Vivanco JM, Edison AS (2009) Bacterial attraction and quorum sensing inhibition in *Caenorhabditis elegans* exudates. J Chem Ecol 35:878–892

Kaufman DG, Franz CM (1993) Biosphere 2000: protecting our global environment. HarperCollins College Publishers, New York

Kirk JL, Beaudette LA, Hart M, Moutoglis P, Klironomos JN, Lee H, Trevors JT (2004) Methods of studying soil microbial diversity. J Microbiol Methods 58:169–188

Klironomos JN (2003) Variation in plant response to native and exotic arbuscular mycorrhizal fungi: underground processes in plant communities. Ecology 84:2292–2301

Kobae Y, Sekino T, Yoshioka H, Nakagawa T, Martinoia E, Maeshima M (2006) Loss of AtPDR8, a plasma membrane ABC transporter of *Arabidopsis thaliana*, causes hypersensitive cell death upon pathogen infection. Plant Cell Physiol 47:309–318

Kobae Y, Hata S (2010) Dynamics of periarbuscular membranes visualized with a fluorescent phosphate transporter in arbuscular mycorrhizal roots of rice. Plant Cell Physiol 51:341-353

Kosslak RM, Bookland R, Barkei J, Paaren HE, Appelbaum ER (1987) Induction of *Bradyrhizobium japonicum* common nod genes by isoflavones isolated from *Glycine max*. Proc Natl Acad Sci USA 84:7428–7432

Kowalchuka GA, Hola WHG, Van Veen JA (2006) Rhizosphere fungal communities are influenced by Senecio jacobaea pyrrolizidine alkaloid content and composition. Soil Biol Biochem 38:2852–2859

Liljeroth E, Burgers SLGE, van Veen VA (1991) Changes in bacterial populations along roots of wheat (*Triticum aestivum* L.) seedlings. Biol Fertil Soils 10:276–280

Liu Z, Lozupone C, Hamady M, Bushman FD, Knight R (2007) Short pyrosequencing reads suffice for accurate microbial community analysis. Nucleic Acids Res 35:e120

Mader P, Fliessbach A, Dubois D, Gunst L, Fried P, Niggli U (2002) Soil fertility and biodiversity in organic farming. Science 296:1694–1697

Maillet F, Poinsot V, Andre O, Puech-Pages V, Haouy A, Gueunier M, Cromer L, Giraudet D, Formey D, Niebel A, Martinez EA, Driguez H, Becard G, Denarie J (2011) Fungal lipochitooligosaccharide symbiotic signals in arbuscular mycorrhiza. Nature 469:58–63

Maxwell CA, Hartwig UA, Joseph CM, Phillips DA (1989) A chalcone and two related flavonoids released from alfalfa roots induce nod genes of *Rhizobium meliloti*. Plant Physiol 91:842–847

Mazzola M, Funnell DL, Raaijmakers JM (2004) Wheat cultivar-specific selection of 2,4-diacetylphloroglucinol-producing fluorescent Pseudomonas species from resident soil populations. Microb Ecol 48:338–348

Micallef SA, Shiaris MP, Colon-Carmona A (2009) Influence of *Arabidopsis thaliana* accessions on rhizobacterial communities and natural variation in root exudates. J Exp Bot 60:1729–1742

Miller RM, Reinhardt DR, Jastrow JD (1995) External hyphal production of vesicular-arbuscular mycorrhizal fungi in pasture and tallgrass prairie communities. Oecologia 103:17–23

Morgan JA, Bending GD, White PJ (2005) Biological costs and benefits to plant–microbe interactions in the rhizosphere. J Exp Bot 56:1729–1739

Mougel C, Offre P, Ranjard L, Corberand T, Gamalero E, Robin C, Lemanceau P (2006) Dynamic of the genetic structure of bacterial and fungal communities at different developmental stages of *Medicago truncatula* Gaertn. cv. Jemalong line J5. New Phytologist 170:165–175

Narasimhan K, Basheer C, Bajic VB, Swarup S (2003) Enhancement of plant-microbe interactions using a rhizosphere metabolomics-driven approach and its application in the removal of polychlorinated biphenyls. Plant Physiol 132:146–153

Oehl F, Sieverding E, Mader P, Dubois D, Ineichen K, Boller T, Wiemken A (2004) Impact of long-term conventional and organic farming on the diversity of arbuscular mycorrhizal fungi. Oecologia 138:574–583

Parniske M (2008) Arbuscular mycorrhiza: the mother of plant root endosymbioses. Nat Rev Microbiol 6:763–775

Perry JN, Moens M (2006) Plant nematology. CABI, Cambridge

Peters NK, Frost JW, Long SR (1986) A plant flavone, luteolin, induces expression of *Rhizobium meliloti* nodulation genes. Science 233:977–980

Phillips DA, Joseph CM, Maxwell CA (1992) Trigonelline and stachydrine released from alfalfa seeds activate nodD2 protein in *Rhizobium meliloti*. Plant Physiol 99:1526–1531

Postma-Blaauw MB, de Goede RG, Bloem J, Faber JH, Brussaard L (2010) Soil biota community structure and abundance under agricultural intensification and extensification. Ecology 91:460–473

Priha O, Grayston SJ, Pennanen T, Smolander A (1999) Microbial activities related to C and N cycling and microbial community structure in the rhizospheres of *Pinus sylvestris*, *Picea abies* and *Betula pendula* seedlings in an organic and mineral soil. FEMS Microbiol Ecol 30:187–199

Raaijmakers JM, Paulitz TC, Steinberg C, Alabouvette C, Moënne-Loccoz Y (2009) The rhizosphere: a playground and battlefield for soilborne pathogens and beneficial microorganisms. Plant Soil 321:341–361

Rangarajan S, Saleena LM, Nair S (2002) Diversity of Pseudomonas spp. isolated from rice rhizosphere populations grown along a salinity gradient. Microb Ecol 43:280–289

Redecker D, Kodner R, Graham LE (2000) Glomalean fungi from the Ordovician. Science 289:1920–1921

Redmond J, Batley M, Djordjevic M, Innes R, Kuempel P, Rolfe B (1986) Flavones induce expression of nodulation genes in *Rhizobium*. Nature 323:632–635

Reganold J, Elliott L, Unger Y (1987) Long-term effects of organic and conventional farming on soil erosion. Nature 330:370–372

Reinhart KO, Callaway RM (2006) Soil biota and invasive plants. New Phytologist 170:445–457

Remy W, Taylor TN, Hass H, Kerp H (1994) Four hundred-million-year-old vesicular arbuscular mycorrhizae. Proc Natl Acad Sci USA 91:11841–11843

Rengel Z (2002) Breeding for better symbiosis. Plant Soil 245:147–162

Rotem J (1994) The genus Alternaria: biology, epidemiology and pathogenicity. American Phytopathological Society Press, St Paul

Ryan PR, Dessaux Y, Thomashow LS, Weller DM (2009) Rhizosphere engineering and management for sustainable agriculture. Plant Soil 321:363–383

Simon L, Bousquet J, Levesque RC, Lalonde M (1993) Origin and diversification of endomycorrhizal fungi and coincidence with vascular land plants. Nature 363:67–69

Sprent JI (2007) Evolving ideas of legume evolution and diversity: a taxonomic perspective on the occurrence of nodulation. New Phytologist 174:11–25

Stein M, Dittgen J, Sanchez-Rodriguez C, Hou BH, Molina A, Schulze-Lefert P, Lipka V, Somerville S (2006) Arabidopsis PEN3/PDR8, an ATP binding cassette transporter, contributes to nonhost resistance to inappropriate pathogens that enter by direct penetration. Plant Cell 18:731–746

Sugiyama A, Vivanco JM, Jayanty SS, Manter DK (2010) Pyrosequencing assessment of soil microbial communities in organic and conventional potato farms. Plant Dis 94:1329–1335

Takagi S (1976) Naturally occurring iron-chelating compounds in oat and rice root-washings. Soil Sci Plant Nutr 22:423–433

Trevors JT (2010) One gram of soil: a microbial biochemical gene library. Antonie Van Leeuwenhoek 97:99–106

Wissuwa M, Mazzola M, Picard C (2009) Novel approaches in plant breeding for rhizosphere-related traits. Plant Soil 321:409–430

Wu T, Chellemi DO, Graham JH, Martin KJ, Rosskopf EN (2008) Comparison of soil bacterial communities under diverse agricultural land management and crop production practices. Microb Ecol 55:293–310

Yang CH, Crowley DE (2000) Rhizosphere microbial community structure in relation to root location and plant iron nutritional status. Appl Environ Microbiol 66:345–351

Yoneyama K, Xie X, Sekimoto H, Takeuchi Y, Ogasawara S, Akiyama K, Hayashi H, Yoneyama K (2008) Strigolactones, host recognition signals for root parasitic plants and arbuscular mycorrhizal fungi, from Fabaceae plants. New Phytologist 179:484–494

Index

G. Witzany and F. Baluška (eds.), *Biocommunication of Plants*,
Signaling and Communication in Plants 14,
DOI 10.1007/978-3-642-23524-5, © Springer-Verlag Berlin Heidelberg 2012

Q

R

T